JP Evans
CLARE HALL

Semiconductor Optics

Springer
Berlin
Heidelberg
New York
Barcelona
Budapest
Hong Kong
London
Milan
Paris
Tokyo

C.F. Klingshirn

Semiconductor Optics

With 270 Figures

Professor. Dr. CLAUS F. KLINGSHIRN
Institut für Angewandte Physik
Universität Karlsruhe
Postfach 6980
76128 Karlsruhe, Germany

ISBN 3-540-58312-2 Springer-Verlag Berlin Heidelberg New York

Library of Congress Cataloging-in-Publication Data. Klingshirn, C.F. (Claus F.), 1944- Semiconductor optics/C.F. Klingshirn. p. cm., Includes bibliographical references and index., ISBN 3-40-58312-2 1. Semiconductors-Optical properties. I. Title. QC611.6.o6K65 1995 537.6′226-dc20 95–1801

Printed in Germany

Typesetting: Thomson Press (I) Ltd., Madras

SPIN: 10048555 54/3140/SPS–5 4 3 2 1 0 – Printed on acid-free paper

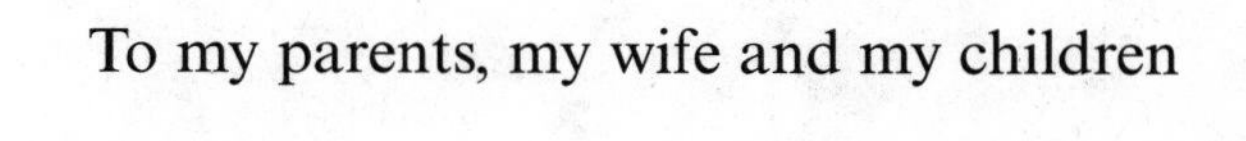

To my parents, my wife and my children

Wahrheit und Klarheit sind komplementär.

E. MOLLWO

This aphorism was coined in the nineteen-fifties by E. MOLLWO, Professor of Physics at the Institut für Angewandte Physik of the Universität Erlangen during a discussion with W. HEISENBERG. The author hopes that, with respect to his book, the deviations from exact scientific truth (Wahrheit) and perfect understandability (Klarheit) are in a reasonable balance.

Preface

One of the most prominent senses of many animals and, of course, of human beings is sight or vision. As a consequence, all phenomena which are connected with light and color, or with the optical properties of matter, have been focal points of interest throughout the history of mankind. Natural light sources such as the Sun, the Moon and stars, or fire, were worshipped as gods or godesses in many ancient religions. Fire, which gives light and heat, was for many centuries thought to be one of the four elements – together with earth, water, and air. In alchemy, which marks the dawn of our modern science, the Sun and the Moon appeared as symbols of gold and silver, respectively, and many people tried to produce these metals artificially. Some time later, Johann Wolfgang von Goethe (1749–1832) considered his "Farbenlehre" as more important than his poetry. In the last two centuries a considerable fraction of modern science has been devoted to the investigation and understanding of light and the optical properties of matter. Many scientists all over the world have added to our understanding of this topic. As representatives of the many we should like to mention here only a few of them: I. Newton (1643–1727), Maxwell (1831–1879), M. Planck (1858–1947), A. Einstein (1879–1955), N. Bohr (1885–1962), and W. Heisenberg (1901–1967).

The aim of this book is more modest. It seeks to elucidate one of the numerous aspects in the field of light and the optical properties of matter, namely the interaction of light with semiconductors, i.e., semiconductor optics. The investigation of the properties of semiconductors has, in turn, its own history, which has been summarized recently by H.J. Queisser[1]. In Queisser's book one can find early examples of semiconductor optics, namely the observation of artificially created luminescence by V. Cascariolo in Bologna at the beginning of the 17th century, or by K.F. Braun (1850–1918), inventor of the "Braun'sche Röhre" (Braun's tube) now usually called CRT (cathode ray tube), at the beginning of this century.

Another root of semiconductor optics comes from the investigation of the optical properties of insulators, especially of the color (Farb- or F-) centers in alkali halides. This story has been written down recently by J. Teichmann[2]. It is inseparably connected with names such as Sir Nelville Mott and A. Smakula, but especially with R.W. Pohl (1884–1976) and his school in Göttingen.

Together with J. Franck (1882–1964) and M. Born (1882–1970) R.W. Pohl was one of the outstanding physicists of the "golden years of physics" at Göttingen before 1933[3]. The present author considers himself a scientific grandson of Pohl, with E. Mollwo (1909–1993), F. Stöckmann and W. Martienssen as the intermediate generation, and he owes to them a large part of his scientific education.

Scientific interest in semiconductor optics comprises both fundamental and applied research. It has been an extremely lively, rapidly developing area of research for the last five decades and more, as can be seen from the contributions to the series of International Conferences on the Physics of Semiconductors[4] and on the Physics of Luminescence[5]. It does not need much of a prophetic gift to predict that semiconductor optics will continue to be a major topic of solid state physics far into the next century. Many applications of semiconductor optics are known from everyday life such as light-emitting diodes (LED) in displays, laser diodes in compact-disk (CD) players, and laser printers.

Karlsruhe, February 1995 C.F. KLINGSHIRN

[1]H.-J. Queisser: *Kristallene Krisen* (Piper, München 1985)

[2]J. Teichmann: *Zur Geschichte der Festkörperphysik–Farbzentrenforschung bis 1940* (Steiner, Wiesbaden 1988)

[3]F. Hund, H. Maier-Leibnitz, E. Mollwo: Eur. J. Phys. **9**, 188 (1988)
E. Mollwo: Physik in unserer Zeit **15**, 110 (1984)
A.D. Beyerchen: *Scientists under Hitler*, (Yale Univ. Press, New Haven 1977)

[4]The Series of Int'l Conferences on the Physics of Semiconductors (ICPS) was started in 1950 in Reading. Proceedings of the more recent ones are
12th ICPS, Stuttgart (1974), ed. by M.H. Pilkuhn (Teubner, Stuttgart 1974)
13th ICPS, Rome (1976), ed. by F.G. Fumi (Tipographia Marves, Rome 1976)
14th ICPS, Edinburgh (1978), ed. by B.L.H. Wilson (The Institute of Physics, Bristol 1979)
15th ICPS, Kyoto (1980), ed. by S. Tanaka, Y. Toyozawa: J. Phys. Soc. Jpn. **49**, Suppl. A (1980)
16th ICPS, Montpellier (1982), ed. by M. Averous: Physica B **117** + **118** (1983)
17th ICPS, San Francisco (1984), ed. by J.M. Chadi, W.A. Harrison (Springer, Berlin Heidelberg 1984)
18th ICPS, Stockholm (1986), ed. by O. Engström (World Scientific, Singapore 1987)
19th ICPS, Warsaw (1988), ed. by W. Zawadzki (The Institute of Physics, Polish Academy of Sciences, 1988)
20th ICPS, Thessaloniki (1990), ed. by E.M. Anastassakis, J.D. Joannopoulos (World Scientific, Singapore 1990)
21st ICPS, Beijing (1992), ed. by Ping Jiang, Hou-Zhi Zheng (World Scientific, Singapore 1993)
22st ICPS, vancouver (1994) to be published

[5]The proceedings of the Series of Int'l Conferences of Luminescence (ICL) are published in J. Lumin. The more recent ones were
ICL, Berlin (1981), ed. by I. Broser, H.-E. Gumlich, R. Broser: J. Lumi. **24/25** (1981)
ICL, Madison (1984), ed. by W.M. Yen, J.C. Wright: J. Lumin. **31/32** (1984)
ICL, Beijing (1987), ed. by Xu Xurong: J. Lumin. **40/41** (1987)
ICL, Lisbon (1990), ed. by S.J. Formosinho, M.D. Sturge: J. Lumi. **48/49** (1990)
ICL, Storrs (1993) ed. by D.S. Hamilton, R.S. Meltzer and M.D. Sturge: J. Lumi. **60/61** (1995)

Acknowledgements

This book is based on various lectures given by the author at the Universities of Karlsruhe, Frankfurt and Kaiserslautern and at some of the Summer Schools in Erice (1981, 1983, ..., 1993. See [1.7]). The sources of the scientific information presented here are partly the references given. Of equal importance, however, is the physics which I learned from my academic teachers and from many collegues and co-workers during fruitful discussions. Without trying to be complete, I should like to mention my academic teachers Prof. Dr. D. Fleischmann, Prof. Dr. H. Volz, Prof. Dr. E. Mollwo (†), Prof. Dr. R. Helbig, Prof. Dr. H. Hümmer (Erlangen) and Prof. Dr. F. Stöckmann and Prof. Dr. W. Ruppel (Karlsruhe). From the collegues I acknowledge with great pleasure fruitful and stimulating discussions with Prof. Dr. H. Haug, Prof. Dr. W. Martienssen, and Prof. Dr. E. Mohler and Dr. Banyai (Frankfurt/Main), Prof. Dr. U. Rössler (Regensburg). Prof. Dr. J.B. Grun, Prof. Dr. B. Hönerlage, and Dr. R. Lévy (Strasbourg), Prof. Dr. E. Göbel and Prof. Dr. S. Schmitt-Rink (†), (Marburg), Prof. Dr. D.S. Chemla (Berkeley), Prof. Dr. S.W. Koch (Marburg), Prof. Dr. I. Broser, Prof. Dr. R. Zimmermann, and Prof. Dr. F. Henneberger (Berlin), Dr. D.A.B. Miller (Holmdel), Dr. I. Bar-Joseph (Rehovot), Prof. Dr. B. Di Bartolo (Boston/Erice) and many others which I cannot mention here by name.

My special thanks are due to all of my students and co-workers who produced in their Diploma, doctorial, or Habilitation theses many of the fine results presented in this book. I also include at this point the funding agencies who supported our research: the Deutsche Forschungsgemeinschaft, the Stiftung Volkswagenwerk, and the Bundesministerium für Forschung und Technologie.

At the final stage of this book project the critical reading of some of the chapters by my coworkers Dr. H. Kalt, Dr. U. Woggon, Dipl.-Phys. W. Langbein was of great help, as was the assistance of Dr. A. Uhrig, Dipl.-Phys. A. Wörner, and Mrs. I. Wollscheidt who helped with the figures, and Mrs. S. Müller who patiently typed the final version.

Last but not least, I should like to thank the publishing house Springer, especially Dr. H. Lotsch for the excellent and fruitful cooperation and for the encouragement to write and complete this book, and Dr. A. Lahee and Dr. V. Wicks for their careful and competent reading.

Contents

1 Introduction

This introductory chapter consists of an outline of the fundamental concepts and ideas on which the text is based, including the rather limited prerequisites so that the reader can follow it and, finally, some hints about its contents.

1.1 Aim and Concepts

The aim of this book is to explain the optical properties of semi-conductors, e.g., the spectra of transmission, reflection and luminescence, or of the complex dielectric function in the infrared, visible and near-ultraviolet part of the electromagnetic spectrum. We want to evoke in the reader a clear and intuitive understanding of the physical concepts and foundations of semiconductor optics and of some of their numerous applications. To this end, we try to keep the mathematical apparatus as simple and as limited as possible in order not to conceal the physics behind mathematics. We give ample references for those who want to enter more deeply into the mathematical concepts [1.1–3].

In this spirit, this present textbook is not only suitable for graduate and postgraduate students of physics, but also for students of neighboring disciplines, such as material science and electronics.

The prerequisites for the reader are an introductory or undergraduate course in general physics and some basic knowledge in atomic physics and quantum mechanics. The reader should know, for example, what the Schrödinger equation is, what the words eigen- (or proper-) state and eigen energy mean, and what quantum mechanics predicts about plane waves, the hydrogen atom or the harmonic oscillator, how to calculate transition probabilities e.g., by Fermi's golden rule. Some basic knowledge of solid state physics will facilitate reading of this book, although the basic concepts will always be outlined here.

At the end of every chapter we give several problems which can be solved with the information given in the text, combined with some basic knowledge of physics, some thinking and some creativity.

1.2 Outline

In the first part of this book (Chaps. 2–17) we shall present the linear optical properties of semiconductors. We start in Chap. 2 with Maxwell's equations and photons and introduce in Chap. 3 the basic concepts of the interaction of light with matter. In Chaps. 4–8 a model system of oscillators is treated with respect to the optical properties which can be expected for such a system. Chapters 9–12 are used to introduce the elementary excitations or quasiparticles in semiconductors, followed by a presentation of the linear optical properties resulting from the interaction of these quasiparticles with light in Chaps. 13–16. Chapter 17 gives a short résumé of the linear optical properties of semiconductors.

We include in Chaps. 9–16 modern concepts of semiconductor optics such as the properties of systems of reduced dimensionality, e.g., quantum wells, or disordered systems which lead to localization. At present, more than 600 different semiconductor materials are known. Many of them and their properties are listed in several volumes of Landolt-Börnstein [1.4]. We shall concentrate here on the most important ones. They are usually tetrahedrally coordinated and comprise, e.g., the group IV elements Si and Ge, the III–V compounds such as GaAs, the IIb–VI semiconductors such as CdS or ZnSe, and the Ib–VII materials such as the Cu halides.

Chapters 18–23 contain the main aspects of the nonlinear optical properties of semiconductors including optical bistability as an example for an application of nonlinear optical properties.

In Chaps. 24, 25, which can be considered as a kind of appendix, we shall outline some experimental techniques of semiconductor spectroscopy and some elements of group theory which are relevant for the description of semiconductor optics.

In the sections on the linear and on the nonlinear optical properties of semiconductors, the main emphasis is placed on those properties which are connected with excitations in the electronic system of semiconductors, since these aspects have obtained the widest interest both in fundamental and applied research as can be seen from an inspection of the conference series mentioned in the preface.

At the end of most chapters a selection of references will be given for further reading which penetrate deeper into the topic, consider some further aspects, or give a more detailed theoretical description. Since the number of original publications, conference proceedings or summer schools on the topics covered here is "close to infinite", it is definitely only possible to cite a very small fraction of them, the choice of which is partly arbitrary and determined by the author's research interest. Furthermore, we shall not give references at all for things which can be considered to belong to the "general education or culture" in physics but we give references to the sources of original data in the figures. These figures have all been redrawn and generally modified for the purpose of this textbook. We apologize for these deficiencies.

The present book thus complements the text books [1.5] and [1.6] which concentrate more on atomic and molecular spectroscopy and on solid state spectroscopy in general. A rather remarkable series of books on various aspects of optical properties of solids, with some emphasis on insulators results from the International Schools on "Atomic and Molecular Spectroscopy" held every two years in Erice (Sicily) [1.7]. Other series, which contain a lot of information on solid state optics are listed in [1.8]. The books cited in [1.5, 9] present various topics of laser spectroscopy of solids, again with emphasis on insulators. Information about the optical properties of metals are found e.g., in [1.10].

1.3 Problems

1. What are the basic conservation laws in nature?
2. Try to remember some of the basic concepts of quantum mechanics:
   - What is the Hamiltonian in classical and in quantum mechanics?
   - Write down the time-independent and the time-dependent Schrödinger equation for a single particle.
   - What are the eigenenergies and eigenfunctions of a one-dimensional harmonic oscillator and of the hydrogen atom?
   - What can you calculate with time-independent perturbation theory?
   - What does Fermi's golden rule say about transition probabilities?

2 Maxwell's Equations and Photons

In this chapter we consider Maxwell's equations and what they reveal about the propagation of light in vacuum and in matter. We introduce the concept of photons and present their density of states. Since the density of states is a rather important property, not only for photons, we approach this quantity in a rather general way. We will use the density of states later also for other (quasi-) particles including systems of reduced dimensionality. In addition, we introduce the occupation probability of these states for various groups of particles.

2.1 Maxwell's Equations

Maxwell's equations can be written in different ways. We use here the macroscopic Maxwell's equations in their differential form. Throughout this book the internationally recommended system of units known as SI (système international) is used. These equations are given in their general form in (2.1a–f), where bold characters symbolize vectors and normal characters scalar quantities.

$$\nabla \cdot \boldsymbol{D} = \rho, \quad \nabla \cdot \boldsymbol{B} = 0 \, , \tag{2.1a,b}$$

$$\nabla \times \boldsymbol{E} = -\dot{\boldsymbol{B}}, \quad \nabla \times \boldsymbol{H} = \boldsymbol{j} + \dot{\boldsymbol{D}} \, , \tag{2.1c,d}$$

$$\boldsymbol{D} = \varepsilon_0 \boldsymbol{E} + \boldsymbol{P}, \quad \boldsymbol{B} = \mu_0 \boldsymbol{H} + \boldsymbol{M} \, . \tag{2.1e,f}$$

The various symbols have the following meanings:

$\boldsymbol{E}$ = electric field strength
$\boldsymbol{D}$ = electric displacement
$\boldsymbol{H}$ = magnetic field strength
$\boldsymbol{B}$ = magnetic induction or magnetic flux density
ρ = charge density
$\boldsymbol{j}$ = electrical current density
$\boldsymbol{P}$ = polarization density of a medium, i.e., electric dipole moment per unit volume

$\boldsymbol{M}$ = magnetization density of the medium, i.e., magnetic dipole moment per unit volume[1]

$\varepsilon_0 \simeq 8.859 \cdot 10^{-12}$ As/Vm is the permittivity of vacuum

$\mu_0 = 4\pi \cdot 10^{-7}$ Vs/Am is the permeability of vacuum

∇ = Nabla-operator, in Cartesian coordinates $\nabla = (\partial/\partial x, \partial/\partial y, \partial/\partial z)$

$\dot{} = \partial/\partial t$ i.e., a dot means differentiation with respect to time.

The applications of ∇ to scalar or vector fields are usually denoted by

$$\nabla \cdot f(\boldsymbol{r}) = \operatorname{grad} f ,$$
$$\nabla \cdot \boldsymbol{A}(\boldsymbol{r}) = \operatorname{div} \boldsymbol{A} ,$$
$$\nabla \times \boldsymbol{A}(\boldsymbol{r}) = \operatorname{curl} \boldsymbol{A} ,$$

and the Laplace operator Δ is defined as

$$\Delta \equiv \nabla^2 .$$

If Δ is applied to a scalar field ρ we obtain

$$\Delta\rho = \frac{\partial^2 \rho}{\partial x^2} + \frac{\partial^2 \rho}{\partial y^2} + \frac{\partial^2 \rho}{\partial z^2} . \tag{2.2}$$

Application to a vector field $\boldsymbol{E}$ results in

$$\Delta \boldsymbol{E} = \begin{pmatrix} \dfrac{\partial^2 E_x}{\partial x^2} + \dfrac{\partial^2 E_x}{\partial y^2} + \dfrac{\partial^2 E_x}{\partial z^2} \\ \dfrac{\partial^2 E_y}{\partial x^2} + \dfrac{\partial^2 E_y}{\partial y^2} + \dfrac{\partial^2 E_y}{\partial z^2} \\ \dfrac{\partial^2 E_z}{\partial x^2} + \dfrac{\partial^2 E_z}{\partial y^2} + \dfrac{\partial^2 E_z}{\partial z^2} \end{pmatrix} . \tag{2.3}$$

Further rules for the use of ∇ and of Δ and their representations in other than Cartesian coordinates (polar or cylindrical coordinates) are found in compilations of mathematical formulae [2.1].

Equations (2.1a, b) show that free electric charges ρ are the sources of the electric displacement and that the magnetic induction is source-free. Equations (2.1b, c) demonstrate how temporally varying magnetic and electric fields generate each other. In addition, the $\boldsymbol{H}$ field can be created by a macroscopic current density $\boldsymbol{j}$. Equations (2.1e, f) are the material equations in their general form. From them we learn that the electric displacement is given by the sum of electric field and polarization, while the magnetic flux density is given by the sum of magnetic field and magnetization. Some authors prefer not to differentiate between $\boldsymbol{H}$ and $\boldsymbol{B}$. This leads to difficulties, as can be easily seen from the fact that $\boldsymbol{B}$ is source-free (2.1b) but $\boldsymbol{H}$ is not, as follows from the inspection of the fields of every simple permanent magnet.

---

[1] Some authors prefer to use $\boldsymbol{M}' = \boldsymbol{M}\mu_0^{-1}$ and thus $\boldsymbol{B} = \mu_0(\boldsymbol{H} + \boldsymbol{M}')$. We prefer the notation of (2.1e, f) for symmetry arguments.

2.2 Electromagnetic Radiation in Vacuum

In vacuum the following conditions are fulfilled

$$\boldsymbol{P}=0; \quad \boldsymbol{M}=0; \quad \rho=0; \quad \boldsymbol{j}=0\,. \tag{2.4}$$

With the help of (2.1e, f) this simplifies (2.1c, d) to

$$\nabla\times\boldsymbol{E}=-\mu_0\dot{\boldsymbol{H}} \quad \text{and} \quad \nabla\times\boldsymbol{H}=\varepsilon_0\dot{\boldsymbol{E}}\,. \tag{2.5a, b}$$

Applying $\nabla\times$ to (2.5a) and $\partial/\partial t$ to (2.5b) yields

$$\nabla\times(\nabla\times\boldsymbol{E})=-\mu_0\nabla\times\dot{\boldsymbol{H}} \quad \text{and} \quad \nabla\times\dot{\boldsymbol{H}}=\varepsilon_0\ddot{\boldsymbol{E}}\,. \tag{2.6}$$

From (2.6) we find with the help of the properties of the ∇ operator

$$-\mu_0\varepsilon_0\ddot{\boldsymbol{E}}=\nabla\times(\nabla\times\boldsymbol{E})=\nabla(\nabla\cdot\boldsymbol{E})-\nabla^2\boldsymbol{E}\,. \tag{2.7}$$

With (2.4), (2.3) and (2.1a) we see that

$$(\nabla\boldsymbol{E})=0 \tag{2.8}$$

and (2.7) reduces to the usual wave equation, written here for the electric field

$$\nabla^2\boldsymbol{E}-\mu_0\varepsilon_0\ddot{\boldsymbol{E}}=0\,. \tag{2.9}$$

Solutions of this equation are all waves of the form

$$\boldsymbol{E}(\boldsymbol{r},t)=\boldsymbol{E}_0 f(\boldsymbol{kr}-\omega t)\,. \tag{2.10}$$

$\boldsymbol{E}_0$ is the amplitude, f is an arbitrary function whose second derivate exists. The wave vector $\boldsymbol{k}$ and the angular frequency ω obey the relation

$$\frac{\omega}{k}=\left(\frac{1}{\mu_0\varepsilon_0}\right)^{1/2}=c \quad \text{with} \quad k=|\boldsymbol{k}|=2\pi/\lambda_{\text{v}}\,. \tag{2.11}$$

In the following we use for simplicity only the term "frequency" for $\omega=2\pi/T$ where T is the temporal period of the oscillation.

In (2.11), c is the vacuum speed of light and λ_{v} is the wavelength in vacuum. From all possible solutions of the form (2.10) we shall concentrate in the following on the most simple ones, namely on plane harmonic waves, which can be written as

$$\boldsymbol{E}(\boldsymbol{r},t)=\boldsymbol{E}_0\exp[\mathrm{i}(\boldsymbol{kr}-\omega t)]\,. \tag{2.12}$$

For all waves (not only those in vacuum), the phase and group velocities v_{ph} and v_{g} are given by

$$v_{\text{ph}}=\frac{\omega}{k}; \quad v_{\text{g}}=\frac{\partial\omega}{\partial\boldsymbol{k}}=\operatorname{grad}_{\boldsymbol{k}}\omega\,, \tag{2.13}$$

where v_{ph} gives the velocity with which a certain phase propagates, (e.g., a maximum of a monochromatic wave) while v_{g} gives the speed of the center of mass of a wave packet with middle frequency ω and covering a small frequency interval $d\omega$ as shown schematically in Figs. 2.1a, b, respectively. The formulas (2.13) are of general validity. The $\mathrm{grad}_{\boldsymbol{k}}$ on the r.h.s. of (2.13) means a differentiation with respect to $\boldsymbol{k}$ in the sense of $\nabla_{\boldsymbol{k}} = (\partial/\partial k_x, \partial/\partial k_y, \partial/\partial k_z)$ and has to be used instead of the more simple expression $\partial\omega/\partial k$ in anisotropic media. For the special case of electromagnetic radiation in vacuum we find from (2.11, 13)

$$v_{\mathrm{ph}} = v_{\mathrm{g}} = c = (\mu_0 \varepsilon_0)^{-1/2} \tag{2.14}$$

Now we want to see what constraints are imposed by Maxwell's equations on the various quantities such as $\boldsymbol{E}_0$ and $\boldsymbol{k}$. Inserting (2.10) or (2.12) into (2.8) gives

$$\nabla \cdot \boldsymbol{E} = \mathrm{i}\boldsymbol{E}_0 \cdot \boldsymbol{k} \exp[\mathrm{i}(\boldsymbol{kr} - \omega t)] = 0 \ . \tag{2.15}$$

This means that

$$\boldsymbol{E}_0 \perp \boldsymbol{k} \tag{2.16}$$

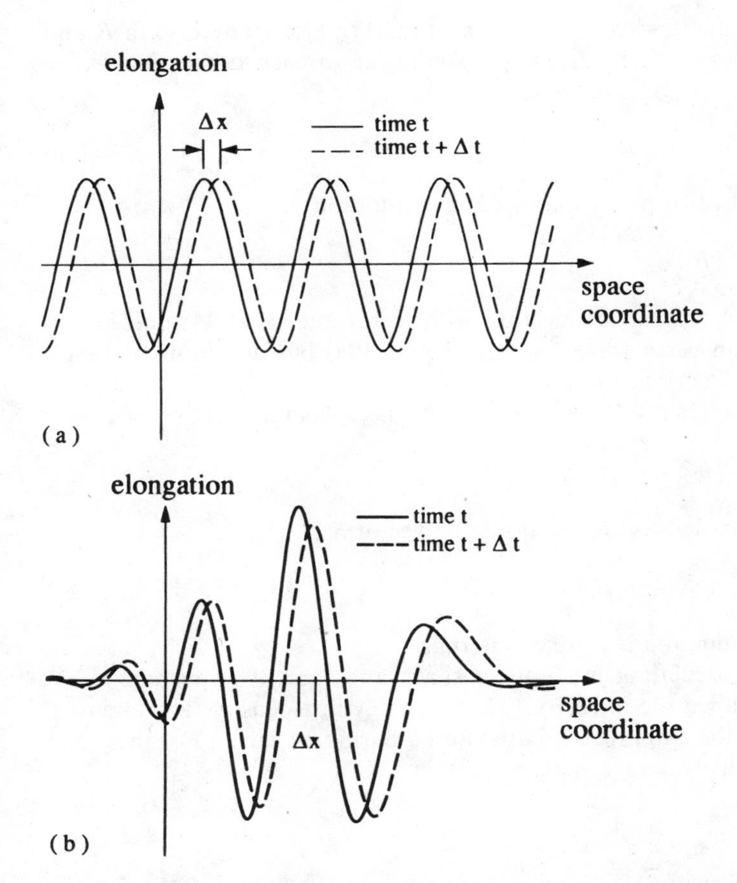

Fig. 2.1a, b. A harmonic wave (**a**) and a wave packet (**b**) shown at two different times t and $t + \mathrm{d}t$ to illustrate the concepts of phase and group velocity, respectively

or, in other words, the electromagnetic wave is transverse in $\boldsymbol{E}$. What can we learn from Maxwell's equations for the other fields? From (2.5) we have for plane waves

$$\boldsymbol{H} = (\omega\mu_0)^{-1}\boldsymbol{k} \times \boldsymbol{E} = \boldsymbol{H}_0 \exp[(\mathrm{i}(\boldsymbol{kr} - \omega t)]$$

with

$$\boldsymbol{H}_0 = (\omega\mu_0)^{-1}\boldsymbol{k} \times \boldsymbol{E}_0 \, . \tag{2.17}$$

Furthermore we have with (2.1e, f) and (2.4)

$$\boldsymbol{D} = \boldsymbol{D}_0 \exp[\mathrm{i}(\boldsymbol{kr} - \omega t)] = \varepsilon_0 \boldsymbol{E}_0 \exp[\mathrm{i}(\boldsymbol{kr} - \omega t)] \, ,$$

$$\boldsymbol{B} = \boldsymbol{B}_0 \exp[\mathrm{i}(\boldsymbol{kr} - \omega t)] = \omega^{-1}\boldsymbol{k} \times \boldsymbol{E}_0 \exp[\mathrm{i}(\boldsymbol{kr} - \omega t)] \, . \tag{2.18}$$

The electromagnetic wave is, according to (2.17), also transverse in $\boldsymbol{B}$ and the electric and magnetic fields are perpendicular to each other, that is, we have in general

$$\boldsymbol{D} \perp \boldsymbol{k} \perp \boldsymbol{B} \perp \boldsymbol{D} \, . \tag{2.19a}$$

In vacuum and isotropic media one has in addition

$$\boldsymbol{E} \parallel \boldsymbol{D} \quad \text{and} \quad \boldsymbol{H} \parallel \boldsymbol{B} \, . \tag{2.19b}$$

As we shall see later in connection with (2.15) and (2.43, 44) one has in matter usually transverse waves, which obey (2.19a) but additionally, longitudinal waves exist under certain conditions.

The momentum density Π of the electromagnetic field is given by

$$\Pi = \boldsymbol{D} \times \boldsymbol{B}, \quad \Pi \parallel \boldsymbol{k} \tag{2.20}$$

and the energy flux density by the Poynting vector $\boldsymbol{S}$

$$\boldsymbol{S} = \boldsymbol{E} \times \boldsymbol{H} \tag{2.21}$$

with $\boldsymbol{S} \parallel \Pi$ in vacuum and isotropic materials.

$\boldsymbol{S}$ is a rapidly oscillating function of space and time. The average value $<\boldsymbol{S}>$ is usually called the intensity I or the energy flux density. The intensity is proportional to the amplitude squared for all harmonic waves. For the plane monochromatic waves treated here, we obtain

$$<\boldsymbol{S}> = \frac{1}{2}|\boldsymbol{E}_0 \times \boldsymbol{H}_0| = \frac{1}{2}\frac{1}{c\mu_0}\boldsymbol{E}_0^2 = \frac{1}{2}\frac{c^2}{\mu_0}\boldsymbol{B}_0^2 = \frac{1}{2}\,\varepsilon_0^{-1}\boldsymbol{H}_0^2 \, . \tag{2.22}$$

Equations (2.20) and (2.21) are also valid in matter.

2.3 Electromagnetic Radiation in Matter; Linear Optics

Now we treat Maxwell's equations in matter. Doing so we have in principle to use the equations in their general from (2.1). However we will still make some assumptions which are reasonable for semiconductors: we assume that there are no macroscopic free space charges and that we have a nonmagnetic material. Actually, all matter has some diamagnetism. But this is a rather small effect of the order of 10^{-6} so it can be neglected for our purposes. Paramagnetic and especially ferromagnetic contributions can be significantly larger for low frequencies. However, even these contributions diminish rapidly for higher frequencies. Consequently the assumption of a nonmagnetic material is a good approximation in the visible range of the electromagnetic spectrum even for ferromagnetic materials. Furthermore, the more common semiconductors are not ferro-, ferri- or antiferromagnetic and have only a small concentration of paramagnetic centres with negligible influence on the optical properties. The only exceptions are semiconductors which contain a considerable amount of e.g., Mn or Fe ions as does $Zn_{1-y}Mn_ySe$. We refer the reader to [2.2] and references therein for this class of materials.

The current term $\boldsymbol{j}$ in (2.1d) deserves some more consideration. The current is driven by the electric field

$$\boldsymbol{j} = \sigma \boldsymbol{E} , \tag{2.23}$$

where σ is the conductivity. For intrinsic or weakly doped semiconductors, the carrier density is small and consequently σ is as well. Then the following inequality holds

$$|\boldsymbol{j}| = |\sigma \boldsymbol{E}| \ll |\boldsymbol{D}| . \tag{2.24}$$

In the following we will consider this case and neglect $\boldsymbol{j}$ in (2.1d). For heavily doped semiconductors (2.24) is no longer valid and σ will have some influence on the optical properties at least in the infrared (IR). We come back to this situation in connection with plasmons in Chaps. 12 and 13.

The basic material equation still left in comparison with the vacuum case is now (2.1e) $\boldsymbol{D} = \varepsilon_0 \boldsymbol{E} + \boldsymbol{P}$.

If we proceed with this equation again in the manner of (2.5–9) the result is

$$\nabla^2 \boldsymbol{E} - \mu_0 \varepsilon_0 \ddot{\boldsymbol{E}} = \mu_0 \ddot{\boldsymbol{P}} . \tag{2.25}$$

Equation (2.25) states the well-known fact that every dipole $\boldsymbol{p}$ and every polarization $\boldsymbol{P}$ with a non vanishing second derivative in time radiates an electromagnetic wave.

As long as we have no detailed knowledge about the relationships between $\boldsymbol{D}, \boldsymbol{E}$ and $\boldsymbol{P}$ we cannot go beyond (2.25). Now we make a very important assumption. We assume a linear relationship between $\boldsymbol{P}$ and $\boldsymbol{E}$:

$$\frac{1}{\varepsilon_0}\boldsymbol{P} = \chi\boldsymbol{E} \tag{2.26a}$$

or

$$\boldsymbol{D} = \varepsilon_0(1+\chi)\boldsymbol{E} = \varepsilon\varepsilon_0\boldsymbol{E} \tag{2.26b}$$

with

$$\varepsilon = \chi + 1 \, . \tag{2.27}$$

This linear relation is the reason why everything that is treated in the following Chaps. 3–17 is called linear optics. A linear relation is what one usually assumes between two physical quantities as long as one does not have more precise information. In principle we can also consider (2.26a) as an expansion of $\boldsymbol{P}(\boldsymbol{E})$ in a power series in $\boldsymbol{E}$ which is truncated after the linear term. We come back to this aspect in Chap. 18.[2] The quantities ε and χ are called the dielectric function and the susceptibility, respectively. They can be considered as linear response functions.

Both quantities depend on the frequency ω and on the wave-vector $\boldsymbol{k}$, and they both have a real and an imaginary part as shown for ε.

$$\varepsilon = \varepsilon(\omega, \boldsymbol{k})\,; \quad \chi = \chi(\omega, \boldsymbol{k}) = \varepsilon(\omega, \boldsymbol{k}) - 1 \, , \tag{2.28}$$

$$\varepsilon(\omega, \boldsymbol{k}) = \varepsilon_1(\omega, \boldsymbol{k}) + \mathrm{i}\varepsilon_2(\omega, \boldsymbol{k}) \, . \tag{2.29}$$

The frequency dependence is dominant and will be treated first in Chaps. 4–7. We drop the $\boldsymbol{k}$ dependence for the moment but come back to it in connection with spatial dispersion in Chap. 6. In Chap. 8 we discuss the properties of ε as a function of frequency and wave vector or as a function of time and space.

The value of $\varepsilon(\omega)$ for $\omega \simeq 0$ is usually called the dielectric constant.

In general ε and χ are tensors. For simplicity we shall consider them to be scalar quantities if not stated otherwise, e.g., in connection with birefringence in Sect. 3.1.7.

Using the linear relations of (2.26) we can transform (2.25) into

$$\nabla^2\boldsymbol{E} - \mu_0\varepsilon_0\varepsilon(\omega)\ddot{\boldsymbol{E}} = 0 \, . \tag{2.30a}$$

If magnetic properties are to be included, a corresponding linear approach would lead to

$$\nabla^2\boldsymbol{E} - \mu_0\mu(\omega)\varepsilon_0\varepsilon(\omega)\ddot{\boldsymbol{E}} = 0 \, , \tag{2.30b}$$

---

[2] A constant term in this power expansion such as $\boldsymbol{P} = \boldsymbol{P}_0 + \chi\boldsymbol{E}$ would describe a spontaneous polarization of matter which occurs e.g., in pyro- or ferro-electric materials. With arguments similar to the ones given for ferromagnetics we can neglect such phenomena in the discussion of the optical properties of semiconductors.

where $\mu(\omega)$ is the magnetic susceptibility. As outlined above we have in the visible for most semiconductors $\mu(\omega) \simeq 1$.

As for (2.25) the solutions of (2.30) are again all functions of the type

$$\boldsymbol{E} = \boldsymbol{E}_0 f(\boldsymbol{kr} - \omega t) , \tag{2.31}$$

or for our present purposes,

$$\boldsymbol{E} = \boldsymbol{E}_0 \exp[\mathrm{i}(\boldsymbol{kr} - \omega t)] . \tag{2.32}$$

The relationship between $\boldsymbol{k}$ and ω is however now significantly different from (2.11). It follows again from inserting the ansatz (2.31 or 32) into (2.30) and now reads:

$$\frac{c^2 \boldsymbol{k}^2}{\omega^2} = \varepsilon(\omega) . \tag{2.33}$$

This relation appears in Chap. 5 again under the name "polariton equation". It can also be written in other forms:

$$k = \frac{\omega}{c} \varepsilon^{1/2}(\omega) = \frac{2\pi}{\lambda_\mathrm{v}} \varepsilon^{1/2}(\omega) = k_\mathrm{v} \varepsilon^{1/2}(\omega) , \tag{2.34}$$

where λ_v and k_v refer to the vacuum values.

For the square root of ε we introduce for simplicity a new quantity $\tilde{n}(\omega)$ which we call the complex index of refraction

$$\tilde{n}(\omega) = n(\omega) + \mathrm{i}\kappa(\omega) = \varepsilon^{1/2}(\omega) . \tag{2.35}$$

The equations (2.11) and (2.33–35) can be interpreted in the following way. In vacuum an electromagnetic wave propagates with a wave vector $\boldsymbol{k}_\mathrm{v}$ which is real and given by (2.11). In matter, light propagates with a wave vector $\boldsymbol{k}$ which can be a complex quantity given by (2.34), or, with the help of (2.35), by

$$k = \frac{\omega}{c} \tilde{n}(\omega) = \frac{\omega}{c} n(\omega) + \mathrm{i} \frac{\omega}{c} \kappa(\omega) = \frac{2\pi}{\lambda_\mathrm{v}} \tilde{n}(\omega) = k_\mathrm{v} \tilde{n} . \tag{2.36}$$

We should notice that k is for complex $\tilde{n}$ not simply $|\boldsymbol{k}|$ since $|\boldsymbol{k}|$ is always a positive, real quantity. Here k means just neglecting the vector character of $\boldsymbol{k}$ but k can still be a real, imaginary or complex quantity according to (2.36). The direction of $\boldsymbol{k}$ is still parallel to $\boldsymbol{D} \times \boldsymbol{B}$ as in (2.20).

Writing the plane wave explicitly we have:

$$\boldsymbol{E}_0 \exp[\mathrm{i}(\boldsymbol{kr} - \omega t)] = \boldsymbol{E}_0 \exp\left\{\mathrm{i}\left[\frac{\omega}{c} n(\omega) \hat{\boldsymbol{k}}\boldsymbol{r} - \omega t\right]\right\} \exp\left[-\frac{\omega}{c} \kappa(\omega) \hat{\boldsymbol{k}}\boldsymbol{r}\right] , \tag{2.37}$$

where $\hat{\boldsymbol{k}}$ is the unit vector in the direction of $\boldsymbol{k}$, i.e., in the direction of propagation.

Obviously $n(\omega)$ describes the oscillatory spatial propagation of light in matter; it is often called the refractive index in connection with Snells' law of refraction. This means that the wavelength λ in a medium is connected with the wavelength λ_v in vacuum by

$$\lambda = \lambda_v n^{-1}(\omega) . \tag{2.38}$$

In (2.37) $\kappa(\omega)$ describes a damping of the wave in the direction of propagation. This effect is usually called absorption or, more precisely, extinction. We given the precise meaning of these two quantities in Sect. 3.1.5. Here we compare (2.37) with the well-known law of absorption for the light intensity I of a parallel beam propagating in z-direction

$$I(z) = I(z=0)\,\mathrm{e}^{-\alpha z} , \tag{2.39}$$

where $\alpha(\omega)$ is usually called the absorption coefficient, especially in Anglo-Saxon literature. In German literature $\alpha(\omega)$ is also known as "Absorptionskonstante" (absorption constant) and dimensionless quantities proportional to $\kappa(\omega)$ are called "Absorptionskoeffizient" or "Absorptions index" (absorption coefficient or absorption index). So some care has to be taken regarding what is meant by one or the other of the above terms.

Bearing in mind that the intensity is still proportional to the amplitude squared (2.22), a comparison between (2.37) and (2.39) yields

$$\alpha(\omega) = \frac{2\omega}{c}\,\kappa(\omega) = \frac{4\pi}{\lambda_v}\,\kappa(\omega) . \tag{2.40}$$

The phase velocity of light in a medium is now given by (2.13)

$$v_{\mathrm{ph}} = \frac{\omega}{\mathrm{Re}\{k\}} = c n^{-1}(\omega) . \tag{2.41}$$

For the group velocity we can get rather complicated dependencies originating from

$$v_{\mathrm{g}} = \frac{\partial \omega}{\partial k} . \tag{2.42}$$

We return to this aspect later.

2.4 Transverse, Longitudinal and Surface Waves

The only solution of (2.15) for light in vacuum is a transverse electromagnetic wave (2.19). This solution exists for light in matter as well. However (2.15) has now with the use of (2.26) the form

$$\nabla\cdot\boldsymbol{D} = \nabla\varepsilon_0\,\varepsilon(\omega)\boldsymbol{E} = 0 . \tag{2.43}$$

Apart from the above-mentioned transverse solution with $\boldsymbol{E} \perp \boldsymbol{k}$ there is a new solution which does not exist in vacuum ($\varepsilon_{\mathrm{vac}} \equiv 1$), namely

$$\varepsilon(\omega) = 0 .$$

This means that we can find longitudinal solutions at the frequencies at which $\varepsilon(\omega)$ vanishes. We call these frequencies correspondingly ω_L and note

$$\varepsilon(\omega_L)=0; \quad \boldsymbol{E} \| \boldsymbol{k} \text{ is possible} \tag{2.44}$$

Now let us consider the other fields for this longitudinal wave in matter. From (2.26) we see immediately that we have for the longitudinal modes

$$\boldsymbol{D}=0 \quad \text{and} \quad \boldsymbol{E}=-\frac{1}{\varepsilon_0}\boldsymbol{P}\,. \tag{2.45}$$

In matter, the Maxwell's equation $\nabla \times \boldsymbol{E} = -\dot{\boldsymbol{B}}$ is still valid. This leads for plane waves in nonmagnetic material to

$$\boldsymbol{H}_0=(\omega\mu_0)^{-1}\boldsymbol{k}\times\boldsymbol{E}_0\,. \tag{2.46}$$

For the longitudinal wave it follows from (2.44) that

$$\boldsymbol{H}_0=0 \quad \text{and} \quad \boldsymbol{B}=\mu_0\boldsymbol{H}=0\,. \tag{2.47}$$

The longitudinal waves which we found in matter are not electromagnetic waves but pure polarization waves with $\boldsymbol{E}$ and $\boldsymbol{P}$ opposed to each other with vanishing $\boldsymbol{D}, \boldsymbol{B}$ and $\boldsymbol{H}$.

Until now we were considering the properties of light in the bulk of a medium. The boundary of this medium will need some extra consideration e.g., the interface between vacuum (air) and a semiconductor. This interface is crucial for reflection of light and we examine this problem in Sects. 3.1, 4.4 and 6.2. Here we only want to state that the boundary conditions allow a surface mode, that is, a wave which propagates along the interface and has field amplitudes which decay exponentially on both sides. These waves are also known as surface polaritons for reasons discussed in more detail in Chap. 7.

2.5 Photons and Some Aspects of Quantum Mechanics and of Dispersion Relations

Maxwell's equations are the basis of the classical theory of light. They describe problems like light propagation and the diffraction at a slit or a grating e.g., in the frame of Huygen's principle.

In the interaction of light with matter, its quantum nature becomes apparent, e.g., in the photoelectric effect which shows that a light field of frequency ω can exchange energy with matter only in quanta $\hbar\omega$. Therefore, the proper description of light is in terms of quantum mechanics or of quantum electrodynamics. However, we shall not go through these theories here in detail nor do we want to address the aspects of quantum statistics of coherent and incoherent light sources, but we present in the following some of their well-known results and refer the reader to the corresponding literature [2.3] to [2.7] for a comprehensive discussion.

The electromagnetic fields can be described by their potentials $\boldsymbol{A}$ and ϕ by

$$\boldsymbol{E} = -\operatorname{grad}\phi - \dot{\boldsymbol{A}}; \quad \boldsymbol{B} = \nabla \times \boldsymbol{A} . \tag{2.48}$$

The so-called vector potential $\boldsymbol{A}$ is not exactly defined from (2.48). A gradient of a scalar field can be added. We can choose e.g., the so-called Coulomb gauge

$$\nabla \cdot \boldsymbol{A} = 0 . \tag{2.49}$$

In this case ϕ is the usual electrostatic potential obeying the Poisson equation:

$$\nabla^2 \phi = -\frac{\rho}{\varepsilon_0 \varepsilon(\omega)} . \tag{2.50}$$

In vacuum we still have $\rho = 0$ and we assume the same for the description of the optical properties of matter.

Now we should carry out the procedure of second quantization, for simplicity again for plane waves. A detailed description of how one begins with Maxwell's equations and arrives at photons within the framework of second quantization is beyond the scope of this book see [2.3–7]. On the other hand we want to avoid that the creation and annihilation operators appear like a "deus ex machina". Therefore we try at least to outline the procedure.

First we have to write down the classical Hamilton function H which is the total energy of the electromagnetic field using $\boldsymbol{A}$ and ϕ. Then we must find some new, suitable quantities $p_{\boldsymbol{k},s}$ and $q_{\boldsymbol{k},s}$ which are linear in $\boldsymbol{A}$ and which fulfill the canonic equations of motion

$$\frac{\partial H}{\partial q_{\boldsymbol{k},s}} = -\dot{p}_{\boldsymbol{k},s} , \quad \frac{\partial H}{\partial p_{\boldsymbol{k},s}} = \dot{q}_{\boldsymbol{k},s} , \tag{2.51}$$

and are thus canonically conjugate variables. Here $\boldsymbol{k}$ is the wave vector of our plane electromagnetic or $\boldsymbol{A}$-wave and s the two possible transverse polarizations. The Hamilton function reads in these variables:

$$H = \frac{1}{2} \sum_{\boldsymbol{k},s} (p_{\boldsymbol{k},s}^2) + \omega_{\boldsymbol{k}}^2 \, q_{\boldsymbol{k},s}^2 . \tag{2.52}$$

This is the usual form of the harmonic oscillator. The quantization condition

$$p_{\boldsymbol{k},s} \, q_{\boldsymbol{k},s} - q_{\boldsymbol{k},s} \, p_{\boldsymbol{k},s} = \frac{\hbar}{\mathrm{i}} \tag{2.53}$$

for all $\boldsymbol{k}$ and $s = 1, 2$ gives then the well-known result for the harmonic oscillator: The electromagnetic radiation field has for every $\boldsymbol{k}$ and polarization s energy steps

$$E_{\boldsymbol{k}} = \left(n_{\boldsymbol{k}} + \frac{1}{2} \right) \hbar \omega_{\boldsymbol{k}} \quad \text{with} \quad n_{\boldsymbol{k}} = 0, 1, 2, \ldots \tag{2.54}$$

It can exchange energy with other systems only in units of $\hbar\omega$. These energy units or quanta are called photons. The term $\hbar\omega/2$ in (2.54) is the zero-point energy of every mode of the electromagnetic field.

The so-called particle-wave dualism, that is, the fact that light propagates like a wave showing, e.g., diffraction or interference and interacts with matter via particle-like quanta, can be solved by the simple picture that light is an electromagnetic wave, the amplitude of which can have only discrete values so that the energy in the waves just fulfills (2.54).

From the above introduced, or better, postulated quantities $p_{k,s}$ and $q_{k,s}$ we can derive by linear combinations operators $a^{\dagger}_{k,s}$ and $a_{k,s}$ with the following properties: If $a_{k,s}$ acts on a state which contains $n_{k,s}$ quanta of momentum $\boldsymbol{k}$ and polarization s it produces a new state with $n_{k,s}-1$ quanta. Correspondingly, $a^{\dagger}_{k,s}$ increases $n_{k,s}$ by one. We call therefore $a_{k,s}$ and $a^{\dagger}_{k,s}$ annihilation and creation operators, respectively. Since the operators $a_{k,s}$ and $a^{\dagger}_{k,s}$ describe bosons (see below), their permutation relation is

$$a_{k,s}a^{\dagger}_{k,s} - a^{\dagger}_{k,s}a_{k,s} = 1 \ . \tag{2.55a}$$

This holds for equal $\boldsymbol{k}$ and s and is zero otherwise.

The operator $a^{\dagger}_{k,s}a_{k,s}$ acting on a photon state gives the number of photons $n_{k,s}$ times the photon state and is therefore called the number operator. Summing over all possible $\boldsymbol{k}$-values and polarizations s gives finally the Hamilton operator

$$H = \sum_{\mathbf{k},s} \hbar\omega_{k,s} a^{\dagger}_{k,s} a_{k,s} \ . \tag{2.55b}$$

It is clear to the author that the short outline given here is not sufficient to explain the procedure to a reader who is not familiar with it. However, since the intent is not to write a textbook on quantum electrodynamics, we want to stress here only that the electromagnetic radiation field in vacuum can be brought into a mathematical form analogous to that of the harmonic oscillator, and that quantum mechanics gives for every harmonic oscillator the energetically equidistant terms of (2.54).

The harmonic oscillator is one of the fundamental systems, which has been investigated in physics and is understood in great detail. In theoretical physics a problem can be considered as "solved" if it can be rewritten in the form of the harmonic oscillator. Apart from the electromagnetic radiation field in vacuum, we will come across some other systems which are treated in this way. For those readers who are not familiar with the concept of quantization and who wish to study the procedure in a quiet hour by themselves, we recommend [2.3] to [2.7] or the references given therein.

Here are some more results: The two basic polarizations of single quanta of the electromagnetic field, – of the photons – are left and right circular σ^- and σ^+. A linearly polarized wave can be considered as a coherent superposition of a left and right circularly polarized one. The term coherent means that two light beams have a fixed-phase relation relative to each other. The component of the angular momentum s in the direction of the quantization axis which is parallel to $\boldsymbol{k}$ is for photons thus

$$s_{\parallel} = \pm\hbar \; . \tag{2.56}$$

This means that photons have integer spin and are bosons (2.5a). The third possibility $s_{\parallel} = 0$ expected for spin one particles is forbidden, because longitudinal electromagnetic waves do not exist (2.19) at least in vacuum.

Photons in thermodynamic equilibrium are described by Bose-statistics. The occupation probability f_{BE} of a state with frequency ω is given by

$$f_{\mathrm{BE}} = [\exp(\hbar\omega/k_{\mathrm{B}}T) - 1]^{-1} \; , \tag{2.57}$$

where T is the temperature and k_{B} is Boltzmann's constant.

The chemical potential μ which could appear in (2.57) is zero, since the number of photons is not conserved.

An approach to describe non-thermal photon fields e.g. luminescence by a non-vanishing μ is found in [2.8].

The momentum $\boldsymbol{p}$ of a photon with wave vector $\boldsymbol{k}$ is given, as for all quanta of harmonic waves, by

$$\boldsymbol{p} = \hbar\boldsymbol{k} \; . \tag{2.58}$$

To summarize, we can state that photons are bosons with spin $\pm\hbar$, energy $\hbar\omega$ and momentum $\hbar\boldsymbol{k}$ which propagate according to the wave equations.

A very important property of particles in quantum mechanics is their dispersion relation. By this we mean the dependence of energy E or frequency ω on the wave vector $\boldsymbol{k}$ i.e., the $E(\boldsymbol{k})$ or $\omega(\boldsymbol{k})$ relation. For photons in vacuum we find the classical relation given already in (2.11)

$$E = \hbar\omega = \hbar ck \; . \tag{2.59}$$

The dispersion relation for photons in vacuum is thus a linear function with slope $\hbar c$ as shown in Fig. 2.2. Correspondingly we find again both for phase and group velocity with (2.13)

$$v_{\mathrm{ph}} = v_{\mathrm{g}} = c \; . \tag{2.60}$$

2.6 Density of States and Occupation Probabilities

A quantity which is crucial in quantum mechanics for the properties of particles is their density of states. It enters, e.g., in Fermi's golden rule which allows one to calculate transition probabilities. We want to discuss this problem in a general way for systems of different dimensionalities $d = 3, 2$ and 1. We shall need these results later on for low-dimensional semiconductor structures. The discussion of the density of states, especially in various dimensions, is not so commonly treated as the harmonic oscillator, and so we shall spend some time on this problem and delve more into details. At the end of this

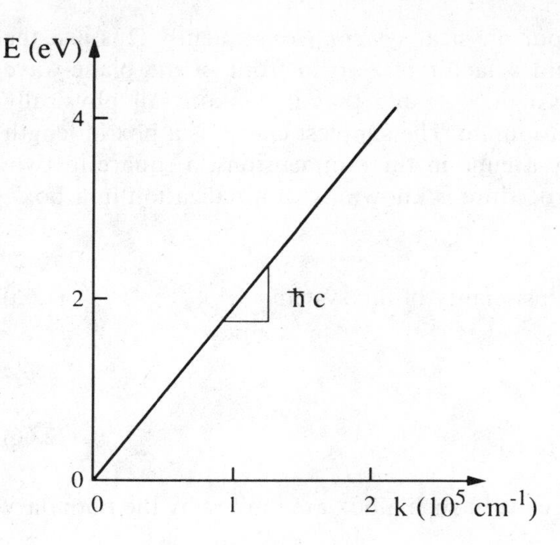

Fig. 2.2. The dispersion relation of photons in vacuum

section we shall also state the occupation probability in thermodynamic equilibrium for classical particles, for fermions and bosons.

If we consider a particle which is described by a wave function[3] $\phi(\boldsymbol{r})$ then the probability w to find it in a small element of space $\mathrm{d}\tau = \mathrm{d}x\,\mathrm{d}y\,\mathrm{d}z$ around $\boldsymbol{r}$ is

$$w(\boldsymbol{r})\,\mathrm{d}\tau = \phi^*(\boldsymbol{r})\phi(\boldsymbol{r})\,\mathrm{d}\tau\ . \tag{2.61}$$

Since the particle has to be somewhere in the system, $w(\boldsymbol{r})$ has to be "normalized", that is,

$$\int_{\text{system}} w(\boldsymbol{r})\,\mathrm{d}\tau = \int_{\text{system}} \phi^*(\boldsymbol{r})\phi(\boldsymbol{r})\,\mathrm{d}\tau = 1\ . \tag{2.62}$$

Here, the functions $\phi(\boldsymbol{r})$ are of the form exp $(\mathrm{i}\boldsymbol{kr})$. For normalization purposes a factor has to be added

$$\phi(\boldsymbol{r}) = \Omega^{-1/2}\exp(\mathrm{i}\boldsymbol{kr})\ . \tag{2.63}$$

The normalization condition (2.62) results in

$$\Omega^{-1}\int_{\text{system}} \exp(-\,\mathrm{i}\boldsymbol{kr})\exp(\mathrm{i}\boldsymbol{kr})\,\mathrm{d}\tau = \Omega^{-1}\int_{\text{system}} \mathrm{d}\tau = \Omega^{-1}\,V_{\text{system}} = 1\ , \tag{2.64}$$

[3] The letter ϕ has been already used for the electrostatic potential e.g., in (2.50). Since there are more different physical quantities than letters of the alphabet, we sometimes use the same letter for different quantities, but from the context it should be clear what is meant.

where V_{system} is the volume of our physical system. Consequently Ω is just the volume of the system. To avoid a factor of zero in front of the plane-wave term, one assumes that the system is so big that it contains all physically relevant parts, but that it is not infinite. The simplest choice is a box of length L, or, more precisely speaking, a cube in three dimensions, a square in two, and an interval in one. This procedure is known as "normalization in a box". Consequently we have

$$V_{\text{system}} = L^d \quad \text{with} \quad d = \text{dimensionality of the system} \tag{2.65}$$

and

$$\Omega^{1/2} = L^{d/2} \quad \text{for} \quad d = 3, 2, 1 \; . \tag{2.66}$$

The wave vectors which can exist in such a box are limited by the boundary conditions.

If we assume that we have an infinitely high potential barrier around the box, then the wavefunction must have nodes at the walls (Fig. 2.3a).

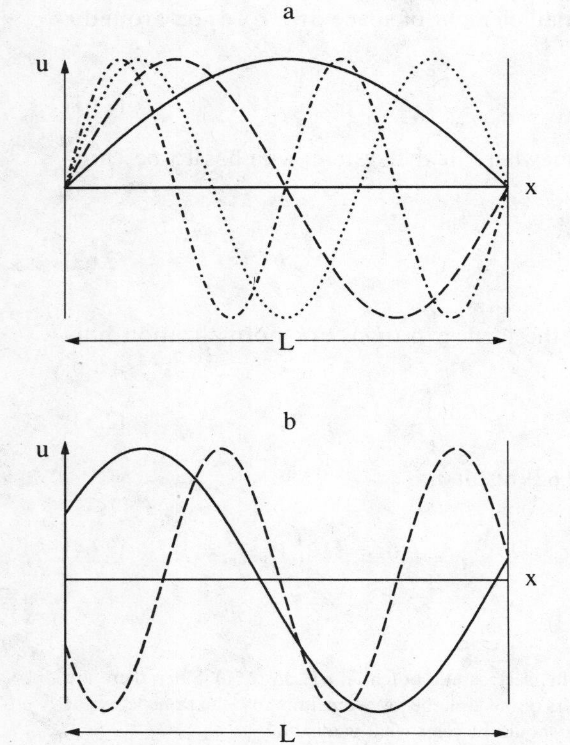

Fig. 2.3a, b. Plane waves which have nodes at the boundaries (**a**) or which obey periodic boundary conditions (**b**)

Consequently the components k_i of $\boldsymbol{k}$ must fulfill

$$k_i = n_i \frac{\pi}{L}; \quad n_i = 1, 2, 3 \ldots; \quad i = 1, \ldots, d\,, \tag{2.67}$$

where the index i runs over all dimensions.

Such a wave is a standing wave, i.e., a coherent superposition of two waves with $\boldsymbol{k}$ and $-\boldsymbol{k}$ and equal amplitudes. In the following we must consider therefore only positive values of $\boldsymbol{k}$. The various modes are distributed equally spaced over the k_i-axes with a spacing Δk_i

$$\Delta k_i = \frac{\pi}{L}\,. \tag{2.68}$$

In other words, every state (or mode) needs a volume V_k in $\boldsymbol{k}$-space given by

$$V_k = \left(\frac{\pi}{L}\right)^d\,. \tag{2.69}$$

Another approach is to impose periodic boundary conditions. Then the plane wave should have equal amplitude and slope on opposite sides of the cube according to Fig. 2.3b. In this case one can fill the infinite space by adding boxes in all dimensions and one finds:

$$k_i = n_i' \frac{2\pi}{L}; \quad n_i' = \pm 1, \pm 2, \pm 3 \ldots \tag{2.70}$$

This means for Δk_i

$$\Delta k_i' = \frac{2\pi}{L}\,. \tag{2.71}$$

In contrast to the case of standing waves, we now have to consider positive and negative values of n_i separately. This procedure results finally in the same density of states.

As a consequence we find that plane waves have in Cartesian coordinates in $\boldsymbol{k}$-space a constant density on all axes.

Often one wants to know the number of states in a shell between k and $k + \mathrm{d}k$ independent of the direction of $\boldsymbol{k}$. This question can be answered by introducing polar coordinates in $\boldsymbol{k}$-space. The differential volume $\mathrm{d}V_k$ of a shell of thickness dk in a d-dimensional $\boldsymbol{k}$-space is given by

$$\begin{aligned} \mathrm{d}V_k &= 2dk \quad \text{for} \quad d = 1\,, \\ \mathrm{d}V_k &= 2\pi k\, dk \quad \text{for} \quad d = 2\,, \\ \mathrm{d}V_k &= 4\pi k^2 dk \quad \text{for} \quad d = 3\,. \end{aligned} \tag{2.72}$$

Depending on the boundary condition we have to take into account only positive (2.67), or positive and negative (2.70), values of $\boldsymbol{k}$ or n_i.

The number $\hat{D}(\boldsymbol{k})$ of states in $\boldsymbol{k}$-space found between k and $k+dk$ in polar coordinates is given by dividing dV_k by the volume for each state and by multiplying by g_s. The quantity g_s considers degeneracies such as the spin degeneracy. For photons we have $g_s = 2$ according to the σ^+ and σ^- polarizations (see above). The results are

$$\hat{D}(k)dk = g_s \frac{L}{\pi} dk \quad \text{for} \quad d=1\,,$$

$$\hat{D}(k)dk = g_s \frac{L^2}{2\pi} k dk \quad \text{for} \quad d=2\,,$$

$$\hat{D}(k)dk = g_s \frac{L^3}{2\pi^2} k^2 dk \quad \text{for} \quad d=3\,. \tag{2.73}$$

The derivation of this result is depicted for $d=2$ in Fig. 2.4.

If we neglect constant prefactors and divide by dk we find

$$\hat{D}(k) \propto g_s L^d k^{d-1}\,, \quad d=1,2,3\,\ldots\,. \tag{2.74}$$

If we consider not the number of states in the box of volume L^d but the density of states $D(k)$ per unit of space (e.g., per cm^3 or m^3) the term L^d in (2.74) disappears yielding

$$D(k) \propto g_s k^{d-1} \tag{2.75}$$

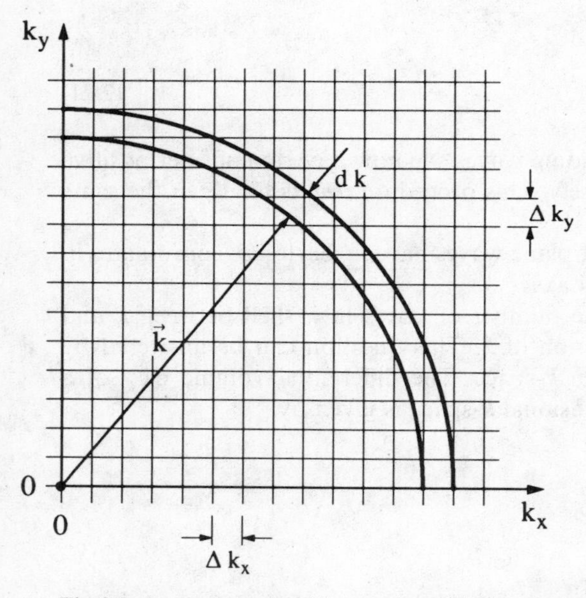

Fig. 2.4. A two-dimensional $\boldsymbol{k}$-space in which the states are equally spaced on the k_x-and k_y-axes to derive (2.73) for $d=2$

This result has to be expected since the density of states per unit volume must be independent of the size of the box which we have in mind provided the box is sufficiently large.

We want to stress here that we assumed only plane waves but did not make any specific assumptions about which type of particles are represented by these plane waves – photons, electrons etc. Therefore this result is valid for all particles described by plane waves.

The next step is now to calculate the density of states on the energy axis, i.e.,

$$D(E)\,\mathrm{d}E\,. \tag{2.76}$$

This quantity gives the number of states in the energy interval from E to $E+\mathrm{d}E$. To calculate this quantity we need the specific dispersion relation $E(\boldsymbol{k})$ and its inverse $\boldsymbol{k}(E)$ as seen from the identity (2.77):

$$D(E)\mathrm{d}E = D[k(E)]\frac{\mathrm{d}k}{\mathrm{d}E}\cdot\mathrm{d}E = D[k(E)]\frac{1}{|\mathrm{grad}_k E(\boldsymbol{k})|}\,\mathrm{d}E\,. \tag{2.77}$$

The term on the right-hand side of (2.77) gives the generalized equation which is also valid for anistropic cases.

In particular for photons in vacuum we have with (2.59)

$$k=\frac{E}{\hbar c}=\frac{\omega}{c};\quad \frac{\mathrm{d}k}{\mathrm{d}E}=\frac{1}{\hbar c}\,. \tag{2.78}$$

Inserting this result in (2.77), for the case $d=3$ we find

$$D(E)\mathrm{d}E=\frac{E^2}{\pi^2(\hbar c)^3}\,\mathrm{d}E$$

or

$$D(\omega)\mathrm{d}\omega \propto \omega^2\mathrm{d}\omega\,. \tag{2.79}$$

The next quantity, which we need is the occupation probability of the states discussed above. We restrict ourselves in the following to thermodynamic equilibrium. There are three types of statistics which can be considered:

For classical, distinguishable particles, Boltzmann statistics apply:

$$f_{\mathrm{B}}=\exp[-(E-\mu)/k_{\mathrm{B}}T]\,. \tag{2.80a}$$

For bosons, i.e., indistinguishable particles with integer spin, photons being an example, one must use the Bose–Einstein statistics

$$f_{\mathrm{BE}}=\{\exp[(E-\mu)/k_{\mathrm{B}}T]-1\}^{-1}\,. \tag{2.80b}$$

Fermions, or indistinguishable particles with half-integer spin e.g., electrons obey the Fermi–Dirac statistics f_{FD}

$$f_{\mathrm{FD}}=\{\exp[(E-\mu)/k_{\mathrm{B}}T]+1\}^{-1}\,. \tag{2.80c}$$

The Boltzmann constant is k_B and the chemical potential is μ which gives the average energy necessary to add one more particle to the system. For fermions μ is also known as the Fermienergy E_F. The probability to find a particle in the interval from E to $E + dE$ is then given by the product of the density of states $D(E)$ and the occupation probability f

$$D(E)\, f(E, T, \mu)\, dE\ . \tag{2.81}$$

In Fig. 2.5 we plot f_B, f_{BE} and f_{FD} as a function of $(E-\mu)/k_B T$.

The Boltzmann statistics show the well-known exponential dependence. The Fermi–Dirac statistics never exceed one, realizing thus Pauli's exclusion principle. The Bose–Einstein statistics have a singularity for $E = \mu$. This gives rise to Bose – Einstein condensation, or in other words, a macroscopic population of a single state, if μ falls in a region with a finite density of states. In this case the species with energies $E = \mu$ and those with $E > \mu$ must be considered separately. Furthermore it is obvious from Fig. 2.5 that f_{BE} and f_{FD} converge to f_B for $(E-\mu)/kT > 1$.

The chemical potential μ is zero for quanta whose number is not conserved, for e.g., photons or phonons. We introduce this topic in Chap. 9.

If the number N or density n of particles in a system is known, as is the case for electrons at non-relativistic energies, then μ is well defined by (2.82).

$$\int D(E)\, f(E, \mu, T)\, dE = n\ , \tag{2.82}$$

which says that the density of particles is equal to the integral over the product of the density of states and the probability that a state is occupied.

As an example, we apply now the above statements to photons in a three-dimensional box in thermodynamic equilibrium. With (2.79b) and (2.80b), (2.83) is obtained.

$$N(\omega)d\omega = D(\omega) f_{BE}(\omega, T)d\omega \propto \omega^2 \left[\exp\left(\frac{\hbar\omega}{k_B T}\right) - 1\right]^{-1} d\omega\ . \tag{2.83}$$

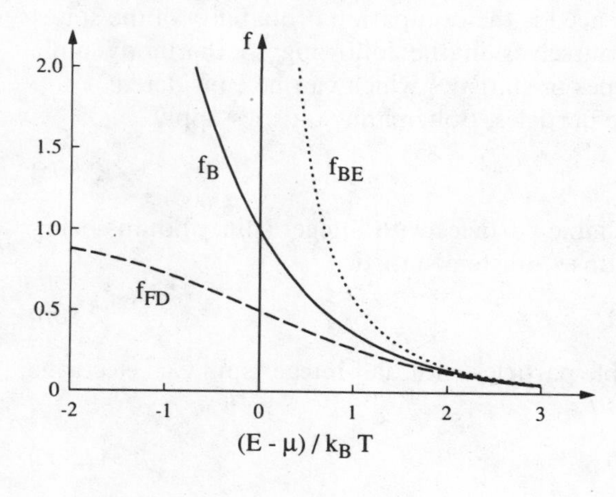

Fig. 2.5. The Boltzmann, Bose–Einstein and Fermi–Dirac distribution functions versus $(E-\mu)/ k_B T$

For the energy content $U(T)$, the result is

$$U(T) = \hbar \int_0^\infty \omega N(\omega) \mathrm{d}\omega \propto \int_0^\infty \omega^3 \left[\exp\left(\frac{\hbar\omega}{k_B T}\right) - 1 \right]^{-1} \mathrm{d}\omega \,. \tag{2.84}$$

Equation (2.83) is nothing other than Planck's law of black-body radiation. By substituting x for the variable $\hbar\omega/k_B T$ in (2.84) we immediately find the Stefan–Boltzmann T^4 law

$$U(T) \propto T^4 \int_0^\infty x^3 (\exp x - 1)^{-1} \mathrm{d}x = T^4 A \,, \tag{2.85}$$

where A is a constant.

2.7 Problems

1. The intensity of the sunlight falling on the earth is, for normal incidence and before its passage through the atmosphere, about 1.5 kW/m^2. Calculate the electric-field strength.

2. Pulsed high power lasers can be easily focussed to a power density I of 100 MW/cm^2. Calculate the E and B fields. Compare them with the electric field in an H atom at a distance of one Bohr radius, and the magnetic field on the surface of the earth, respectively.

3. Calculate the number of photon modes in the visible part of the spectrum ($\approx 400\,\mathrm{nm} \lesssim \lambda_v \lesssim 800\,\mathrm{nm}$) in a box of 1 cm^3.

4. Calculate the momentum and energy of a photon with $\lambda_v = 500\,\mathrm{nm}$. At which acceleration voltage has an electron the same momentum?

5. Show qualitatively the $\boldsymbol{B}$, $\boldsymbol{H}$ and $\boldsymbol{M}$ fields of a homogeneously magnetized, brick-shaped piece of iron.

6. Check whether the maximum of $N(\omega)$ in (2.83) shifts in proportion to T (Wien's law), originally formulated as $\lambda_{max} \propto T^{-1}$.

3 Interaction of Light with Matter

In the next two sections we present some basic interaction processes of light with matter from two different points of view. First we consider matter as a homogeneous medium described by the complex dielectric function $\varepsilon(\omega)$ or by the complex index of refraction $\tilde{n}(\omega)$ (Sect. 3.1). We concentrate especially on the reflection and transmission of light at the plane interface between two media. As an especially simple case we investigate the boundary of matter and vacuum. In the later section 3.2 we will discuss the interaction of the radiation field with individual atoms. In this case quantum mechanics must be used. We will employ what we have explored in Chap. 2.

3.1 Macroscopic Aspects for Solids

3.1.1 Boundary Conditions

Let us start with the macroscopic description of the optical properties of semiconductors. In Fig. 3.1 we show the wave vectors and field amplitudes in the vicinity of the interface between two media for two linear polarizations. In Fig. 3.1a the electric field $\boldsymbol{E}_\mathrm{i}$ of the incident beam is polarized parallel to the plane of incidence, which is defined by the wave vector of the incident light $\boldsymbol{k}_\mathrm{i}$ and the normal to the plane interface $\boldsymbol{e}_\mathrm{n}$. As we will see later from the boundary conditions, the wave vectors and the electric fields of transmitted and reflected beams (indices tr and r, respectively) are in the same plane; the magnetic fields according to (2.19) and (2.43) perpendicular to it. In Fig. 3.1b we have just the opposite situation for **E** and **B**.

One often assumes for simplicity that the medium I is vacuum (or air), i.e., $\varepsilon_\mathrm{I}(\omega) = \tilde{n}_\mathrm{I}(\omega) \equiv 1$. We do not use this approximation here but we still assume that media I and II are isotropic. This means $\varepsilon_\mathrm{I}(\omega)$ and $\varepsilon_\mathrm{II}(\omega)$ are scalar functions and $\tilde{n}(\omega)$ does not depend on orientation. Phenomena which appear if we drop this assumption are dealt with later. Furthermore, we assume that there is only one reflected and one transmitted beam. This assumption seems trivial and it is indeed for reflection. In transmission there may be more than one propagating beam, as we shall discuss in Chap. 5. What we first want to know are the dependences of the angles α_r and α_tr on α_i, i.e., the laws of reflection and refraction. We then want to know the coefficients of reflection r

and transmission t of the interface between media I and II. We can define these coefficients for the field amplitudes E_0.

$$r_\parallel = \frac{E_{0r}}{E_{0i}}; \quad t_\parallel = \frac{E_{0tr}}{E_{0i}} \tag{3.1a}$$

for the configuration of Fig. 3.1a or

$$r_\perp = \frac{E_{0r}}{E_{0i}}; \quad t_\perp = \frac{E_{0tr}}{E_{0i}} \tag{3.1b}$$

for the configuration of Fig. 3.1b.

However, what is usually measured is the reflectivity R and transmittivity T of an interface for the intensities. We have

$$R_{\perp,\parallel} = |r_{\perp,\parallel}|^2 \tag{3.2}$$

because incident and reflected beams propagate in the same medium.

In order to calculate all the quantities given above we need the corresponding number of equations. They can be deduced from Maxwell's equations as boundary conditions which must be fulfilled at the interface. To do so, we need two general laws of vector analysis which are known as the laws of Gauß and of Stokes, respectively. They read for a given vector field $\boldsymbol{A}$ [2.1].

$$\int_{\text{volume}} \nabla \cdot \boldsymbol{A} \, d\tau = \oint_{\text{surface}} \boldsymbol{A} \cdot d\boldsymbol{f} \tag{3.3a}$$

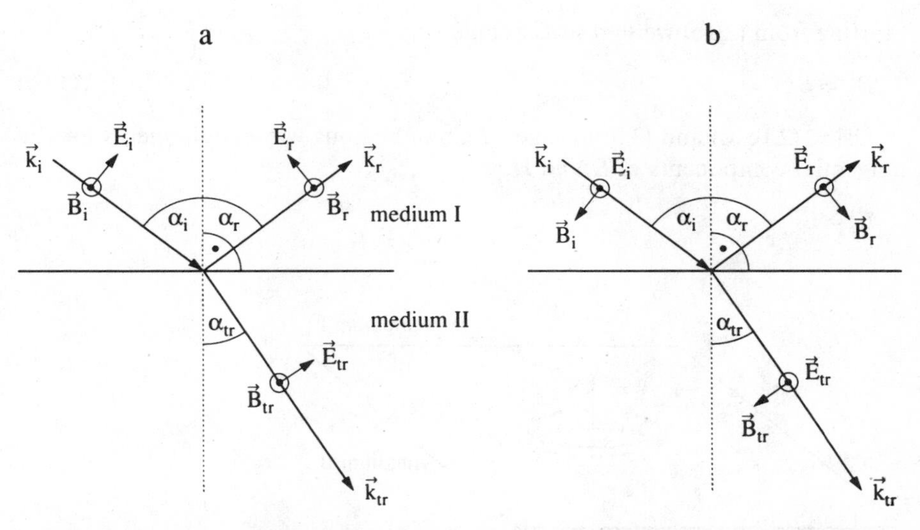

Fig. 3.1a, b. The $\boldsymbol{E}$ and $\boldsymbol{B}$ fields and the wave vectors $\boldsymbol{k}$ for incident, transmitted and reflected beams at an interface between two isotropic media for two different, orthogonal, linear polarizations **(a)** and **(b)** respectively

and

$$\int_{\text{surface}} (\nabla \times \boldsymbol{A}) \cdot \mathrm{d}\boldsymbol{f} = \oint_{\text{line}} \boldsymbol{A}\, \mathrm{d}\boldsymbol{s} . \tag{3.3b}$$

Starting with (2.1a), with the help of (3.3a), we obtain

$$\int_V \operatorname{div} \boldsymbol{D}\, \mathrm{d}\tau = \oint \boldsymbol{D} \cdot \mathrm{d}\boldsymbol{f} = \int_V \rho\, \mathrm{d}\tau . \tag{3.4}$$

We choose the integration volume in the form of a tiny (differentially small) cylinder which contains the interface and has its top and bottom in the media I and II, respectively (Fig. 3.2). Furthermore it is assumed that the ratio of the height to the radius of this cylinder is also infinitesimally small, so that the contribution to the whole integral from the lateral surface of the cylinder is negligible. Then the middle and right-hand-side terms of (3.4) yield

$$(\boldsymbol{D}_{\mathrm{I}} - \boldsymbol{D}_{\mathrm{II}}) \cdot \mathrm{d}\boldsymbol{f} = (D_{\mathrm{n,I}} - D_{\mathrm{n,II}})\, \mathrm{d}f = \rho_{\mathrm{s}}\, \mathrm{d}f , \tag{3.5}$$

where the index n means the normal component and ρ_s a surface charge density.

Since we assumed that there are no free charges ρ at all (Sect. 2.3) and consequently no surface charges, the right-hand side of (3.5) vanishes and we find as a boundary condition that the normal component of $\boldsymbol{D}$ is continuous across the interface:

$$D_{\mathrm{n}}^{\mathrm{I}} = D_{\mathrm{n}}^{\mathrm{II}} . \tag{3.6a}$$

Starting from (2.1b) we find in the same way

$$B_{\mathrm{n}}^{\mathrm{I}} = B_{\mathrm{n}}^{\mathrm{II}} . \tag{3.6b}$$

Using (2.1c, d) and (3.3b) we get in an analogous way requirements for the tangential components of $\boldsymbol{E}$ and $\boldsymbol{H}$.

$$E_{\mathrm{t}}^{\mathrm{I}} = E_{\mathrm{t}}^{\mathrm{II}} \tag{3.7a}$$

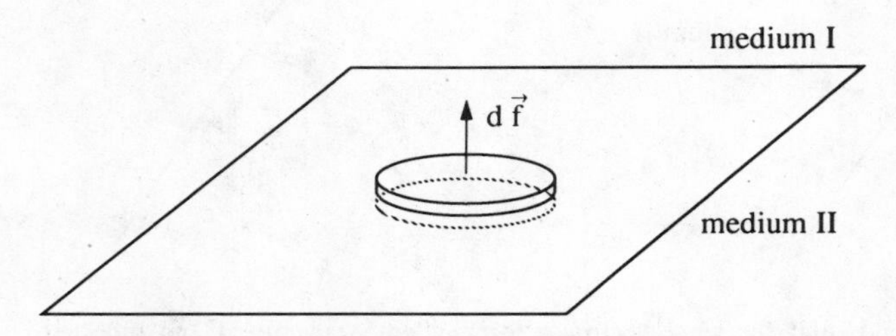

Fig. 3.2. Schematic drawing of the infinitesimally small cylinder used to deduce with (3.4) the boundary condition (3.6)

and

$$H_t^{\mathrm{I}} = H_t^{\mathrm{II}} . \tag{3.7b}$$

Equations (3.6) and (3.7) represent the boundary conditions for electric and magnetic fields. Actually only two of them are independent, the other ones follow directly with the linear approaches (2.26). This is enough to calculate for a given incident beam the properties of the reflected and refracted ones. In order to do so the boundary conditions must be applied to a specific problem. For the configuration of Fig. 3.1 this reads for the incident, reflected and transmitted electric fields

$$\boldsymbol{e}_{\mathrm{n}} \times \boldsymbol{E}_{\mathrm{i}} - \boldsymbol{e}_{\mathrm{n}} \times \boldsymbol{E}_{\mathrm{r}} = \boldsymbol{e}_{\mathrm{n}} \times \boldsymbol{E}_{\mathrm{tr}} , \tag{3.8}$$

since the vector product of the various $\boldsymbol{E}_s$ with the unit vector $\boldsymbol{e}_{\mathrm{n}}$ normal to the interface generate just the tangential components. Using another one of the set of equations (3.7,8) allows one to calculate the properties of the reflected and transmitted beams. This procedure involves some basically simple but lengthy algebra and does not give further insight into the physics. In accordance with the concept of this book, we consequently skip these calculations which can be found in the literature, (See e.g. [3.1] to [3.3] and references therein) but present the results giving some cross-links to other physical approaches to obtain them.

3.1.2 Laws of Reflection and Refraction

The first, not too surprising, result from the above-mentioned procedure is

$$\omega_{\mathrm{i}} = \omega_{\mathrm{r}} = \omega_{\mathrm{tr}} . \tag{3.9}$$

This means all three beams have the same frequency. This becomes clear from classical physics, as we shall see in Chap. 4 since atoms oscillate with frequency ω_i under the influence of the incident field and can therefore radiate, according to the linear approach (2.26), only at this frequency. The relation (3.9) is also intelligible from the point of view of quantum mechanics, bearing in mind the law of energy conservation and the fact that a single photon has energy $\hbar\omega$ and can be either reflected or refracted.

The next results are the laws of reflection and Snell's law of refraction. The first one states

$$\alpha_{\mathrm{i}} = \alpha_{\mathrm{r}} \tag{3.10a}$$

and;

$$\boldsymbol{k}_{\mathrm{i}}, \boldsymbol{k}_{\mathrm{r}} \text{ and } \boldsymbol{e}_{\mathrm{n}} \text{ are in one plane,} \tag{3.10b}$$

namely in the above-introduced plane of incidence.

The second one reads

$$\frac{\sin\alpha_i}{\sin\alpha_{tr}} = \frac{n_{II}}{n_I}; \quad \boldsymbol{k}_i, \boldsymbol{k}_{tr} \text{ and } \boldsymbol{e}_n \text{ in one plane}. \tag{3.10c}$$

In Fig. 3.1 we manifest the situation for $n_I < n_{II}$, i.e., the refraction from an optically thinner into an optically thicker medium. In the opposite case, one reaches a critical angle α_i^c for which $\alpha_{tr} = 90°$ given by the condition

$$\alpha_i^c = \arcsin\frac{n_{II}}{n_I}. \tag{3.11}$$

For $\alpha_i \geqslant \alpha_i^c$ there is a totally reflected beam but no longer a transmitted one. However, the boundary conditions (3.6) and (3.7) require finite field amplitudes in medium II. Inspection of the boundary conditions shows that a so-called evanescent wave exists in medium II which propagates parallel to the surface. Its field-amplitudes decay exponentially in the direction normal to the interface over a distance of a few wavelengths, as shown schematically in Fig. 3.3a. The reflected wave has under these conditions the same intensity as the incident one. Correspondingly the phenomen is known as total (internal) reflection.

If medium II has only a thickness of the order of a wavelength and is then covered by material I again, then the evanescent wave couples into this medium giving rise to a propagating refracted wave (Fig. 3.3b). Consequently the intensity of the reflected wave decreases. This phenomenon is called attenuated, or frustrated, total reflection (ATR) or the optical tunnel effect in analogy to the quantum-mechanical tunnel effect.

The laws (3.10, 11) can be also deduced from the principle of Maupertius or Fermat, which says that for geometrical optics the optical path length, i.e., the product of the geometrical path length and the refractive index n between two

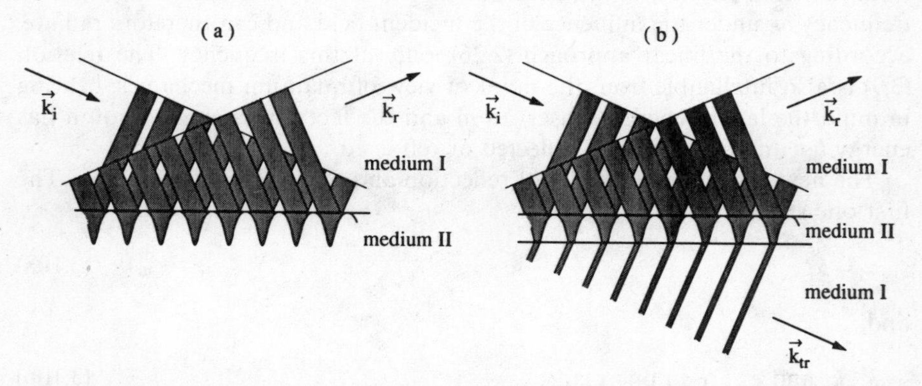

Fig. 3.3a, b. The evanescent wave in the case of total internal reflection (**a**) and the arrangement for the "optical tunneling" effect or attenuated total reflection (**b**)

points A and B is an extremum, generally a minimum. This is shown schematically for the case of refraction in Fig. 3.4. From all in principle possible ways to travel from A to B, the light propagation is along the one for which (3.12) holds.

$$\delta \int n \mathrm{d}s = \delta(n_{\mathrm{I}} \overline{AC} + n_{\mathrm{II}} \overline{CB}) = 0 \;, \tag{3.12}$$

i.e., the variation δ of the optical path length vanishes. Equation (3.10) can be deduced from (3.12).

3.1.3 Noether's Theorem and Some Aspects of Conservation Laws

A third way of deriving (3.9, 10) relies on the law of momentum conservation. Since conservation laws are essential in other fields as well as in physics (See problem 1 of Chap. 1) we shall dwell here on them for some time.

We start with the theorem of E. Noether, which is usually not taught in standard physics courses, though it is of great importance. In simple words it says:

A conservation law follows from every invariance of the Hamilton operator H.

We are not going to prove this statement. Instead we give some well known applications.

If H is invariant against infinitesimal translations in time $\mathrm{d}t$, i.e., if H does not depend explicitly on time, then the total energy E of the system described by H is conserved

$$H(t) = H(t + \mathrm{d}t) \longrightarrow E = \text{const.} \tag{3.13a}$$

If H is invariant against an infinitesimal translation along an axis x, then the x-component of the total momentum $\boldsymbol{p}$ is conserved

$$H(x) = H(x + \mathrm{d}x) \longrightarrow p_x = \text{const.} \tag{3.13b}$$

If H is invariant against an infinitesimal rotation $\mathrm{d}\phi$, e.g., around an axis z, that is, $\mathbf{d}\boldsymbol{\phi} = (0, 0, \mathrm{d}\phi)$ then the z-component of the angular moment $\boldsymbol{L}$ is

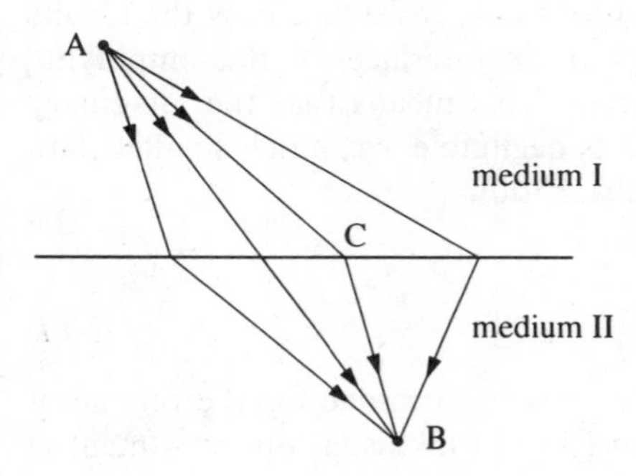

Fig. 3.4. Various possible paths of a light beam travelling from A to B

conserved

$$H(\mathbf{r}) = H(\mathbf{r} - \mathbf{r} \times \mathbf{d}\phi_z) \longrightarrow L_z = \text{const.} \tag{3.13c}$$

The axis along which (3.13c) is valid is called the quantization axis.

For the problem of light reflection and refraction at an interface, (3.13a) is still valid, resulting in energy conservation which we used, e.g., in connection with (3.9). Concerning (3.13b) the problem is only invariant against infinitesimal translation parallel to the interface. Correspondingly only the momentum parallel to this interface is conserved. We learned with (2.58) that the momentum of photons (and of all other free particles) is $\hbar \boldsymbol{k}$. Consequently at the interface the conservation laws must be fufilled.

$$k_{\mathrm{i}\|} = k_{\mathrm{r}\|}\,, \tag{3.14a}$$

$$k_{\mathrm{i}\|} = k_{\mathrm{tr}\|}\,. \tag{3.14b}$$

Since incident and reflected beams propagate in the same medium, the lengths of the wave vectors are equal, too.

$$|\boldsymbol{k}_{\mathrm{i}}| = |\boldsymbol{k}_{\mathrm{r}}|\,, \tag{3.15}$$

The only solution for (3.14a) is then obviously the law of reflection (3.10). For the relation of $\boldsymbol{k}_{\mathrm{i}}$ and $\boldsymbol{k}_{\mathrm{t}}$ we find accordingly in addition to (3.15)

$$k_j = k_{\mathrm{vac}}\, n_j = \frac{\omega}{c} n_j\,; \quad j = \mathrm{I, II}\,. \tag{3.16}$$

The simultaneous solution of (3.15) and of (3.16) gives just (3.10c).

When we describe damping by a complex wave vector (2.36, 37) then the above conservation laws applied to the real, i.e., oscillatory parts of $\boldsymbol{k}$. For clarity, the situation is depicted (again) in Fig. 3.5.

The conservation law (3.13c) still holds for a quantization axis perpendicular to the interface.

3.1.4 Reflection and Transmission at an Interface and Fresnel's Formulae

Continuing to exploit the boundary conditions (3.6, 7) we give now the results for the transmission t and the reflection r of an interface for the simplifying assumption that both media are transparent. This means that the imaginary part κ of the complex index of refraction $\tilde{n}$ is negligible, i.e., much smaller than the real part n. This is the case of weak absorption.

$$|\kappa| \ll |n| \simeq 1 \quad \text{weak absorption,}$$

$$|\kappa| \gg |n| \simeq 1 \quad \text{strong absorption.} \tag{3.17}$$

The resulting equations are known as the Fresnel formulae for the regime of weak absorption. They read, according to [3.1–3] (as usual our treatment is

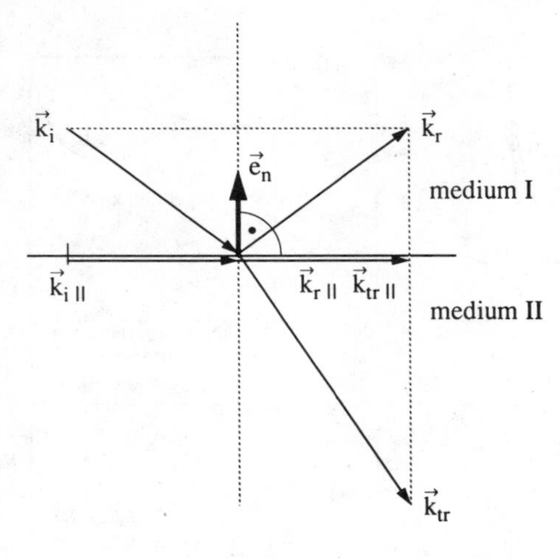

Fig. 3.5. The momenta $\hbar k$ of incident, reflected and refracted beams at an interface

for non-magnetic materials)

$$
\begin{aligned}
r_\perp &= \frac{n_\text{I}\cos\alpha_\text{i} - n_\text{II}\cos\alpha_\text{t}}{n_\text{I}\cos\alpha_\text{i} + n_\text{II}\cos\alpha_\text{i}} = -\frac{\sin(\alpha_\text{i}-\alpha_\text{tr})}{\sin(\alpha_\text{i}+\alpha_\text{tr})}\,, \\
r_\| &= \frac{n_\text{II}\cos\alpha_\text{i} - n_\text{I}\cos\alpha_\text{tr}}{n_\text{I}\cos\alpha_\text{i} + n_\text{II}\cos\alpha_\text{i}} = \frac{\tan(\alpha_\text{i}-\alpha_\text{tr})}{\tan(\alpha_\text{i}+\alpha_\text{tr})}\,, \\
t_\perp &= \frac{n_\text{I}\cos\alpha_\text{i}}{n_\text{I}\cos\alpha_\text{i} + n_\text{II}\cos\alpha_\text{i}} = \frac{2\sin\alpha_\text{t}\cos\alpha_\text{i}}{\sin(\alpha_\text{i}+\alpha_\text{tr})}\,, \\
t_\| &= \frac{2n_\text{I}\cos\alpha_\text{i}}{n_\text{I}\cos\alpha_\text{tr} + n_\text{II}\cos\alpha_\text{i}} = \frac{2\sin\alpha_\text{i}\cos\alpha_\text{i}}{\sin(\alpha_\text{i}+\alpha_\text{t})\cos(\alpha_\text{i}-\alpha_\text{tr})}\,.
\end{aligned}
\tag{3.18}
$$

The relation between α_i and α_tr according to (3.10c) is used to progress from one set of formula to the other. The signs in (3.18) depend on the way in which we defined the field amplitudes in Fig. 3.1. This is, however, not a serious problem since it is usually not possible to measure directly the field amplitudes in the optical regime but only the intensities $I = \langle S \rangle$. [See (2.21)]

We display in Fig. 3.6 formulae (3.18) graphically and the phase shift between the various reflected components, assuming that the incident ones are in phase. Furthermore we show the results for strong absorption not covered by (3.18).

The experimentally accessible quantities are R and T which can be calculated from (2.22, 3.2).

We discuss first the reflectivity R. For the orientation $R_\perp$, R increases monotonically with α_i. The limiting values for $\alpha_\text{i} = 0$ and $\alpha_\text{i} = 90°$ are given in (3.19).

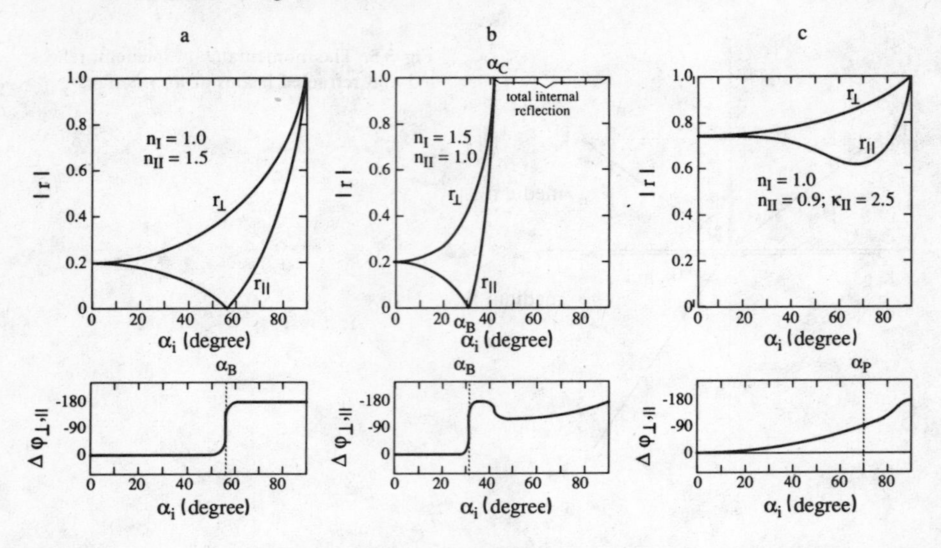

Fig. 3.6a–c. The reflection coefficient r for the (electric) field amplitudes according to (3.18) and the phase difference between the reflected amplitudes for the orientations $r_\perp$ and $r_\parallel$ and reflection at an optically thicker (**a**) and thinner (**b**) medium and for reflection at a strongly absorbing medium (**c**). According to [3.3]

For weak absorption ($n \gg \kappa$)

$$R(\alpha_i = 0) = \left(\frac{n_{II} - n_I}{n_{II} + n_I}\right)^2 \tag{3.19a}$$

For $n_I = 1$ the answer, again for weak absorption, is

$$R(\alpha_i = 0) = \left(\frac{n_{II} - 1}{n_{II} + 1}\right)^2 ; \tag{3.19b}$$

and for strong absorption and $n_I = 1$

$$R(\alpha_i = 0) = \frac{(n_{II} - 1)^2 + \kappa_{II}^2}{(n_{II} + 1)^2 + \kappa_{II}^2} , \tag{3.19c}$$

while for grazing incidence

$$R(\alpha_i = 90°) = 1 \tag{3.19d}$$

in all cases.

In contrast $R_\parallel$ goes through a minimum at a certain angle α_B with

$$R_\parallel(\alpha = \alpha_B) = 0 . \tag{3.20}$$

The angle α_B is known as Brewster's angle or the polarization angle. For $\alpha = \alpha_B$ only the component polarized perpendicularly to the plane of incidence is reflected. So this angle can be used to polarize light, if unpolarized light is directed to the interface. Note that the transmitted beam is not strictly polarized, but has only some preference for the orientation parallel to the plane of incidence.

The condition which comes from (3.18) for $r_{\parallel}=0$ is

$$n_{\rm II}\cos\alpha_{\rm i}=n_{\rm I}\cos\alpha_{\rm t} \quad \text{or} \quad \tan(\alpha_{\rm i}+\alpha_{\rm t})=\infty\ . \tag{3.21}$$

Equation (3.21) has, apart from the trivial solution (no interface $\rightarrow n_{\rm I}=n_{\rm II}$; $\alpha_{\rm i}=\alpha_{\rm t}$), the solution

$$\alpha_{\rm i}+\alpha_{\rm t}=90^{\circ}\ , \tag{3.22}$$

i.e., the reflected and refracted beams propagate perpendicularly to each other. This fact can be easily understood. As we shall see in Chap. 4 the reflected beam is radiated from the forced oscillations of the atoms close to the surface which forms the optically thicker medium (assuming for the moment that medium I is vacuum). Since dipoles do not radiate in the direction of their axis, and since the polarization in the medium is perpendicular to $\boldsymbol{k}_{\rm t}$ (transverse wave) we find directly (3.22).

For the case $n_{\rm II}<n_{\rm I}$ we also find the critical angle $\alpha_{\rm c}$ for total internal reflection in Fig. 3.6b which we mentioned already earlier.

If we send light on the interface polarized differently than $E_{\parallel}$ or $E_{\perp}$ to the plane of incidence, we can decompose it always into two components with the above orientations, we calculate their reflected or transmitted amplitudes with (3.18) or Fig. 3.6 and superpose them again, taking into account the relative phase shifts given in Fig. 3.6. In the general case of a phase shift different from 0° or 180° the reflected light will be elliptically polarized for a linearly polarized incident beam. However, in experimental investigations of the optical properties of semiconductors, one tries to avoid this additional complication, usually by choosing the simplest geometries.

For strong absorption, R does not reach zero for any polarization (Fig. 3.6c), and starts already for $\alpha_{\rm i}=0$ rather close to one. This leads to a statement which may seem contradictory in itself at first glance: strongly absorbing materials absorb only a small fraction of the incident light. The solution is clear, since the bigger fraction is reflected. The smaller fraction which actually enters the medium is absorbed however over a short distance. In a weakly absorbing medium, the major portion of the light is transmitted through the surface and may be completely absorbed if the medium is thick enough. Indian ink is in the sense of (3.17) a weakly absorbing medium, metals are strongly absorbing over wide spectral ranges and have R close to unity.

With increasing $\alpha_{\rm i}$, $R_{\perp}$ increases monotonically while $R_{\parallel}$ goes through a shallow minimum as seen in Fig. 3.6c. The principle angle of incidence $\alpha_{\rm p}$ is defined by some authors as the $\alpha_{\rm i}$ for which the slopes of the curves $R_{\parallel}(\alpha_{\rm i})$ and $R_{\perp}(\alpha_{\rm i})$ are equal. Other authors prefer to use the minimum of $R_{\perp}$ as definition of $\alpha_{\rm p}$. The difference is marginal. The phase shift between the two components is just $\pi/2$ for $\alpha_{\rm p}$.

A consequence of the smooth variation of the phase shift with α_i is the fact that linearly polarized light impinging on a metallic mirror is usually elliptically polarized after reflection except for the simple orientations $E_\parallel$ and $E_\perp$. This fact should be remembered when building an optical setup in the lab.

To conclude this subsection, we shall shortly consider transmission through a single interface. For a lossless interface, energy conservation requires for the incident, reflected and transmitted powers P of the light

$$P_i = P_r + P_{tr} , \tag{3.23a}$$

where the power is defined as energy per unit of time.

Despite (3.23a), $|r|^2$ and $|t|^2$ do not add up to unity, since these quantities give information about the reflected and transmitted light intensities. This quantity gives, as stated already earlier, the energy flux density, i.e., the energy per unit of time and of area. Since the cross-sections of the incident and reflected beams are equal, but different from the transmitted one for $\alpha_i \neq 0$ as shown in Fig. 3.7, a corresponding correction factor has to be added to $T = |t|^2$ to fulfill (3.23a):

$$T_{\parallel,\perp} = \frac{I_{tr}}{I_i}\frac{\cos\alpha_{tr}}{\cos\alpha_i} = |t_{\parallel,\perp}|^2 \frac{\cos\alpha_{tr}}{\cos\alpha_i} . \tag{3.23b}$$

For more details about transmission and reflection at a plane interface see [3.1–3] or the references given therein.

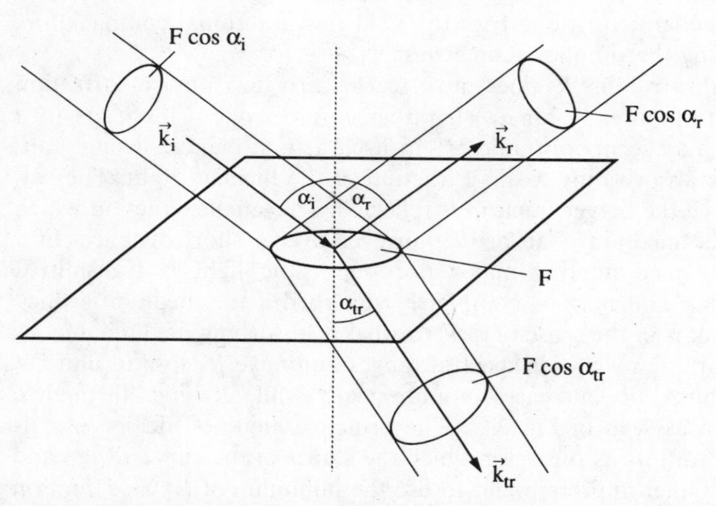

Fig. 3.7. The relation of the cross sections of incident, transmitted and reflected beams at a plane interface between two media

3.1.5 Extinction and Absorption of Light

Until now we have considered mainly what happens in the immediate surroundings of a plane interface between two media. Now we shall spend a few minutes on the propagation of a beam within a medium, continuing the discussion in Sect. 2.3 in connection with (2.37).

If light propagates in a medium other than vacuum, its intensity or field amplitude decreases in most cases with increasing distance, even if we (unphysically) assume a strictly parallel beam and neglect all diffraction losses connected with its finite diameter. In reality, both assumptions above can be fulfilled only to a good approximation but never in a strict, mathematical sense. The decay is usually exponential with increasing distance d [See (2.37–40)].

$$I = I_0 e^{-\alpha(\omega)d} . \tag{3.24}$$

If energy is pumped into a suitable material and in a suitable way $\alpha(\omega)$ may even become negative for a certain range of frequencies and consequently light is amplified. We call these materials active, laser or inverted materials in contrast to passive materials with $\alpha \geqslant 0$.

The attenuation of light according to (3.25) is called "extinction". It comprises two groups of phenomena:

$$\alpha_{\text{extinction}}(\omega) = \alpha_{\text{absorption}}(\omega) + \alpha_{\text{scattering}}(\omega) \tag{3.25}$$

Extinction is the more comprehensive term. It enters in the damping Γ in Chap. 4 or in the phase relaxation time T_2 discussed in more detail in Chap. 22 via $\gamma = 2\hbar T_2^{-1}$ and contains two contributions. Absorption is the transformation of the energy of the light field into other forms of energy like heat, chemical energy or electromagnetic radiation which is not coherent and generally also frequency shifted with respect to the incident beam. This latter phenomenon is usually called (photo-)luminescence. The other contribution to extinction is attenuation by (coherent) scattering of light. If the scattering particles do not show absorption in the visible and have typical sizes large compared to the wavelength of the light λ, the material usually looks white. Examples are the powder of ZnO (just to start with a semiconductor), ground sugar and salt, clouds, snow, the foam of beer, milky quartz or the bark of a birch tree. The reason is that the scattering of light at the interfaces by reflection and refraction is roughly wavelength independent and thus the same for all colors.

If the particles are small compared to λ, the scattering at these particles becomes wavelength dependent. Often one finds an ω^4 law

$$\frac{I_{\text{scatter}}}{I_{\text{incident}}} \propto \omega^4 . \tag{3.26}$$

This relation follows from a combination of (2.25) and (2.22). It explains that the sunlight propagating through clear atmosphere preferentially loses the high frequencies, i.e., short wavelength or blue parts of its spectrum by scattering from N_2, O_2 and other molecules of the atmosphere. Consequently the sun itself looks yellow to red depending on the thickness of air through which the sunlight has to travel and the sky appears blue from the scattered light.

If the scattering or absorbing particles are diluted and do not interact with each other, one finds proportionality between their concentration n_p and α

$$\alpha(\omega) = n_p \alpha_s(\omega) , \tag{3.27}$$

where α_s is the specific extinction constant. Equation (3.27) is also known as Beer's law.

Though there is evidently a rather clear definition of the terms "extinction" and "absorption", one uses often in "every day" language in the lab and also in many books including this one the word absorption instead of extinction, sometimes for convenience, and sometimes because it is not always clear which group of phenomena is responsible for the attenuation of a light beam along its path through matter.

3.1.6 Transmission Through a Slab of Matter and Fabry–Perot Modes

We discuss now in connection with Figs. 3.8 and 3.9 the transmission and reflection of a plane-parallel slab of matter of geometrical thickness d with ideal, lossless surfaces. The surrounding material I is air or vacuum ($n_I = 1$, $\kappa_I = 0$). The total transmission $\hat{T}$ or reflection $\hat{R}$ does not only depend on material II and on α_i but also on the properties of the incident light field, e.g., on its polarization and on its coherence length l_c, i.e., the distance over which there is a fixed phase-relation. We can discuss here only some limiting cases.

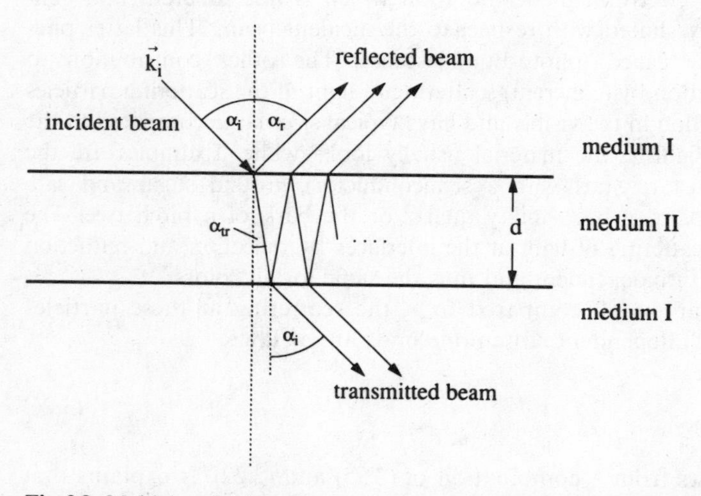

Fig. 3.8. Multiple reflection of an incident light beam in a plane-parallel slab of matter

For strong absorption in the sense discussed in connection with (3.18) it is easily possible to detect the beam reflected from the front surface. The transmitted beam is strongly attenuated for $d \gg \lambda \simeq 0.5\,\mu\text{m}$, i.e., for an optical density $\alpha(\omega)d \gg 1$. Consequently $\hat{R}$ reduces to R given for normal incidence by (3.19c). T is difficult to measure since sufficiently thin samples are often not easily available and sometimes have optical properties different from the bulk material. Similar statements for $\hat{R}$ are true for $\alpha(\omega)\,d \gg 1$ even if $\alpha(\omega)$ is small. This can be expressed by

$$\hat{R}(\omega) = R(\omega) \quad \text{for} \quad \alpha(\omega)\,d \gg 1\,. \tag{3.28}$$

The most convenient regime in which to measure $\alpha(\omega)$ is $1 \leqslant \alpha(\omega)\,d \leqslant 3$. The reflection has to be taken into account only once at the front and rear surfaces since multiply reflected beams are very weak due to the absorption. We find with Fig. 3.8

$$\hat{T}(\omega) = [1 - R^{\mathrm{I}\to\mathrm{II}}(\omega, \alpha_{\mathrm{i}})]\exp[-\alpha(\omega)\,d\cos^{-1}\alpha_{\mathrm{t}}]\,[1 - R^{\mathrm{II}\to\mathrm{I}}(\omega, \alpha_{\mathrm{t}})]\,. \tag{3.29a}$$

This simplifies for normal incidence to (3.29b) bearing in mind that $R^{\mathrm{I}\to\mathrm{II}} = R^{\mathrm{II}\to\mathrm{I}}$ for weakly absorbing material (See (3.19a).)

$$\hat{T}(\omega) \simeq [1 - R(\omega)]^2\,\mathrm{e}^{-\alpha(\omega)d}\,. \tag{3.29b}$$

For the conditions of (3.28) we can write

$$\hat{R}(\omega) \simeq R(\omega) + [1 - R(\omega)^2]R(\omega)\,\mathrm{e}^{-2\alpha(\omega)d} \simeq R(\omega)\,. \tag{3.29c}$$

For materials with an optical density $\alpha(\omega)d \leqslant 1$ things become more complicated again, because we must consider multiple reflection. The behavior depends strongly on the relation of the optical pathlength and the coherence length l_{c}. For $dl_{\mathrm{c}}^{-1} \gg 1$ we have to add intensities resulting in (in this case, for normal incidence)

$$\hat{T}(\omega) \simeq \frac{[1 - R(\omega)]^2\mathrm{e}^{-\alpha(\omega)d}}{1 - R^2(\omega)\mathrm{e}^{-2\alpha(\omega)d}} \simeq [1 - R(\omega)]^2\mathrm{e}^{-\alpha(\omega)d} \quad \text{for} \quad \alpha d \lesssim 1\,. \tag{3.29d}$$

For long coherence lengths, the field amplitudes interfere with approriate phases and the two plane-parallel interfaces form a Fabry–Perot resonator. There are two limiting cases: in one case all partial waves reflected at the two surfaces interfere constructively in the resonator. This condition is fulfilled if an integer number m of half waves fits in the resonator, expressed mathematically as

$$\lambda_m = \frac{d2n(\omega)}{m} \quad \text{or} \quad \omega_m = \frac{m\pi c}{dn(\omega_m)} \quad \text{or} \quad k_m = m\,\frac{\pi}{d} \quad \text{with} \quad m = 1, 2, 3 \ldots. \tag{3.30}$$

In this case we have a large field amplitude in the resonator which may surpass even the amplitude of the incident beam, a total transmission $\hat{T}$ close to unity and correspondingly a weak total reflection $\hat{R}$ (even if R is close to unity!). In the opposite case of mainly destructive interference of the partial waves in the resonator, we find just the opposite situation, that is, $\hat{R} \leqslant 1$; $\hat{T} \ll 1$. Such a device is called etalon or Fabry–Perot resonator (or FP resonator).

The general formula for the FP resonator reads approximately [3.4]

$$\hat{T} = \frac{A}{1 + F\sin^2\delta}, \quad \hat{R} = \frac{B + F\sin^2\delta}{1 + F\sin^2\delta} \tag{3.31a}$$

with

$$F = \frac{4R_\alpha}{(1-R_\alpha)^2},$$
$$A = \frac{e^{-\alpha d}(1-R_F)(1-R_B)}{(1-R_\alpha)^2}, \tag{3.31b}$$
$$B = \frac{R_F(1-R_\alpha/R_F)}{(1-R_\alpha)^2},$$

R_F: reflectivity of front surface of the Fabry–Perot resonator;
R_B: reflectivity of back surface of the Fabry–Perot resonator;
$R_\alpha = (R_F R_B)^{1/2} e^{-\alpha d}$; $\delta = n(\omega)\, k^{vac}\, d = n(\omega)\, \omega d/c$.

One often has $R_F = R_B = R$.

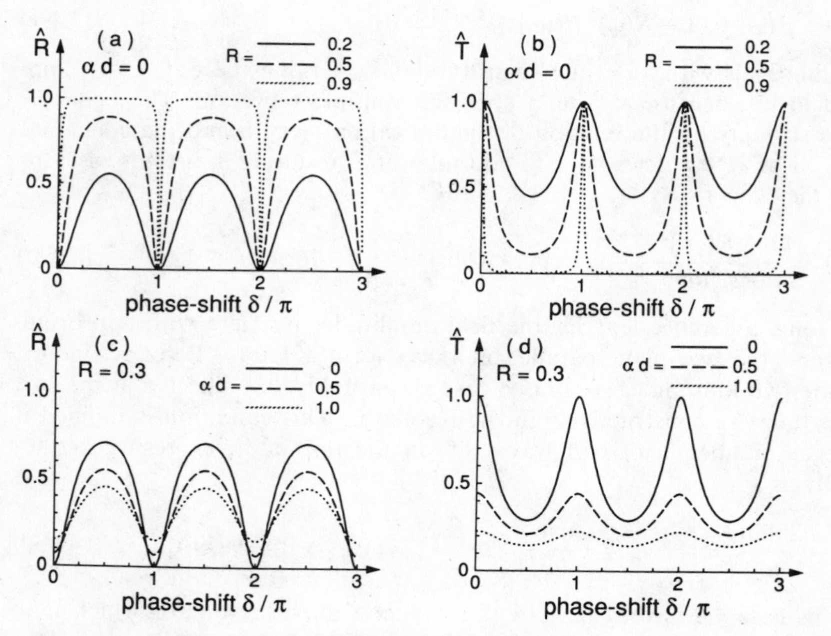

Fig. 3.9a–d. Transmission $\hat{T}$ **(b, d)** through and reflection $\hat{R}$ **(a, c)** from a Fabry–Perot resonator as a function of the phase shift δ for one half round trip for various values of the reflectivity R of a single surface or the optical density αd of the medium in the resonator

For a lossless medium $\alpha = 0$, and normal incidence these expressions reduce to

$$\hat{T} = \frac{1}{1 + F' \sin^2\delta} , \tag{3.32a}$$

$$\hat{R} = \frac{F' \sin^2\delta_2}{1 + F' \sin^2\delta} = 1 - \hat{T} . \tag{3.32b}$$

with the phase shift per round trip δ given in (3.31) and the finesse F' given by

$$F' = \frac{4R}{1 - R^2} . \tag{3.33}$$

In Fig. 3.9 we show $\hat{T}$ and $\hat{R}$ as a function of δ for various values of R and α. For vanishing damping $\hat{T}(\omega)$ reaches unity and $\hat{R}(\omega)$ zero for the conditions of (3.32). Increasing F' makes the FP resonances narrower. $\hat{T}$ and $\hat{R}$ always add up to unity. For finite damping or α this is no longer the case. $\hat{T}$ remains below unity and the height of the resonance decreases with decreasing F', i.e., increasing α for constant values of R. A more detailed treatment of the Fabry–Perot and of related problems like dielectric single and multilayer coatings is beyond the scope of this book and [3.1–5] are suggested for further reading.

3.1.7 Birefringence and Dichroism

Until now we have assumed, for the sake of simplicity, that our sample is isotropic, i.e., that $\varepsilon(\omega)$ is a scalar function and that means with (2.20, 21 and 26) that $\boldsymbol{D}$ and $\boldsymbol{E}$ as well as $\boldsymbol{S}$ and Π are parallel. Later on we shall in general use this assumption again. Here we want to have a short look at what happens if we have an anisotropic material. Indeed, many crystals are anisotropic, including the hexagonal wurtzite structure of several semiconductors. Even cubic crystals can show a weak anisotropy for a finite wave vector $\boldsymbol{k} \neq 0$ since cubic symmetry is lower in symmetry than spherical symmetry, see also Chap. 25. This latter aspect will not be considered for the moment. In the mechanical model which we shall treat in Chap. 4, we can already understand such anisotropies if we assume that some oscillators can be excited (i.e., elongated) only in one direction, e.g., in the x-direction but not in the others. Such an oscillator would react only on the component of an incident electric field polarized $\boldsymbol{E} \| \hat{\mathbf{x}}$. In the microscopic model the same approach means that the oscillator strength f introduced also in Chap. 4 depends on the direction of polarization and is, e.g., finite for light polarized parallel to a crystallographic axis and zero perpendicular to it. Indeed it is already sufficient that the oscillator strength is different for different orientations of the polarization with respect to the crystallographic axis in order to obtain birefringence.

To describe such situations it is necessary to remember that the dielectric function $\varepsilon(\omega)$ is generally a tensor. It describes the connection between the two vectors $\boldsymbol{D}$ and $\boldsymbol{E}$.

Usually one tries to align the cartesian coodinates for $\varepsilon(\omega)$ in a simple way with respect to the crystallographic axes. In uniaxial systems one identifies the z-axis with the crystallographic $\boldsymbol{c}$-axis and the $x-y$ plane with the (usually almost isotropic) plane perpendicular to $\boldsymbol{c}$.

If transformed on these main axes, the tensor $\varepsilon(\omega)$ has therefore in the main diagonal two equal elements

$$\varepsilon_{xx}(\omega) = \varepsilon_{yy}(\omega) \neq \varepsilon_{zz}(\omega) \tag{3.34a}$$

and zeros otherwise.

For biaxial systems of even lower symmetry, such simple connections are often no longer possible and one finds

$$\varepsilon_{xx}(\omega) \neq \varepsilon_{yy}(\omega) \neq \varepsilon_{zz}(\omega) \neq \varepsilon_{xx}(\omega) \; . \tag{3.34b}$$

Since all of the important semiconductors crystallize either in cubic systems (diamond structure with point-group O_h, zincblende structure T_d) or hexagonal ones (wurtzite structure C_{6v}) we will not go below uniaxial symmetry and refer the reader for these problems to books on crystal optics [3.1–3] or on crystallography [3.6]. The meaning of "point groups" will be explained Chap. 25.

The consequences of the tensor character of $\varepsilon(\omega)$ are birefringence and dichroism. We briefly outline both effects below. Dichroism means literally that a crystal has two different colors depending on the direction of observation. In a more general sense one describes with the word dichroism every dependence of the absorption spectra on the direction of polarization. In Fig. 3.10 we show schematically transmission spectra for a dichroitic, uniaxial

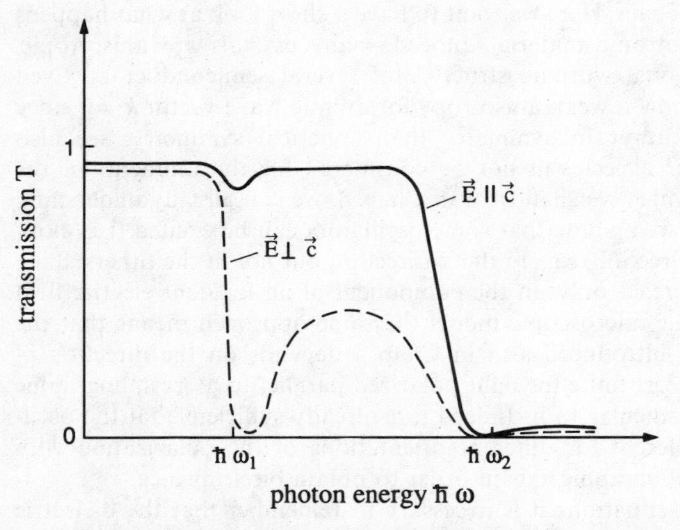

Fig. 3.10. Schematic drawing of the transmission spectra of a dichroitic material for two different polarizations of light with respect to the crystallographic axis

material of a certain thickness. The sample is transparent for both polarizations below $\hbar\omega_1$. The resonance at $\hbar\omega_1$ is assumed to couple more strongly to the light field (i.e., to have larger "oscillator strength") for the orientation $\boldsymbol{E} \perp c$ than for $\boldsymbol{E} \parallel \boldsymbol{c}$. Above $\hbar\omega_2$ light is absorbed almost completely for both orientations. The dichroitic region obviously lies betweem $\hbar\omega_1$ and $\hbar\omega_2$.

In some materials this region covers a wide spectral range, in some cases the whole visible part of the spectrum. In such a case this material can be used as a polarizer. As an example polaroid films contain long organic molecules which are oriented parallel to each other by the stretching of the film during the manufacturing process. These molecules absorb radiation polarized parallel to the chain, and transmit for the perpendicular orientation over most of the visible spectrum. Another material which is known for its dichroism are some colored varieties of tourmaline. In many semiconductors the dichroitic spectral range is rather narrow and amounts often only to a few tens of meV. These materials are, of course, of no use as commercial polarizers, but the investigation of the dichroism gives very important information on the symmetries and selection rules of the resonances. We will see some examples of this in Chap. 14.

If we assume that the eigenfrequencies and/or oscillator strengths of some resonance(s) depend on polarization, then we know immediately from the Kramers-Kronig relations, presented in Chap. 8, that the real part of the refractive index $n(\omega)$ depends also on the orientation of $\boldsymbol{E}$ relative to $\boldsymbol{c}$, i.e., the material is birefringent. We can even state that every dichroitic material must show birefringence and that birefringent material must have some spectral range in which dichroism occurs. For uniaxial materials (e.g., crystals with uniaxial symmetry C_{6v} or as a prototype calcite) an incident light beam can always be decomposed into two components of the electric field polarized parallel and perpendicular to the main section. The main section is the plane defined by the crystallographic axis and the incident wave vector. The beam polarized perpendicular to the main section is called the ordinary (o) beam. Its refraction is described by Snells" law and the refractive index $\tilde{n}$ (ω) is independent of orientation. This fact can be understood since the ordinary beam is always polarized perpendicular to the $\boldsymbol{c}$-axis and we assume that uniaxial materials are isotropic in the plane $\perp \boldsymbol{c}$. This is strictly correct only for vanishing wave vectors (and corresponds just to the situation for the dipole approximation in Sect. 3.2.2) and to a very good approxmation for small but finite $\boldsymbol{k}$ values. The so-called extraordinary beam (eo), the polarization of which falls in the main section, has components $\boldsymbol{E} \parallel \boldsymbol{c}$ and $\boldsymbol{E} \perp \boldsymbol{c}$ the weights of which depend on the angle $\gamma = \angle(\boldsymbol{k}, \boldsymbol{c})$. It is not surprising that the refractive index experienced by the extraordinary beam depends an γ, since the relative coupling to the oscillators active for the orientations $\parallel \boldsymbol{c}$ or $\perp \boldsymbol{c}$ changes with γ. For a general direction of incidence an unpolarized (or elliptically polarized) beam will be decomposed into two beams polarized perpendicular to each other – the ordinary and the extraordinary ones – which will be separated in space, as shown schematically in Fig. 3.11. This is the concept, which allows us

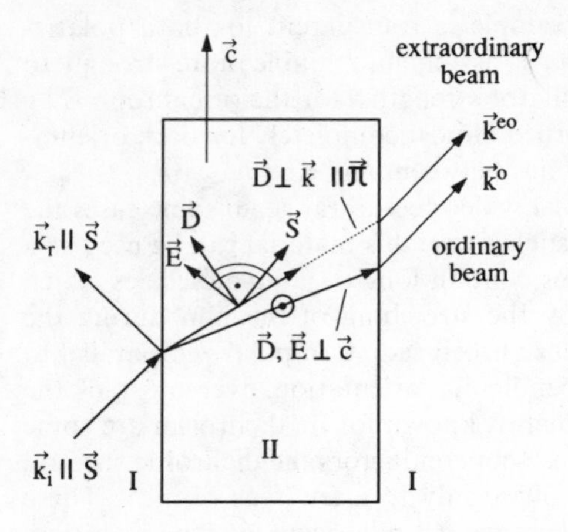

Fig. 3.11. Birefringence for an unpolarized beam falling on a birefringent material at a given angle with the crystallographic axis *c* parallel to the interface

to use birefringent materials as polarizers. We note here already that the wave vector $\boldsymbol{k}$ and the Poynting vector $\boldsymbol{S}$ of the eo beam are not necessarily parallel to each other. The reason will be given in connection with Fig. 3.13. In the case of Fig. 3.11 the refractive index of the ordinary beam is greater than of the extraordinary one. This situation is called positive-uniaxial birefringence, the opposite situation corrrespondingly negative-uniaxial birefringence.

There are two limiting orientations which result in rather clear and simple situation. Therefore these orientations are usually investigated in semiconductor optics. One situation is $\boldsymbol{k} \parallel \boldsymbol{c}$. In this case the $\boldsymbol{E}$ field can be only perpendicular to $\boldsymbol{c}$, this means one observes the ordinary beam only, independent of the polarization of the incident beam. The other clear orientation is $\boldsymbol{k} \perp \boldsymbol{c}$. In this case one can choose by a polarizer the orientation $\boldsymbol{E} \perp \boldsymbol{c}$ for the ordinary beam or $\boldsymbol{E} \parallel \boldsymbol{c}$ for the extraordinary beam, but in the latter situation, the $\boldsymbol{E}$ field acts only on oscillators which can be elongated parallel to $\boldsymbol{c}$.

Oblique incidence on a surface cut parallel or perpendicular to $\boldsymbol{c}$ or normal incidence on a surface cut under an arbitrary angle (Fig. 3.12) with respect to $\boldsymbol{c}$ are much more complicated to evaluate concerning the spectra of reflection or transmission. The worst situation is, of course, oblique incidence on a plane at an arbitrary angle to the $\boldsymbol{c}$-axis. Scientists working on semiconductor optics usually try to avoid these situations, scientists devoting their work to crystal optics find it challenging and even prefer biaxial systems to others.

A situation which allows us to discuss various aspects of birefringence is perpendicualr incidence on a plane at an oblique angle with respect to $\boldsymbol{c}$. We shall dwell a few minutes on this topic.

The experimental result is shown in Fig. 3.12. The incident beam is split into two when entering the birefringent material. The ordinary beam continues to propagate normal to the surface as expected from Snells' law (3.10b) or from the conservation of momentum parallel to the surface (3.14). The extraordinary one seems to violate these two rules. Since a violation of the law of momentum

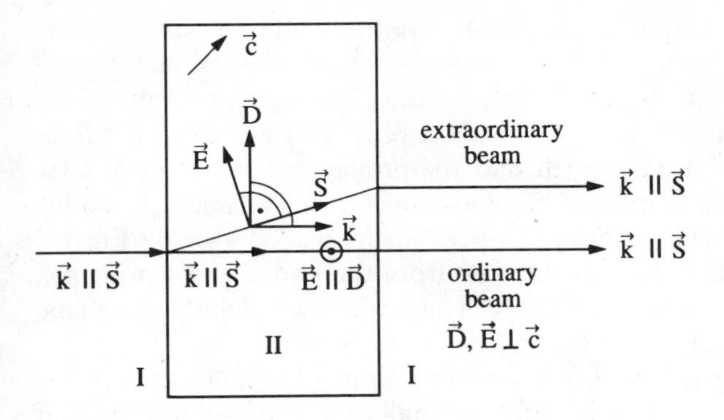

Fig. 3.12. Same as Fig. 3.11 but for normal incidence on a surface cut under an arbitrary angle with respect to $\boldsymbol{c}$

conservation would be very serious, not only for physics, we have to look more closely at this problem.

There are two rather simple ways to present birefringence. One, which we shall outline towards the end of this subsection is in terms of the indicatrix. In the other method, one uses in polar coordinates a plot which gives the phase velocity $v_{\mathrm{ph}} = cn^{-1}(\omega)$. This is basically the inverse of the real part of $\tilde{n}(\omega)$ as a function of the direction of propagation. In a uniaxial system, the figures produced when we include all directions are a sphere for the ordinary beam and a figure with rotational symmetry with respect to $\boldsymbol{c}$ for the extraordinary

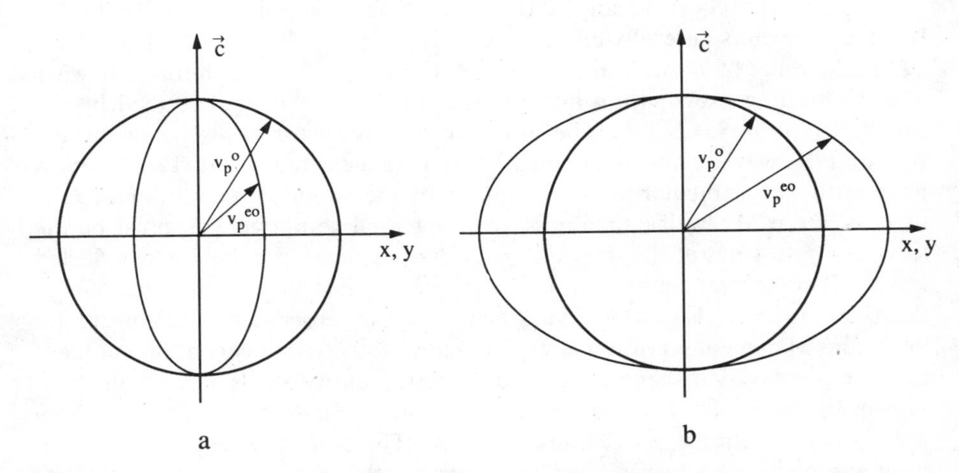

Fig. 3.13a, b. The phase velocity of light in an optically uniaxial material shown for one frequency ω in polar coordinates for negative (**a**) and positive (**b**) birefringence for the ordinary (o) and extraordinary (eo) beams

one. They touch for propagation of light parallel to the crystallographic axis as shown in Fig. 3.13 since there are only o-waves for this orientation, as mentioned above. Now recall Huygen's principle for the propagation of light which says that every point illuminated by an incident primary wave front becomes the source of a secondary wavelet such that the primary wavefront at a later time results from the superposition of the amplitudes of all wavelets. In addition, we must discard the back-travelling waves in the way shown by Fresnel and Kirchhoff which would appear otherwise from the above given principle. With this amendment we can contruct wavefronts when we identify the shape of the wavelets with Fig. 3.13.

Shown in Fig. 3.12 are the $\boldsymbol{S}, \Pi, \boldsymbol{k}, \boldsymbol{E}$ and $\boldsymbol{D}$ vectors. The magnetic vector $\boldsymbol{B}$ and $\boldsymbol{H}$ are parallel to each other and normal to the paper and are, in this context, of no further interest since we are dealing with nonmagnetic material.

In Fig. 3.14a the situation is shown for the *o*-beam. The wavelets are spheres, the resulting wave front is parallel to the vacuum-medium interface. The vectors of energy-flux density $\boldsymbol{S} = \boldsymbol{E} \times \boldsymbol{H}$, of moment density $\Pi = \boldsymbol{D} \times \boldsymbol{B}$ and the wavevector $\boldsymbol{k}$ ($\hat{=}$ momentum $\hbar \boldsymbol{k}$) are parallel. (Diffraction effects caused by the finite beam diameter are neglected here, though they are obviously also described by Huygens' principle.) The situation for the *eo*-beam is presented in Fig. 3.14b. The wave front produced by the superposition (or interference) of the wavelets, and constructed as the tangent to the wavelets, is still parallel to the interface. This wave front describes the $\boldsymbol{D}$ field because we know from Maxwell's equations that the boundary condition for $\boldsymbol{D}$ is that the normal component D_n is continuous over the interface. In our case, $D_n = 0$ on both sides. As a consequence, the classical momentum density $\boldsymbol{\Pi}$ and the momentum $\hbar \boldsymbol{k}$ of the light quanta are still perpendicular to the interface, as required by the conservation of the momentum component parallel to the interface, which is here obviously zero.

On the other hand, we can see that the whole wave front is shifting sideways with continuing propagation into the medium. This shift is described by the poynting vector $\boldsymbol{S} = \boldsymbol{E} \times \boldsymbol{H}$. The direction of this vector is just given by the origin of the wavelet and the point where the tangent touches it. The $\boldsymbol{E}$ field is necessarily perpendicular to $\boldsymbol{S}$. As required by the boundary condition for $\boldsymbol{E}$ as deduced from Maxwell's equations, the tangential component $\boldsymbol{E}_t$ must be the same on both sides of the interface (including incident, refracted and reflected beams). The normal component of $\boldsymbol{E}$ can change, and that is what happens in the orientation of Fig. 3.14b. To summarize, we observe that there are no violations of any conservation laws. The tangential components of the momentum are conserved at the interface and for this quantity Snells' law is still valid. However, the direction of energy propagation given by $\boldsymbol{S}$ changes, but there are no conservation laws for this direction. The law of energy conservation itself has, of course, to be fulfilled, this means in this case that the total amount of energy falling per unit time on the interface equals the sum of transmitted and reflected energies.

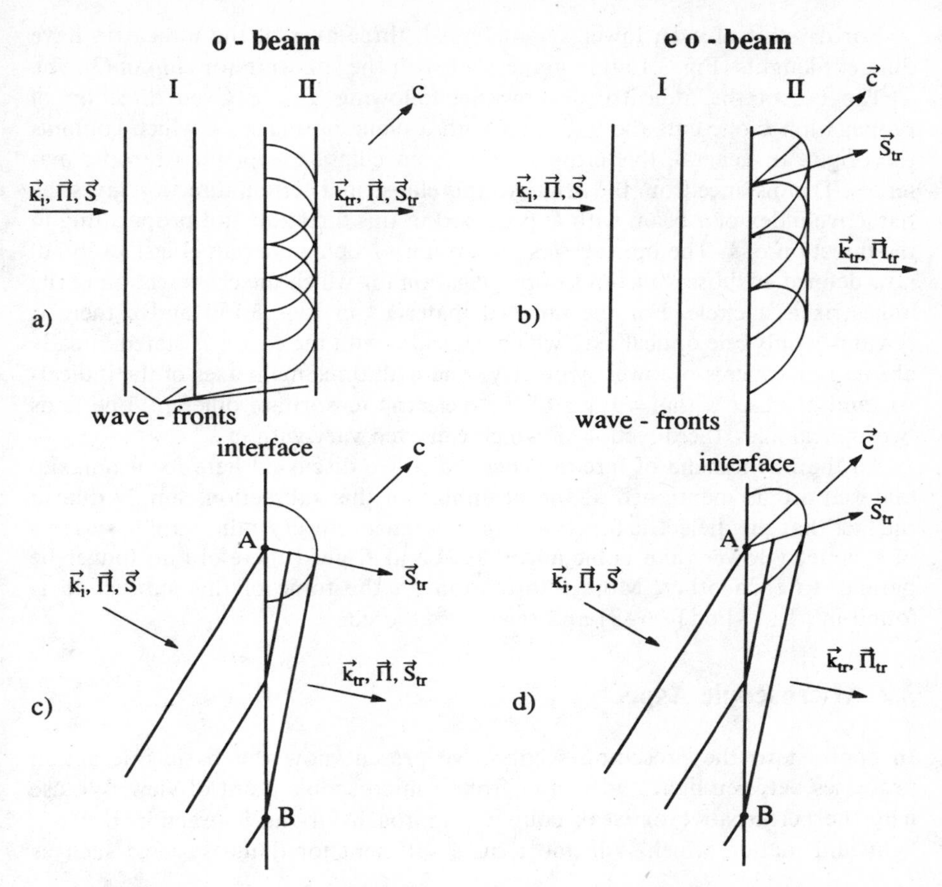

Fig. 3.14a–d. The construction of the wave fronts of the ordinary (**a**, **c**) and the extraordinary (**b**, **d**) beams for various geometries using Fig. 3.13

Figure 3.14c and d finally show schematically the general situation i.e., oblique incidence on a surface cut at an arbitrary angle with respect to ***c***. The construction simply uses the size of the wavelets around point A when the incident wave front just reaches the interface at point B.

The basic idea of the representation of birefringence by the indicatrix is explained in connection with Fig. 3.15. We plot from the origin lines in all directions with a length equal to the refractive index $n(\omega)$ of a wave with ***D*** polarized in this direction. In doing so, we get a sphere for an isotropic material (Fig. 3.15a) and ellipsoids with rotational symmetry for uniaxial materials. In Figs. 3.15b and c we show the situation for positively and negatively birefrigent uniaxial crystals, respectively. The rotation axis coincides in this case with the crystallographic ***c***-axis.

For crystals of even lower symmetry, all three axes of the indicatrix have different lenghts (Fig. 3.15d) in agreement with the situation for $\varepsilon(\omega)$ in (3.35b).

The use of the indicatrix is now the following: For a given direction of propagation $\boldsymbol{k}$ one cuts the indicatrix with a plane normal to $\boldsymbol{k}$ which contains the origin. In general, this cross section is an ellipse except for isotropic materials. The distance from the origin to this ellipse in a certain direction gives the refractive index of a beam with $\boldsymbol{D}$ polarized in this direction and propagating in the direction of $\boldsymbol{k}$. The optical axes are for anisotropic materials (Figs. 3.15b–d) now defined as those directions of propagation for which the cross section of the indicatrix is a circle. For the uniaxial materials in Figs. 3.15b and c there is obviously only one optical axis, which coincides with the $\boldsymbol{c}$-axis as stated already above. For systems of lower symmetry, where all three main axes of the indicatrix (and all three $\varepsilon_{ii}(\omega) i = x, y, z$ of the dielectric tensor) are different, one finds two optical axes, the direction of which can even vary with ω.

All the phenomena of birefringence which we discussed here for a uniaxial material are as mentioned at the beginning of this subsection, simply due to the fact that the dielectric function $\varepsilon(\omega)$ is a tensor for crystallographic systems of symmetry lower than cubic and that $\boldsymbol{D}$ and $\boldsymbol{E}$ need therefore no longer be parallel to each other. More information on the topic of this subsection is found in [3.1–3] or [3.6–8] and references therein.

3.2 Microscopic Aspects

In contrast to the preceding section, we present now the basic interaction processes between light and matter from a microscopic point of view. We use here the perturbative or weak coupling approach for the interaction between light and matter, which is in most cases sufficient for dilute systems such as

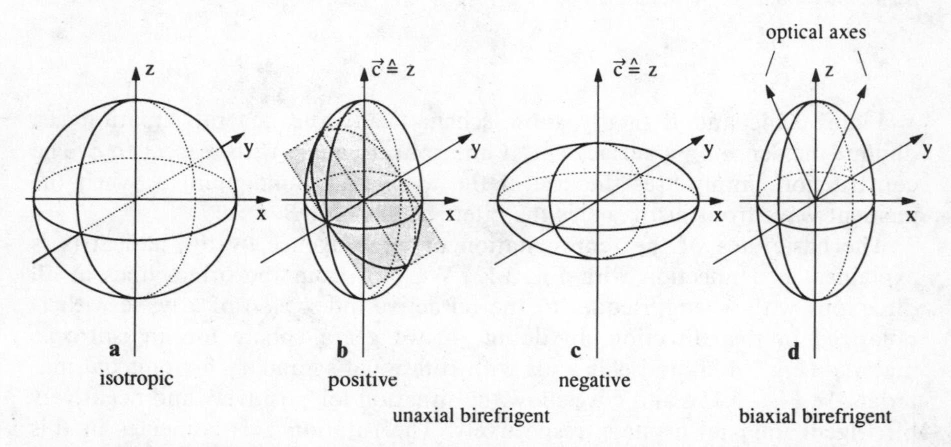

Fig. 3.15a–d. The indicatrix for isotropic (**a**) uniaxial (**b** and **c**) and biaxial materials (**d**) according to [3.6]

gases. For solids the strong coupling approach is often necessary, which leads to the concept of polaritons and which is introduced in chap. 5. We describe first in words the basic interaction mechanisms between light and matter, namely absorption, spontaneous and stimulated emission (Sect. 3.2.1), then we proceed to the treatment of linear optical properties in the framework of perturbation theory (Sect. 3.2.2). Since these topics are also treated in many textbooks (e.g., [1.1, 5, 6] or [2.3–7]) it is not necessary to go into too much detail here.

3.2.1 Absorption, Stimulated and Spontaneous Emission, Virtual Excitation

For simplicity we assume that we have a certain number of two-level "atoms" as shown in Fig. 3.16. Every atom has one electron which can be either in the ground or in the excited state. Later we will extend the model from a two-level system to bands in semiconductors, but the basic interaction processes remain the same.

In Fig. 3.16a an incident photon hits an atom in its ground state. With a certain probability the photon is annihilated and the electron gains enough energy to reach the excited state. For reasons of energy conservation, the photon has to fulfill the condition

$$\hbar\omega = E_{ex} - E_g , \tag{3.35}$$

where $E_{ex} - E_g$ is the energy difference between the ground and excited states. We call this process absorption in agreement with the definition in Sect. 3.1.5 if the energy of the photon is soon converted into other forms of energy, that is, if the electron undergoes some scattering processes, which destroy its coherence to the incident light field and eventually returns to its ground state and looses its energy e.g., as phonons i.e., as heat or as a photon which is not coherent with the incident one.

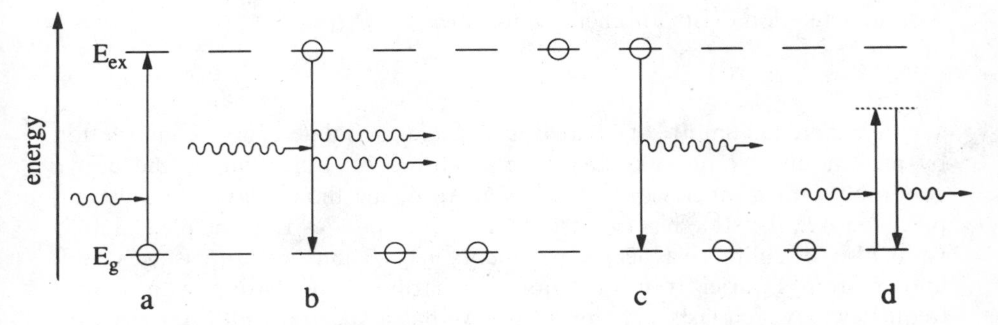

Fig. 3.16a–d. Schematic representations of various interaction processes of light with matter namely absorption (**a**), stimulated emission (**b**), spontaneous emission (**c**) and virtual excitation (**d**)

If an incident photon hits an atom with its electron in the excited state it can induce with a certain probability a transition of the electron from the excited to the ground state. In this process a second photon is created which is identical in momentum, energy, polarization and phase to the incident one. This process is called induced or stimulated emission. This process can be used to amplify a photon field. It is therefore the basic mechanism for all lasers (Light Amplification by Stimulated Emission of Radiation). Absorption and stimulated emission are closely related events.

An electron in the excited state can also with a certain probability reenter the ground state by itself, either by emitting a photon (Fig. 3.16c) or by loosing the transition energy through phonons or collisions. In the present context, the first mechanism is of interest. It is called spontaneous emission or spontaneous radiative recombination, while the second possibility is known as non-radiative recombination. Spontaneous emission can also be understood in another way. In Sect. 2.5 we saw in connection with (2.54) that photons are similar to harmonic oscillators and have consequently a zero-point energy. This zero-point energy exists for all photon modes. It cannot be absorbed because a harmonic oscillator does not have any states below the zero-point energy, but it can induce a transition in the way as discussed in connection with Fig. 3.16b. So we can consider spontaneous emission as a process induced by the zero-point vibrations of the electro-magnetic field, which are also called vacuum fluctuations.

The last process presented here is virtual excitation. Understanding this phenomenon often causes some problems for students. So we develop this topic slowly and try to explain it from various points of view in context with Fig. 3.16d. Virtual excitation means the creation of a state with the same wave function as the excited state, but with an energy which is different from the eigenenergy of this excited state. This process becomes possible through the uncertainty principle of quantum mechanics which can be written in space and momentum coordinates.

$$\Delta x_i \Delta p_i \geqslant h \quad \text{for} \quad i = 1, 2, 3 . \tag{3.36a}$$

A similar relation exists for energy and time

$$\Delta E \, \Delta t \approx \hbar . \tag{3.36b}$$

We need here (3.36b). It says that it is possible to violate energy conservation by an amount ΔE up until a maximum time Δt which fulfills the above condition. Or, in other words, if we want to define the energy with a certain precision ΔE, the state has to exist at least for a time Δt. In principle (3.36b) is valid also in simple classical wave theory (e.g., acoustics) and is very well known from Fourier transformations. A harmonic oscillation with central frequency ω, which lasts only for a time Δt, has a spectral width $\Delta\omega$ given by

$$\Delta\omega \, \Delta t \geqslant 1 . \tag{3.36c}$$

The connection between (3.37b) to (3.37c) comes simply from the relation

$$E = \hbar\omega \ . \tag{3.37}$$

If we send now a photon with energy $\hbar\omega'$ to the atom, we can excite the electron for a maximum time Δt given by (3.36b) or for our specific case by

$$\Delta t \approx \hbar[|(E_{ex} - E_g) - \hbar\omega'|]^{-1} \ . \tag{3.38}$$

At the latest after the time Δt has elapsed, the excited state must collapse. The simplest way to collapse is to emit a photon identical to the one which caused the virtual excitation. This "new" photon has however a certain phase delay with respect to the incident photon, because the energy was stored for a maximum time Δt in the atom. As a consequence, an electromagnetic wave propagates with a lower phase-velocity through an ensemble of atoms than through vacuum. The same effect is described phenomenologically by the refractive index $n(\omega)$. (See (2.13) and (2.41).) So we get a first hint of how $n(\omega)$ can be understood and calculated in quantum mechanics. Obviously Δt increases if we approach the resonance condition in (3.38) and consequently wave propagation through a material will deviate more strongly from that through vacuum. This is indeed the case, as we shall see in Sect. 4.3.

If the virtually excited state emits a photon $\hbar\omega'$ in a direction different from the incident one, we have a scattering process as discussed in connection with (3.26). If $\hbar\omega$ approaches the resonan energy, this scattering process is also known as resonance fluorescence.

In connection with this scattering, we may ask how can light propagate at all in a clear or transparent medium? The answer is that we have in dense media many scattering atoms or centers in the coherence volume of light, independent of whether it is light from an incandescent lamp, a laser or another source. As a consequence, every scattered wave finds another one which has a phase difference, resulting in destructive interference. The only way that all scattered waves interfere constructively is just the usual propagating wave. The explanation for blue sky, which we gave in Sect. 3.1.5 fulfills, apart from the condition that the diameter of the scattering centers is small compared to the wavelength of the scattered radiation, another condition, namely that there are only a few scattering centers (i.e., gas molecules) per coherence volume of sunlight, so that the mutually destructive interference of the scattered waves is not complete.

If the virtually excited state disappears under simultaneous emission of a photon and the creation or annihilation of a phonon (i.e., a quantum of the lattice vibrations, as in Sect. 9.6), energy conservation for the emitted photon $\hbar\omega_R$ implies

$$\hbar\omega_R = \hbar\omega' \pm \hbar\Omega_{phonon} \ . \tag{3.39}$$

This phenomenon is called Raman scattering for optical phonons, and Brillouin scattering for acoustic phonons. The "–" sign gives the Stokes and the "+" sign the anti-Stokes emission. Similar processes are also possible with more than one phonon. More details will be given in Sect. 13.3.

From the few phenomena outlined briefly above which involve virtual excitation, it is obvious that this mechanism is of some importance for the optical properties of matter. Therefore we want to examine it from another point of view and outline the well-known classical analog of the virtually excited states. In addition this analogy gives some justification for the calculation of the dielectric function used in Chap. 4.

Virtual excitation in quantum mechanics corresponds to a driven or forced oscillation in classical mechanics. If we have an oscillator of eigen frequency ω_0 (corresponding to the energetic differences $E_{ex} - E_g$ in quantum mechanics) and if we excite it with an external frequence ω, it will oscillate with frequency ω after a short damped transient feature of oscillations with ω_0. The amplitude of these steady oscillations increases with decreasing detuning $|\omega - \omega_0|$ depending on the properties of the oscillator, e.g., its damping. This increase of the amplitude of the classical oscillator corresponds to the increase of Δt in (3.39) in the picture of virtual excitation, and it is qualitatively understandable that we will get the strongest deviations of $\varepsilon(\omega)$ or $\tilde{n}(\omega)$ from the vacuum value ($\varepsilon = \tilde{n} = 1$) in the vicinity of the resonance ω_0. We elaborate this concept in detail in Chap. 4.

However, before doing so, we shall demonstrate how the various transitions shown in Fig. 3.16 and some others can be treated quantitatively in quantum mechanics by perturbation theory.

3.2.2 Perturbative Treatment of the Linear Interaction of Light with Matter

In this section we present first the Hamiltonian of the system elaborating the perturbation terms. Then we outline shortly how a perturbation causes transitions between various eigenstates. Finally we join these two things together ending with an understanding of the theoretical description of absorption and stimulated or spontaneous transitions.

The Hamiltonian of the total system consisting of the electron states in Fig. 3.16 of the two level atoms (or of the bands of the semiconductor), the radiation field and the interaction of these two systems can be written as

$$H = H_{el} + H_{rad} + H_{interac} . \tag{3.40}$$

In the picture of second quantization outlined in Sect. 2.5 $H_{interac}$ contains terms which describe, e.g., the annihilation of a photon and of an electron in the ground state and the creation of an electron in the excited state for the process shown in Fig. 3.16a, weighted with a factor which contains the transition matrix element.

The exact solution of the total Hamiltonian leads to the polariton concept (Chap. 5) in the field of linear optics and describes among other things the changes of the electronic states introduced by the presence of the radiation in nonlinear optics for which we will see some examples in Sects. 19.3 and 19.4

For our present purposes we follow an approach which is widely used and which treats the radiation field as a small perturbation, that is, we assume that the eigenstates φ_n and the eigenenergies $E_{\rm n}$ of $H_{\rm el}$ do not change much in the presence of the electromagnetic field, and that the eigenstates of $H_{\rm rad}$ are the photons described already in Sect. 2.5. The approximation which we use now is known as the semiclassical treatment of radiation. It consists of replacing the canonical conjugate momentum $\boldsymbol{p}$ in the Hamilton function by

$$\boldsymbol{p} \to \boldsymbol{p} - e\boldsymbol{A} \,, \tag{3.41}$$

where $\boldsymbol{A}$ is the vector potential (2.48).

If we replace $\boldsymbol{p}$ by its operator

$$\boldsymbol{p} = \frac{\hbar}{\mathrm{i}} \nabla \tag{3.42}$$

the single particle Hamiltonian reads

$$H = \frac{1}{2m}\left(\frac{\hbar}{\mathrm{i}}\nabla - e\boldsymbol{A}\right)^2 + V(\boldsymbol{r}) \tag{3.43}$$

including any electrostatic potential into $V(\boldsymbol{r})$. Making use of the Coulomb guage (2.49) we can evaluate (3.43) to obtain

$$H = -\frac{\hbar^2}{2m}\nabla^2 + V(\boldsymbol{r}) - \frac{e}{m}\boldsymbol{A}\frac{\hbar}{\mathrm{i}}\nabla + \frac{e^2}{2m}\boldsymbol{A}^2 \,, \tag{3.44}$$

$$H = H_{\rm el} - \frac{e}{m}\boldsymbol{A}\frac{\hbar}{\mathrm{i}}\nabla + \frac{e^2}{2m}\boldsymbol{A}^2 \,, \tag{3.45}$$

$$H = H_{\rm el} + H^{(1)} + H^{(2)} \,. \tag{3.46}$$

In (3.46) there are two perturbation terms $H^{(1)}$ and $H^{(2)}$. If we assume that $\boldsymbol{A}$ and thus the light intensity are small, and in the regime of linear optics they are small by definition, then $H^{(1)}$ is a perturbation term of first order and $H^{(2)}$ is small of second order. Consequently $H^{(1)}$ must be used in first-order perturbation theory. In the second-order approximation we have to use $H^{(1)}$ in second-order perturbation theory and $H^{(2)}$ in first-order perturbation theory, etc. We shall come back to this latter aspect in Chap. 18.

In order to arrive at Fermi's golden rule for the transition rate ω_{ji} from an initial state i (e.g., the ground state g in Fig. 3.16) to another state

j (e.g., the excited state ex in Fig. 3.16) one uses the time-dependent Schrödinger equation

$$H\psi = \mathrm{i}\hbar\,\frac{\partial \psi}{\partial t} \tag{3.47a}$$

with the stationary solutions when $H = H_0$

$$\psi_n(\boldsymbol{r}, t) = \varphi_n(\boldsymbol{r})\mathrm{e}^{-\mathrm{i}(E_n/\hbar)t} \,. \tag{3.47b}$$

For the solution in the presence of a perturbation $H^{(1)}$ we make the ansatz

$$\psi(\boldsymbol{r}, t) = \sum_n a_n(t)\varphi_n(\boldsymbol{r})\,\mathrm{e}^{-\mathrm{i}(E_n/\hbar)t} \,. \tag{3.48}$$

We assume that the perturbation is switched on at $t = 0$. Before the system is in state i, i.e.,

$$\left.\begin{aligned} a_\mathrm{i}(t) &= 1 \\ a_{n\neq\mathrm{i}}(t) &= 0 \end{aligned}\right\} \quad \text{for} \quad t \leqslant 0 \,. \tag{3.49}$$

For $t > 0$ the $a_{n\neq\mathrm{i}}(t)$ start to grow and under these conditions the transition rate ω_{ij} of Fermi's golden rule becomes

$$\omega_{ij} = \frac{2\pi}{\hbar}\,|H_{ij}^{(1)}|^2 D(E) \,, \tag{3.50a}$$

where $D(E)$ is the density of the final states modified by momentum conservation if applicable. $H_{ij}^{(1)}$ is the transition matrix element given by

$$H_{ij}^{(1)} = \int \psi_j^*(\boldsymbol{r}) H^{(1)} \psi_i(\boldsymbol{r})\,\mathrm{d}\tau \,. \tag{3.50b}$$

For a non-degenerate two level system $D(E)$ is simply one per atom. The square of the transition matrix element $|H_{ij}|^2$ is known as the transition probability. Later on we will assume, for simplicity of writing, that some constant factors as the term $2\pi/\hbar$ are incorporated in this $|H_{ij}|^2$. Transition probabilities are given apart from some coefficients by the square of the respective transition matrix elements of (3.50) in the case of first order perturbation and by the terms (3.51) for second order.

The transition rate ω_{ij} is proportional to the transition probability multiplied by the square of the amplitude of the perturbation $H^{(1)}$, i.e., here by $|\boldsymbol{A}_0|^2 \sim I$.

If the first-order perturbation term (3.50b) vanishes, then according to what we stated above the second-order contribution reads

$$\omega_{ij}^{(2)} = \frac{2\pi}{\hbar}\left|\sum_{k\neq i,j} \frac{H_{jk}^{(1)} H_{ki}^{(1)}}{E_i - E_k} + H_{ij}^{(2)}\right|^2 D(E) \,. \tag{3.51}$$

We restrict ourselves for the moment to the first order according to (3.50) and discuss the term $H_{ij}^{(1)}$ in some more detail for the perturbations of (3.45).

With the vector potential $\boldsymbol{A}$

$$\boldsymbol{A}=\boldsymbol{A}_0 \mathrm{e}^{\mathrm{i}(\boldsymbol{kr}-\omega t)} \tag{3.52a}$$

we find, e.g., for the absorption process (3.50)

$$\omega_{\mathrm{g}\to\mathrm{ex}}=\frac{2\pi}{\hbar}\left|\frac{-e\hbar}{\mathrm{i}m}\boldsymbol{A}_0\cdot\int\varphi_{\mathrm{ex}}^{*}(\boldsymbol{r})\,\mathrm{e}^{\mathrm{i}(E_{\mathrm{ex}}/\hbar)t}\boldsymbol{e}_A\,\mathrm{e}^{\mathrm{i}(\boldsymbol{kr}-\omega t)}\nabla\varphi_{\mathrm{g}}(\boldsymbol{r})\mathrm{e}^{-\mathrm{i}(E_{\mathrm{g}}/\hbar)t}\mathrm{d}\tau\right|^2 D(E_{\mathrm{g}}+\hbar\omega)$$

$$\propto|\boldsymbol{A}_0<\varphi_{\mathrm{ex}}|H^{(1)}|\varphi_{\mathrm{g}}>|^2 D(E_{\mathrm{g}}+\hbar\omega)=:A_0^2|H_{\mathrm{eg}}^{(1)}|^2\,D(E_{\mathrm{g}}+\hbar\omega), \tag{3.52b}$$

where $\boldsymbol{e}_A$ is the unit vector in the direction of A. We also give in (3.52) a generally used abreviation for the integral. A significant transition rate occurs only if the time dependent exponential functions vanish, or, mathematically

$$E_{\mathrm{ex}}-E_{\mathrm{g}}-\hbar\omega=0\,. \tag{3.53a}$$

This is again the law of energy conservation. If the φ_i have plane-wave character and are described by a wave vector $\boldsymbol{k}$, a similar argument results in $\boldsymbol{k}$ conservation.

$$\hbar\boldsymbol{k}_{\mathrm{ex}}-\hbar\boldsymbol{k}_{\mathrm{g}}-\hbar\boldsymbol{k}=0\,. \tag{3.53b}$$

This is not the case for the two-level atoms discussed here but is true for most of the eigenstates of an ideal semiconductor.

We see that the transition rate is proportional to A_0^2 and thus to the light intensity $I=\langle S\rangle$ or the density of photons $N_{\mathrm{ph}}(\omega)$ in a certain mode:

$$\omega_{ij}\propto A_0^2|H_{\mathrm{eg}}^{(1)}|^2\propto I|H_{\mathrm{eg}}^{(1)}|^2\propto N_{\mathrm{ph}}|H_{\mathrm{eg}}^{(1)}|^2\,. \tag{3.54}$$

By partial integration using the fact that the eigenfunctions form an othonormal set, or that the $H^{(1)}$ is Hermitian adjoint, or by the argument of the microscopic reversibility of a transitions from state $i\to j$ and from $j\to i$ induced by some perturbation $H^{(1)}$, we find that

$$|\textstyle\int\varphi_j^* H^{(1)}\varphi_i\,\mathrm{d}\tau|^2=|\textstyle\int\varphi_i^* H^{(1)}\varphi_j\,\mathrm{d}\tau|\,. \tag{3.55}$$

We see that the probabilities for induced emission and absorption are the same and that the rates differ only by factors containing the number of atoms in the upper and lower states. Spontaneous emission has to be treated in the sense mentioned above as emission stimulated by the zero field. We return to this aspect in a moment. First the interaction operator $H^{(1)}$ should be simplified to reach the so-called dipole approximation.

We note that the radius of an atom ($r\simeq 0.1$ nm) and the distance between neighboring atoms in a solid ($a\simeq 0.3$ nm) are small compared to the wavelength in the visible ($\lambda \simeq 500$ nm). Therefore there is practically no phase shift of the electromagnetic radiation over one atom or between one atom and its neighbors. Thus we can expand the term e^{ikr} in (3.52) in a power series and stop after the constant term

$$e^{ikr} = 1 + \frac{ik\boldsymbol{r}}{1!} + \frac{ik\boldsymbol{r}^2}{2!} + \cdots \simeq 1 . \tag{3.56}$$

This is the first step towards the dipole approximation. It means, that the momentum of the photon $\hbar \boldsymbol{k}$ in (3.53b) is negligible.

The matrix element $H_{ij}^{(1)}$ still contains the momentum operator $p = \frac{\hbar}{i}\nabla$

$$H_{ij}^{(1)} \sim \langle \varphi_i | \boldsymbol{p} | \varphi_i \rangle =: \langle \boldsymbol{p}_{ij} \rangle .$$

With the semiclassical relation

$$\frac{\hbar}{i}\nabla = p = m\dot{\boldsymbol{r}} \tag{3.57}$$

and some plausibility of arguments [2.3], we find that

$$\int \varphi_j^* \frac{\hbar}{i} \nabla \varphi_i \, d\tau = m \frac{i}{\hbar}(E_i - E_j) \int \varphi_j^* \boldsymbol{r} \varphi_i \, d\tau = m\omega \int \varphi_j^* \boldsymbol{r} \varphi_i \, d\tau . \tag{3.58}$$

For a detailed derivation of this relation actually some knowledge of the wave functions is required (For details see [2.7].)

We note that in this so-called dipole approximation (3.56 = 59) the transition rate is given by

$$\frac{1}{D(E)} w_{ij} \sim I\omega |\boldsymbol{e}_A \langle e\boldsymbol{r}_{ij} \rangle|^2 = iI\omega |H_{ij}^D|^2 . \tag{3.59}$$

This result can also be obtained in a more intuitive way if we remember that the energy of a dipole $e\boldsymbol{r}$ in an electric field $\boldsymbol{E} = \dot{\boldsymbol{A}}$ is given by

$$e\boldsymbol{r} \cdot \boldsymbol{E} = H^{(1)} . \tag{3.60}$$

Using this approach in combination with (3.50) yields directly (3.59).

From now on we will call $|H_{ij}^D|^2$ the dipole-transition probability and the operator $e\boldsymbol{r}$ the dipole operator H^D.

Transitions using higher terms in (3.44, 51, 56) correspond to quadrupole, octupole and higher order transitions.

To conclude this section, we calculate the net rate of the transitions shown in Fig. 3.16a–c.

We assume that we have a density of photons N_{ph} which populate only one mode in the sense used for the calculation of the density of states in Sect. 2.6 that is, all photons have the same wave vector $\boldsymbol{k}$, polarization $\boldsymbol{e}_A$ and energy $\hbar\omega$.

Furthermore $\hbar\omega$ fulfills the energy conservation law according to (3.54). The density of identical two-level atoms is N_A where a fraction α_g is in the ground state and correspondingly $(1-\alpha_g)$ are in the excited state. The net rate of the change of N_{ph} with time is then given using (3.54, 59)

$$\frac{\partial N_{ph}}{\partial t} = -N_A \alpha_g N_{ph} |H^D_{g\to e}|^2 + N_A(1-\alpha_g)(1+N_{ph})|H^D_{e\to g}|^2 . \tag{3.61}$$

The first term on the r.h.s. describes the absorption of photons, the second one the spontaneous and stimulated emission in the factor $(1+N_{ph})$.

From (3.55) we see that

$$|H^D_{g\to e}|^2 = |H^D_{e\to g}|^2 = |H^D|^2 \tag{3.62}$$

and hence

$$\frac{1}{|H^D|^2}\frac{\partial N_{ph}}{\partial t} = N_{ph} \cdot N_A(1-2\alpha_g) + N_A(1-\alpha_g) . \tag{3.63}$$

The first term on the right-hand side depends linearly on N_{ph} and describes the net rate of absorption and stimulated emission. The second terms gives the spontaneous emission since it is independent of N_{ph}.

There is net absorption for $\alpha_g > 1/2$ (absorption coefficient $\alpha(\omega) > 0$ or $\partial N_{ph}/\partial t < 0$ and amplification or optical gain for $\alpha_g < 1/2[\alpha(\omega) < 0]$, i.e., for gain, more than half of the atoms have to be in the upper state. This situation cannot be reached in thermal equilibrium, but only under the influence of a suitable source of pump power. Usually one or more additional energy levels are required (three- and four-level lasers). This fact can be easily elucidated with the following argument. We start with a situation where all atoms are in the ground state i.e., $\alpha_g\ (t=0)=1$. If we send for $t>0$ a photon field with frequency ω fulfilling (3.52a) into the system we initially have absorption since $\alpha(\omega)$ is given by

$$-\alpha(\omega) \sim N_A(1-2\alpha_g) . \tag{3.64}$$

With increasing time and pump power $\alpha(\omega)$ decreases because α_g decreases. For the situation $\alpha_g = 1/2$ the absorption vanishes and the material becomes transparent. This means no more pumping is possible to reach $\alpha_g > 1/2$. In Chaps. 19, 20 and especially 21 we will reexamine the above considerations using the proper statistics introduced in Sect. 2.6.

3.3 Problems

1. Consider the interface between vacuum (or air) and glass ($n = 1.45$) at a wavelength λ_{vac} of 0.5 µm.

Calculate for the angle of incidence $\alpha_i = 45°$ the incident, reflected, and transmitted wave vectors, and the transmitted and reflected intensities for both polarizations.

Calculate Brewster's angle for the transition air → glass and glass → air and the angle for the onset of total internal reflection.

2. Find a piece of polarizing material (polaroid) and observe the light reflected from a nicely polished floor or scattered from the blue sky using different orientations of the light propagation and of the polarization. Do not look into the sun! Try to explain your findings.

3. Play with a piece of clear calcite and the polaroid.

4. Find in a textbook the definition and meaning of Einstein's coefficients.

4 Ensemble of Uncoupled Oscillators

The optical properties of matter are determined by the coupling of various types of oscillators in matter to the electromagnetic radiation field. In other words, an incident electromagnetic field will cause these oscillators to perform driven or forced oscillations. The amplitude of these driven oscillations depends on the angular frequency ω of the incident field, on the eigenfrequency ω_0 of the oscillators, on the coupling strength f between electromagnetic field and oscillator, and on its damping γ. In semiconductors these oscillators or resonances include optical phonons, excitons, and plasmons. They will be explained in some detail in Chaps. 9–12. We can anticipate that many basic features of the optical properties of these resonances are similar. Therefore it is reasonable to discuss first, in a general way, the optical properties of an ensemble of model oscillators. By using the results of Chaps. 4–8 we shall obtain in Chaps. 13–17 a quite simple and straightforward access to the optical properties of semiconductors.

It turns out that a treatment of the optical properties of an ensemble of model oscillators in terms of classical mechanics and electrodynamics yields results which are, in many respects, very close to reality. This is especially true for the spectra of the complex dielectric function or refractive index, or of the spectra of reflection and transmission. All four are closely connected (Chap. 8). We shall therefore follow this classical approach for some while, and explain at the appropriate places what modifications appear if quantum mechanics is applied.

We will now consider the optical properties of an ensemble of oscillators. We begin with the simplest case of uncoupled oscillators and refine the concept in various steps up to Chap. 8.

4.1 Equations of Motion and the Dielectric Function

We assume that we have an ensemble of identical uncoupled harmonic oscillators. For simplicity we choose a periodic one-dimensional array in the direction of light propagation with a lattice-constant a as shown in Fig. 4.1a. These harmonic oscillators all have the same eigenfrequency ω'_0. If we neglect damping for the moment, then ω'_0 can be expressed in a mechanical model by the mass m and the force constant β of the springs as

$$\omega_0'^2 = \beta m^{-1} \ . \tag{4.1}$$

If we elongate the oscillators in phase (Fig. 4.1b) the whole ensemble oscillates with ω_0' and the same will be true if we excite neighboring oscillators in antiphase (Fig. 4.1c). The first case corresponds to

$$\lambda = \infty \quad \text{or} \quad k = 0 \,, \tag{4.2a}$$

and the second one to

$$\lambda_{\min} = 2a \quad \text{or} \quad k_{\max} = \frac{\pi}{a} \,. \tag{4.2b}$$

The point $k = 0$ is usually called the Γ-point in $\boldsymbol{k}$-space and the condition (4.2b) gives the boundary of the first Brillouin zone for a simple linear chain or cubic lattice. The values given by (4.2b) are the shortest physically meaningful wavelength and the largest $\boldsymbol{k}$-vector in our system. The eigenfrequency will also be ω_0' for all λ or $\boldsymbol{k}$-values in between and so we get the horizontal dispersion relation for our system in Fig. 4.2a, i.e., the width of the band of

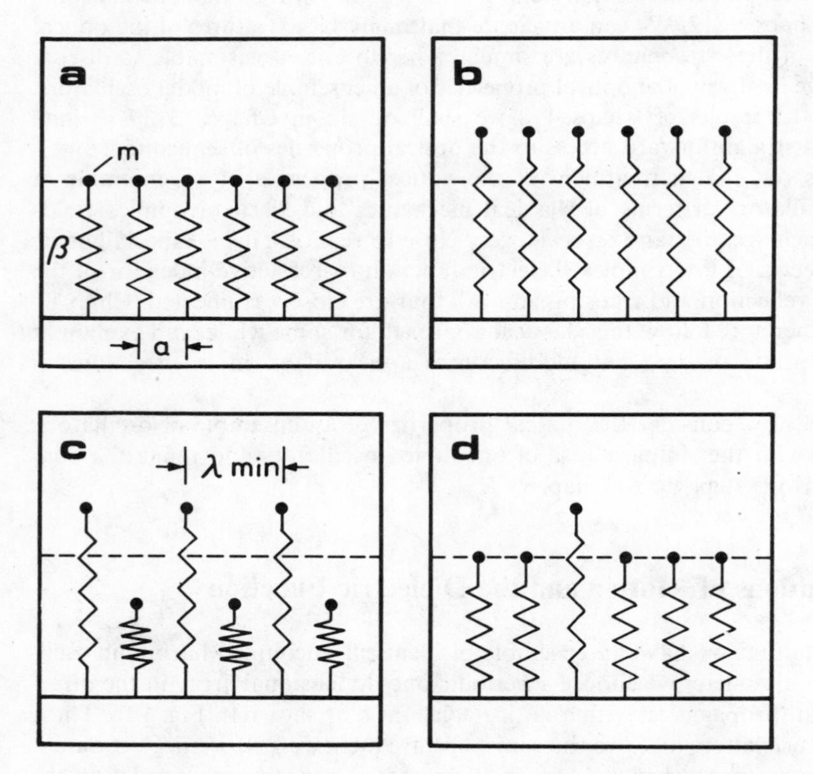

Fig. 4.1a–d. A part of a periodic array of uncoupled oscillators in their equilibrium position (**a**); elongated with a wavelength $\lambda \Rightarrow \infty$ (**b**); with the shortest physically meaningful wavelength $\lambda_{\min}$ (**c**), and a non-propagating wave packet (**d**). After [4.1]

eigenfrequencies is zero (so-called flat band). A wave packet created in our system by elongating only one oscillator (Fig. 4.1d) will not propagate, since there is no coupling to the neighboring oscillators. In agreement we find from (2.13) that the group velocity $\partial\omega/\partial k$ is zero.

What we have seen here is an example of the general rule that in an ensemble of identical oscillators [or of atoms or of other (quasi-) particles] a finite coupling between neighbors results in a finite bandwidth and a non-vanishing group velocity while vanishing coupling results in vanishing band-width and group velocity (see also Fig. 4.2)

$$\begin{aligned}&\text{zero coupling} \longleftrightarrow \text{zero bandwidth} \longleftrightarrow v_g = 0\,,\\&\text{finite coupling} \longleftrightarrow \text{finite bandwidth} \longleftrightarrow v_g \neq 0\,.\end{aligned} \tag{4.3}$$

We will discuss the implications of relaxing the assumption "uncoupled" later on, proceeding to the more realistic assumption of coupled oscillators in Chap. 6.

In the next step we couple the independent oscillators to the electric field of the electromagnetic radiation given by

$$\boldsymbol{E} = (E_0, 0, 0)\exp[\mathrm{i}(k_z z - \omega t)]\,. \tag{4.4a}$$

This means that the light wave propagates in the z-direction and is polarized in the x-direction, parallel to the elongation of the oscillators. By considering the oscillator at $z = 0$ or making use of the dipole approximation of (3.59, 60)

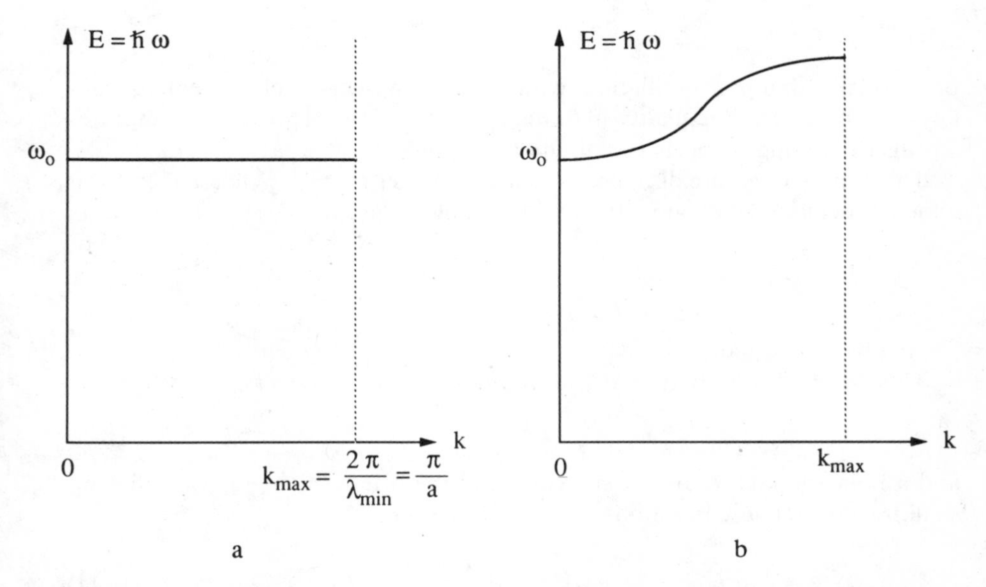

Fig. 4.2a, b. The dispersion relation of an ensemble of uncoupled (**a**) and of coupled (**b**) oscillators

(i.e., $a \ll \lambda$) we can drop the spatial variation in (4.4a) to obtain

$$\boldsymbol{E} = (E_0, 0, 0)\mathrm{e}^{-\mathrm{i}\omega t} . \tag{4.4b}$$

For the coupling, we assume that the mass m of every oscillator carries a charge e. For neutrality reasons, we need then a charge $-e$ fixed at the equilibrium position of every mass. $\boldsymbol{E}$ will then exert a force on the oscillator, and the elongation x is connected with an electric dipole moment via

$$p_x = ex . \tag{4.5}$$

This or similar models and/or the resulting equations for $\varepsilon(\omega)$, go by various names such as Lorentz or Selmaier oscillator, Helmhotz–Ketteler or Kramers–Heisenberg formula.

The equation of motion is given by

$$m\ddot{x} - \gamma m\dot{x} + \beta x = eE_0\mathrm{e}^{-\mathrm{i}\omega t} , \tag{4.6}$$

where we have included a phenomenological damping constant γ. Equation (4.6) is a linear, inhomogeneous differential equation of second order. The general solution is a sum of the general solution of the corresponding homogeneous equation and of a special solution of the inhomogeneous one. Correspondingly we use the following ansatz:

$$x(t) = x_0 \exp[-\mathrm{i}(\omega_0'^2 - \gamma^2/4)^{1/2} t] \exp(-t\gamma/2) + x_\mathrm{p}\mathrm{e}^{-\mathrm{i}\omega t} . \tag{4.7}$$

The first term on the right-hand side is the solution of the homogeneous equation and describes a transient feature. For

$$\omega_0'^2 - \gamma^2/4 > 0 , \tag{4.8}$$

one finds a damped oscillation with a damping-dependent eigenfrequency $(\omega_0'^2 - \gamma^2/4)^{1/2}$. The inequality (4.8) defines the regime of weak damping. For stronger damping one gets essentially an exponentially decaying term.

This transient feature disappears in any case after $t > \gamma^{-1}$. What is then left is a forced oscillation with amplitude x_p. Inserting the ansatz (4.7) into (4.6) we find

$$x_\mathrm{p} = \frac{eE_0}{m}(\omega_0'^2 - \omega^2 - \mathrm{i}\omega\gamma)^{-1} , \tag{4.9}$$

i.e., the usual resonance term.

This oscillation is connected with a dipole moment at every oscillator of

$$p_x = ex_\mathrm{p} \tag{4.10}$$

and a polarizability $\hat{\alpha}(\omega)$. The hat on $\hat{\alpha}$ and the prime on ω_0' indicate that we shall introduce two corrections here and in Sect. 4.2.

$$\hat{\alpha}(\omega) = \frac{ex_\mathrm{p}}{E_0} = \frac{e^2}{m}(\omega_0'^2 - \omega^2 - \mathrm{i}\omega\gamma)^{-1} . \tag{4.11}$$

If we use a three-dimensional array of oscillators with density N, the result is the same and we shall get a preliminary polarization density $\boldsymbol{P}$

$$\boldsymbol{P} = N\hat{\alpha}\boldsymbol{E}_0 = \frac{Ne^2}{m}(\omega_0'^2 - \omega^2 - \mathrm{i}\omega\gamma)^{-1}\boldsymbol{E}_0 . \tag{4.12}$$

From (4.12) and (2.26) we get the following expressions for the dielectric displacement $\boldsymbol{D}$, the dielectric function $\varepsilon(\omega)$, and the susceptibility $\chi(\omega)$:

$$\boldsymbol{D} = \varepsilon_0\boldsymbol{E} + \boldsymbol{P} = \varepsilon_0\left[1 + \frac{Ne^2}{m\varepsilon_0}(\omega_0'^2 - \omega^2 - \mathrm{i}\omega\gamma)^{-1}\right]\boldsymbol{E} \tag{4.13}$$

and

$$\varepsilon(\omega) = 1 + \frac{Ne^2}{m}(\omega_0'^2 - \omega^2 - \mathrm{i}\omega\gamma)^{-1} = \chi(\omega) + 1 , \tag{4.14}$$

where N is the number of oscillators per unit volume. Now we want to address the two corrections to the above set of equations in order to get our final expression for $\varepsilon(\omega)$.

4.2 Corrections Due to Quantum Mechanics and Local Fields

The term Ne^2m^{-1}, e.g., in (4.12) gives the coupling strength of the electromagnetic field to the oscillators in our mechanical model.

In quantum mechanics, this coupling is given by the transition matrix element squared. For dipole-allowed transitions this reads, as mentioned already in Sect. 3.2.2, Eqs. (3.58–60):

$$|H_{ij}^{\mathrm{D}}|^2 = |\langle j|H^{\mathrm{D}}|i\rangle|^2 . \tag{4.15}$$

where i and j stand for initial and final state and H^{D} for the dipole operator. For dipole-forbidden transitions, magnetic dipole or electric quadrupole matrix elements can become relevant. They are usually orders of magnitude smaller. There are different conventions for introducing the transition matrix element into the dielectric function:

Some authors define a dimensionless quantity $\hat{f}$ which is proportional to $|H_{ij}|^2$, call it the oscillator strength, and use it to multiply the term Ne^2m^{-1}. Others call the whole numerator oscillator strength f'; i.e.,

$$f' \propto |H_{ij}^{\mathrm{D}}|^2\, Ne^2m^{-1} . \tag{4.16}$$

We follow the second way for simplicity of notation, but stress that there is no physical difference between the methods. We obtain

$$\varepsilon(\omega) = 1 + \frac{f'}{\omega_0'^2 - \omega^2 - \mathrm{i}\omega\gamma} . \tag{4.17}$$

For the next correction to (4.17) we have to consider what is the electric field $\boldsymbol{E}$ [e.g., in (4.6)] acting on the oscillators. For dilute systems, $\boldsymbol{E}$ is just the external incident field and we can use (4.17) as it is. For dense systems, i.e., systems with a high density N of oscillators, the local field E^{loc} acting on the oscillators consists of two parts, namely the external field and the field created by all the other dipoles. Taking into account this effect leads for cubic materials to the so-called Clausius–Mosotti or Lorenz–Lorentz equation, which relates the polarizability $\hat{\alpha}$ to $\varepsilon(\omega)$ through

$$\frac{\varepsilon(\omega)-1}{\varepsilon(\omega)+2}=\hat{\alpha}(\omega)=\frac{\frac{1}{3}N\frac{e^2}{m}}{\omega_0'^2-\omega^2-\mathrm{i}\omega\gamma}\,. \tag{4.18a}$$

Obviously (4.18a) recovers the form (4.17) for dilute systems (small N) for which $\hat{\alpha}$ is small and $\varepsilon(\omega)$ deviates only a little from unity. If this approximation is not valid, $\varepsilon(\omega)$ can be rewritten in the form (4.17) but with a shifted eigenfrequency

$$\omega_0^2=\omega_0'^2-\frac{Ne^2}{3m\varepsilon_0}=\omega_0'^2-f'/3\,. \tag{4.18b}$$

A similar procedure is also valid for crystal symmetries other than the cubic one, resulting in

$$\varepsilon(\omega)=1+\frac{f'}{\omega_0^2-\omega^2-\mathrm{i}\omega\gamma}\,. \tag{4.19}$$

This formula now incorporates local field effects and the quantum mechanical transition probabilities. The new eigenfrequency ω_0 is the only physically relevant one, and the one that appears in the experiments.

4.3 Spectra of the Dielectric Function and of the Complex Index of Refraction

As we shall see later, a semiconductor contains not only one type of oscillator and one resonance frequency ω_0, but many of them – like phonons, excitons etc. In order to take care of this fact we have to sum over all resonances leading to

$$\varepsilon(\omega)=1+\sum_j\frac{f_j'}{\omega_{0j}^2-\omega^2-\mathrm{i}\omega\gamma_j}\,. \tag{4.20}$$

This is essentially the so-called Helmholtz–Ketteler formula or Kramers–Heisenberg dielectric function.

We now want to discuss the contribution of an isolated resonance at $\omega_{0j'}$ in (4.20) which allows some simplification. For closely spaced resonances the whole formula (4.20) has to be used. However, we note that the contribution of a single resonance at $\omega_{0j'}$ is constant for $\omega \ll \omega_{0j'}$ with a contribution to $\varepsilon(\omega)$

$$\frac{f'_{j'}}{\omega_{0j'}^2} = \text{const} \tag{4.21}$$

and tends to zero for $\omega \gg \omega_{0j'}$.

This means that in the vicinity of a resonance we can neglect the contributions from all lower resonances $\omega_{0j} \ll \omega_{0j'}$ and the constant contributions of all higher resonances $\omega_{0j} \gg \omega_{0j'}$ can be summarized in a so-called background dielectric constant ε_b. Obviously ε_b is unity for the highest resonance in the sum of (4.20). So we finally get the simplified expression in the spectral surroundings of $\omega_{0j'}$

$$\varepsilon(\omega) = \varepsilon_b \left(1 + \frac{f_{j'}}{\omega_{0j'}^2 - \omega^2 - \mathrm{i}\omega\gamma_{j'}} \right) \tag{4.22a}$$

with

$$f_{j'} = f'_{j'} \varepsilon_b^{-1} \; .$$

In the following we drop the index j'. Equation (4.22a) can be separated into real and imaginary parts

$$\varepsilon(\omega) = \varepsilon_b \left(1 + \frac{f(\omega_0^2 - \omega^2)}{(\omega_0^2 - \omega^2)^2 + \omega^2\gamma^2} + \mathrm{i} \frac{\omega\gamma f}{(\omega_0^2 - \omega^2)^2 + \omega^2\gamma^2} \right)$$

$$= \varepsilon_1(\omega) + \mathrm{i}\varepsilon_2(\omega) \; . \tag{4.22b}$$

In Fig. 4.3 we show the real and imaginary parts of $\varepsilon(\omega)$. For negligible damping $\gamma \to 0$ we find a pole in $\mathrm{Re}\{\varepsilon(\omega)\}$, and $\mathrm{Im}\{\varepsilon(\omega)\}$ converges to a δ function at ω_0. Finite damping results in a broadening of $\mathrm{Im}\{\varepsilon(\omega)\}$ to the Lorentzian lineshape of (4.22b) and a smooth connection of the two branches of $\mathrm{Re}\{\varepsilon(\omega)\}$.

We concentrate now on the case of small damping. One of the two frequencies of special interest for $\varepsilon(\omega)$ is the eigenfrequency ω_0, which is connected with the singularity. The other one corresponds to the point at which $\mathrm{Re}\{\varepsilon(\omega)\}$ crosses zero. Going back to (2.15) and (2.43) we find

$$\nabla \cdot \boldsymbol{D} = \varepsilon_0 \nabla \cdot \varepsilon(\omega) \cdot \boldsymbol{E} = 0 \; . \tag{4.23}$$

As already discussed, this equation is usually used to argue that electromagnetic waves are transverse, since $\nabla \cdot \boldsymbol{E} = 0$ is zero for this case. The other solution $\varepsilon(\omega = \omega_L) = 0$ gives the frequency of a longitudinal mode. This mode is a pure polarization mode with $\varepsilon_0 \boldsymbol{E} = -\boldsymbol{P}$, i.e., antiparallel polarization and electric field (see Sect. 2.4).

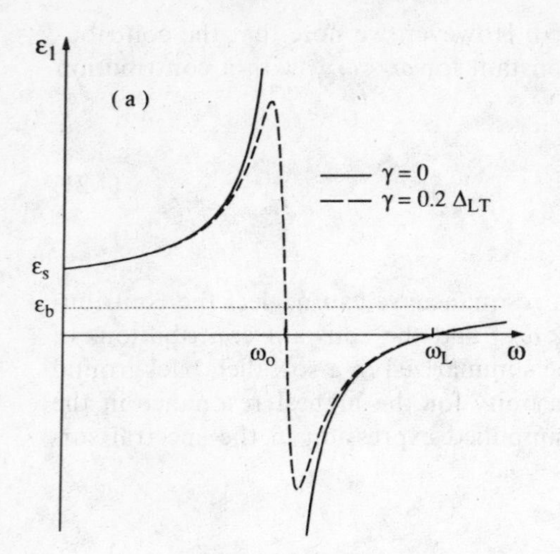

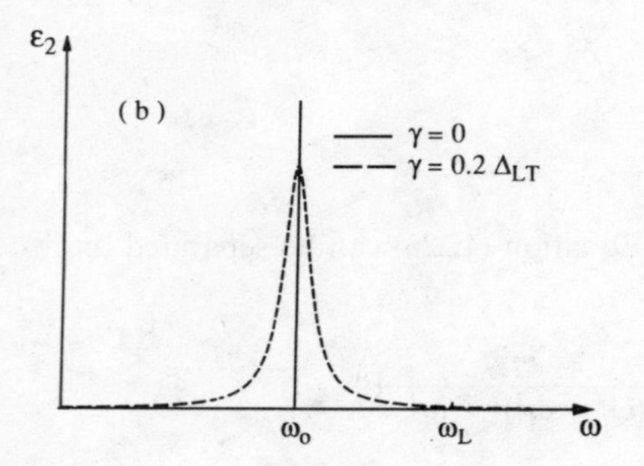

Fig. 4.3a,b. The real (**a**) and imaginary (**b**) parts of the dielectric function for zero and finite damping

For $\gamma \Rightarrow 0$ we find

$$\varepsilon(\omega = \omega_{\mathrm{T}} = \omega_0) = \infty \,,$$

$$\varepsilon(\omega = \omega_{\mathrm{L}}) = 0 \,. \tag{4.24}$$

The relation between these two frequencies is given by

$$\omega_{\mathrm{L}}^2 - \omega_{\mathrm{T}}^2 = f \sim |H_{ij}^{\mathrm{D}}|^2 \,, \tag{4.25}$$

and we can make the following statements. A finite oscillator strength f is necessarily connected with a finite longitudinal–transverse splitting energy Δ_{LT} and vice versa.

$$|H_{ij}|^2 \neq 0 \Leftrightarrow f \neq 0 \Leftrightarrow \Delta_{\mathrm{LT}} = \hbar(\omega_{\mathrm{L}} - \omega_{\mathrm{T}}) \neq 0 \,. \tag{4.26}$$

If we call the roughly constant value of $\varepsilon(\omega)$ below ω_T the static dielectric constant ε_s and the value above ε_b, as already mentioned, we find

$$\varepsilon_s = \varepsilon_b(1 + f/\omega_0^2) \tag{4.27}$$

and the Lyddane–Sachs–Teller relation

$$\frac{\varepsilon_s}{\varepsilon_b} = \frac{\omega_L^2}{\omega_0^2} > 1 \quad \text{for} \quad f > 0\,. \tag{4.28}$$

Furthermore we can make the slightly tricky statement that the background dielectric constant ε_b of a resonance ω_{0j} is simultaneously the static dielectric constant for the next higher resonance ω_{0j+1}.

Finally we can define the term "small damping" already used several times above by

$$\gamma < \hbar^{-1}\Delta_{LT} = \omega_L - \omega_T \tag{4.29}$$

and frequencies ω far above or below ω_0 means

$$\hbar|\omega - \omega_0| \gg \Delta_{LT}\,. \tag{4.30}$$

Now we concentrate on the complex index of refraction $\tilde{n}(\omega) = n(\omega) + \mathrm{i}\kappa(\omega)$, which is connected to the dielectric function via (2.35)

$$\tilde{n}(\omega) = \varepsilon^{1/2}(\omega) \tag{4.31}$$

and consequently

$$\varepsilon_1(\omega) = n^2(\omega) - \kappa^2(\omega)$$

$$\varepsilon_2(\omega) = 2n(\omega)\kappa(\omega)$$

or

$$n(\omega) = \left(\frac{1}{2}\{\varepsilon_1(\omega) + [\varepsilon_1^2(\omega) + \varepsilon_2^2(\omega)]^{1/2}\}\right)^{1/2}$$

$$\kappa(\omega) = \left(\frac{1}{2}\{-\varepsilon_1(\omega) + [\varepsilon_1^2(\omega) + \varepsilon_2^2(\omega)]^{1/2}\}\right)^{1/2}. \tag{4.32}$$

In Fig. 4.4 we show the real and imaginary parts of $\tilde{n}$ for vanishing and for finite γ.

When approaching the resonance from the low-frequency side, one sees that both $\mathrm{Re}\{\tilde{n}\} = n$ and the real part of k increase drastically. Between ω_0 and ω_L we find

$$\omega_0 \leqslant \omega \leqslant \omega_L \begin{cases} n = 0 & \text{for} \quad \gamma = 0 \\ n \ll 1 & \text{for} \quad \gamma \neq 0 \\ \kappa > n & \end{cases} \tag{4.33}$$

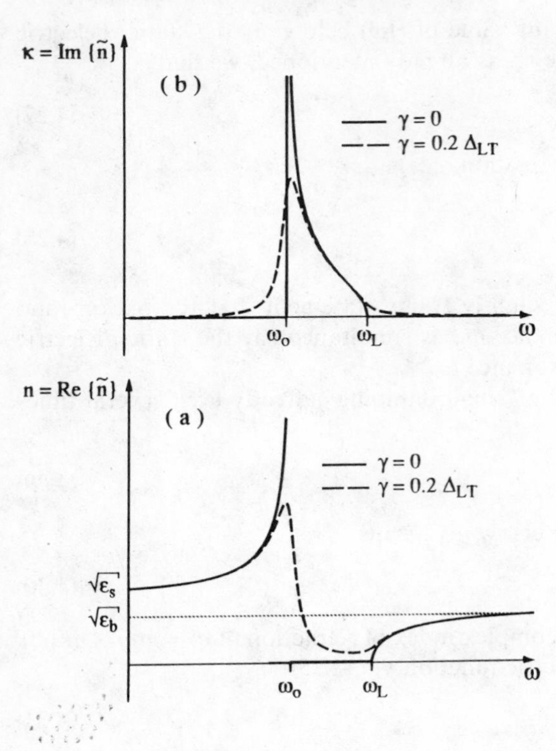

Fig. 4.4a, b. Real (**a**) and imaginary (**b**) parts of the complex index of refraction for zero and finite damping

while κ starts with a singularity at $\omega = \omega_0$ for $\gamma = 0$ and then drops to small values for ω approaching ω_L.

This means that for $\gamma = 0$ there is no propagating, i.e., no spatially oscillating or wave–like solution in the medium for the spectral region addressed in (4.33). Instead we have only a spatially exponentially decaying amplitude similar to the type known for total reflection in the medium with the lower index of refraction. For finite γ we get a small real part of $\tilde{n}$, which means that some light can penetrate into the medium, but this light is damped over a distance shorter than the wavelength in the medium since we have $\kappa > n$, i.e., "strong absorption" according to (3.17).

4.4 The Spectra of Reflection and Transmission

With a knowledge of $\tilde{n}(\omega)$ we can now discuss the spectra of reflection and transmission. We start with the reflectivity $R(\omega)$ of a single interface between vacuum (or air) and the medium. We discuss only the situation of normal incidence [see Sect. 3.1.4 and eq. (3.19)].

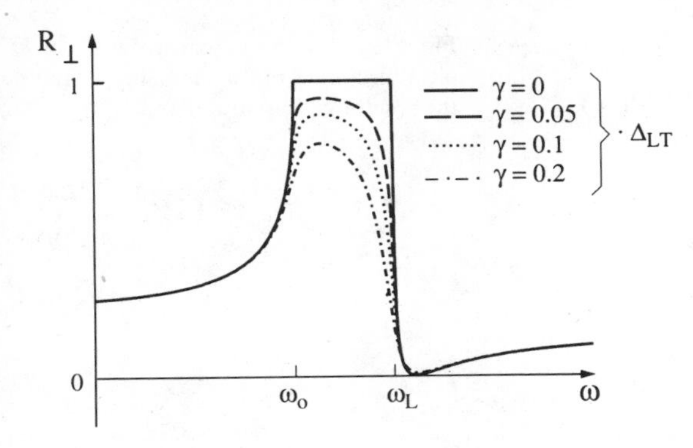

Fig. 4.5. The reflection spectrum of a single resonance with zero and finite damping

Oblique incidence is described by the Fresnel formula (3.18).

$$R(\omega)=\frac{I_{\rm r}}{I_{\rm i}}=\frac{[n(\omega)-1]^2+\kappa^2(\omega)}{[n(\omega)+1]^2+\kappa^2(\omega)} \tag{4.34}$$

$R(\omega)$ is plotted in Fig. 4.5. First we discuss the situation for $\gamma=0$. When approaching the resonance from the low frequency side, R starts with an almost constant value determined from (3.19b) and (4.27, 28):

$$R=\frac{(\varepsilon_{\rm s}^{1/2}-1)^2}{(\varepsilon_{\rm s}^{1/2}+1)^2} \quad \text{for} \quad \omega_0-\omega \gg \frac{1}{\hbar}\Delta_{\rm LT} \ . \tag{4.35a}$$

Above $\omega_{\rm L}$, the reflectivity converges towards a lower constant value

$$R=\left(\frac{\varepsilon_{\rm b}^{1/2}-1}{\varepsilon_{\rm b}^{1/2}+1}\right)^2 \quad \text{for} \quad \omega-\omega_0 \gg \hbar^{-1}\Delta_{\rm LT} \ . \tag{4.35b}$$

Just below ω_0 the reflectivity R increases and reaches the value $R=1$ for $\omega=\omega_0$. Between ω_0 and $\omega_{\rm L}$ we have $R=1$. For $\omega>\omega_{\rm L}$ R drops rapidly and reaches zero at the frequency where $n(\omega)=1$. Then R increases again towards a constant value given by (4.35b). We see now that all light is reflected in the region where we have no propagating mode in the medium in agreement with the discussion of (4.33). Such a band is also called stop-band because the light is "stopped" and sent back.

For finite γ the reflection spectrum is smoothed out due to the finite values of $n(\omega)$ for $\omega_0 \leqslant \omega \leqslant \omega_{\rm L}$ and of $\kappa(\omega)$ for $\omega \leqslant \omega_0$ and for $\omega_{\rm L}<\omega$.

The spectral region between ω_0 and $\omega_{\rm L}$ is often also called "Reststrahl bande" for the following reason. If we send a light beam with a broad, essentially flat, spectral distribution of the intensity around ω_0 onto a sample

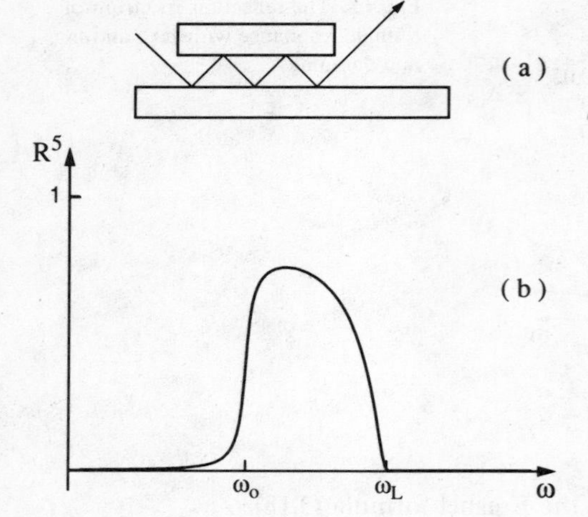

Fig. 4.6a, b. An arrangement for multiple reflection from a medium (**a**) and the resulting "Reststrahlbande" (**b**)

and allow it to be reflected several times, e.g., in the configuration given in Fig. 4.6, then significant intensity will remain only in the region $\omega_0 \leqslant \omega \leqslant \omega_L$. It should be noted that we do not assume any coherent superposition of the beams, i.e., the Reststrahlbande comes only from $\varepsilon(\omega)$ and has nothing to do with Fabry–Perot resonators (Sect. 3.1.6) or a Lummer–Gehrke plate.

Up to now we have discussed the reflectivity of a single interface. In experiments it is usually easier to handle semiconductor plates with two surfaces and a geometrical thickness d rather than samples filling a semi-infinite half-space. To handle this problem, we use the results of Sect. 3.1.6.

In the vicinity of the resonance the reflection spectrum will remain the same as in Fig. 4.5 since multiple reflection is suppressed due to the large value of κ and thus of the absorption coefficient (2.40)

$$\alpha(\omega) = 2\frac{\omega}{c}\kappa(\omega) . \tag{4.36}$$

Away from the resonance we have to consider for the total transmission $\hat{T}$ contributions due to multiple reflection at the two surfaces. There are two cases, as already discussed in Sect. 3.1.6; if d is longer than the coherence length of the light source, or if the two surfaces of the slab are not exactly plane-parallel (e.g., due to steps on the surfaces or due to a small angle between them) we have to add intensities. This results in a total transmission $\hat{T}$ and reflection $\hat{R}$ of the slab given by (3.29).

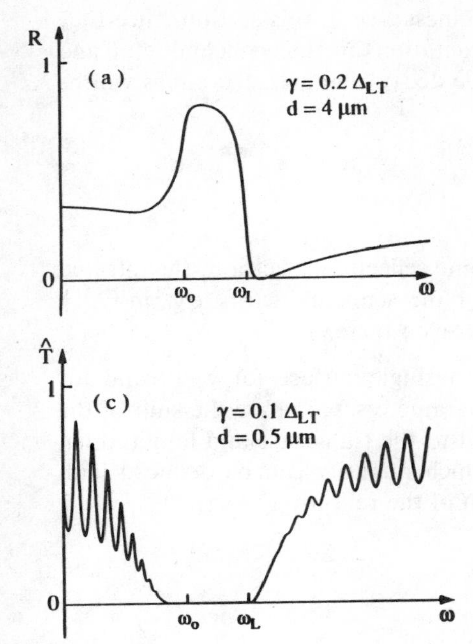

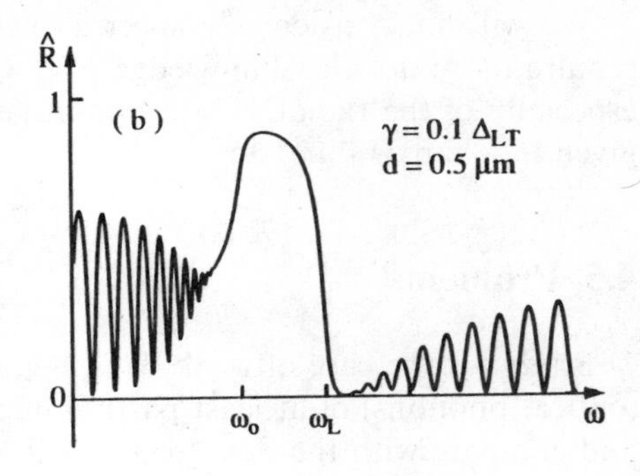

Fig. 4.7a–c. The reflectivity $\hat{R}$ of a thin slab of matter in the vicinity of a resonance for the cases where intensities (**a**) or field amplitudes (**b**) have to be added and the transmission (**c**) for the case where amplitudes add

In Fig. 4.7a we show $\hat{R}$ for this situation. We start with a reflectivity given by adding those of front and rear side. Approaching the resonance, $\hat{R}$ drops first slightly because the absorption reduces the influence of the rear side. The reflection around the resonance remains the same as above in Fig. 4.5. If the sample becomes transparent above the resonance, we again get a contribution from the rear side to the reflectivity.

In Figs. 4.7b and c we give an example for the $\hat{T}(\omega)$ and $\hat{R}(\omega)$ spectra for the second case when the sample forms a FP resonator using the set of equations (3.31, 32). We see that the finesse F is reduced close to the resonance due to increasing absorption. The spectral spacing between the resonances decreases as the resonance is approached from lower frequencies, due to the steep increase of $n(\omega)$ or of $\boldsymbol{k}$. Above resonance, the FP mode structure reappeares.

Every directional dependence of the quantities ω_0, f and γ which describe a resonance will lead to a directional dependence of $n(\omega)$ and $\kappa(\omega)$, i.e., to birefringence and dichroism, discussed in Sect. 3.1.7. In Chap. 6 we shall see that a dependence of the dispersion $\omega_0(\boldsymbol{k})$ can lead to a weak birefringence, and indeed this effect is observed. For a cubic crystal with point group T_d (see Chap. 25) one finds that light propagating with $\boldsymbol{k}$ parallel to the [110] direction may display slightly different optical properties when the electric field is polarized in the directions [$\bar{1}$10] and [001].

We will not consider the spectra of luminescence in this section since they require more detailed knowledge of the excitations in the semiconductor and especially of the radiative and nonradiative decay channels. Examples will be given in Chaps. 14 and 15.

4.5 Problems

1. Study for the case of weak damping, some reflection spectra in the infrared (optical phonons) of at least partly ionic bound semiconductors, e.g., in [1.4], and compare with the data for ω_0 and ω_L given there.

2. Calculate the spectra of reflection for a single surface for weak and for strong damping with otherwise constant parameters. Compare the shift of the reflection maxima and minima with respect to the transverse and longitudinal eigenenergies ω_0 and ω_L, respectively. Which quantity can be deduced with reasonable accuracy from a first inspection of the reflection spectra?

5 The Concept of Polaritons

Here we want to discuss in more detail what is actually propagating when "light" travels through matter. In vacuum the situation was quite clear on our present level of understanding in Sect. 2.2. Light in vacuum is a transverse electromagnetic wave, the quanta of which are known as photons.

There are two levels at which one can describe the interaction of light with matter. One is the so-called perturbative treatment or weak coupling case. In this approach, which we used in Sect. 3.2, the electromagnetic field and the excitations of the matter are treated as independent quantities. As shown in Fig. 3.16a a photon is absorbed and the matter goes from the ground state to the excited state, and that's it. This approach is sufficient for many purposes but, if we look closer, we see that this is not the whole story. The optically excited state of the matter is necessarily connected with some polarization $\boldsymbol{P}$. Otherwise the transition would be optically forbidden, i.e., it would not couple to the electromagnetic field via the dipole-operator [see (3.59)]. On the other hand, we know that every oscillating polarization emits an electromagnetic wave [see (2.25)] which may act back onto the incident electromagnetic field. This interplay will lead us in the following to the strong coupling limit between light and matter and to the concept of polaritons, see e.g. [5.1–6] and references therein. In later chapters we shall see that many of the experimentally observed phenomena can be described quantitatively only in the strong coupling limit. In this chapter we even give an example which shows that this can be true for gases, such as Na vapor.

5.1 Polaritons as New Quasiparticles

Due to the relation (2.26) repeated here:

$$\boldsymbol{P} = \varepsilon_0[\varepsilon(\omega) - 1]\boldsymbol{E} \; , \tag{5.1}$$

the electric field in matter is always accompanied by a polarization wave. This statement is true as long as $\varepsilon(\omega)$ or $\tilde{n}(\omega)$ deviate from one, and holds in the whole spectral range from $\omega = 0$ up to the highest eigenfrequencies of the electronic system of a semiconductor. These highest eigenfrequencies are actually situated in the X-ray region, for example, the K or L absorption edges. Above this region $\varepsilon(\omega)$ and $\tilde{n}(\omega)$ approach unity from below according to

(4.19). In other words, we can state that, for the whole spectral range discussed in this book, i.e., for the IR, VIS and UV, light travelling in a solid is always a mixture of an electromagnetic wave and a "mechanical" polarization wave and this is not only true for semiconductors, but also for other dense media like metals, insulators and liquids and close to resonance even for gases. As we shall later see, these polarization waves include motions of different ions in a semiconductor relative to each other, two particle, i.e., electron–hole pair excitations, and collective motions of the electron cloud with respect to the nuclei and the inner filled electron shells. Later (Chaps. 9–12), we shall discuss these excitations in more detail and see that they can also be quantized to form quasiparticles with energy $\hbar\omega$ and momentum $\hbar \boldsymbol{k}$ in a similar way as for photons. The names of these quanta, or quasiparticles, are phonons, excitons and plasmons, respectively. We can therefore repeat the above statement about what is propagating as "light" in matter, namely that it is a mixture of photons and other quasiparticles that describe the quanta of the polarization field.

In the mechanical model used in Chap. 4 we would describe this phenomenon as follows: an incident electromagnetic wave excites the oscillators. This oscillation is connected with a polarization which itself radiates again an electromagnetic wave; this in turn excites the oscillators, etc.

Anyone who is not familiar with the concept of second quantization can simply skip the following treatment. There will be no problems in understanding the rest of the book. It must just be remembered that light propagating in matter is a mixture of an electromagnetic and a polarization wave. This mixed wave can be quantized and the energy quanta are known as polaritons. They are the quasiparticles of "light" in matter.

The Hamiltonian, which, in the picture of second quantization, describes the interacting system of the photons and other quasiparticles (or excitations of matter) reads [5.4]

$$H = \sum_k \hbar\omega_k a_k^+ a_k + \sum_{k'} E(\boldsymbol{k}') B_{k'}^+ B_{k'} + \mathrm{i}\hbar \sum_k g_k (B_k^+ a_k + \mathrm{h.c.}) \,. \tag{5.2}$$

The first two terms on the right-hand side contain the number operators $a_k^+ a_k$ and $B_{k'}^+ B_{k'}$ of the photons and of the other quasiparticles (see also (2.55b)) which the also bosons, and represent the Hamiltonian of the non-interacting systems. The third term describes the interaction, e.g., the annihilation of a photon a_k and creation of another quasiparticle B_k^+ (under momentum conservation) or vice versa. The prefactors g_k simply contain the transition matrix elements H_{ij} discussed in Sect. 3.2.2. If we use these terms as a perturbation, we arrive back at the weak-coupling limit.

The crucial point now is that the whole Hamiltonian (5.2) can be diagonalized by a proper choice of linear combinations p_k of creation and annihilation operators of photons and of the quasiparticles representing the matter. In the following, we merely outline briefly this Bogoliubov-transformation-like procedure. For details the reader is referred to [5.4].

The above procedure brings the Hamiltonian of (5.2) into the following form:

$$H = \sum_k E_k p_k^+ p_k \tag{5.3}$$

with suitable coefficients u_k and v_k:

$$p_k = u_k B_k + v_k a_k \ . \tag{5.4}$$

The p_k and p_k^+ are the annihilation and creation operators for the quanta of the mixed state of photon and polarization wave, which are consequently called polaritons.

5.2 The Dispersion Relation of Polaritons

All wave-like excitations can be described by two quantities, namely their (angular) frequency ω which is connected with the quantum energy simply by $E = \hbar\omega$ and their wave vector $\boldsymbol{k}$ which gives the (quasi-)momentum $\hbar\boldsymbol{k}$. The relation which connects ω and $\boldsymbol{k}$ is usually called dispersion relation $E(\boldsymbol{k})$ or $\omega(\boldsymbol{k})$.

The dispersion relations $E(\boldsymbol{k})$ which we have encountered until now were very simple. For photons in vacuum it was a straight line with slope $\hbar c$ (Fig. 2.2) and for the polarization wave of uncoupled oscillators a horizontal line (Fig. 4.2a). The relation between ω and $\boldsymbol{k}$ for polaritons, i.e., for the light quanta in matter, can be derived from classical physics and agrees with the results of the quantum-mechanical treatment outlined above.

We remember that the wave vector in matter $\boldsymbol{k}$ is connected with the wave vector in vacuum $\boldsymbol{k}_\mathrm{v}$ by the complex refractive index $\tilde{n}(\omega)$ (2.36). To get rid of the vector character, we consider the squares

$$\boldsymbol{k}^2 = k^2 = \tilde{n}^2(\omega) k_\mathrm{v}^2 \ . \tag{5.5}$$

Now we also recall (2.35, 11) saying that $\tilde{n}^2(\omega) = \varepsilon(\omega)$ and $\boldsymbol{k}_\mathrm{v}^2 = (2\pi/\lambda_\mathrm{v})^2 = (\omega/\lambda)^2$ and obtain again (2.33)

$$\frac{c^2 k^2}{\omega^2} = \varepsilon(\omega) \ . \tag{5.6}$$

This is the so-called polariton equation. On the other hand, we know $\varepsilon(\omega)$ which is given in the vicinity of a single resonance by (4.20, 22).

Putting (5.6) and (4.22) together we find

$$\frac{c^2 k^2}{\omega^2} = \varepsilon_\mathrm{b}\left(1 + \frac{f}{\omega_0^2 - \omega^2 - \mathrm{i}\omega\gamma}\right) . \tag{5.7}$$

This is an implicit representation of $\omega(\boldsymbol{k})$ for the polaritons. For the simplest case, namely vanishing damping γ and no dependence of ω_0 or f on k, it is quite easy to calculate $\boldsymbol{k}(\omega)$ and $\omega(\boldsymbol{k})$. We do not give the formulas here because they bring no further physical insight, but in Fig. 5.1 we give the dispersion relation for the case just mentioned including only one resonance. For several resonances the r.h.s. of (4.22) has to be replaced by (4.20). The dispersion relation starts for $\omega = 0$ and $\boldsymbol{k} = 0$ as a straight line. This part is called the lower polariton branch (LPB). Since the dispersion of photons in vacuum is also a straight line (but with a different slope!), the dispersion relation is said to be "photon-like" as long as it is a straight line.

It bends over when we approach the resonance frequency ω_0, and in this region the polariton dispersion is called phonon-like or exciton-like, depending on whether the resonance corresponds to a phonon or an exciton, respectively. Between the transverse and longitudinal eigenfrequencies there is no propagating mode for the present approximation of uncoupled oscillators, i.e., we have again the stop-band or Reststrahlbande discussed in Sect. 4.4. There is a longitudinal branch, which does usually not couple to the electromagnetic field. At $\hbar\omega_L$ the upper polariton branch (UPB) begins. This bends upwards again displaying a photon-like behaviour, but now with a slope $\hbar c \varepsilon_b^{-1/2}$ compared to $\hbar c \varepsilon_s^{-1/2}$ for the LPB. Between ω_T and ω_L, $\boldsymbol{k}$ is purely imaginary since $\tilde{n}(\omega)$ is imaginary in this range; the consequences for the optical properties were discussed already in Sect. 4.4 above.

Acutally, the dispersion relation shown in Fig. 5.1 is not so surprising as it looks at first glance. If we take the spectral dependences of $n(\omega)$ and of $\kappa(\omega)$, i.e., of the real and imaginary parts of $\tilde{n}(\omega)$, from Fig. 4.4, turn the ω-axis from the x-direction into the y-direction and multiply $n(\omega)$ and $\kappa(\omega)$ by ωc^{-1}, i.e., essentially by a straight line through the origin, according to

$$\mathrm{Re}\{k\} = n(\omega)\omega c^{-1}; \quad \mathrm{Im}\{k\} = \kappa(\omega)\omega c^{-1}, \tag{5.8}$$

we obtain Fig. 5.1.

The dispersion relation of the polariton can also be deduced from the quantum-mechanical "non-crossing rule". This non-crossing rule says roughly the following: There are two energy levels E_1 and E_2, which depend on some

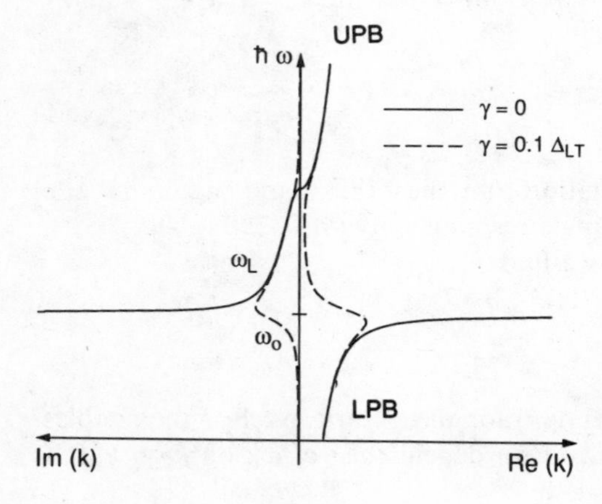

Fig. 5.1. The polariton dispersion in the vicinity of a single resonance for vanishing (*solid line*) and finite damping (*dashed line*)

parameter p. This parameter can be the wave vector, a particle density, a constant electric or magnetic field, a strain field, etc. We assume that these two levels cross as a function of the parameter as sketched in Fig. 5.2 by the dashed lines. If there is any coupling between these two levels, then the cross-over point disappears, and the two levels repel each other in the way shown by the solid lines in Fig. 5.2. The splitting at the former cross-over point is just proportional to the coupling strength between the two levels.

If we relax the assumption of vanishing damping, the polariton dispersion is changed as shown in Fig. 5.1 by the dashed line. Now the propagating modes below ω_T and above ω_L experience damping too. Between ω_T and ω_L there appears a region with negative group velocity $d\omega/dk < 0$. At first glance, this situation seems unphysical in the sense that a light pulse moves out of a sample ($v_g < 0$) when we send it onto the sample. A first way out of this dilemma is given by the strong absorption which is necessarily connected with the region of negative v_g. It allows one to interpret the negative v_g in this spectral region as a wave that is damped out faster than it can propagate.

A more satisfactory interpretation can be found later in Sect. 6.1, where we take into account spatial dispersion. For this more general case, the problem of negative v_g essentially disappears. For a discussion of the fact, that both the phase- and group velocities can exceed for some frequencies the vacuum speed of light, see e.g. [5.7].

To conclude this section we stress that the concept of polaritons is not an academic problem but occurs basically in all interaction processes between light and matter as long as the mean distance between the oscillators is smaller than the coherence length of the incident light. If we simply look out of the window, we rely on the polariton concept in the sense that light does not propagate as a pure electromagnetic wave in the glass, but as a mixture of electromagnetic and polarization waves, the quanta of which are the polaritons.

In Fig. 5.3 we show, with the help of experimental data from [5.8], that the polariton concept is important even in the vicinity of strong resonances in a gas. A continuous spectrum around the yellow Na line is sent in Fig. 5.3a through a homogeneous vapor of Na. The absorption is clearly seen in the spectrum on the r.h.s. (On the l.h.s. the $\hbar\omega$ axis is normal to the paper). If the

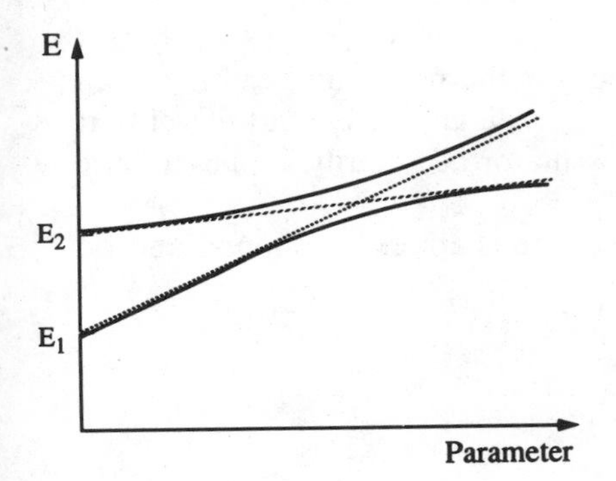

Fig. 5.2. A sketch to illustrate the quantum-mechanical non-crossing rule for two non-interacting (*dashed line*) and interacting (*solid line*) energy levels

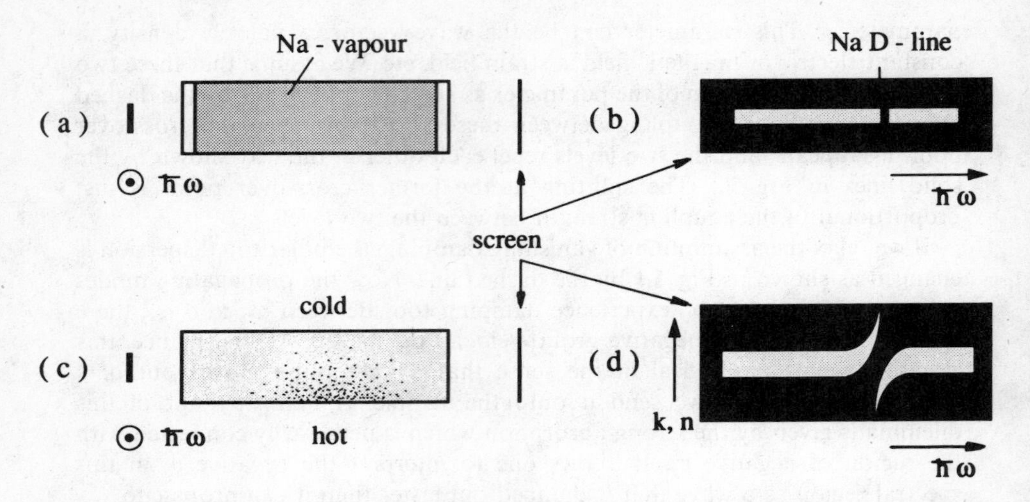

Fig. 5.3a–d. Absorption and dispersion of Na vapor in the vicinity of the yellow sodium line. Schematic arrangement for the absorption (**a**) in a homogeneous vapor and the resulting absorption spectrum (**b**); deflection in a "prism" of Na vapor (**c**) and the dispersion relation (**d**). The energy axis is normal to the paper in (**a**) and (**c**); (**b**) and (**d**) are photographic positives. According to [5.8]

Na vapor is brought into a prism-like shape by heating the tube from below and cooling it above, one clearly observes a deflection of the spectrum in the "Na vapor prism" around the resonance position. When small, the deflection is approximately proportional to $n(\omega)$ or $k(\omega)$. The structure on the r.h.s. of Fig. 5.3b thus directly represents the upper and lower polariton branches of Fig. 5.1, i.e., light is also propagating in Na vapor as a mixture of an electromagnetic and a polarization wave. Further away from the resonance n is so close to unity that to within experimental error no deflection of the spectrum is observable.

5.3 Problems

What dispersion relations would you expect for the polariton resulting from oscillators with the dispersion relations of Figs. 4.2a and b? Do not forget that a finite coupling between photons and the oscillator necessarily implies a finite Δ_{LT}.

Check if you were right when you come to the chapters on phonon and on exciton polaritons.

6 Coupled Oscillators and Spatial Dispersion

For our model system we have assumed until now zero coupling between neighboring oscillators as shown in Fig. 4.1. The consequence of this assumption was a dispersion relation which is simply a horizontal line (Fig. 4.2a). We now relax this assumption by using a concept introduced in [6.1] and elaborated in detail in [6.2–8] and consider a more realistic coupling between neighboring oscillators realized, e.g., by weak springs, as shown schematically in Fig. 6.1. The most important consequence of this coupling is that the eigenfrequency is now a function of $\boldsymbol{k}$ as shown in Fig. 4.2b. For very long wavelength, i.e., $\boldsymbol{k} \rightarrow 0$, neighboring oscillators are still in phase and the coupling springs are not elongated. Therefore they still oscillate with the same frequency. For decreasing wavelength, the coupling springs are elongated and increase the "effective" spring constant. As a consequence, ω_0 increases with increasing $\boldsymbol{k}$. The resulting band width 2B indicated in Fig. 4.2b is directly proportional to the coupling strength between neighboring oscillators. A wave packet created by elongating one or a few oscillators, as in Fig. 4.1d, will now propagate with a finite group velocity $v_g = \mathrm{d}\omega/\mathrm{d}k$ and will show some dispersion. This means that the width of the spatial envelope of the wave packet will increase with time as indicated in Fig. 2.1b. This phenomenon can be easily observed by throwing a stone onto the still surface of a lake. The expanding ring-like wave structure shows a drastic increase of the width of its envelope function during propagation. The detailed shape of $\omega(\boldsymbol{k})$ depends on the physical nature of the oscillators and the coupling mechanism.

For our model system, ω_0 increases with $\boldsymbol{k}$ and we shall see that this is also true, e.g., for excitons. For optical phonons one has usually a decrease of the eigenfrequency with increasing $\boldsymbol{k}$. These points will be presented in more detail in Chaps. 9, 11, 13 and 14. For the following discussion of our model system we shall use the dispersion relation of Fig. 4.2b, but the conclusions will also be qualitatively valid for other dependences of $\omega_0(\boldsymbol{k})$.

The fact that the eigenfrequency ω_0 of some excitation of a solid depends on $\boldsymbol{k}$ is often called "spatial dispersion", for reasons given later. However, the $\omega(\boldsymbol{k})$ dependence of the photons themselves, or of the polaritons, is not usually classed as spatial dispersion.

Since we have used the word dispersion now in various connections, we shall summarize the meanings here:

The term "dispersion relation" or simply "dispersion" means the relation $E(\boldsymbol{k})$ or $\omega(\boldsymbol{k})$ for all wave-like excitations independent of the functional

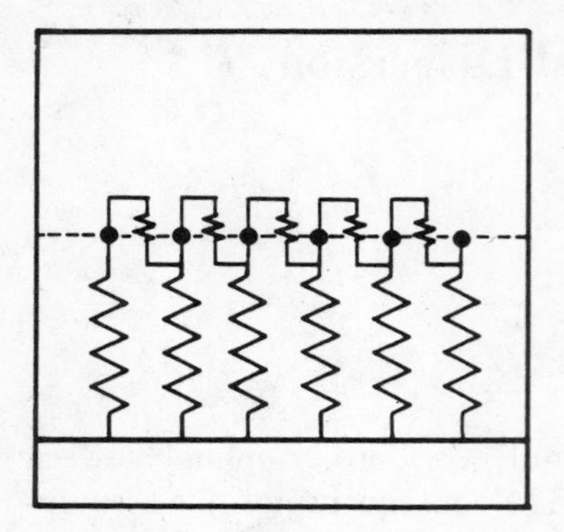

Fig. 6.1. An ensemble of coupled oscillators. Compare with Fig. 4.1. After [4.1]

dependence. It can be simply a horizontal line, a linear or parabolic relation, or something more complicated. Every excitation which has a wave-like character has a dispersion relation.

The term "spatial dispersion" means that the eigenfrequency of one of the elementary excitations in a solid depends on $\boldsymbol{k}$ and is not just a horizontal line (Fig. 4.2a, b).

In technical optics, dispersion often refers more specifically to the dependence of the refractive index n on the wavelength. Materials such as the glass used to make lenses, generally have a decreasing refractive index with increasing wavelength i.e., $\mathrm{d}n/\mathrm{d}\lambda < 0$. This behavior is called "normal dispersion" and is the usual behavior in the transparent spectral region, while an abnormal dispersion, i.e., $\mathrm{d}n/\mathrm{d}\lambda > 0$ is limited to strongly absorbing regions (Fig. 4.4).

Finally, the word dispersion is also used for the fact that the envelope of a wavepacket, e.g., of a short light pulse in matter, becomes spatially broader with time.

The last two meanings of the word dispersion are actually consequences or special examples of the general definition of the term given first. If not stated otherwise, we use the term dispersion (-relation) in this book to mean the $E(\boldsymbol{k})$ or $\omega(\boldsymbol{k})$ relation.

6.1 Dielectric Function and the Polariton States with Spatial Dispersion

The dielectric function given in (4.22) for the simple case of an isolated resonance has to be modified if we want to take spatial dispersion effects into account. The eigenfrequency ω_0 has to be replaced by $\omega_0(\boldsymbol{k})$ and the oscillator strength f and the damping γ may also depend on $\boldsymbol{k}$ resulting in

$$\varepsilon(\omega, \boldsymbol{k}) = \varepsilon_{\mathrm{b}}\left(1 + \frac{f(\boldsymbol{k})}{\omega_0^2(\boldsymbol{k}) - \omega^2 - \mathrm{i}\omega\gamma(\boldsymbol{k})}\right) \tag{6.1}$$

The most significant change is the fact that ε is now a function of two independent variables, ω and $\boldsymbol{k}$.

Along with the transverse eigenfrequency $\omega_0(\boldsymbol{k})$, the longitudinal eigenfrequency ω_L defined as $\varepsilon(\omega_L)=0$ also becomes $\boldsymbol{k}$-dependent, i.e., $\omega_L=\omega_L(\boldsymbol{k})$. The same is true for the longitudinal-transverse splitting Δ_{LT} which is connected with $f(\boldsymbol{k})$. In principle all of the above quantities have to be given as a function of $\boldsymbol{k}$.

What one usually does is to give – if possible – an analytic expression for $\omega_0(\boldsymbol{k})$ and to still consider f and γ as $\boldsymbol{k}$-independent. Though there is clear experimental evidence that f and γ depend on $\boldsymbol{k}$ (see e.g. [6.6] and references therein), these dependences are usually less critical for the correct description of the optical properties of semiconductors than the $\boldsymbol{k}$-dependence of ω_0 and will be neglected in the following. For $\omega_0(\boldsymbol{k})$ we again use for simplicity a parabolic relation, but we stress once more that the consequences are qualitatively similar for other relations. We have

$$\varepsilon(\omega,\boldsymbol{k})=\varepsilon_b\left(1+\frac{f}{\omega_0^2+2\omega_0 A\boldsymbol{k}^2-\omega^2-\mathrm{i}\omega\gamma}\right). \tag{6.2a}$$

with

$$\omega_0^2(\boldsymbol{k})=(\omega_0+A\boldsymbol{k}^2)^2=\omega_0^2+2\omega_0 A\boldsymbol{k}^2+A^2\boldsymbol{k}^4\approx\omega_0^2+2\omega_0 A\boldsymbol{k}^2$$
$$\text{for}\quad |\boldsymbol{k}|\ll\pi/a\,. \tag{6.2b}$$

The approximation used in (6.2b) is usually valid for massive and for effective mass particles (see Sect. 10.4).

To determine the dispersion relation of the polariton we must again combine the polariton equation with the dielectric function resulting again in an implicit relation for $\omega(\boldsymbol{k})$:

$$\frac{c^2\boldsymbol{k}^2}{\omega^2}=\varepsilon_b\left(1+\frac{f}{\omega_0^2+2\omega_0 A\boldsymbol{k}^2-\omega^2-\mathrm{i}\omega\gamma}\right). \tag{6.3}$$

We consider first the case of vanishing damping in Fig. 6.2a and start with the real part of $\boldsymbol{k}$. The transverse lower polariton branch starts photon-like and then bends over to asymptotically approach the parabolic dispersion relation of the resonance. If there is a constant Δ_{LT}, i.e., $f\neq f(\boldsymbol{k})$, then the longitudinal branch (not shown here) starts at ω_L and is then essentially parallel to the exciton-like part of the LPB. At ω_L the transverse upper polariton branch also begins, going over into a photon-like (i.e., linear) dispersion relation with slightly steeper slope, as shown already in Fig. 5.1a. The imaginary part of $\boldsymbol{k}$ starts at ω_L and bends downwards to reach assymptotically a curve which is produced by reflecting the transverse eigenfrequency $\omega_0(\boldsymbol{k})$ through the point $(\omega_0, \boldsymbol{k}=0)$. This means that the UPB has a purely imaginary continuation below ω_L.

If we include a small but finite damping, we end up with the situation shown in Fig. 6.2b. It can be seen that the LPB and the UPB now extend over

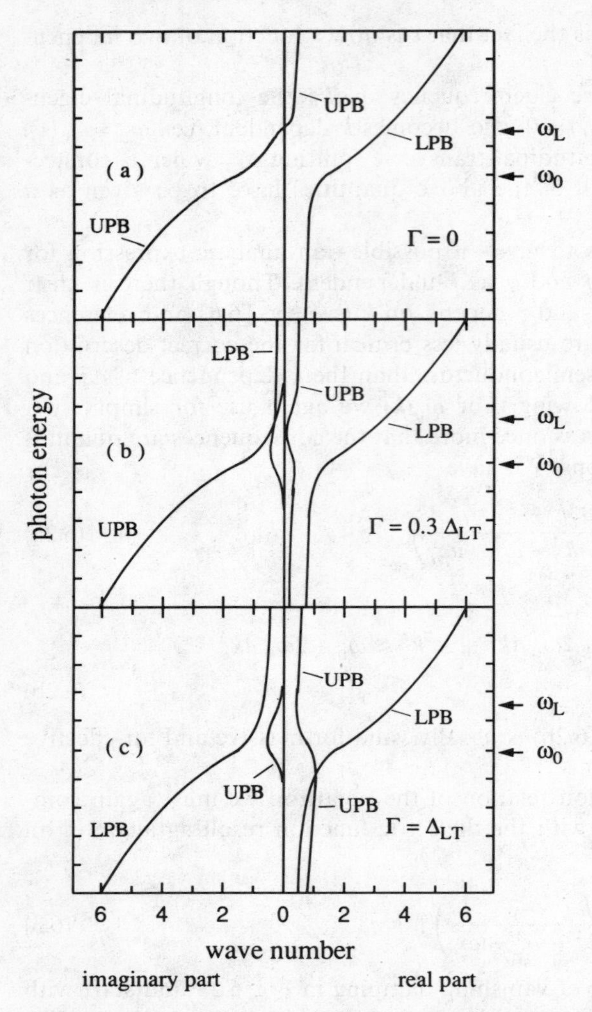

Fig. 6.2a–c. The real and imaginary parts of a polariton dispersion relation in the vicinity of a resonance with spatial dispersion for vanishing **(a)** small ($\hbar\Gamma < \Delta_{LT}$) **(b)** and strong ($\hbar\Gamma \approx \Delta_{LT}$) damping **(c)**. After [6.7]

the whole energy range. They have both a real and an imaginary part at all frequencies. The imaginary part of the LPB is very small below ω_0, peaks between ω_0 and ω_L, and decreases gradually for higher energies. The UPB has a very small imaginary part above ω_L which increases continuously below, but there is also a small real part of the UPB below ω_L. The problem of negative group velocity in the spectral region of the resonance, mentioned already in Sect. 5.2, has become less severe and appears only in the form of a small hump in the dispersion relation for finite damping. For the case of strong damping shown in Fig. 6.2c the role of the LPB is gradually taken over by the UPB below ω_0 [6.7], and one approaches a situation similar to the case without spatial dispersion (compare with Fig. 5.1b). In other words this means that one can describe the optical properties of a strongly damped resonance by neglecting the influence of spatial dispersion. However, it should be stressed, that this effect is still present in principle, but the strong damping is what dominates the optical properties.

6.2 Reflection and Transmission and Additional Boundary Conditions

From the polariton dispersion with spatial dispersion shown in Fig. 6.2 we can easily recover the real and imaginary parts of $\tilde{n}$ by just reversing the procedure given in (5.8). We will not spend time on this procedure but discuss directly the optical spectra, especially the consequences of spatial dispersion on the reflection as compared to the situation of Fig. 4.5 where spatial dispersion was still neglected. The two most important points are, first, that there is no more stop-band, i.e., there is at least one propagating mode for every frequency ω (with and without damping) and, second, that in some spectral regions there is more than one propagating mode. This second point is especially obvious for $\omega > \omega_L$.

We now want to discuss the consequences of these two new phenomena arising from spatial dispersion on the optical spectra, especially on the reflection spectrum.

The fact that we have at least one propagating mode – generally with real and imaginary part – for all frequencies even between ω_T and ω_L, means that the reflectivity for (normal) incidence no longer reaches unity, even for the case of vanishing damping, as sketched in Fig. 6.3.

The fact that we have more than one mode (propagating or evanescent, i.e., $\mathrm{Im}(\boldsymbol{k}) \ll \mathrm{Re}(\boldsymbol{k})$ or $\mathrm{Im}(\boldsymbol{k}) \gg \mathrm{Re}(\boldsymbol{k})$, respectively in the solid for one frequency means that the two boundary conditions deduced from Maxwell's equations in connection with (3.6, 7) and Figs. 3.1, 2 are no longer sufficient. For a given incident wave we could deduce the amplitudes of one reflected and one transmitted wave. If there are two or even more states in the medium at the same frequency coupling to the incident field, we need one or more additional boundary conditions (abc). To make the situation clear, we show in Fig. 6.4 the wave vectors for such a case for normal and oblique incidence. The incident and reflected beams obey the usual law of reflection, their components parallel to the surface are equal. The same is true for the transmitted beams, but the total length of the wave vectors on the LPB and the UPB are different, in agreement with Fig. 6.2. Obviously the two beams are travelling in different spatial directions and this is why the $\boldsymbol{k}$-dependence of ω_0 is called "spatial dispersion". Though Fig. 6.4 has some similarity with the picture for birefringence (3.12) we point out that the reasons are quite different. In

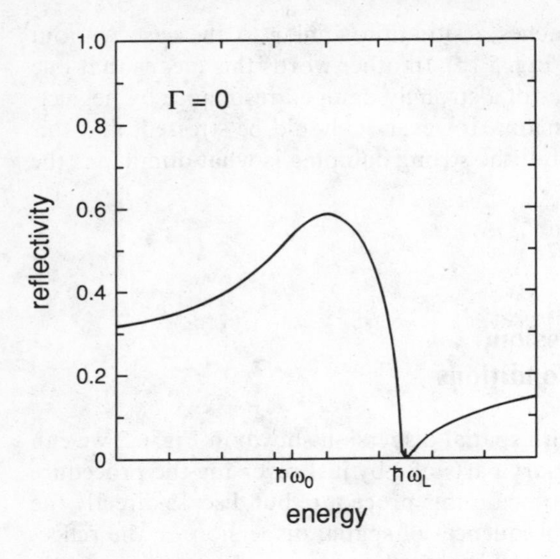

Fig. 6.3. Sketch of the reflection spectrum in the vicinity of a resonance with spatial dispersion and without damping. Compare with Fig. 4.5. After [6.8]

Fig. 3.12 the o and eo beams have orthogonal polarization and the phenomenon needs uniaxial or lower symmetry. Spatial dispersion occurs even for cubic symmetry and the two beams are polarized in the same direction. Spatial dispersion also occurs for crystals of lower symmetry which then may show birefringence in addition. In this case the dispersion curves of Fig. 6.2 have to be drawn twice with different parameters for the o and the eo beam.

As already mentioned, the abc cannot be deduced from Maxwell's equations. Their capacity is exhausted with one reflected and one transmitted beam. Since the complex index of refraction around the resonance is rather different for the LPB and the UPB, which therefore contribute differently to the reflection spectrum according to (3.20), the abc should contain information about the "branching ratio", i.e., which fractions of the incident beam couple in the medium to the LPB and to the UPB as a function of frequency.

The abc are somewhat arbitrary (we shall explain later on why) and are based mainly on arguments of physical plausibility. On the vacuum side of the interface the polarization is trivially zero. To avoid an unphysical discontinuity in the polarization, one possible abc is that the polarization of the medium must be zero at the interface

$$\boldsymbol{P}(z=0)=0\,. \tag{6.4a}$$

Another argument says that the polarization should vary smoothly across the interface, implying that the derivative with respect to the normal direction has to vanish, resulting in

$$\left.\frac{\mathrm{d}\boldsymbol{P}}{\mathrm{d}z}\right|_{z=0}=0\,. \tag{6.4b}$$

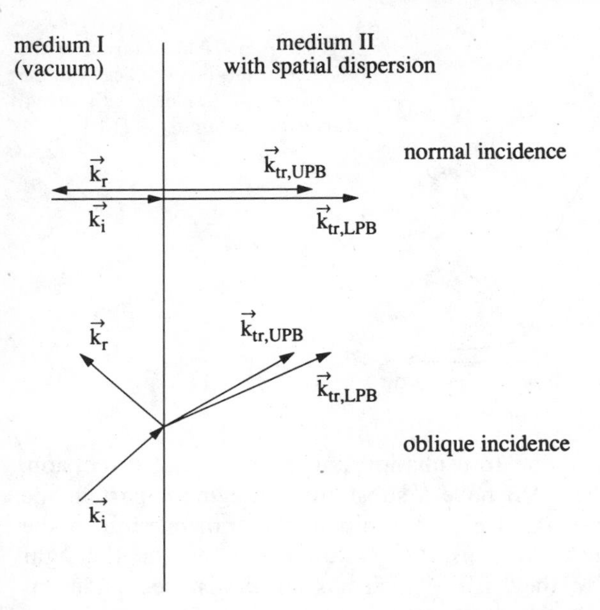

Fig. 6.4. The wave vectors of a resonance with spatial dispersion for $\omega > \omega_L$. Note that there is more than one propagating wave in the medium

Some authors favor a linear combination of the two conditions:

$$P|_{z=0} + \beta \left. \frac{\partial P}{\partial z} \right|_{z=0} = 0 \quad \text{with} \quad -1 \leqslant \beta \leqslant +1 \, . \tag{6.4c}$$

The reflection spectrum shown in Fig. 6.3 is actually calculated for an exciton resonance using the material parameters of CdS and the abc (6.4a), the so-called Pekar–Hopfield abc. See also Chap. 14 and [6.1–8] for further details of the abc.

It turns out that experimentally observed spectra, e.g., of exciton resonances, can be fitted with all the above-mentioned abc, but with slightly different values for the other parameters, such as f and γ, that describe the resonance.

As a rule of thumb, one can state for all abc and weak damping that the light propagating in matter at frequencies ω sufficiently below the transverse eigenfrequency and above the longitudinal one travels almost completely on the LPB and on the UPB, respectively. "Sufficiently" means in this context

$$\hbar|\omega - \omega_{T,L}| \gtrsim 10\Delta_{LT} \, . \tag{6.5}$$

The crucial spectral range where spatial dispresion and the problem of abc are of importance is thus the resonance and its vicinity.

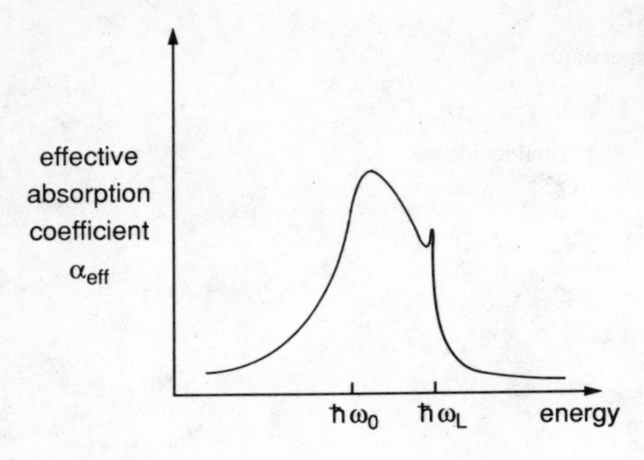

Fig. 6.5. A possible spectrum of the effective absorption coefficient in the vicinity of a resonance with spatial dispersion. According to [6.8]

Now let us have a look at the transmission, including spatial dispersion. Since both the LPB and the UPB have a substantial imaginary part in the vicinity of the resonance, we again expect a dip in the transmission in the region around the resonance. At considerably lower frequencies, the light couples almost completely to the LPB – which has an almost negligible imaginary part in this region – so that the sample is transparent. The same is true significantly above ω_L for the UPB, at least if there are no other resonances. The shape of the effective absorption coefficient $\alpha(\omega)$ or of $\kappa(\omega)$ can look somewhat different from Fig. 4.4 eventually developing a spike at the longitudinal eigenfrequency, as indicated in Fig. 6.5. In addition, one can see from Fig. 6.2 that the imaginary parts of $\boldsymbol{k}$ and thus κ or α are different for the two polariton branches. The amplitudes and light intensities of each polariton branch therefore decay exponentially with thickness, but the sum of both, which is the only experimentally accessible quantity, may show a non-exponential decay with sample thickness. Furthermore the damping may in some cases be higher close to the surface compared to the bulk of the sample due to imperfections introduced into the lattice close to the surface.

Strong damping, i.e., $\hbar\gamma \geqslant \Delta_{LT}$, see (6.5), reduces the importance of the influence of spatial dispersion on the optical spectra as already mentioned above. The resonance in the reflection spectrum is then already so strongly washed out by damping that the details, whose description relies on spatial dispersion, are no longer observable.

It is physically not completely satisfying that the choice of the abc is to some extent arbitrary (6.4). The parameters of the resonance deduced with the use of different abc are (slightly) different. On the other hand, the value of quantities such as f cannot depend on the abc chosen by the physicist running the fitting program. The answer is that the problem of the abc is an artificial one. The dielectric function $\varepsilon(\omega, \boldsymbol{k})$ describes the optical properties in the bulk of the sample. In our derivation of $\varepsilon(\omega, \boldsymbol{k})$ in Chap. 4 no surface was included. Later on, however, we use this dielectric function to describe the optical properties of the interface between two semi-infinite half-spaces, one usually being vacuum, the other the semiconductor under consideration. The price we have to pay for this "improper" use of the dielectric function is the problem of the abc. If we were to calculate the optical properties of a half-space from the outset, the problem of the abc could be avoided. Indeed, some calculations have used this idea [6.2]. However, this procedure has its own problems. For example, the band-structure and the exciton states have to be calculated for the half-space, which means that we can in principle no longer use Bloch's theorem for the direction normal to the interface. This causes complications which can be overcome only with difficulty and by using various approximations and simplifications. Therefore most authors prefer to use the bulk dielectric function and some of the abc to evaluate the optical spectra. More information about the problem of abc and the rather lengthy formulas for calculating the spectra of reflection can be found in [6.1–8] and references therein.

6.3 Real and Imaginary Parts of Wave Vector and Frequency

Until now, we have assumed that we can describe the light wave or polariton wave propagating in matter by a real frequency and a wave vector which has a real and an imaginary part according to the complex index of refraction. In principle, however, one could take $\boldsymbol{k}$ to be real and introduce a complex frequency $\tilde{\omega}$ by including the damping term $-\mathrm{i}\omega\gamma$ as in (4.14, 22):

$$\tilde{\omega}=\omega-\mathrm{i}\gamma\ . \tag{6.6}$$

At first glance, there is no reason to prefer one approach over the other. Actually both approaches are possible and it is the experiments performed that decide which model is the more appropriate one.

If we shine a monochromatic wave with well-defined frequency ω(e.g., a spectrally narrow laser beam) on the sample, or if we select such a frequency with a monochromator, then we have the situation of (4.14), i.e., a forced oscillation, and here we have to use a purely real ω, but a complex $\boldsymbol{k}$ to describe the decrease in amplitude of the polariton wave as it travels through the crystal.

If, on the other hand, we could by some means create at a certain time (e.g., $t=0$), a polariton wave in the sample with constant amplitude everywhere and let it evolve for $t>0$, then the amplitude would decay with time, but would remain the same everywhere in space. This is just the situation described by a real $\boldsymbol{k}$ and a complex $\tilde{\omega}$. The reader might think that a wave with constant amplitude from $-\infty$ to $+\infty$ in space is highly unphysical, but in fact this approximation is as valid as that of a strictly monochromatic wave, which necessarily endures from $t=-\infty$ to $t=+\infty$. Everything with a finite temporal duration has a finite spectral halfwidth.

These are just consequences of the fact that the time t and frequency ω domains and the space $\boldsymbol{r}$ and wave vector $\boldsymbol{k}$ domains are connected with each other by one and three-dimensional Fourier transforms, respectively.

Multiplication of ω or $\boldsymbol{k}$ with $\hbar$ then immediately gives the "uncertainty relations" for energy and time or momentum and space (3.37).

Since the experimental situation discussed first is much more frequently used than the second one, we will restrict ourselves for the rest of this book to the situation of complex $\boldsymbol{k}$ and real ω. Bearing in mind, however, that there is a third uncertainty relation for truth and understandability of a text as stated at the beginning of the book, we should inform the reader that, in principle, one needs to use both a complex $\boldsymbol{k}$ and a complex ω since all excitations usually have a finite lifetime or phase relaxation time. Chapter 22 gives further details on this topic.

6.4 Problems

1. Sketch the dispersion of a polariton resonance with spatial dispersion and an oscillator strength which increases with $\boldsymbol{k}$. (Assume for simplicity zero damping). Does Δ_{LT} then also depend on $\boldsymbol{k}$?

2. Sketch the dispersion of the polariton for two close lying resonances A and B, with and without spatial dispersion for an order of the energies at $\boldsymbol{k}=0$ $\hbar\omega_0^A < \hbar\omega_{\mathrm{L}}^A < \hbar\omega_0^B < \hbar\omega_{\mathrm{L}}^B$. Is it possible for a single orientation of the polarization to have the sequence $\hbar\omega_0^A < \hbar\omega_0^B < \hbar\omega_{\mathrm{L}}^A < \hbar\omega_{\mathrm{L}}^B$?

7 Surface Polaritons

We would now like to say a few words about the surface polaritons already mentioned briefly in Sect. 2.4

7.1 Definition and Dispersion

Surface polaritons are also quanta or quasiparticles of the mixed state of an electromagnetic and a polarization wave. They are distinguished by the fact that they can only propagate along the interface between two different media. The amplitudes decay exponentially with distance from the interface, as shown schematically in Fig. 7.1, i.e., surface polaritons are evanescent waves on both sides of the interface, in contrast to the one-sided evanescent wave in the case of total internal reflection of Fig. 3.3a. For every volume polariton there exists a surface polariton.

We want now to discuss the conditions for which surface polaritons can exist. For simplicity we restrict ourselves to the case of vanishing damping and no spatial dispersion: $\gamma = 0$ and $A = 0$ in (6.2). We assume that the interface is formed by an essentially non-dispersive medium I described by a constant real index of refraction $n_\mathrm{I}^2 = \varepsilon_\mathrm{I}$ on one side (for vacuum $n_\mathrm{I} = 1$) and the medium under consideration with $\varepsilon(\omega) = \tilde{n}^2(\omega)$ on the other (medium II). If the surface polariton cannot propagate into medium I or II, there must be some physical reasons preventing this decay.

As a first condition we may state that in medium II there are no propagating waves between the transverse and longitudinal eigenfrequencies ω_T and ω_L, as discussed for example in connection with Figs. 4.4, 4.5 and 5.1a. The propagation into medium I can be excluded if the wave vector $\boldsymbol{k}_\mathrm{s}$ of the surface polariton is larger then $\boldsymbol{k}_\mathrm{I}$ of a wave propagating in medium I. Under such a condition the conservation law for $\boldsymbol{k}$ parallel to the interface results with Fig. 7.1b in

$$\boldsymbol{k}_\mathrm{s}^2 + \boldsymbol{k}_\perp^2 = \boldsymbol{k}_\mathrm{I}^2 \quad \text{and} \quad \boldsymbol{k}_\mathrm{s}^2 > \boldsymbol{k}_\mathrm{I}^2 \Rightarrow \boldsymbol{k}_\perp^2 < 0 \,. \tag{7.1}$$

The r.h.s of (7.1) simply says that $\boldsymbol{k}_\perp$ is purely imaginary and this is what we need for an evanescent wave normal to the interface.

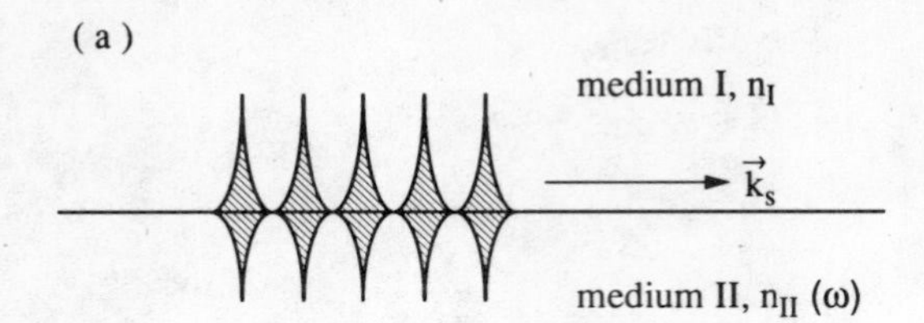

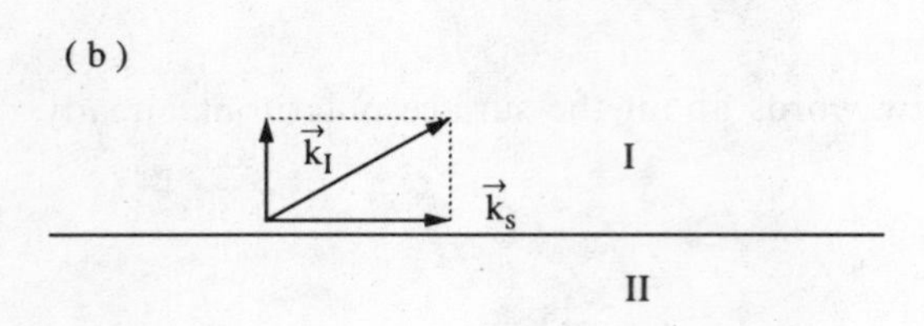

Fig. 7.1a, b. Schematic drawing of the decay of the amplitudes of a surface polariton propagating along an interface (**a**) and a sketch to illustrate the derivation of (7.1) (**b**)

We can summarize these arguments by saying that surface polaritons can be expected in the spectral region given by

$$\omega_T \leqslant \omega_s \leqslant \omega_L \ , \tag{7.2a}$$

$$\boldsymbol{k}_s \geqslant n_I \omega_c^{-1} \ , \tag{7.2b}$$

or

$$\mathrm{Re}\{\varepsilon_{II}(\omega)\} < 0 \quad \text{and} \quad |\mathrm{Re}\{\varepsilon_{II}(\omega)\}| > \varepsilon_I \ . \tag{7.3}$$

The dispersion relation of surface polaritons $\omega_s(\boldsymbol{k}_s)$ can be deduced for the present assumptions from the boundary conditions given by Maxwell's equations. We will not go through the procedure here, but merely give the result and refer the reader for its derivation to the literature [7.1–3].

$$\boldsymbol{k}_s = \left(\frac{\varepsilon_I - \varepsilon_{II}(\omega)}{\varepsilon_I + \varepsilon_{II}(\omega)}\right)^{1/2} \frac{\omega}{c} \ . \tag{7.4}$$

We should note that $\varepsilon_{II}(\omega)$ is negative in the region of (7.2a). In order to get a real value of $\boldsymbol{k}_s$ we evidently have, in addition to (7.2a, b), to fulfill (7.3).

The polarization of the surface polaritons is as follows. If the interface is the x–y-plane and the surface wave propagates in the x-direction, i.e., $\boldsymbol{k}_s \| x$, then the electric field is in the x–z-plane and the magnetic induction is along the y-axis.

In Fig. 7.2 we show the dispersion of the surface polariton for a resonance without (a), and with (b), spatial dispersion. In the latter case the calculated dispersion relation $\omega_s(\boldsymbol{k}_s)$ is slightly influenced by the abc used. For more details see [7.1–3].

To conclude this section we should briefly stress one point. Since surface polaritons cannot propagate into medium I (generally vacuum) they cannot be created by shining light of an appropriate frequency on the sample. The same

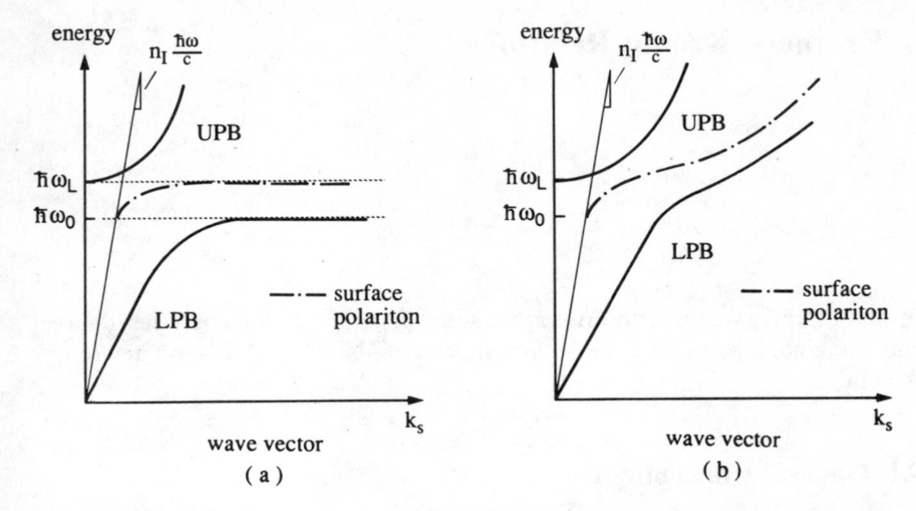

Fig. 7.2a, b. The dispersion of the surface polariton for a resonance without (**a**), and with (**b**), spatial dispersion. The damping is assumed to be negligible in both cases

is true for the other side. As a consequence, it is not possible to excite surface polaritons directly. A frequently used method involves attenuated total reflection. This technique will be outlined briefly in Chap. 14.

7.2 Problems

1. Find out something about the method of attenuated total reflection (ATR) from Chap. 14. Compare this method with the "optical tunnel effect" of Fig. 3.3b.

2. Apart from the use of ATR methods, it is possible to excite surface polaritons optically if a periodic structure, i.e., a grating, is formed at the interfaces. What is the principle behind this?

Compare this with the statements about momentum conservation in Sects. 3.1.3 and 9.2.

8 Kramers–Kronig Relations

In this Chapter we want to investigate some general relations between the real and imaginary parts of $\tilde{n}$ or ε. For more details see [8.1, 2] and references therein.

8.1 General Concept

We stated in Sect. 2.3 that the susceptibility χ and the dielectric function $\varepsilon = \chi + 1$ are response functions of the medium which describe the response (in this case the polarization) to a stimulus (in this case the incident electric field) for the special case of an incident monochromatic wave with frequency ω. We now leave ω-and $\boldsymbol{k}$-space for a moment and go to t and $\boldsymbol{r}$ space, i.e., to time and real space.

The most general expression for a linear response function is

$$\frac{1}{\varepsilon_0}\boldsymbol{P}(\boldsymbol{r},t) = \int_{-\infty}^{+\infty}\int_{-\infty}^{+\infty} \chi(\boldsymbol{r},\boldsymbol{r}',t,t')\boldsymbol{E}(\boldsymbol{r}',t')\,\mathrm{d}t'\mathrm{d}\boldsymbol{r}' \,. \tag{8.1}$$

This means that the polarization $\boldsymbol{P}$ at point $\boldsymbol{r}$ and time t depends on the electric field at all other places and at all times.

We are now going to simplify (8.1) in various steps.

First we assume that the sample is homogeneous in time, i.e., its properties do not depend on t explicitly. Then χ depends only on the time difference $t - t'$. Since our medium consists of atoms, it is not homogenous in space, but if we assume that all wavelengths present in $\boldsymbol{E}(\boldsymbol{r}',t')$ are much longer than the lattice constant, then an analogous approach holds for $\boldsymbol{r} - \boldsymbol{r}'$. This is essentially the same assumption as made in the dipole approximation (3.2).

Equation (8.1) then transforms into

$$\frac{1}{\varepsilon_0}\boldsymbol{P}(\boldsymbol{r},t) = \int_{-\infty}^{+\infty}\int_{-\infty}^{+\infty} \chi(\boldsymbol{r}-\boldsymbol{r}',t-t')\boldsymbol{E}(\boldsymbol{r}',t')\mathrm{d}t'\,\mathrm{d}\boldsymbol{r}' \,. \tag{8.2}$$

The response function χ is said to be non-local, i.e., the polarization at $\boldsymbol{r}$ also depends on the electric field at other places $\boldsymbol{r}'$. This is just the phenomenon of spatial dispersion, as becomes clear by inspecting the model of Fig. 6.1. Let us now assume that there is no spatial dispersion, i.e., we go from Fig. 6.1 back to Fig. 4.1. The $\boldsymbol{r} - \boldsymbol{r}'$ dependence of the susceptibility then reduces to a δ-function

$\delta(\boldsymbol{r}-\boldsymbol{r}')$, i.e., to a local response function. Integration over this function is trivial and we can ignore the $\boldsymbol{r}$ dependence for a homogeneous medium without spatial dispersion (see above) resulting in

$$\frac{1}{\varepsilon_0}\boldsymbol{P}(t)=\int_{-\infty}^{+\infty}\chi(t-t')\,\boldsymbol{E}(t')\,\mathrm{d}t' \,. \tag{8.3}$$

Now we use a very important physical argument, namely causality. This argument is, however, strictly valid only in classical physics. This means that the response $\boldsymbol{P}$ cannot come before the stimulus and thus

$$\chi(t-t')\equiv 0 \quad \text{for} \quad t'>t \tag{8.4}$$

or

$$\frac{1}{\varepsilon_0}\boldsymbol{P}(t)=\int_{-\infty}^{t}\chi(t-t')\boldsymbol{E}(t')\,\mathrm{d}t' \,. \tag{8.5}$$

We next execute a Fourier transform

$$\boldsymbol{P}(\omega)=\int_{-\infty}^{+\infty}\boldsymbol{P}(t)\,\mathrm{e}^{\mathrm{i}\omega t}\,\mathrm{d}t \,, \tag{8.6a}$$

$$\boldsymbol{E}(\omega)=\int_{-\infty}^{+\infty}\boldsymbol{E}(t)\,\mathrm{e}^{\mathrm{i}\omega t}\,\mathrm{d}t \,, \tag{8.6b}$$

$$\chi(\omega)=\int_{-\infty}^{+\infty}\chi(t-t')\,\mathrm{e}^{\mathrm{i}\omega(t-t')}\,\mathrm{d}t \,. \tag{8.6c}$$

Inserting (8.5) into (8.6a) results in

$$\frac{1}{\varepsilon_0}\boldsymbol{P}(\omega)=\int_{-\infty}^{+\infty}\mathrm{e}^{\mathrm{i}\omega t}\left[\int_{-\infty}^{t}\chi(t-t')\boldsymbol{E}(t')\,\mathrm{d}t'\right]\mathrm{d}t \,. \tag{8.7}$$

Introducing

$$1=\mathrm{e}^{-\mathrm{i}\omega t'}\mathrm{e}^{\mathrm{i}\omega t'} \tag{8.8}$$

in the inner integral and rearranging the terms gives

$$\frac{1}{\varepsilon_0}\boldsymbol{P}(\omega)=\int\boldsymbol{E}(t')\mathrm{e}^{\mathrm{i}\omega t'}\left[\int\chi(t-t')\mathrm{e}^{\mathrm{i}\omega(t-t')}\,\mathrm{d}t\right]\mathrm{d}t'=\chi(\omega)\boldsymbol{E}(\omega) \,. \tag{8.9}$$

This is identical to (2.26).

With the knowledge of (8.9) we can apply Cauchy's theorem, which connects the real and imaginary parts of the Fourier transforms of analytic functions. This theorem leads us to

$$\varepsilon_1(\omega) - 1 = \frac{1}{\pi} P \int_{-\infty}^{+\infty} \frac{\varepsilon_2(\omega')}{\omega' - \omega} \mathrm{d}\omega'$$

and

$$\varepsilon_2(\omega) = -\frac{1}{\pi} \int_{-\infty}^{+\infty} \frac{\varepsilon_1(\omega') - 1}{\omega' - \omega} \mathrm{d}\omega', \tag{8.10}$$

where P in front of the integral means the principal value.
Equation (8.10) can be rewritten as

$$\varepsilon_1(\omega) - 1 = \mathrm{Re}\{\chi(\omega)\} = \frac{2}{\pi} P \int_0^{\infty} \frac{\omega' \varepsilon_2(\omega)}{\omega'^2 - \omega^2} \mathrm{d}\omega'$$

and

$$\varepsilon_2(\omega) = \mathrm{Im}\{\chi(\omega)\} = -\frac{2\omega}{\pi} P \int_0^{\infty} \frac{\varepsilon_1(\omega')}{\omega'^2 - \omega^2} \mathrm{d}\omega' . \tag{8.11}$$

Similar relations hold for the real and imaginary parts of $\tilde{n}(\omega)$:

$$n(\omega) = 1 + \frac{1}{\pi} P \int_{-\infty}^{+\infty} \frac{\kappa(\omega')}{\omega' - \omega} \mathrm{d}\omega'$$

and

$$\kappa(\omega) = -\frac{1}{\pi} P \int_{-\infty}^{+\infty} \frac{n(\omega') - 1}{\omega' - \omega} \mathrm{d}\omega' \tag{8.12}$$

or for the phase $\phi(\omega)$ and amplitude $\rho(\omega)$ of the reflectivity $r(\omega)$ given for normal incidence by

$$r_\perp(\omega) = \frac{n(\omega) - 1 + \mathrm{i}\kappa}{n(\omega) + 1 + \mathrm{i}\kappa} = \rho(\omega)\,\mathrm{e}^{\mathrm{i}\phi(\omega)} . \tag{8.13}$$

Compare (8.13) with the expression for $R_\perp$ in (3.19).

$$\ln\{\rho(\omega)\} = \frac{1}{\pi} P \int_{-\infty}^{+\infty} \frac{\phi(\omega')}{\omega' - \omega} \mathrm{d}\omega' ,$$

$$\phi(\omega) = -\frac{1}{\pi} P \int_{-\infty}^{+\infty} \frac{\ln \rho(\omega')}{\omega' - \omega} \mathrm{d}\omega' = \frac{-2\omega}{\pi} P \int_0^{\infty} \frac{\ln \rho(\omega')}{\omega'^2 - \omega^2} \mathrm{d}\omega' . \tag{8.14}$$

The relations (8.10–12, 14) are known as Kramers–Kronig relations. They are of very general nature and rely only on causality and locality of the response. The two most important consequences are first, that if $\varepsilon_1(\omega)$ or $n(\omega)$ deviate in some frequency range from 1, then there must necessarily be absorption structures somewhere, i.e., $\varepsilon_2(\omega) \neq 0$ or $\kappa(\omega) \neq 0$ and vice versa and, second, that if either the real or imaginary part of $\varepsilon(\omega)$, $\tilde{n}(\omega)$ or $r(\omega)$ is known over the whole spectral range, then the other part can be calculated. Due to the denominator it is in practice sufficient to know the real or imaginary part only over a finite and not too small region around ω to be able to calculate the other part.

If spatial dispersion i.e. a non local response are included, the Kramer–Kronig relations become more complicated. This topic is beyond the scope of this book and for further details we refer the reader to [8.1–5] and references therein.

We now leave the subject of ensembles of oscillators and proceed to consider the elementary excitations characteristic of semiconductors. They will later replace the model oscillators considered so far.

8.2 Problem

If you are interested in the analysis of complex functions, derive the Kramers–Kronig relations from the properties of analytic complex functions $f(z)$ with $z \in C$ and give the restrictions imposed on $f(z)$.

9 Lattice Vibrations and Phonons

In this chapter we start to discuss topics that are specific to semiconductors. We shall inspect the elementary excitations and quasi particles in semiconductors in Chaps. 9 to 12. These will be needed to describe and understand the linear optical properties in Chaps. 13–17. More details about these elementary excitations are found in textbooks on solid state physics; see for example, [9.1–7].

9.1 Adiabatic Approximation

If we want to describe a semiconductor, all we have to do, in principle is to solve the Schrödinger equation for the problem. This depends on the coordinates of the ion cores, consisting of the nucleus and the tightly bound electrons in the inner shells and the outer or valence electrons with coordinates $\boldsymbol{R}_j$ and $\boldsymbol{r}_i$ and masses M_j and m_0, respectively. The Hamiltonian reads:

$$H = -\frac{\hbar^2}{2}\sum_{j=1}^{M}\frac{1}{M_j}\Delta_{\boldsymbol{R}_j} - \frac{\hbar^2}{2m_0}\sum_{i=1}^{N}\Delta_{\boldsymbol{r}_i} + \frac{1}{4\pi\varepsilon_0}$$
$$\times\left(\sum_{j>j'}\frac{e^2 Z_i Z_{i'}}{|\boldsymbol{R}_j-\boldsymbol{R}_{j'}|} + \sum_{i>i'}\frac{e^2}{|\boldsymbol{r}_i-\boldsymbol{r}'_i|} + \sum_{i,j}\frac{e^2 Z_j}{|\boldsymbol{R}_j-\boldsymbol{r}_i|}\right). \qquad (9.1)$$

Z_j is the effective charge of the ion core j and the indices j and i run over all M ion cores and N electrons, respectively.

We want to stress here that of the four fundamental interactions so far known, namely strong, electromagnetic, weak, and gravitational interaction, only the electromagnetic is of importance for the typical properties of semiconductors such as the transport and optical properties discussed in this book. Within the electromagnetic interaction we restrict ourselves here to the electric ones (including exchange interaction) since electric interactions are usually much stronger than magnetic ones, basically since electric interactions begin with monopole – monopole (i.e., Coulomb) interaction, whereas magnetic interactions start only with dipole – dipole interactions, due to the absence of magnetic monopoles [see (2.1b)]. The wavefunction solving (9.1) depends on all coordinates $\boldsymbol{R}_j$ and $\boldsymbol{r}_i$ including spins.

$$\phi(\boldsymbol{r}_i, \boldsymbol{R}_j) = E\phi(\boldsymbol{r}_i, \boldsymbol{R}_j) \qquad (9.2)$$

Since the indices j and i running from one to M and N, respectively, both count of the order of 10^{23} particles per cm^3 of semiconductor, it is obvious that there is at present no realistic chance of solving (9.1, 2), though a proper solution would, in principle, contain all information about a given semiconductor. If we do not want to get stuck at this point we must use some approximations to simplify (9.1). The most important one is the so-called adiabatic or Born-Oppenheimer approximation. It starts from the fact that the mass of an ion core is three to five orders of magnitude heavier than a free electron, i.e.,

$$M_j \simeq 1836 \cdot A_j m_0 \tag{9.3}$$

where A_j is the mass number of ion j. Since the electric forces that bind the outer electrons to the atom, and which can be described by a force constant β, are comparable to the ones which bind neighboring atoms or ions, we can easily see, even from classical arguments, that the highest resonance frequencies Ω with which ions move are much lower than the corresponding values ω for electrons

$$\Omega \simeq (\beta M_j^{-1})^{1/2} \ll \omega = (\beta m_0^{-1})^{1/2} \ . \tag{9.4}$$

Consequently, the electrons can practically instantaneously follow the motion of the ion cores, but not vice versa. This is the essence of the adiabatic approximation. On this basis we can separate ϕ $(\boldsymbol{r}_i\ \boldsymbol{R}_j)$ into a product of a wavefunction which depends only on the $\boldsymbol{R}_j$ and describes the motion of the ion cores, and another one which gives the wavefunction of the electron system depending on the momentary values of the $\boldsymbol{R}_j$. In a next step we will further assume that all ions are at their equilibrium positions resulting finally in

$$\phi(\boldsymbol{r}_i, \boldsymbol{R}_i) = \phi(\boldsymbol{r}_i)\,\phi(\boldsymbol{R}_i) \ , \tag{9.5}$$

and treating the interaction between electrons and the deviation of the ions from their equilibrium position in perturbation theory.

Before we start to inspect both factors of (9.5) we shall briefly outline how we describe a periodic lattice.

9.2 Lattices in Real and Reciprocal Space

In most cases we shall consider crystalline semiconductors. Disordered systems will be mentioned explicitly. Crystalline solids have a periodic spatial arrangement of atoms, i.e., they show long-range order. We can define in such a case three non-coplanar elementary translation vectors $\boldsymbol{a}_i (i = 1, 2, 3)$ with the property that if we start at a special atom, e.g., a Ga atom in a GaAs crystal, we reach an identical atom if we move by a vector $\boldsymbol{R}$ given by

$$\boldsymbol{R} = n_i \boldsymbol{a}_1 + n_2 \boldsymbol{a}_2 + n_3 a_3 \tag{9.6}$$

with $n_i = 0, \pm 1, \pm 2 \ldots$

The vector $\boldsymbol{R}$ is called a translation vector of the lattice. If we shift the lattice by $\boldsymbol{R}$ it comes to a position which is identical to the starting one.

The vectors $\boldsymbol{a}_i$ define a parallelepiped which is called the unit cell (see also Figs. 9.1, 2). The whole volume of a crystal is completely filled with identical unit cells. The unit cell and the vectors $\boldsymbol{a}_i$ are called primitive if the unit cell has the minimum possible volume. This definition is not unique as we explain for a two-dimensional cubic lattice in Fig. 9.1, where we show a non-primitive unit cell and two primitive ones. By convention, a special primitive unit cell is agreed upon. In our case the one defined by $\boldsymbol{a}_1$ and $\boldsymbol{a}_2$.

The vectors $\boldsymbol{R}$ evidently form for an infinite crystal an Abelian group which is called the translational group (Chap. 25). The positions of the atoms in the unit cell are given by the so-called basis. In Fig. 9.1 the basis consists of two atoms, one atom A at (0, 0) and one atom B at $(1/2\boldsymbol{a}_1, 1/2\boldsymbol{a}_2)$. The translation vectors $\boldsymbol{a}_i$ and the basis is all that we need to describe a periodic lattice.

Apart from the translational group there is another type of symmetry operation which transforms the lattice into itself, but for which at least one point is kept fixed. These symmetry operations also form a group which is called the point group. The elements of this group are for example reflections at mirror planes, rotations around axes with two-, three-, four- or six-fold symmetry or the inversion through the origin (Chap. 25). The combination of translational and point group is the space group. For three-dimensional semiconductors there are a total of 32 point groups and 280 space groups. The most important point groups for semiconductors are O_h (e.g., diamond, Si, Ge) T_d (zinc-blende-type lattices, e.g., GaAs, InP, ZnSe, ZnTe, CuCl) and C_{6v} (wurtzite type lattices, e.g., CdS, ZnO, CdSe, GaN). The first two are cubic and the third is hexagonal. We give in Fig. 9.2 unit cells for the three lattice types. For O_h and T_d one generally uses either a non-primitive cubic unit cell or the primitive one. They contain 8 and 4 atoms, respectively. The O_h lattice consists of two identical face-centered cubic lattices shifted by 1/4 of the space diagonal to the

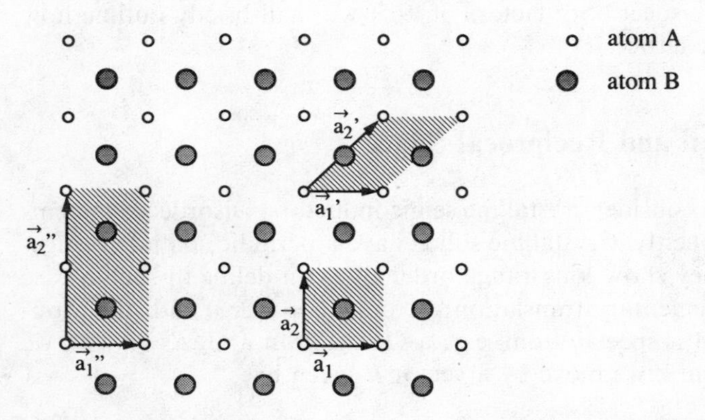

Fig. 9.1. Two primitive (r.h.s) and a non-primitive (l.h.s.) unit cells in a two-dimensional cubic lattice with a basis consisting of two different atoms per unit cell

cubic unit cell. In T_d one has the same principle; however, one of the two sub-lattices consists of atoms A the other of atoms B. Finally C_{6v} is a hexagonal lattice with a polar crystallographic $\boldsymbol{c}$-axis. In all three cases one atom is surrounded tetrahedrally by four others. The difference between T_d and C_{6v} is the position of next-nearest neighbors only. It is recommended that the reader visualizes these differences using some crystal models. The chemical binding of the semiconductors is covalent for the elements (C, Si, Ge) with sp^3 hybridization, and acquires an increasing and finally dominant ionic admixture when going to the III–V, II^b–VI and I^b–VII compounds.

Now we want to introduce the so-called reciprocal lattice. It is defined by its elementary translation vectors $\boldsymbol{b}_i$ in the same way as the lattice in real space. The $\boldsymbol{b}_i$ are given by:

$$\boldsymbol{b}_1 = \frac{2\pi}{V_{uc}} \boldsymbol{a}_2 \times \boldsymbol{a}_3 \tag{9.7}$$

and cyclic permutations of the indices V_{uc} is the volume of the unit cell given by

$$V_{uc} = \boldsymbol{a}_1 (\boldsymbol{a}_2 \times \boldsymbol{a}_3) \ . \tag{9.8}$$

A general translation vector of the reciprocal lattice is usually called $\boldsymbol{G}$

$$\boldsymbol{G} = l_1 \boldsymbol{b}_1 + l_2 \boldsymbol{b}_2 + l_3 \boldsymbol{b}_3 \quad l_i = 0, \pm 1, \pm 2, \ldots \quad i = 1, 2, 3 \ . \tag{9.9}$$

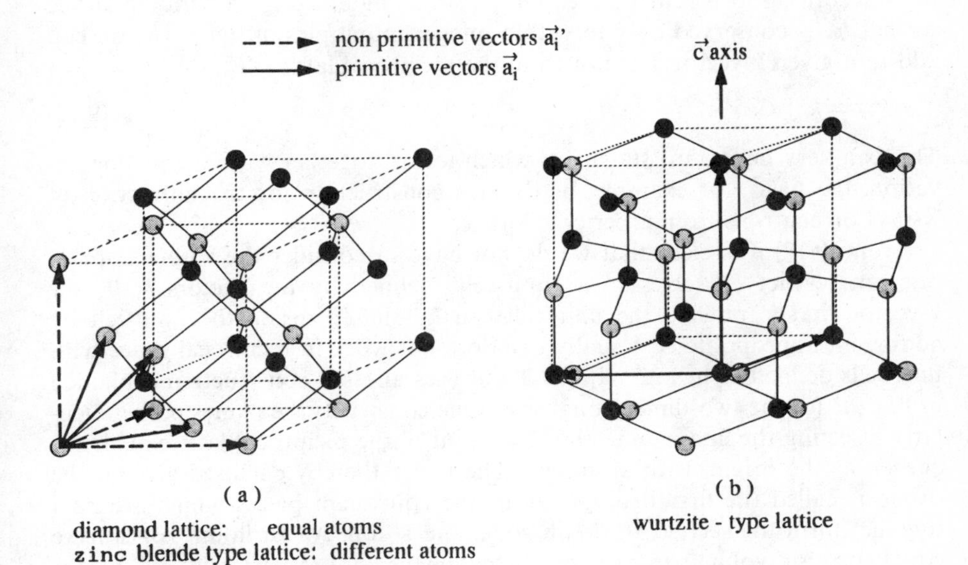

Fig. 9.2a, b. The unit cells for the diamond and zinc-blende-type lattice (**a**) and of the wurtzite lattice (**b**)

Without trying to be complete, we give some properties of the reciprocal lattice and its connections with the real one.

Every periodic function in real space which is sufficiently smooth and has a periodicity given by $f(\boldsymbol{r}+\boldsymbol{R})=f(\boldsymbol{r})$ and $\boldsymbol{R}$ defined by (9.6) can be written as a Fourier series summing over all vectors of the reciprocal lattice

$$f(\boldsymbol{r})=\sum_{\boldsymbol{G}} f_{\boldsymbol{G}}\,\mathrm{e}^{\mathrm{i}\boldsymbol{G}\boldsymbol{r}} \tag{9.10}$$

with

$$f_{\boldsymbol{G}}=\boldsymbol{V}_{\mathrm{uc}}^{-1}\int_{\mathrm{uc}} f(\boldsymbol{r})\,\mathrm{e}^{-\mathrm{i}\boldsymbol{G}\boldsymbol{r}}\,\mathrm{d}\tau\ .$$

The scalar product of $\boldsymbol{R}$ and $\boldsymbol{G}$ always fulfills

$$\boldsymbol{R}\cdot\boldsymbol{G}=2\pi m;\quad m=0,\pm 1,\pm 2\ldots\ . \tag{9.11}$$

As a consequence, we can choose to describe effects occurring in periodic lattices in real space or in reciprocal space. The latter is the appropriate space for wave vectors $\boldsymbol{k}$ or (quasi-)momenta $\hbar\boldsymbol{k}$. The "translation" from one space into the other is given by the three-dimensional Fourier series of (9.10).

In a crystal lattice we no longer have invariance with respect to infinitesimal translations in space (Sect. 3.1.3) but only invariance with respect to translations by integer multiples of $\boldsymbol{a}_i$. The conservation law for the momentum $\hbar\boldsymbol{k}$ which follows from an invariance with respect to infinitesimal translations according to Noether's theorem (3.14b) is modified for a periodic lattice so that $\hbar\boldsymbol{k}$ is conserved only to within integer mupltiples of the $\boldsymbol{b}_i$ i.e., we can add to a given $\boldsymbol{k}$-vector a vector from the reciprocal lattice $\boldsymbol{G}$:

$$\boldsymbol{k}\Longleftrightarrow\boldsymbol{k}+\boldsymbol{G}\ . \tag{9.12}$$

This is a very important statement which forms, together with energy conservation, the basis, for example, of Ewald's construction for the diffraction of X-rays or neutrons from a periodic lattice.

From (9.12) it is clear that we do not have to consider the whole $\boldsymbol{k}$-space, but can restrict ourselves to a "unit-cell" defined by the vectors $\boldsymbol{b}_i$. Every $\boldsymbol{k}$-vector that is outside the unit cell can be shifted inside the unit cell by adding an appropriate $\boldsymbol{G}$. Usually one does not work in reciprocal space with unit cells defined as in Fig. 9.1 or 9.2, but uses another construction explained in Fig. 9.3 for the two-dimensional case. One constructs the planes perpendicularly bisecting the lines connecting one point of the reciprocal lattice, which is chosen as the origin, with all others. The figure thereby enclosed around the origin is called the first Brillouin zone; the equivalent pieces which are next together form the second Brillouin zone, and so on. All Brillouin zones have equal area or volume in two or three dimensions, respectively. All higher Brillouin zones can be shifted into the first one by adding appropriate $\boldsymbol{G}$ vectors. The Brillouin zones also form a type of elementary cells, but

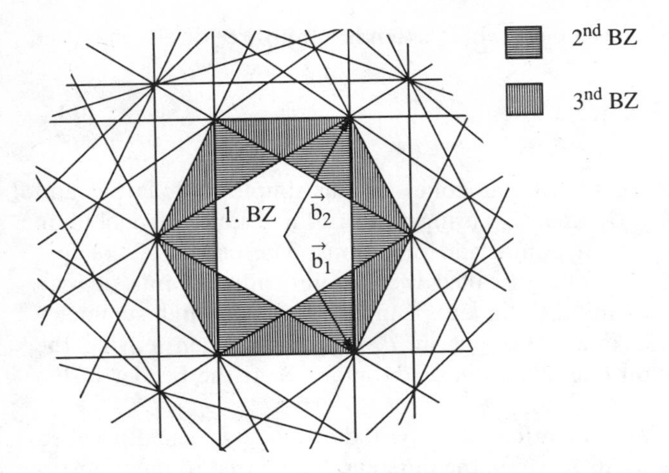

Fig. 9.3. The first Brillouin zones of a two-dimensional, hexagonal lattice

constructed according to Fig. 9.3 and not according to Fig. 9.1. The first cell constructed in real space according to Fig. 9.3 is known as a Wigner-Seitz cell. The names of points and lines of high symmetry in the first Brillouin zone are indicated in Fig. 9.4.

For a simple cubic lattice with

$$\boldsymbol{a}_1 = (a,0,0), \quad \boldsymbol{a}_2 = (0,a,0), \quad \boldsymbol{a}_3 = (0,0,a), \tag{9.13}$$

the $\boldsymbol{b}_i$ are also orthogonal with

$$\boldsymbol{b}_1 = \left[\frac{2\pi}{a},0,0\right], \quad \boldsymbol{b}_2 = \left[0,\frac{2\pi}{a},0\right], \quad \boldsymbol{b}_3 = \left[0,0,\frac{2\pi}{a}\right], \tag{9.14}$$

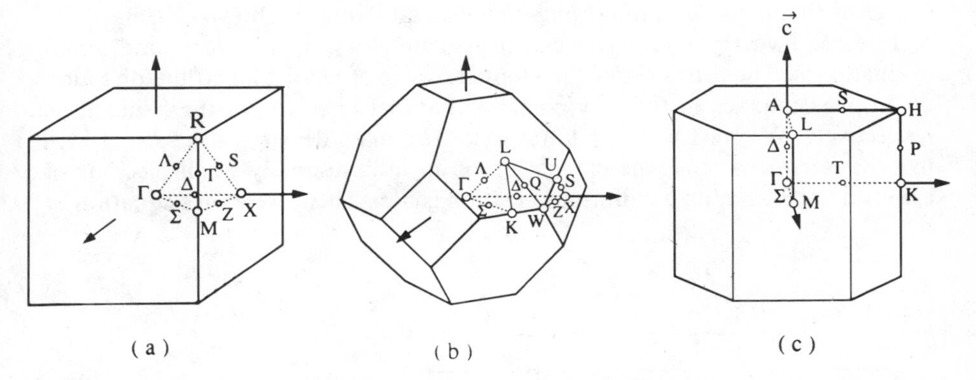

Fig. 9.4a–c. The first Brillouin zones for a simple cubic lattice (**a**), of the diamond and zinc-blende-type structures (point groups O_h and T_d, respectively) (**b**) and of the hexagonal wurtzite-type structure (C_{6v}) (**c**)

and the first Brillouin zone is a cube which extends in all three directions from

$$-\frac{\pi}{a} \leqslant k_i \leqslant +\frac{\pi}{a}, \quad i = x, y, z \, . \tag{9.15}$$

In Fig. 9.4 we give the first Brillouin zones for this simple cubic lattice and for the point groups T_d, O_h and C_{6v} using the primitive unit cell including the notation for some special points and directions. The center of the first Brillouin zone $\boldsymbol{k} = (0, 0, 0)$ is always called the Γ-point, other points of high symmetry are labelled by capital Latin letters and directions of high symmetry by capital Greek letters. As an example: in T_d symmetry, when leaving the Γ-point in the Σ-direction one arrives in at the point K at the border of the first Brillouin zone.

The quantity $\hbar\boldsymbol{k}$ of an excitation in a periodic lattice is usually called quasi-momentum if one wants to stress the difference compared to the momentum $\boldsymbol{p} = \hbar\boldsymbol{k}$ of a free particle in vacuum, e.g., a photon or an electron, where, in contrast to (9.12) no reciprocal lattice vector may be added. Actually it is possible to make a transition from one case to the other: if the lattice constant a goes to zero, the system regains translational invariance with respect to infinitessimally small shifts in real space. On the other hand, the $\boldsymbol{b}_i$ go to infinity in this limit (9.7) and the first Brillouin zone fills the whole $\boldsymbol{k}$-space, so that reciprocal lattice vectors become physically meaningless.

9.3 Vibrations of a String

In sections 9.3 to 9.6 we treat the lattice vibrations and the resulting quanta, the phonons, in the way adopted in many textbooks, i.e., we start with a homogeneous string, proceed to monatomic and diatomic chains and finally arrive at the three-dimensional solid.

Let us first consider a quasi one-dimensional string, as shown schematically in Fig. 9.5. Two types of waves can propagate along it, transverse and longitudinal ones. The direction of the elongation is perpendicular to the direction of propagation, i.e., in the x–y plane, or parallel to it, i.e., in the z-direction, respectively. We start with the latter case. The mass density of the string is ρ, its cross-section A, and, the elongation of an infinitesimally small piece dz of the string at z from its equilibrium position is $u(z)$. Then, Newton's equation of

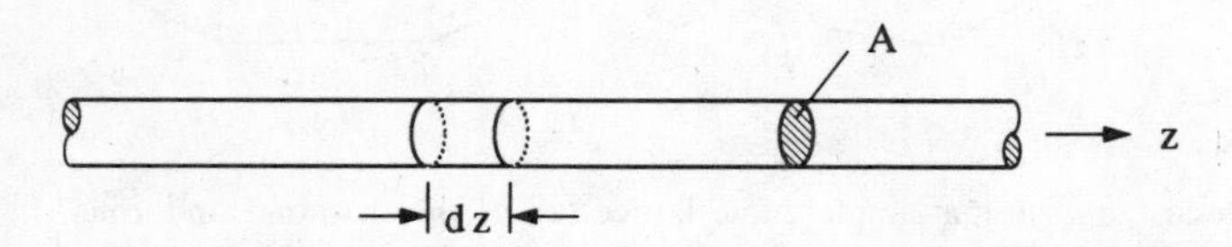

Fig. 9.5. A piece of a string as a model to explain the derivation of (9.19)

motion reads

$$\mathrm{dm}\frac{\partial^2 u}{\partial t^2} = \rho A \cdot \mathrm{d}z \cdot \frac{\partial^2 u}{\partial t^2} = F \, . \tag{9.16}$$

The force F is connected to the elasticity modulus E via

$$F = A \cdot E \, \frac{\partial^2 u}{\partial z^2} \, . \tag{9.17}$$

The appearance of the second derivative in (9.17) is for some people surprising, bearing in mind Hooke's law. However, we must consider that the stress σ is indeed given by

$$\sigma(z) = E \, \frac{\partial u}{\partial z} \, . \tag{9.18}$$

If the stress is the same on both sides of the infinitesimal element of length $\mathrm{d}z$, the resulting forces at z and $z + \mathrm{d}z$ compensate each other to zero. The restoring force F is therefore given by $\mathrm{d}\sigma/\mathrm{d}z$ leading to (9.17).

Putting (9.16) and (9.17) together leads to the standard harmonic wave equation

$$\rho \, \frac{\partial^2 u}{\partial t^2} = E \, \frac{\partial^2 u}{\partial z^2} \, . \tag{9.19}$$

With the ansatz

$$u = u_0 \exp\left[\mathrm{i}(kz - \omega t)\right] \tag{9.20}$$

for a plane wave we find the dispersion relation for longitudinal waves

$$\omega_{\mathrm{L}} = (E/\rho)^{1/2} k \, . \tag{9.21}$$

This is a linear relation as shown in Fig. 9.6. Consequently phase and group velocity are constant and equal, namely, with (2.13):

$$v_{\mathrm{ph}}^{\mathrm{L}} = v_{\mathrm{g}}^{\mathrm{L}} = (E/\rho)^{1/2} \, . \tag{9.22}$$

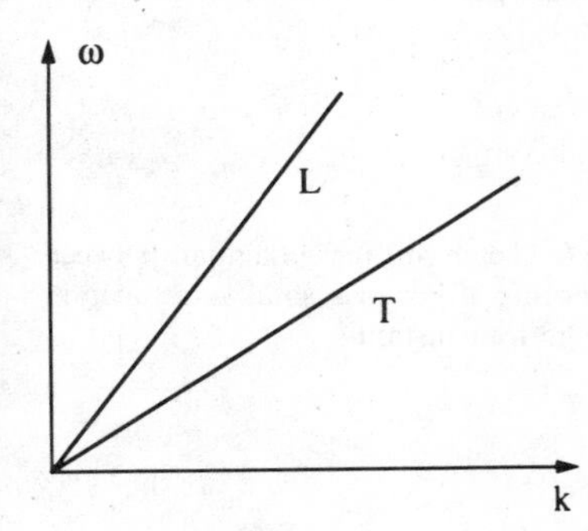

Fig. 9.6. The dispersion relation of waves on a homogeneous string

For the two degenerate, transverse waves we find in a similar way

$$\omega_{\mathrm{T}} = (G/\rho)^{1/2}\, k \tag{9.23}$$

or

$$v_{\mathrm{ph}}^{\mathrm{T}} = v_{\mathrm{g}}^{\mathrm{T}} = (G/\rho)^{1/2}\,, \tag{9.24}$$

where G is the shear or torsion modulus.

Since it is known from the theory of elasticity that

$$G \leqslant E\,, \tag{9.25}$$

we find

$$v_{\mathrm{ph}}^{\mathrm{T}} \leqslant v_{\mathrm{ph}}^{\mathrm{L}}\,, \tag{9.26}$$

a result comparable to (4.28).

9.4 Linear Chains

We now should consider the regime of validity of the above calculation. We assumed a homogeneous string, neglecting the fact that a solid is made up from atoms. Therefore the above approximation can only be valid for wavelengths much longer than the lattice constant or for wave vectors close to the center of the first Brillouin zone, i.e.,

$$\lambda \gg a \quad \text{or} \quad k \ll \frac{\pi}{a}\,. \tag{9.27}$$

For shorter wavelengths we have to consider the atomic structure of solids. The interaction potential between neighboring atoms as a function of the lattice constant a looks approximately like Fig. 9.7 for all types of binding, covalent, ionic or metallic. For sufficiently large lattice constants there is no interaction between the atoms, i.e., $V = 0$; then comes an attractive regime (without which there would be no solids); and this is finally followed by a steep repulsive increase due to Pauli's exclusion principle when the filled

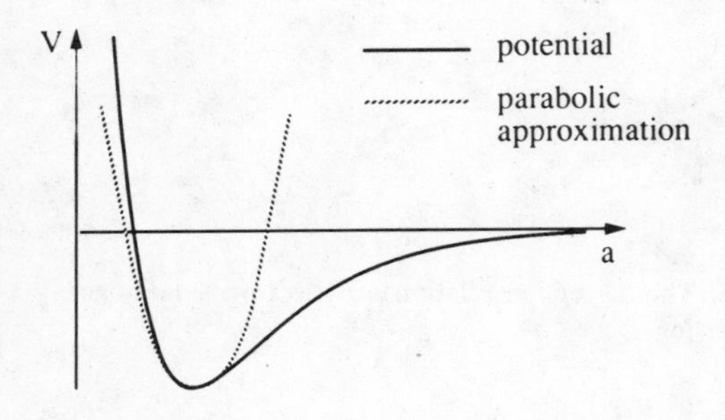

Fig. 9.7. Sketch of the potential between neighboring atoms in a solid as a function of the lattice constant

inner shells of neighboring atoms start to overlap. Different analytic approximations to $V(a)$ are known for example the Born-Mayer or Lennard-Jones potentials. These details have at present no relevance for us. We note that a crystal left to itself will come to a state close to the energetic minimum, i.e., to the equilibrium position a_0. In the vicinity of a_0, $V(a)$ can be approximated by a parabola, that is by a harmonic potential, shown by the dashed line. It is at least qualitatively clear that this harmonic approximation is valid only very close to a_0. For larger deviations from a_0 siginificant anharmonicities (i.e., deviations from the harmonic potential) have to be expected. The anharmonicities are characteristic for lattice vibrations and manifest themselves, among other things, in the thermal expansion of solids and in phonon–phonon interaction.

For the moment, however, the harmonic approximation is good enough and we consider a linear model soild in which every atom with mass M is connected to its neighbors by a "spring" with a force constant D, representing the harmonic potential, leading to the linear-chain model of Fig. 9.8a in which we indicate also the lattice constant a. Evidently we have a basis consisting of one atom per primitive unit cell. At this point it is important to stress the difference between the models of Fig. 6.1 and 9.8. In Fig. 6.1 we had independent oscillators, and a weak coupling between them was introduced only to simulate spatial dispersion. Here, in Sect. 9.4, the coupling springs from the only forces acting on the atoms. We introduce now the displacement of atom u_n from the equilibrium position and obtain the equation of motion, again for the longitudinal mode

$$M\frac{\partial^2 u_n}{\partial t^2} = D[(u_{n+1} - u_n) - (u_n - u_{n-1})] = D(u_{n+1} - 2u_n + u_{n-1}) . \tag{9.28}$$

Instead of the second differential quotient in the homogeneous approximation we now get a second order difference equation.

(a) M a D n - 1 n n + 1

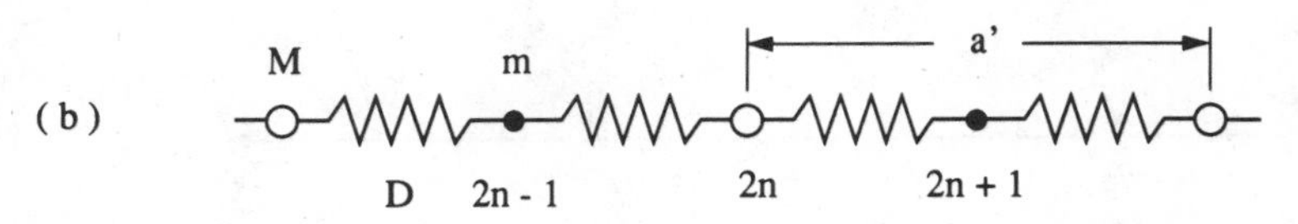

Fig. 9.8. The linear chain model for the cases of one atom per unit cell (**a**) and two atoms per unit cell (**b**)

As a solution of (9.28) we again try a plane wave which reads, in this discrete case,

$$u_n = u_{n,0}\exp[\mathrm{i}(kna - \omega t)]\,,$$
$$u_{n\pm1} = u_{n\pm1,0}\exp\{\mathrm{i}[k(n\pm1)a - \omega t]\}\,.$$

For a plane wave we conclude that the amplitudes of the various atoms are equal, i.e.,

$$u_{n,0} = u_{n+1,0} = u_{n-1,0}\,. \tag{9.29}$$

Inserting (9.29) into (9.28) gives

$$-M\omega^2 = D(\mathrm{e}^{-\mathrm{i}ka} - 2 + \mathrm{e}^{\mathrm{i}ka}) = -2D(1-\cos ka) \tag{9.30}$$

or

$$\omega = \left(\frac{4D}{M}\right)^{1/2}\left|\sin\frac{ka}{2}\right|\,. \tag{9.31}$$

The dispersion relation according to (9.31) is shown in Fig. 9.9 together with the phase and group velocities.

As can be expected from the discussion in Sect. 9.2, the dispersion relation outside the first Brillouin zone repeats just what is inside, or, in other words, the branches inside and outside can be shifted into each other by adding or subtracting reciprocal lattice vectors $l\cdot 2\pi/a$, where l is a positive or negative integer.

The fact that there is nothing new outside the first Brillouin zone can be easily elucidated for the case of lattice vibrations in connection with Fig. 9.10.

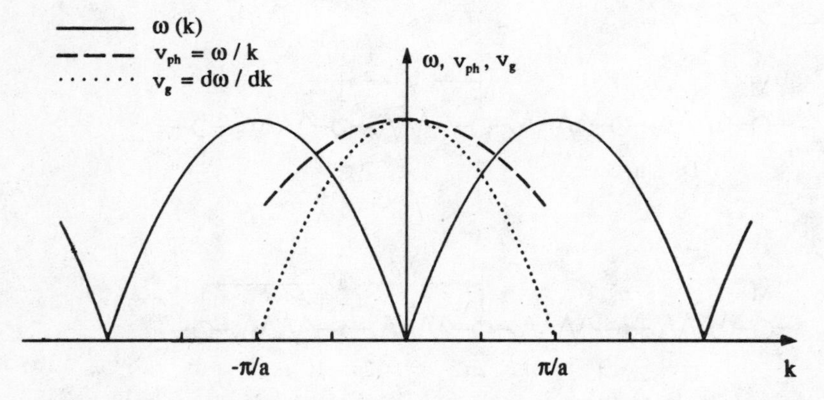

Fig. 9.9. The dispersion relation and the phase and group velocities of the longitudinal vibrations of the monatomic chain of Fig. 9.8a

If adjacent atoms are displaced in antiphase, we get the shortest physically meaningful wavelength (solid line) $\lambda_{\min}$ by

$$\lambda_{\min} = 2a \rightarrow k_{\max} = \frac{2\pi}{\lambda_{\min}} = \frac{\pi}{a} . \tag{9.32}$$

Of course we can define a shorter wavelength as indicated by the dashed line resulting here in $k = 3\pi/a$. But this definition is physically meaningless since we have no atoms at the positions between $z = na$ and $z = (n+1)a$. On the other hand, $k = 3\pi/a$ corresponds to π/a by just adding $G = 2\pi/a$.

We give specify v_{ph} and v_{g} in the first Brillouin zone. Except at $k = 0$ the two quantities are no longer equal and change with k. Consequently a wave packet will become broader during propagation, i.e., it will show "dispersion" in the sense discussed in Chap. 6. The fact that the dispersion relation is horizontal at the border of the first Brillouin zone in the direction normal to this border is generally the case.

Since most semiconductors have more than one atom per primitive unit cell, we now address this situation now with Fig. 9.8b where we evidently have a basis consisting of two atoms with masses M and m. The lattice constant is now a'.

Using the nomenclature of Fig. 9.8b we get the following equations of motion in analogy to (9.28):

$$M \frac{\partial^2 u_{2n}}{\partial t^2} = D(u_{2n+1} - 2u_{2n} + u_{2n-1}) ,$$

$$m \frac{\partial^2 u_{2n+1}}{\partial t^2} = D(u_{2n+2} - 2u_{2n+1} + u_{2n}) . \tag{9.33}$$

Using again an ansatz

$$u_{2n} = u_{2n,0} \exp[\mathrm{i}(2nak - \omega t)] ,$$

$$u_{2n+1} = u_{2n+1,0} \exp\{\mathrm{i}[(2n+1)ak - \omega t]\} , \tag{9.34}$$

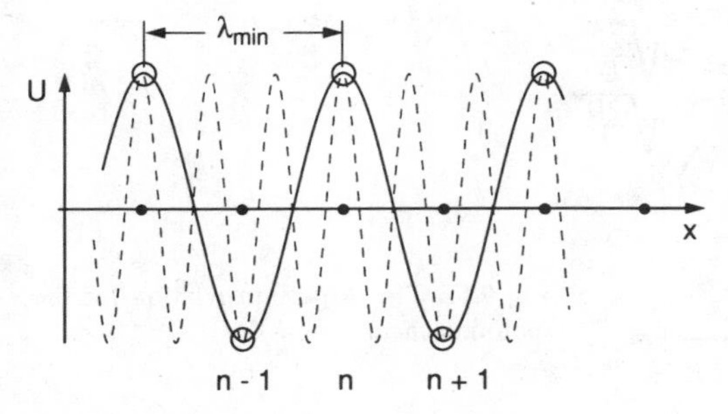

Fig. 9.10. A vibration of the linear chain of Fig. 9.8a with a wave vector inside and outside the first Brillouin zone

and noting again that the amplitudes of equal masses are equal

$$
\begin{aligned}
u_{2n,0} &= u_{2n+2,0} = A_M \,, \\
u_{2n+1,0} &= u_{2n-1,0} = A_m \,,
\end{aligned} \tag{9.35}
$$

we get the following linear set of equations

$$
\left.
\begin{aligned}
(2D - \omega^2 M) A_M - 2D\cos(ka) A_m &= 0 \,, \\
-2D\cos(ka) A_M + (2D - \omega^2 m) A_m &= 0 \,.
\end{aligned}
\right\} \tag{9.36}
$$

These have a non–trivial solution (i.e., one other than $A_M = A_m = 0$) only if the determinant of the coefficients vanishes. The dispersion relation resulting from the corresponding secular equation reads:

$$
\omega^2 = D\left(\frac{1}{m} + \frac{1}{M}\right) \pm D\left[\left(\frac{1}{m} + \frac{1}{M}\right)^2 - \frac{4}{Mm}\sin^2\frac{ka'}{2}\right]^{1/2} \tag{9.37}
$$

The dispersion relation has now two branches, as shown in Fig. 9.11, where we give also the values at some special points. The lower branch is usually called the acoustic branch since sound waves propagate according to its modes. The upper branch is called the optical one, for reasons given below.

We can enter the solution (9.37) into (9.36) and calculate the ratio A_M/A_m. The procedure is straightforward but lengthy and so we present the result only graphically in Fig. 9.12 in agreement with our statement in Sect. 1.1, and discuss it in connection with Fig. 9.13. For the acoustic branch, the two different atoms are displaced in the same direction, Fig. 9.13a. For very long wavelengths (i.e., $k \simeq 0$) the amplitudes are equal. Actually the case $k = 0$ corresponds to a simple displacement of the whole crystal. For increasing k the amplitude of the heavy mass M gets larger than that of m for $M > m$, and, at the boundary of the first Brillouin zone, only the heavy masses oscillate (Fig. 9.13b), resulting in an eigenfrequency $(2D/M)^{1/2}$ as indicated in Fig. 9.11.

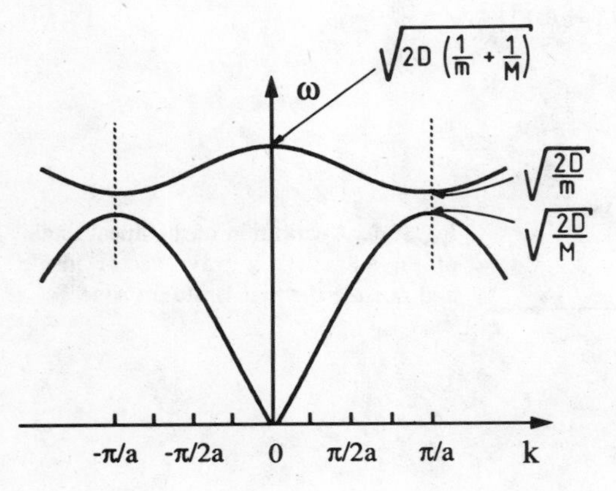

Fig. 9.11. The dispersion relation for the diatomic chain

For the optical branch, the two atoms are displaced in anti-phase (Fig. 9.13c). At the boundary of the first Brillouin zone only the light masses oscillate (Fig. 9.13d).

If the two different atoms carry an electric charge, i.e., if the binding is at least partly ionic, then the oscillation according to Fig. 3.13c is connected with an oscillating electric dipole. This allows it to couple to the electromagnetic light field at least for the transverse eigenmodes and this is why these oscillations are called "optical" modes. We come back to this aspect in Chap. 13.

Here we have discussed in detail only the longitudinal modes, but it is obvious that both for the monatomic and the diatomic chain, for every wave vector and every branch, two (degenerate) transverse oscillations also exist.

The dispersion relation of Fig. 9.11 can also be deduced in another way starting from the one of Fig. 9.9, i.e., from the monatomic chain which we repeat in Fig. 9.14a. Now we imagine that we paint the atoms of the monatomic chain in two different alternating colors, but without changing their physical properties. As a consequence we have increased the length of the primitive unit cell by a factor of two and the new lattice constant a' is given by

$$a' = 2a \tag{9.38}$$

and this in turn reduces the length of the first Brillouin zone by one-half as shown in Fig. 9.14a, b. Consequently, we can shift the outer parts by vectors of the new reciprocal lattice into the first Brillouin zone. This situation is shown in Fig. 9.14b. Since the atoms are still identical, the two branches cross at the border of the first Brillouin zone. If we now also introduce differences in the physical properties of the atoms, for example giving them different masses, then we end up with the situation of Fig. 9.14c which is identical to Fig. 9.11. We shall use this set of arguments again in connection with superlattices later on. First we want to extend this discussion of the lattice vibrations of three-dimensional systems.

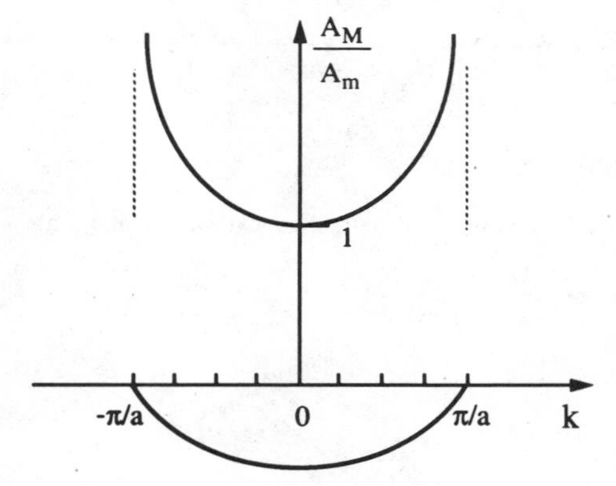

Fig. 9.12. The ratio of the amplitudes A_M/A_m of the two different masses for the two dispersion branches of Fig. 9.11 as a function of k

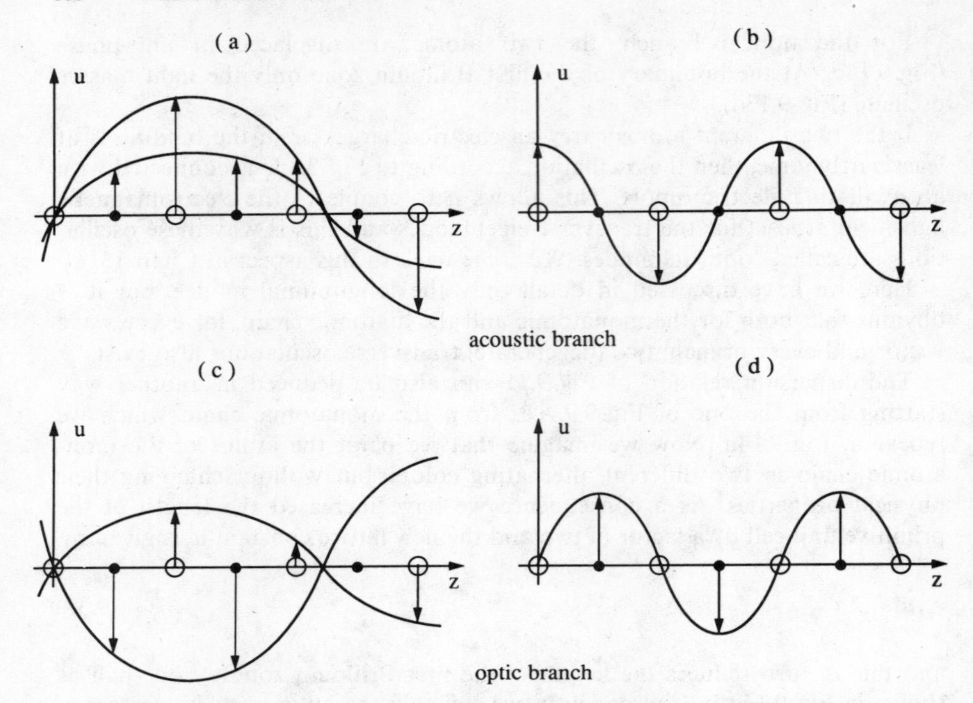

Fig. 9.13a–d. Sketch of the displacements of the atoms on the acoustic (**a, b**) and the optical branches for two different wave vectors

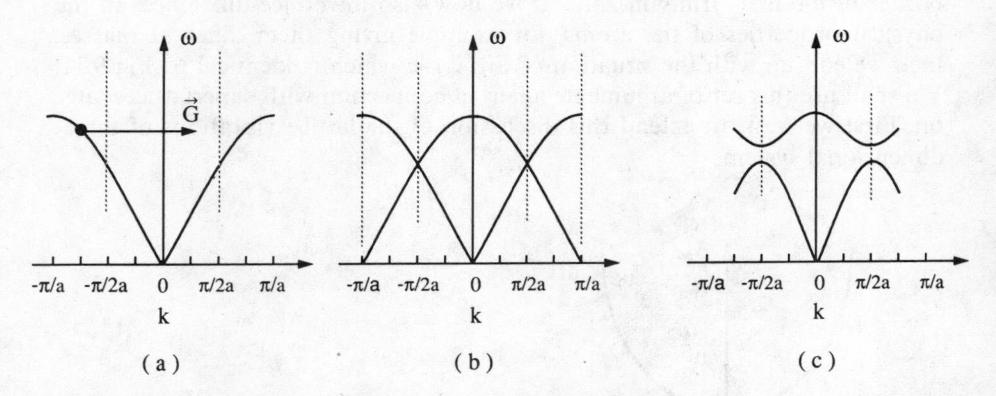

Fig. 9.14a–c. The transition from the monatomic (**a**) diatomic (**c**) chain. (See text for explanation)

9.5 Three-Dimensional Lattices

If we consider a three-dimensional crystal, not too much changes with respect to the chain model of Sect. 9.4, at least for the dispersion relations. The atoms are connected with nearest (and possibly next-nearest) neighbors in the three-dimensional lattice. The set of equation analogous to (9.28) or (9.33) will become correspondingly more complex, but the result will be qualitatively the same. There are still the acoustic branches and, in addition, optical ones if we have more than one atom in the primitive unit cell. There are always three acoustic branches, namely one longitudinal and two transverse ones for every $\boldsymbol{k}$-vector and $3s-3$ optic ones:

$$\left.\begin{array}{ll}\text{number of acoustic branches:} & 3\\ \text{number of optical branches:} & 3s-3\end{array}\right\} \tag{9.39}$$

where s is the number of atoms per primitive unit cell.

If the crystal is anisotropic the dispersion relations will be different for different directions in the Brillouin zone and the degeneracy between the two transverse modes for every $\boldsymbol{k}$ may be lifted. We show such a situation schematically in Fig. 9.15.

9.6 Quantization of Lattice Vibrations: Phonons and the Concept of Quasiparticles

If we look again at (9.28) or (9.33) for the $u_n(t)$ we see that these equations are rather similar to the equation of motion for an harmonic oscillator

$$M\frac{\partial^2 v_n}{\partial t^2} = -Dv_n\,. \tag{9.40}$$

The only difference are the terms with indices $n \pm 1$, i.e., the off-diagonal terms in the language of matrix representation. It is now possible to find appropriate

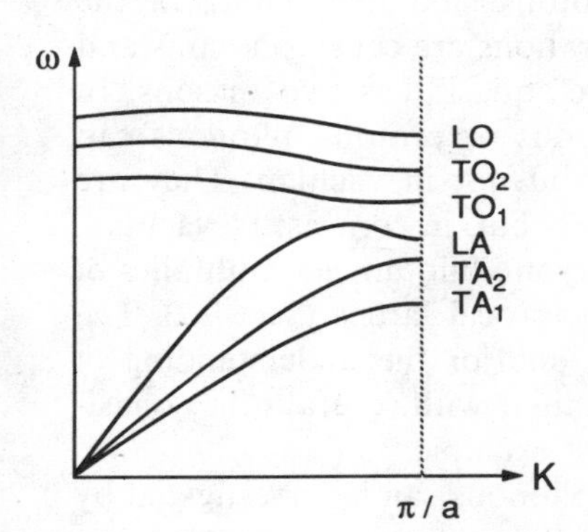

Fig. 9.15. Sketch of the dispersion relation of lattice vibrations for a three-dimensional anisotropic lattice with partly ionic binding

linear combinations of the u_n.

$$v_n = \sum_{n'} a_{n'} u_{n'} \tag{9.41}$$

such that the v_n obey equations like (9.40) or (2.52) or, in other words, to diagonalize the problem.

On the other hand a quantum mechanical treatment of the harmonic oscillator leads to discrete energy levels given by

$$E_n = \hbar\omega_n\left(n + \frac{1}{2}\right), \quad n = 1, 2, 3 \ldots . \tag{9.42}$$

It is now important to note that the dispersion relation shown for example in Fig. 9.15 is not influenced by a linear transformation of the coordinates.

As a consequence we may say that the lattice vibrations consist of quanta according to (9.42) for each wave vector $\boldsymbol{k}$ and branch i. The total energy of the lattice vibrations can be written as

$$E = \sum_{k,i} \hbar\omega_{k,i}\left(n_{k,i} + \frac{1}{2}\right) \tag{9.43}$$

in analogy to (2.55b) or (5.3), where the index i labels the various branches.

In the framework of second quantization it is possible to define creation and annihilation operators $b_{k,i}^{+}$ and $b_{k,i}$ respectively, and the Hamiltonian can then again be written in terms of the number operator, in a similar way to that used already for photons

$$H = \sum_{k,i} \hbar\omega_{k,i}\left(b_{k,i}^{+} b_{ki}^{+} + \frac{1}{2}\right) . \tag{9.44}$$

Obviously there is a close analogy between the quanta or particles of the electromagnetic radiation in vacuum, the photons, and the quanta of the lattice vibrations. The quanta of the lattice vibrations are called phonons and are considered as quasi-particles. The attribute “quasi” has two reasons. In contrast to “real” particles like photons, electrons or protons, phonons can exist only in the (periodic) lattice of a solid and not in vacuum. They are characterized by their energy and momentum $\hbar\boldsymbol{k}$, but, in contrast to vacuum, $\hbar\boldsymbol{k}$ is a “quasi-momentum” which is defined only modulo integer multiplies of the elementary translation vectors $\boldsymbol{b}_i$ of the reciprocal lattice (Sect. 9.2). The concept of quasi-particles is a very important one for the understanding of solids. In the next chapters we shall become familiar with several other quasi-particles in solids. The phonons are just the first example.

The existence and the dispersion relation of phonons can be investigated by inelastic neutron scattering. An incident neutron from a mono-energetic beam with

$$E_\mathrm{i} = \frac{\hbar^2 \boldsymbol{k}_\mathrm{i}^2}{2m_\mathrm{n}} \tag{9.45}$$

is scattered under creation or annihilation of a phonon, resulting in a neutron in the final state with $E_\mathrm{f}, \boldsymbol{k}_\mathrm{f}$, which are given via the conservation laws of energy and quasi-momentum

$$E_\mathrm{f} = \frac{\hbar^2 \boldsymbol{k}_\mathrm{f}^2}{2m_\mathrm{n}} = E_\mathrm{i} \pm \hbar\omega_\mathrm{Phonon}$$

and

$$\boldsymbol{k}_\mathrm{f} = \boldsymbol{k}_\mathrm{i} \pm \boldsymbol{k}_\mathrm{Phonon} + G \tag{9.46}$$

By measuring the properties of the incident and scattered neutrons it is possible to prove the existence of phonons and to determine their dispersion relation.

Figure 9.16 shows the dispersion relation of the phonons in two different semiconductors, Si and CdS, for various directions in $\boldsymbol{k}$-space (see Fig. 9.4). All well-known semiconductors, including the elemental ones like Si and Ge, have more than one atom per unit cell and therefore support both acoustic and optical branches. Si has only covalent binding in contrast to CdS which has a mixed ionic-covalent binding. Therefore the Si atoms do not carry an electric charge and, as a result, even the optical phonons do not couple directly to the radiation field, resulting at $\boldsymbol{k} = 0$ in an oscillator strength $f = 0$ and consequently in $\Delta_\mathrm{LT} = 0$ as a result of (4.26).

In connection with Fig. 9.7 we have introduced a harmonic interaction potential between atoms. Actually there are strong anharmonicities, as already mentioned, which are, among others, due to the fact that the electron distribution changes almost instantaneously with the changing positions $\boldsymbol{R}_j$ of the atoms, c.f. the adiabatic approximation of Sect. 9.1. This fact results in a variation of the "spring-constant" D with the lattice constant a, i.e., in an anharmonicity.

The resulting anharmonicity manifests itself, for example, in scattering processes between phonons. We show some of them schematically in Fig. 9.17. However, one should bear in mind that all combination possibilities are exhausted only when the three-dimensional $\boldsymbol{k}$-space is considered.

In (a) a TO phonon decays under energy and momentum conservation into two acoustic phonons in (b) two transverse acoustic phonons combine to form a longitudinal acoustic phonon. Energy and momentum conservation read e.g. for (a) and (b):

(a) $\quad \hbar\omega_\mathrm{i} = \hbar\omega_\mathrm{f1} + \hbar\omega_\mathrm{f2}, \quad \boldsymbol{k}_\mathrm{i} = \boldsymbol{k}_\mathrm{f1} + \boldsymbol{k}_\mathrm{f2}$

(b) $\quad \hbar\omega_\mathrm{i1} + \hbar\omega_\mathrm{i2} = \hbar_\mathrm{f}, \quad \boldsymbol{k}_\mathrm{i1} + \boldsymbol{k}_\mathrm{i2} = \boldsymbol{k}_\mathrm{f}\,.$

These are so-called n or normal processes.

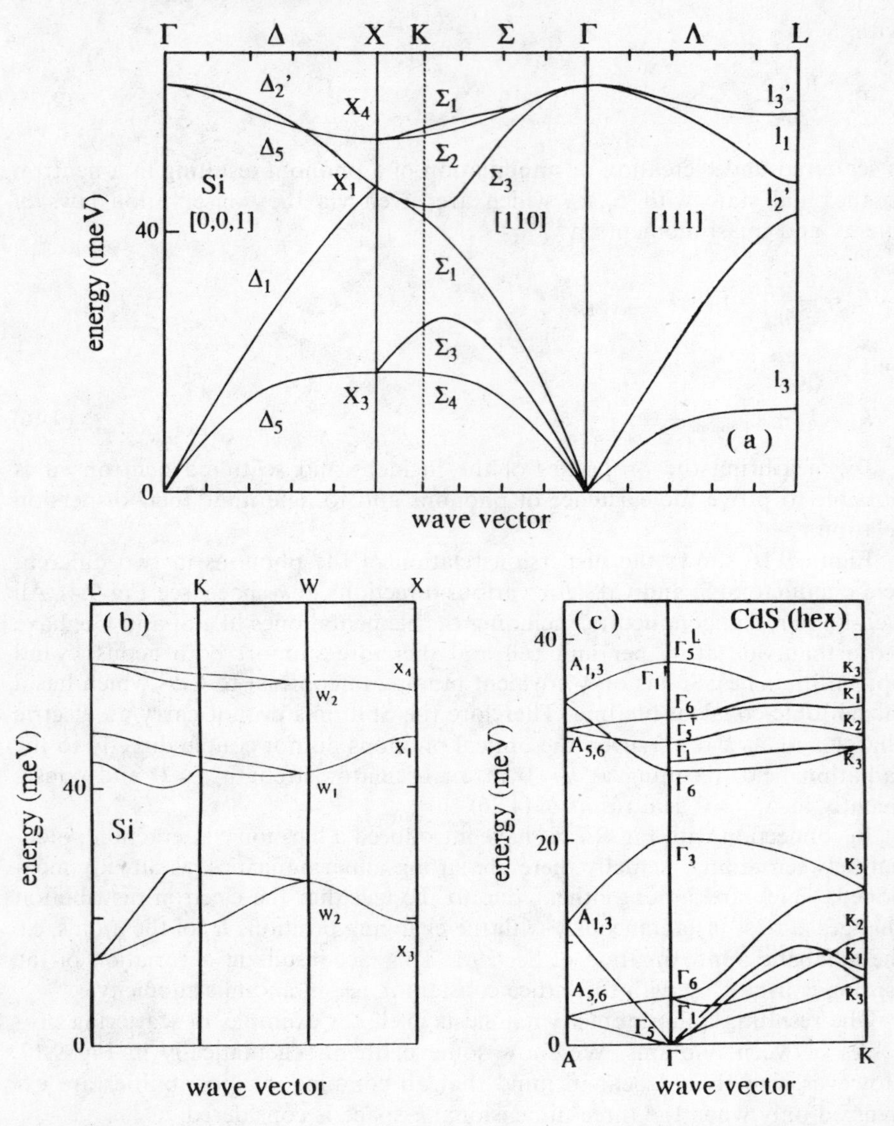

Fig. 9.16a–c. Dispersion relation of phonons in Si and CdS for various directions in $\boldsymbol{k}$-space **a** and **c** show essentially directions in $\boldsymbol{k}$ space, which originate from the Γ point and **b** on the surface of the first Brillouin zone After [9.8]

In the decay process (c) we end up in the second Brillouin zone with one phonon and fold the phonon back with a vector of the reciprocal lattice $\boldsymbol{G}$

$$\hbar\omega_{i1} + \hbar\omega_{i2} = \hbar\omega_f, \quad \boldsymbol{k}_{i1} + \boldsymbol{k}_{f2} = \boldsymbol{k}_f + \boldsymbol{G}\,. \tag{9.47}$$

The situation of (c) is known as a u or Umklapp process.

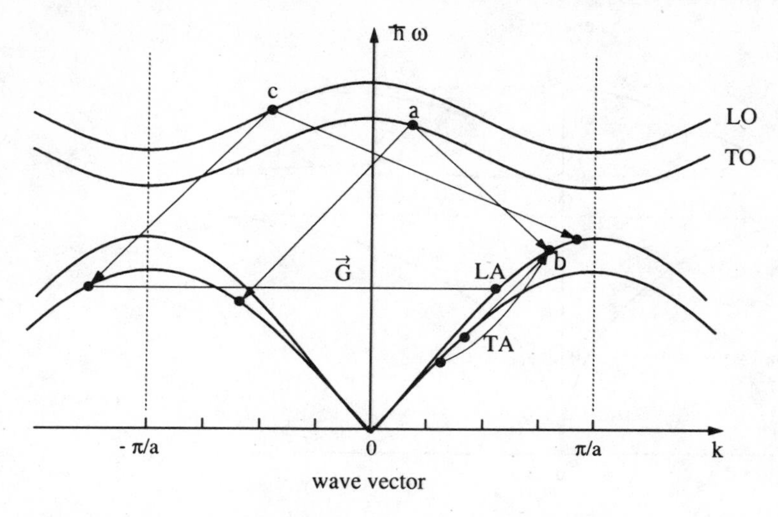

Fig. 9.17. Sketch of possible decay processes of an optical phonon

9.7 The Density of States and Phonon Statistics

Now we want to calculate the density of states of phonons as a first application of what we have learned in Sect. 2.6.

We assume for simplicity that we have two atoms per unit cell and that the resulting three acoustic and optical branches are degenerate. Furthermore we assume an isotropic semiconductor. This results in Fig. 9.18 where we plot on one side of the x-axis $\boldsymbol{k}$ and on the other the density of states $D(\omega)$. As stated in Sect. 2.6 we need the dispersion relation to calculate $D(\omega)$. In the linear part of the acoustic branch we have

$$\omega = v_s k \ , \tag{9.48}$$

where v_s is the constant velocity of sound. This linear relation results immediately in

$$D(\omega) = \text{const} \cdot \omega^2 \tag{9.49}$$

i.e., a parabolic density of states similar to that for photons in vacuum. The difference is only in the proportionality factor of (9.48, 49).

When deviations from the linear dispersion relation start, the calculations get more lengthy and we will not go into details. The denominator on the right of (2.77) tells us, however, that the density of states has in principle a singularity and in practice a steep maximum when we have a horizontal slope of the $\omega(\boldsymbol{k})$ relation, as indicated in Fig. 9.18.

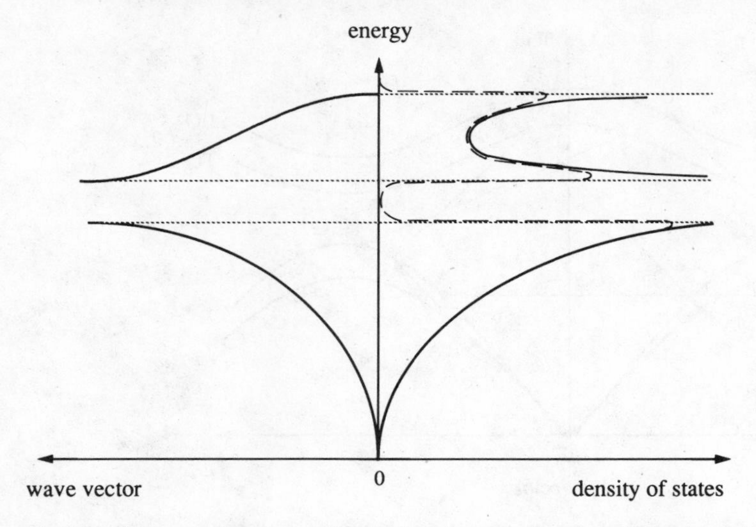

Fig. 9.18. Schematic drawing of the density of states of a three-dimensional, isotropic semiconductor with two atoms per unit cell and degenerate optical and acoustic branches

If we want to know the number of phonons we must integrate over the density of states weighted with the probability that the state is populated.

The commutation relations for phonons obey

$$[b_{k,i}, b^+_{k',i'}] = \partial_{k,k'}\partial_{i,i'} \tag{9.50}$$

where i is the label of the branch. This means that phonons are bosons. So we have in thermal equilibrium to use Bose–Einstein statistics (2.80b)

$$f_{\mathrm{BE}}(\hbar\omega, T) = \{\exp[(E-\mu)/k_{\mathrm{B}}T] - 1\}^{-1}\,. \tag{9.51}$$

Since the number of phonons is not conserved, due e.g., to phonon–phonon interaction (see above), the chemical potential μ is zero (Sect. 2.6).

For the total number of phonons, $N(T)$, we obtain

$$N(T) = \int_0^\infty D(\hbar\omega)\,[\exp(E/k_{\mathrm{B}}T) - 1]^{-1}\,\mathrm{d}\omega \tag{9.52a}$$

and for the energy of the phonon system

$$U(T) = \int_0^\infty \hbar\omega\, D(\hbar\omega)\,[\exp(E/k_{\mathrm{B}}T) - 1]^{-1}\,\mathrm{d}\omega\,. \tag{9.52b}$$

Starting from (9.52b) it is easy to calculate the specific heat of the phonon system

$$c_p \simeq c_v = \frac{\partial U}{\partial T} \tag{9.53}$$

if appropriate approximations are made for $D(\omega)$. Einstein assumed that all phonons have the same frequency ω_E, i.e., he approximated $D(\hbar\omega)$ by a δ-function

$$D(\hbar\omega) = \delta(\omega - \omega_E)\, 3Ns\,, \tag{9.54a}$$

and Debye continued the linear part of the dispersion relation up to a frequency which is also chosen to accommodate all $3Ns$ degrees of freedom of the atoms, where N is the number of unit cells and s the number of atoms per unit cell

$$3Ns = \int_0^{\omega_D} D(\hbar\omega)\,d\omega \quad \text{with} \quad D(\omega) \propto \omega^2\,. \tag{9.54b}$$

For high temperatures (i.e., $k_B T > \hbar\omega_D$ or $\hbar\omega_E$) both approximations give the classical limit, namely the law of Dulong and Petit

$$c_v = 3Nsk_B \tag{9.55}$$

and a continuous approach to zero for $T \to 0$. In the case of Debye's approximation one finds the well-known T^3 law:

$$c_v \sim (T/\theta)^3 \quad \text{with} \quad k_B\theta = \hbar\omega_D \quad \text{and} \quad T < \theta\,. \tag{9.56}$$

Since the discussion of specific heat c_v is outside the scope of this book, we do not take it any further, having introduced it simply to illustrate the applicability of the concept of the density of states, and mention only that the approach in (9.52b, 54b) is, apart from some constants and $\omega_D \to \infty$ identical to the one used to describe Planck's law of blackbody radiation as seen by comparing with (2.84).

Finally, we want to point out the following. Since phonons are bosons, processes which involve phonons can be stimulated in the same way as was discussed in connection with photons in subsection 3.2.1, i.e., we get in transitions involving the emission of a phonon in a certain mode, apart from other terms, a factor like

$$W_{i \to f} \sim (N_{Ph}^k + 1) \tag{9.57}$$

To get a feeling we assume a lattice temperture of 77 K (i.e., liquid N_2) and an acoustic phonon with an energy around 0.2 meV (such phonons will be used in Brillouin scattering in Sect. 13.3) and find for these conditions with (9.51)

$$N_{Ph}^k \simeq 30 \gg 1\,. \tag{9.58}$$

This means that the occupation number is much larger than one and processes which involve the emission of phonons with energies smaller than the thermal energy are stimulated by the phonons. Depending on the process under consideration, it is eventually necessary to consider the reverse process, too, which depends on N_{Ph}^k.

9.8 Phonons in Superlattices

Until now we were mainly considering homogeneous, three-dimensional semiconductors. For the phonons, however, we started with a one-dimensional chain, but we stated that the dimensionality does not have significant influence on the dispersion relation but it does on the density of states, according to what we learned in Sect. 2.6. For the regime of a linear dispersion relation, i.e., for acoustic phonons with not too large wave vectors ($\boldsymbol{k} \ll \pi/a$, see Figs. 9.15–18) we get with (2.77)

$$D(\omega) \sim \omega^{d-1} , \tag{9.59}$$

where d is the dimensionality of the system.

Now we want to address in the context of phonons for the first time a rather modern topic in semiconductor physics, namely superlattices.

A superlattice is a man-made periodic structure which consists of thin alternating layers of two different materials, as shown in Fig. 9.19. The different layers are only a few lattice constants thick and can be prepared by various techniques like molecular-beam epitaxy (MBE), metal-organic chemical vapor-phase deposition (MOCVD), hot-wall epitaxy (HWE) or atomic layer epitaxy (ALE). A description of these methods is beyond the scope of this book, but we give some references for the interested reader, e.g. [9.9]. Especially well-suited materials for growing superlattices are the III–V compounds GaAs, AlAs and their alloys $Al_{1-y}Ga_yAs$, since the lattice constant is almost independent of y. Other systems under investigation involve the two elemental semiconductors Si and Ge or ZnSe and $ZnS_{1-x}Se_x$. Due to their different lattice constants, the fabrication of superlattices is in this case more difficult and the layers are strained (strained layer superlattice).

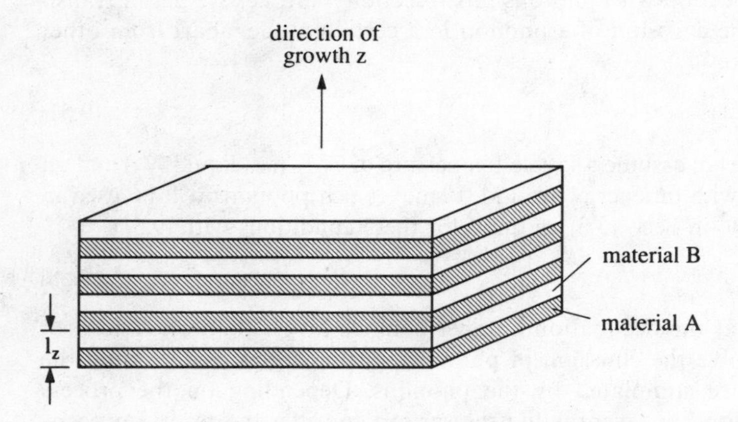

Fig. 9.19. A superlattice consisting of alternating thin layers of two different materials

We now consider the consequences of superlattice formation on the dispersion relation of phonons. For simplicity we consider only one acoustic branch as shown in Fig. 9.20, where the dotted line is the dispersion relation for the three-dimensional homogeneous system in the first Brillouin zone, and the result for a superlattice system is given by solid lines.

Artificially imposing a periodicity l_z in the z-direction (where l_z is necessarily an integer multiple of a or $a/2$ depending on lattice type and orientation) leads to a reduction of the first Brillouin zone in the k_z-direction from $-\pi/a \leqslant k_z \leqslant \pi/a$ to

$$-\frac{\pi}{l_z} \leqslant k_z \leqslant \frac{\pi}{l_z} .$$

All regions which are outside this new, smaller Brillouin zone can be shifted into it by vectors of the new reciprocal lattice

$$G_z = l_3 \frac{2\pi}{l_z} ; \quad l_3 = 0, \pm 1, \pm 2, \ldots \tag{9.60}$$

resulting in the solid curve in Fig. 9.20.

Basically we used the same arguments as in connection with Fig. 9.14. It may happen that some "mini-gaps" open in the dispersion relation at the crossing points in the center and at the boundary of the Brillouin zone. The folding-back of the dispersion curve in the case of a superlattice has consequences for the optical properties of phonons, as will be discussed in Sect. 13.3. For more details of this topic see e.g. [9.10] and references given therein or Chap. 13.

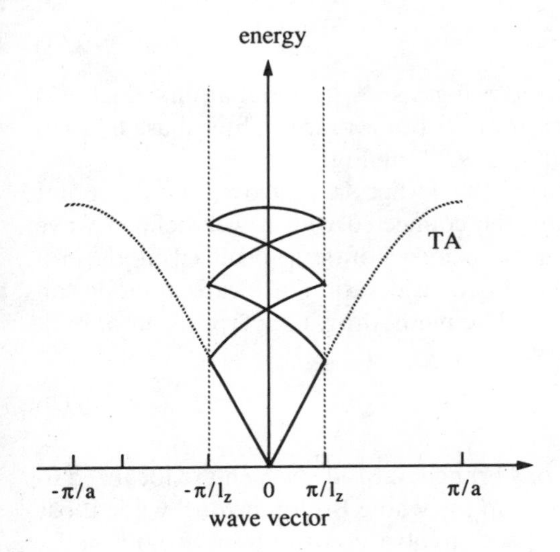

Fig. 9.20. The influence of superlattice formation on the acoustic phonon branch

9.9 Localized Phonon Modes

In our discussion of phonons (and of our model substance in Chaps. 4–6), we assumed until now a perfect arrangement of atoms. In fact, every real semiconductor contains a lot of lattice defects. There are point defects, one-dimensional defects like dislocations, two-dimensional defects like stacking faults and (small angle) grain boundaries, or three-dimensional ones like precipitates. In the present context the most important are the point defects. For an introduction to this topic see [9.1–5]. Point defects include vacancies, interstitital and substitutional atoms. We give some examples for a compound semiconductor *AB* in Fig. 9.21. We come back to this topic when we discuss electron states in semiconductors. What we need here is the fact that a point defect may have a different mass m' and/or spring constant D' as compared to the atom which would be at this place in a perfect lattice.

A consequence of such a point defect is that a localized phonon mode may appear. This is a mode which cannot propagate through the sample with a plane-wave factor as in (9.29 or 9.34). Instead the amplitude has a maximum at the place of the defect and decays exponentially with increasing distance from it. Obvisouly such a mode is localized at or in the vicinity of the point defect.

If the eigenfrequency ω_{loc} of such a localized phonon mode falls into the bands of the intrinsic acoustic or optic modes and couples to them, it will not produce a big effect. Once such a localized mode is excited, it decays rapidly into bulk modes. The situation is different if ω_{loc} falls either in a spectral region where the pure material has no eigenfrequences at all, or couples only weakly to the bulk modes. Then the localized mode can produce, for example, an additional absorption band or Raman satellite. We come to this point later. The situation

$$\omega_{\mathrm{loc}} > \omega_{\mathrm{bulk}}^{\mathrm{LO}} \tag{9.61}$$

can be realized for example by incorporating a substitutional atom, which has approximately the same "spring constant" D but a much lighter mass m' than the atom which it replaces, according to (9.37) or Fig. 9.11.

Since the translation invariance of the lattice is destroyed at the point defect, the eigenfrequency ω_{loc} cannot be connected with a well-defined wave vector. There is, however, a possibility to incorporate a localized mode in a dispersion relation based on the following consideration. A localized mode can be constructed in a Fourier-transform-like method by a superposition of bulk modes with appropriate coefficients:

$$u_{\mathrm{loc}}(\boldsymbol{r}) = \sum_{k,i} a_{k,i} u_{k,i}(\boldsymbol{r}) \, , \tag{9.62}$$

where the index i runs over the various branches. Modes which are localized to one unit cell will need contributions from the whole Brillouin zone, while those which are more extended in real space involve contributions from smaller wave vectors only.

We can now indicate a localized mode in the dispersion relation by a horizontal line covering the region of $\boldsymbol{k}$ values that make substantial contributions to the expansion of (9.62). We show such a situation in Fig. 9.22 where the thickness of the horizontal line is related to the amount of the coefficients $|a_k|$ in (9.62).

It should be noted that every defect is a scattering center for phonons and contributes, together with the anharmonicity mentioned above, to the finite phase relaxation time of phonons.

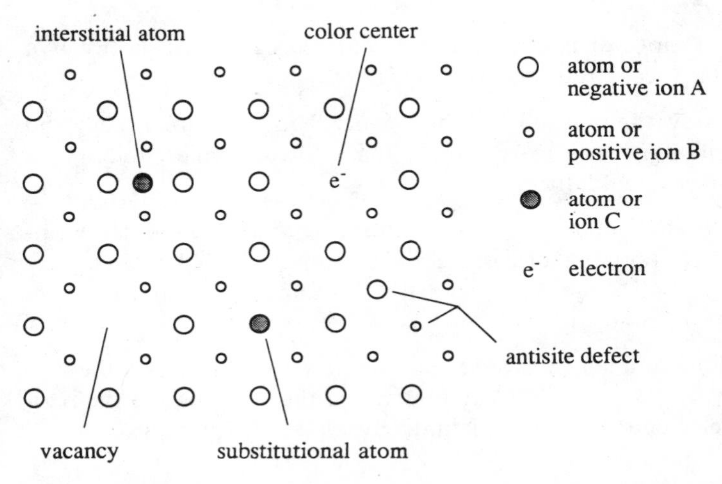

Fig. 9.21. The lattice of a semiconductor AB containing various types of point defects

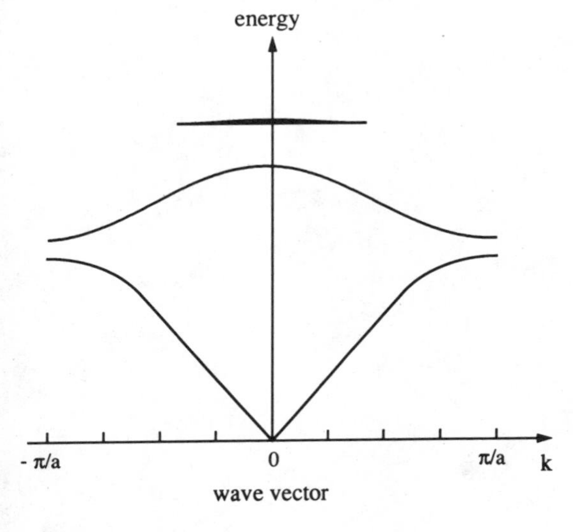

Fig. 9.22. A localized phonon mode represented schematically in a dispersion diagram

To conclude this discussion we would like to stress that even different isotopes act as scattering centers due to their different masses. If, for example, the intrinsic ballistic propagation of phonons is to be investigated, it is desirable to use materials which consist of one isotope only. For more information of this topic see [9.11] and references given therein.

9.10 Problems

1. Inspect or build some lattice models to become familiar with the topics presented in Sect. 9.2.

2. Show that the reciprocal lattice of a face-centered cubic (fcc) lattice is a body centered cubic (bcc) lattice and vice versa.

3. Calculate and draw the unit cell and the Wigner-Seitz cell in real space and the first three Brillouin zones in reciprocal space for a simple cubic and a hexagonal two-dimensional lattice.

4. Study with the data found in [9.8] the transverse and longitudinal eigenfrequencies of optical phonons of a series of semiconductors like ZnO, ZnS, ZnSe, and ZnTe, or of a similar series with the same anion and different cations. What do you conclude?

5. Compare the longitudinal–transverse splitting at $k=0$ for Ge and the corresponding III–V, II–VI and I–VII compounds on the same line of the periodic table of the elements. Can you qualitatively explain the findings?

10 Electrons in a Periodic Crystal Lattice

Let us now return for a moment to (9.1) which describes a crystal. By introducing the phonons we have treated the motion of the atoms. We now assume that the atoms are fixed at their equilibrium positions, i.e., that no phonons are excited. Then the sum over the potentials of all atoms forms a periodic potential for the electrons; but we are still left with a Hamiltonian for about 10^{23} interacting outer electrons per cubic centimeter, which should be properly treated in a many particle formalism. Unfortunately it is extremely difficult to handle this approach. Instead one generally uses a so-called one-electron approximation. The idea is the following: One assumes that the periodically arranged atoms and all interaction potentials between electrons together form a periodic potential $V(\boldsymbol{r})$ with

$$V(\boldsymbol{r}+\boldsymbol{R})=V(\boldsymbol{r})\,. \tag{10.1}$$

Then one calculates the eigenstates for one electron of the corresponding Schrödinger equation and populates these states with electrons according to Fermi–Dirac statistics until all electrons have been accommodated. The potential $V(\boldsymbol{r})$ should ideally be calculated in a self-consistent way. But simpler and more feasible approaches are often used such as a periodic arrangement of screened Coulomb potentials or a "muffin-tin" potential. It is beyond the scope of this book to review the various methods of bandstructure calculations. We just mention that there are basically two approaches to the problem. In one case one starts with free electrons, i.e., plane waves, and introduces a weak periodic potential as a perturbation. These techniques include the nearly-free-electron (NFE), augmented-plane-wave (APW), and orthogonalized-plane-wave (OPW) approaches. In the last two cases, potential terms named pseudo-potentials are introduced to make sure that the calculated wavefunction is orthogonal to the deeper atomic states. The other group of methods begins with the atomic orbitals of the semiconductor atoms, summing up one or more atomic orbitals at every lattice site. These techniques include, for example, the tight-binding approximation and the linear combination of atomic orbitals (LCAO). The first group of methods is more suited for the calculation of conduction-band states, the second for valence-band states. These two terms will be explained below.

For details of bandstructure calculations we refer the reader to the literature [10.1–5], and summarize here just the most important results for the states of an electron in a periodic lattice.

10.1 Bloch's Theorem

Starting from free electrons or from atomic orbitals we find that the electron states in a periodic potential are energetically arranged in energy bands of a certain width, which may be separated by gaps in which no eigenstates exist (Fig. 10.1). The proper eigenstates of a periodic potential are so-called Bloch waves $\phi_{k,i}(\boldsymbol{r})$, with the property

$$\phi_{k,i}(\boldsymbol{r}) = \mathrm{e}^{i\boldsymbol{k}\boldsymbol{r}}\, u_{k,i}(\boldsymbol{r})\,, \tag{10.2}$$

where

$$u_{k,i}(\boldsymbol{r}) = u_{k,i}(\boldsymbol{r}+\boldsymbol{R})\,. \tag{10.3}$$

The $\phi_{k,i}$ are evidently a product of a plane wave and a lattice periodic term $u_{k,i}$ (10.3), where $\boldsymbol{k}$ is the wave vector and i the index of the band, as shown in Fig. 10.2. The eigenenergies in the bands depend both on $\boldsymbol{k}$ and i and are periodic in $\boldsymbol{k}$-space, i.e.,

$$E(\boldsymbol{k}, i) = E(\boldsymbol{k}+\boldsymbol{G}, i)\,. \tag{10.4}$$

In a similar way one finds that

$$\phi_{k,i}(\boldsymbol{r}) = \phi_{k+G,i}(\boldsymbol{r})\,. \tag{10.5}$$

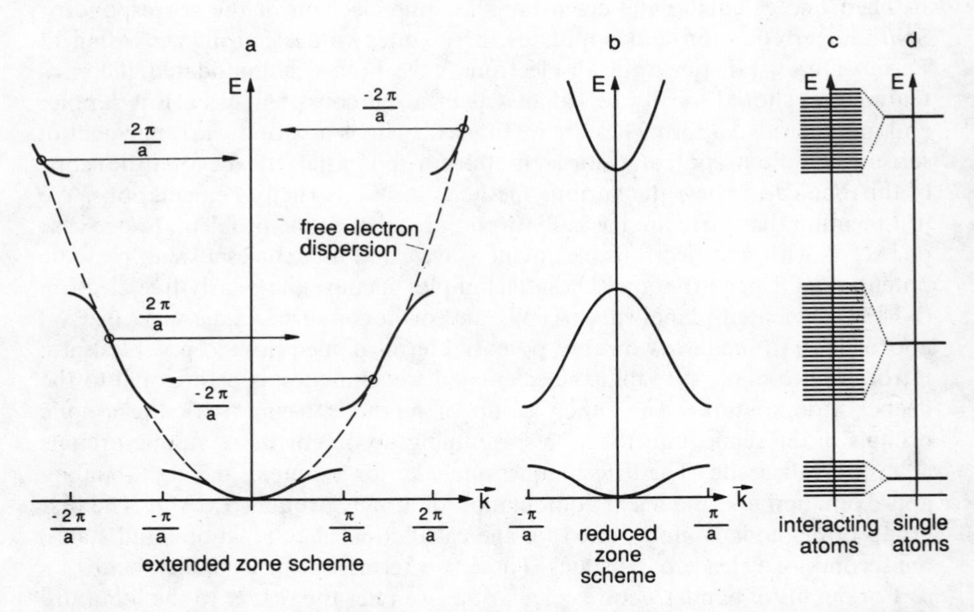

Fig. 10.1a–d. The reduced-zone scheme (**b**) of electronic energy bands in a crystalline solid reached either starting from (nearly) free electrons (**a**) or from atomic orbitals (**c, d**). According to [10.6]

This result once more allows the dispersion relation $E(\boldsymbol{k})$ to be reduced to the first Brillouin zone (Fig. 10.1).

We now want to elucidate the above statements. If we start with free electrons we have the simple, non-relativistic dispersion relation of Fig. 10.1a.

$$E(\boldsymbol{k}) = \frac{\hbar^2 k^2}{2m_0} . \tag{10.6}$$

Introducing a weak periodic potential opens a gap in the dispersion at the boundaries of the Brillouin zones. At the boundary of the first Brillouin zone, the wavelength $\lambda = 2\pi k^{-1}$ matches the lattice constant. We now superimpose two energetically degenerate states, the wave vectors $\boldsymbol{k}_i$ and $\boldsymbol{k}_i'$ of which differ by a certain reciprocal lattice vector $\boldsymbol{G}'$, e.g., $|\boldsymbol{G}'| = 2\pi/a$ in a simple cubic lattice, i.e.,

$$k_i = \frac{\pi}{a}; \quad k_i' = \frac{\pi}{a} - \frac{2\pi}{a} = \frac{\pi}{a} - G_i' = -\frac{\pi}{a}; \quad i = x, y, z \tag{10.7}$$

and we get with (10.4) at the border of the Brillouin zone standing waves with vanishing group velocity, i.e.,

$$v_g = \frac{1}{\hbar} \frac{\partial E}{\partial k_{k_i = \pm\pi/a}} = 0 . \tag{10.8}$$

The appearance of the gap becomes clear with Fig. 10.3, which shows two wavefunctions which have the same wavelength, i.e., the same magnitude of $\boldsymbol{k}$, but different potential energies. One solution has its maxima of the probability $|\phi|^2$ at the potential minima and the other at the maxima. Exactly this difference in potential energy is what causes the gap. Adding suitable vectors $\boldsymbol{G}$ of the reciprocal lattice allows either the periodic continuation of the branches in the extended zone scheme, according to (10.4), or reduction to the first Brillouin zone, as shown in Figs. 10.1b. In the following we shall use this reduced zone scheme unless otherwise stated.

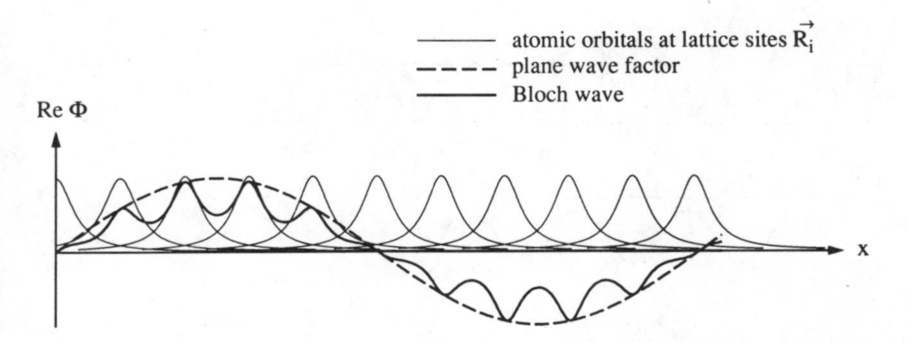

Fig. 10.2. Schematic drawing of the real (or imaginary) part of a Bloch wave in one dimension

If we start from atomic orbitals (Fig. 10.1d) and move the atoms together to form a solid, then the eigenenergies of the electrons split and broaden into a band due to the mutual interaction (i.e., overlap of the wavefunctions) similar to classical coupled oscillators (Fig. 10.1c). Compare also the discussion with Fig 4.2. The width of the band B is directly proportional to the overlap integral J between orbitals of neighboring atoms

$$B = 2Jz \,, \tag{10.9}$$

where z is the coordination number. The periodicity of the arrangement of atoms imposes $\boldsymbol{k}$ as a good quantum number, thus giving the transition from Figs. 10.1c to b.

To elucidate Bloch's theorem further we outline another approach based on symmetry considerations. We assume that $\phi(\boldsymbol{r})$ is a non-degenerate solution of the one-electron periodic Hamiltonian with the potential given in (10.1) and with eigenenergy E, and introduce an operator $T(\boldsymbol{R})$ which shifts the crystal by a lattice vector $\boldsymbol{R}$. Since H is periodic such a translation does not change the problem at all, i.e., H and $T(\boldsymbol{R})$ commute:

$$T(\boldsymbol{R})\,H = H\,T(\boldsymbol{R}) \,. \tag{10.10}$$

This means that there are solutions which are eigenfunctions of both H and $T(\boldsymbol{R})$. These are just the Bloch waves of (10.2, 3).

We continue with (10.10) and get

$$T(\boldsymbol{R})H\phi(\boldsymbol{r}) = HT(\boldsymbol{R})\phi(\boldsymbol{r}) = T(\boldsymbol{R})E\phi(\boldsymbol{r}) = ET(\boldsymbol{R})\phi(\boldsymbol{r}) \,\cdot \tag{10.11}$$

Equation (10.11) says that along with $\phi(\boldsymbol{r})$ the function $T(\boldsymbol{R})\phi(\boldsymbol{r})$ is also an eigenstate with the same energy. Since $\phi(\boldsymbol{r})$ was assumed to be non-degenerate, the two solutions can differ only by a phase factor of unit modulus and the rest must be periodic.

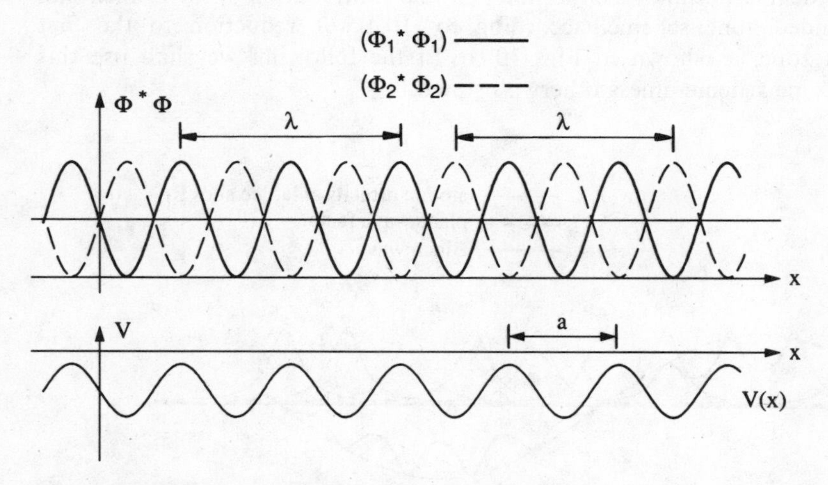

Fig. 10.3. Schematic drawing of two Bloch waves with equal $\boldsymbol{k}$-vector but different potential energies at the boundary of the first Brillouin zone over a periodic potential $v(x)$

We call this periodic part therefore $u(\boldsymbol{r})$ and obtain:

$$T(\boldsymbol{R})\phi(\boldsymbol{r}) = \mathrm{e}^{\mathrm{i}f(\boldsymbol{R})}u(\boldsymbol{r}) \,. \tag{10.12}$$

Since

$$T(\boldsymbol{R}_1)T(\boldsymbol{R}_2) = T(\boldsymbol{R}_1 + \boldsymbol{R}_2) \tag{10.13}$$

holds we can conclude that the function $f(\boldsymbol{R})$ must be linear in $\boldsymbol{R}$. The prefactor in front of $\boldsymbol{R}$ necessarily has the dimension $(\text{length})^{-1}$ and we identify it with $\boldsymbol{k}$, thus reproducing (10.2).

To conclude this section we want to mention that the bands are usually represented in one of two different ways. The first is the dispersion relation (Figs. 10.1b, 10.4a). In the other case one plots the width of the bands and of the gaps between them as a function of the space coordinate $\boldsymbol{r}$ (Fi.g 10.4b). The latter is especially useful to demonstrate spatial inhomogeneities or localized states. We shall meet some examples of both cases later.

10.2 Metals, Semiconductors, Insulators

Having obtained a first insight into the electronic bandstructure in the above section, we now want to make the second step and put all electrons into the bands using, of course, Fermi–Dirac statistics. We further assume zero temperature.

Some of the energetically lower-lying bands will be completely filled. We call bands which are completely filled at $T = 0\,\mathrm{K}$ "valence bands", while all partly filled or empty bands are "conduction bands".

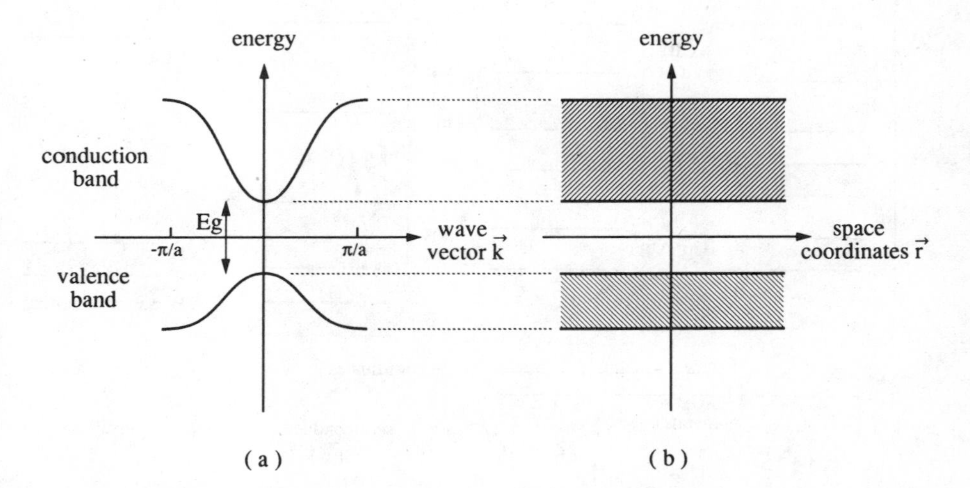

Fig. 10.4a, b. The representation of bands in $\boldsymbol{k}$-space (**a**) and in real space (**b**)

The important region is that around the highest valence and the lowest conduction bands.

If the filling procedure ends in such a way that, at $T = 0$K there are one or more partly-filled conduction bands (Figs. 10.5a, b), we have a metal. This situation arises for example if the atomic orbital which forms the band is itself only partly occupied (e.g., the outer *s*-level of the alkali metals Li, Na...) (Fig. 10.5a) or if a completely filled orbital forms a band which overlaps with a band stemming from an empty atomic orbital, as is the case in the rare earth metals (Ca, Mg...) in Fig. 10.5b. If, on the other hand, the filling procedure gives one or more completely-filled valence bands which are separated by a gap E_g from completely empty conduction bands, we have a semiconductor for

$$0 < E_g \leqslant 4\,\text{eV} \tag{10.14a}$$

and an insulator for

$$E_g \geqslant 4\,\text{eV}\,. \tag{10.14b}$$

The "boarder line" of 4eV is set by convention and is not sharp.

In Table 10.1 we give rough values of E_g for various semiconductors of the group-IV elements, the III–V the II^b–VI and the I^b–VII compounds. For more details see, e.g. [10.7].

Semiconductors with a gap below approximately 0.5 eV, such as InSb, $Cd_{1-y}Hg_yTe$ or the lead salts PbS, PbSe and PbTe, are called narrow-gap semiconductors. If E_g is close to zero, one speaks of semimetals. On the other hand, materials with a gap between approximately 2eV and 4eV are called

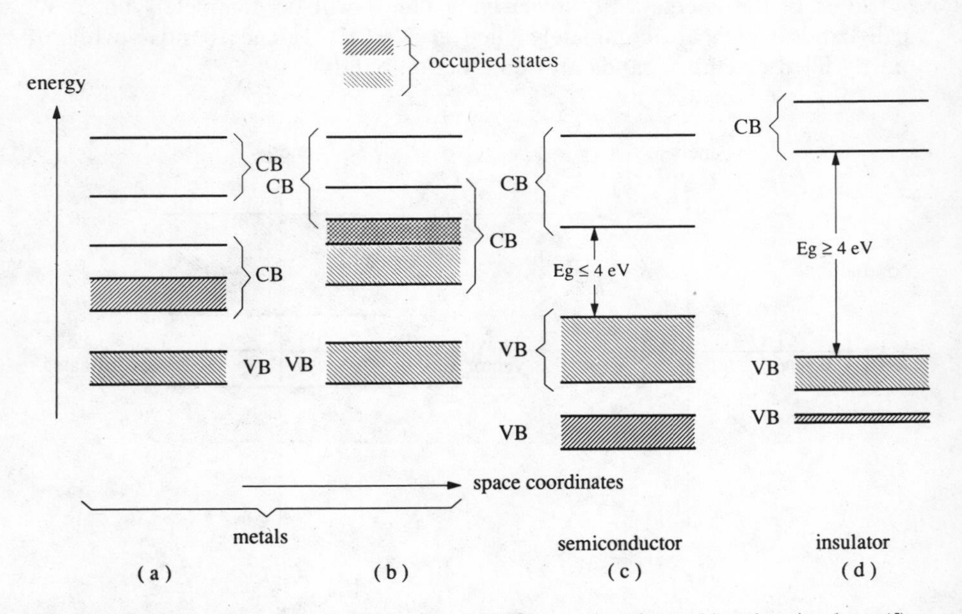

Fig. 10.5a–d. Occupation of the bands for a metal (**a, b**) a semiconductor (**c**) and an insulator (**d**)

Table 10.1. Values of the band gap energy of some group IV, III–V, II–VI and I–VII semiconductors (SC) at low temperature ($T \lesssim 10\,K$), Sy: Symmetry (point group). According to [10.7]

Group IV				III–V				II–VI				I–VII			
SC	Sy	E_g(eV)	dir/indir	SC	Sy	E_g(eV)	dir/indir	SC	Sy	E_g(eV)	dir/indir	SC	Sy	E_g(eV)	dir/indir
C	O_h	5.48	i	AlN	C_{6v}	6.28	d	ZnO	C_{6v}	3.437	d	CuCl	T_d	3.395	d
S_i	O_h	1.17	i	AlP	T_d	2.53	i	ZnS	C_{6v}	3.91	d	CuBr	T_d	3.077	d
Ge	O_h	0.744	i	AlAs	T_d	2.228	i	ZnS	T_d	3.78	d	CuI	T_d	3.115	d
grey Sn	O_h	0	semimetal	AlSb	T_d	1.696	i	ZnSe	T_d	2.82	d	AgCl	T_d	3.249	i
				GaN	C_{6v}	3.503	d	ZnTe	T_d	2.391	d	AgBr	T_d	2.684	i
				GaP	T_d	2.350	i	CdS	C_{6v}	2.583	d	AgI	C_{6v}	3.024	d
				GaAs	T_d	1.518	d	CdSe	C_{6v}	1.841	d				
				GaSb	T_d	0.812	d	CdTe	T_d	1.60	d				

wide-gap semiconductors, for example, CdS, ZnSe, GaN or CuCl. There is obviously a tendency for the gap to decrease with increasing atomic number Z within the rows of the periodic table. Grey tin or HgTe, for example, are semimetals.

Often it is possible to produce solid solutions or alloys between the semiconductors within one row like $Al_{1-y}Ga_y$ As or $CdS_{1-x}Se_x$ for all values of x and y between 0 and 1.

The bandstructure can be basically understood in the following way. In many elemental semiconductors, binding is covalent and arises due to sp^3 hybridization of the uppermost occupied s and p levels. The bonding and antibonding states form the upper valence and the lower conduction bands, respectively. The more ionic semiconductors like the II–VI or I–VII compounds form valence bands from the filled noble-gas shells of the anions, e.g., the 3p levels of S in CdS or the 3p and 3d levels of Cl^- in CuCl, while the conduction band arises from the empty levels of the cations, e.g., the 5s levels of Cd^{2+} in CdS or the 4s levels in Cu^+ in CuCl. Actually there is a continuous transition from covalent to ionic binding when going from the group IV via the III–V and II–VI to the I–VII semiconductors. Additionally, there is a trend of decreasing ionicity with increasing Z within the rows.

10.3 Electrons and Holes in Crystals as New Quasiparticles

As we shall see later, the optical properties of the electronic system of semiconductors are largely determined by transitions of electrons between the upper valence bands and the lower conduction bands.

The bandstructure as presented until now, i.e., in connection with Figs. 10.1–5, describes the so-called $N \pm 1$ particle problem in the following sense: if we consider a semiconductor with a completely filled valence band containing N electrons per cm^3

$$N \simeq 10^{22}\text{–}10^{23}\,\mathrm{cm}^{-3} \tag{10.15}$$

and a completely empty conduction band and add one more electron, we find that this electron can be placed into exactly the conduction band states. If we remove one of the N electrons and ask from which state it came, we find the valence band states.

An obvious step now is to consider the one or few electrons in an otherwise empty conduction band (CB). For an almost filled valence band, however, it is easier to consider the few empty states and their properties instead of the many occupied ones. This idea leads to the concept of "defect-electrons" or "holes". The properties of the hole are connected in the following way (Table 10.2) with the properties of the electron that has been removed from the valence band (VB).

From Table 10.2 we see that the hole has a positive charge and that its wave vector and spin are opposite to those of the electron removed from the

Table 10.2. Properties of a hole in the valence band compared to the properties of the electron that has been removed from the valence band to create the hole

Property	Hole	Removed electron
electric charge q	q_h	$= -q_{re} \approx -1.6 \times 10^{-19} As$
wave vector	$\boldsymbol{k}_h$	$= -\boldsymbol{k}_{re}$
spin	σ_h	$= -\sigma_{re}$
eff. mass	$m_h > 0$	$= -m_{re} < 0$

valence band. The two latter statements are easy to understand. A semiconductor with a completely filled valence band has total momentum and spin equal to zero. If we take one particle out, the remainder acquires for the above quantities values exactly opposite to those of the removed particle. For clarity Figs. 10.6a, b show the bandstructure containing one electron in the conduction band and one hole in the valence band, respectively. The states are equidistant in $\boldsymbol{k}$ [see (2.67)] but we should note that there are usually 10^{22}– 10^{23} states in each band per cm^{-3} and not only the few shown in Fig. 10.6.

The electrons and holes in a semiconductor crystal are quasi-particles. They can exist only in the crystal and not in vacuum, in contrast to normal electrons and positrons with which they have a lot in common, except the magnitude of

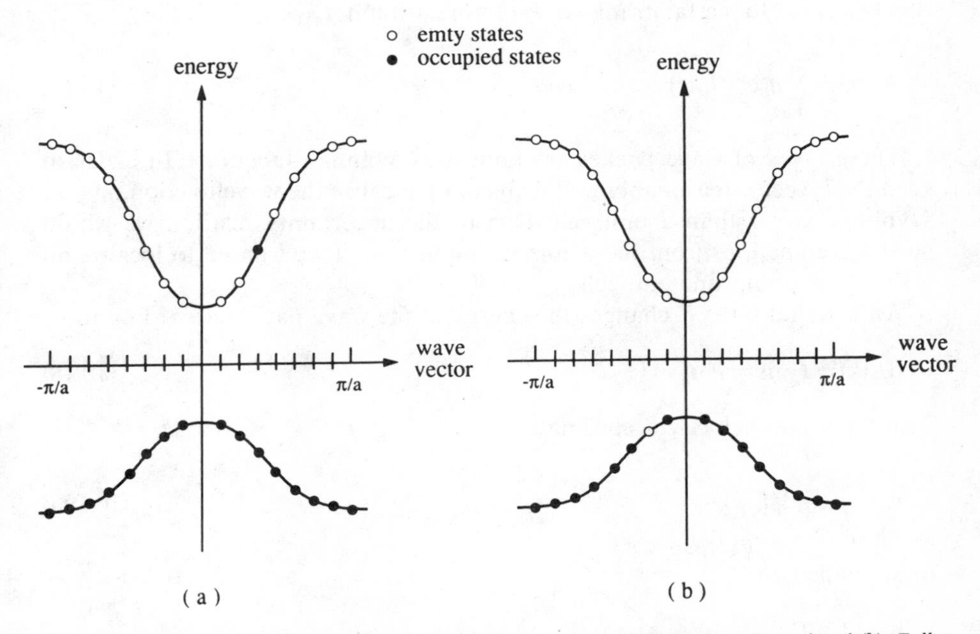

Fig. 10.6a, b. One electron in the conduction band (**a**) and one hole in the valence band (**b**). *Full circles*: occupied states; *open circles*: empty states

the energy gap which is $\simeq 1$ MeV for normal electrons and positrons i.e., twice the rest mass of $511\,\mathrm{keV} = m_0c^2$. The dispersion relations of electrons and holes are different from those of free electrons and positrons which for the non-relativistic case are given by

$$E_{\mathrm{e,P}} = \pm\left(m_0c^2 + \frac{\hbar^2k^2}{2m_0}\right), \tag{10.16}$$

where m_0 is the free electron mass.

The quantity $\hbar\boldsymbol{k}_{\mathrm{e,h}}$ of crystal electron and hole is a quasi-momentum, since it is conserved only modulo reciprocal lattice vectors – see (10.4, 5) – and since the Bloch waves of (10.2, 3) are not proper eigenstates of the momentum operator $\frac{\hbar}{\mathrm{i}}\nabla$.

One should note that the energy of a hole increases if it is brought deeper into the valence band.

10.4 The Effective-Mass Concept

If we want to describe the motion of an electron or hole in a semiconductor under the influence of an external field (e.g., an electric or magnetic field) it is intuitively clear that we ought to consider a wave packet rather than the infinitely extended Bloch waves. To describe such a wave packet we superpose Bloch waves of a certain range of $\boldsymbol{k}$-vectors around a $\boldsymbol{k}_0$

$$\phi_{\boldsymbol{k}_0}(\boldsymbol{r}) = \sum_{\boldsymbol{k}} a_{\boldsymbol{k}} \mathrm{e}^{\mathrm{i}\boldsymbol{k}\cdot\boldsymbol{r}} u_{\boldsymbol{k}}(\boldsymbol{r})\,. \tag{10.17}$$

These types of wave packet are known as Wannier-functions. In order to keep the $\boldsymbol{k}$-vector reasonably well defined, we localize the wavefunction only to a volume larger than a unit cell. Due to the uncertainty relation we would need wavefunctions from the whole Brillouin zone if we wanted to localize an electron to within one unit cell.

An external force $\boldsymbol{F}$ changes the energy of the wave packet according to

$$\mathrm{d}E(\boldsymbol{k}) = \boldsymbol{F}\cdot\mathrm{d}\boldsymbol{s} = \boldsymbol{F}\cdot\boldsymbol{v}_{\mathrm{g}}\mathrm{d}t \tag{10.18}$$

with the group velocity according to (2.13)

$$\boldsymbol{v}_{\mathrm{g}} = \frac{1}{\hbar}\,\mathrm{grad}_{\boldsymbol{k}} E(\boldsymbol{k}) \tag{10.19a}$$

or simplified

$$v_{\mathrm{g}} = \frac{1}{\hbar}\frac{\mathrm{d}E}{\mathrm{d}k} \tag{10.19b}$$

Since, on the other hand, we have

$$\mathrm{d}E(\boldsymbol{k}) = \frac{\mathrm{d}E(\boldsymbol{k})}{\mathrm{d}\boldsymbol{k}} \cdot \mathrm{d}\boldsymbol{k} = \hbar \boldsymbol{v}_{\mathrm{g}} \mathrm{d}\boldsymbol{k} \, , \tag{10.19c}$$

we find from combining (10.18, 10.19c)

$$\hbar \frac{\mathrm{d}\boldsymbol{k}}{\mathrm{d}t} = \boldsymbol{F} = \dot{\boldsymbol{p}} \, . \tag{10.19d}$$

This expression corresponds to Newton's law of motion, but now for the quasi-momenta $\hbar \boldsymbol{k}$ of the crystal electrons or holes.

From (10.19b, d) we get for the acceleration $\boldsymbol{a}$ of the wave packet

$$\boldsymbol{a} = \frac{\mathrm{d}\boldsymbol{v}_{\mathrm{g}}}{\mathrm{d}t} = \frac{1}{\hbar} \frac{\partial^2 E}{\partial k \partial t} = \frac{1}{\hbar} \frac{\partial^2 E}{\partial k^2} \frac{\mathrm{d}\boldsymbol{k}}{\mathrm{d}t} = \frac{1}{\hbar^2} \frac{\partial^2 E}{\partial k^2} \boldsymbol{F} \, . \tag{10.19e}$$

Comparing with the trivial form of (10.19d)

$$\boldsymbol{a} = \frac{1}{m} \boldsymbol{F} \tag{10.19f}$$

we find that the crystal electron and hole move under the influence of external fields through the crystal like a particle, however, with an effective mass given by

$$\frac{1}{m_{\mathrm{eff}}} = \frac{1}{\hbar^2} \frac{\partial^2 E}{\partial k^2} = \frac{1}{\hbar^2} \frac{\partial^2 E}{\partial k_i \partial k_j} \, ; \quad i, j = x, y, z \, . \tag{10.20}$$

For free electrons and positrons (10.20) leads to $m_{\mathrm{eff}} = m_0$.

The right-hand side of (10.20) shows that the effective mass is actually a tensor and can depend on the direction in which the electron or hole moves.

The important point now is that the bands of semiconductors tend to be parabolic in the vicinity of the band extrema, as shown schematically in Fig. 10.7 or in the real band structures discussed below with Figs. 10.9–12. These extrema are most important for the optical and transport properties. The effective masses are constant in these regions. This leads to the so-called effective-mass approximation. Electrons and holes in a semiconductor are simply treated as free particles, but with an effective mass given by (10.20).

We should mention that the mass of an electron is positive if the curvature of the band is positive. Due to the change of the sign of the properties of holes compared to those of the missing electron in the valence band (Table 10.2) the mass of the hole is positive at the maximum where the curvature is actually negative.

The explanation behind the effective masses of electrons and holes is basically the following. The forces which act on an electron are the ones from the other ions and electrons in the crystal and the externally applied ones. For simplicity, we condensed the first of these forces to yield the periodic potential of (10.1) using the one-electron approximation. Then we chose to consider only the external forces acting on the electrons and the price we have to pay is the fact that the electrons and holes react with an effective mass. This mass is fortunately, in some regions at least, constant, allowing the effective-mass approximation mentioned above, but it can change as a function of $\boldsymbol{k}$ and become negative or infinite, as can be seen in Fig. 10.7.

It is important to note that an increasing curvature of a band is necessarily connected with an increasing width of the band. Therefore we find the qualitative relation:

electron easy to move and accelerate $\longleftrightarrow$ low effective mass
$\longleftrightarrow$ large curvature of the band $\longleftrightarrow$ large band width
$\longleftrightarrow$ strong coupling between adjacent atoms and vice versa. (10.21)

Thus we have here another example of the more general discussion given in the introduction to Chap. 6.

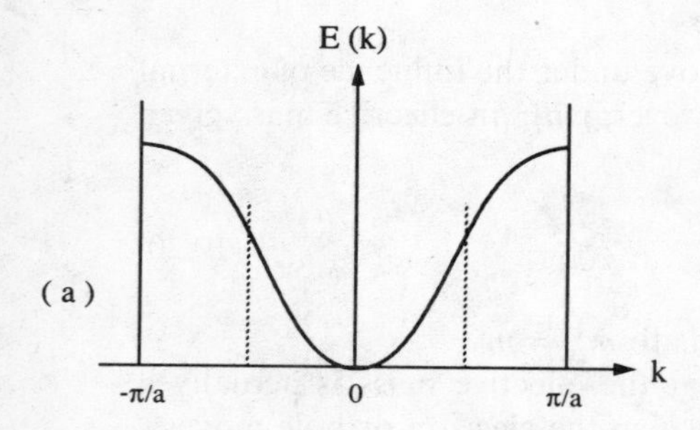

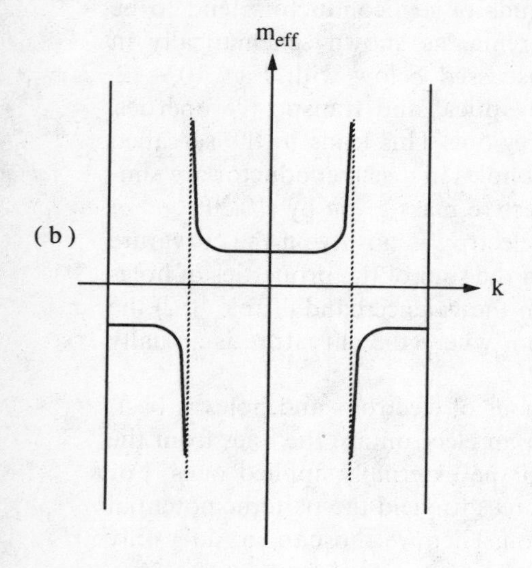

Fig. 10.7 a, b. Schematic dispersion (**a**) of a conduction band and the resulting effective mass (**b**). After [10.8]

In the rest of this book we shall use the effective-mass approximation if not stated otherwise, including the polaron correction discussed in the next section.

We will in future omit the prefix "crystal" when talking about electrons and holes in semiconductors and we indicate the effective masses by indices e and h:

$$m_e = \text{effective mass of electron} ,$$
$$m_h = \text{effective mass of hole} , \tag{10.22}$$

sometimes with an additional index to distinguish different bands.

10.5 The Polaron Concept and Other Electron–Phonon Interaction Processes

Before proceeding to the band structures of real semiconductors, we shall discuss here various aspects of electron–phonon interaction.

If we introduce an electron or a hole into a semiconductor which has at least partially ionic binding, the additional charge carrier will polarize the lattice provided we relax the assumption that the ions are "fixed" at their lattice sites. An electron will attract atoms with a positive charge and repel those with a negative one (Fig. 10.8). For holes, the situation is just the opposite. We can describe this lattice distortion as a superposition of preferentially longitudinal optical phonons, i.e., a free carrier is accompanied by a "phonon cloud". The entity of charge carrier plus phonon cloud is called a polaron. In semiconductors, the radius of the phonon cloud is larger than the lattice constant, resulting in a so-called "large polaron" in contrast to the small polaron which occurs in ionic insulators and may lead to self-localization within a unit cell. The effective mass $m_{e,h}$ of a polaron is greater than that of an

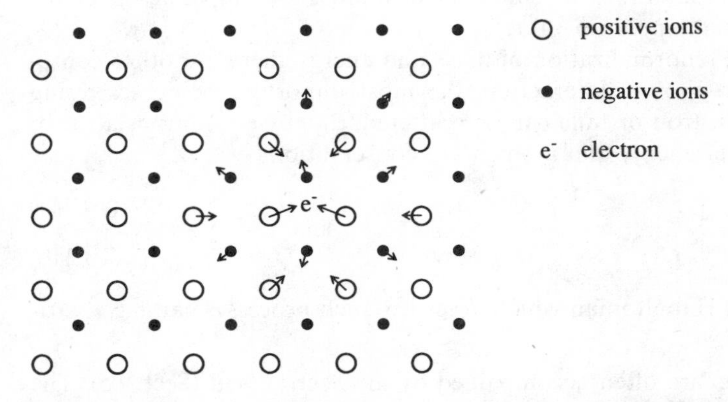

Fig. 10.8. The lattice distortion around a carrier in a (partly) ionic semiconductor illustrating the polaron concept

electron in a rigid lattice $m_{e,h}^{r.l.}$. See e.g. [10.1–3,9].

$$m_{e,h} \simeq m_{e,h}^{r.l.}\left(1 + \frac{\alpha_{e,h}}{6}\right), \tag{10.23a}$$

where the index r.l. stands for "rigid lattice" and α is a dimensionless quantity which describes the Fröhlich coupling of carriers to the LO phonons. One finds

$$\alpha_{e,h} = \frac{e^2}{2\hbar\omega_{LO}}\left(\frac{2m_{e,h}^{r.l.}\omega_{LO}}{\hbar}\right)^{1/2}\left(\frac{1}{\varepsilon_b} - \frac{1}{\varepsilon_s}\right), \tag{10.23b}$$

where ε_s and ε_b are the low-and high-frequency values of the dielectric function below and above the optical phonon resonances, respectively; see also Chaps. 4, 13.

Obviously semiconductors without an ionic bonding contribution have $\varepsilon_b = \varepsilon_s$ [see (4.26, 28)] and thus $\alpha = 0$. For all normal semiconductors one finds

$$\alpha \leqslant 1 \, . \tag{10.23c}$$

Additionally the lattice relaxation leads to a decrease of the width of the gap by an amount ΔE_g with contributions for electrons and holes

$$\Delta E_g^{e,h} = \alpha_{e,h}\hbar\omega_{LO} \, . \tag{10.23d}$$

The radius of the phonon cloud in the polaron r_p is finally given by

$$r_p^{e,h} = (\hbar/2m_{e,h}^{r.l.}\omega_{LO})^{1/2} \, . \tag{10.23e}$$

Basically it is very difficult to "fix the atoms at their lattice sites". Therefore all common experimental techniques to determine the effective masses of electrons and holes, e.g., by cyclotron resonance, or the value of the gap, e.g., by optical spectroscopy, will give polaron values. We therefore continue to use the effective mass approximation and simply bear in mind that all values given for $m_{e,h}$ or E_g are actually polaron values.

Apart from the renormalization of mass and energy there are other consequences of carrier–phonon interaction, the most important being scattering phenomena. An electron or hole can be scattered, e.g., by emitting or absorbing a phonon under energy and momentum conservation.

$$E_e^i = E_e^f \pm \hbar\omega_{Phonon} \, , \tag{10.24a}$$

$$\boldsymbol{k}_e^i = \boldsymbol{k}_e^f \pm \boldsymbol{k}_{Phonon}(+\boldsymbol{G}) \, . \tag{10.24b}$$

The interaction Hamiltonian which describes such processes can have various origins:

- Optical phonons are often accompanied by an electric field (Sect. 9.5). The interaction of carriers with the electric field of preferentially longitudinal optical phonons is known as the Fröhlich interaction.

- Since the width of the gap depends on the lattice constant and on the arrangement of the atoms in the basis, a change of these quantities will influence the bandstructure. On the other hand, a phonon can be considered as a periodic deformation of the arrangement of atoms, and the carriers "feel" the resulting modulation of the bands. The resulting interaction between carriers and phonons is called deformation-potential scattering. The deformation potential scattering occurs for both acoustic and optical phonons.
- Finally, it is known that many non-centrosymmetric crystals show the piezo-electric effect, i.e., the appearance of an electric field as a consequence of strain, i.e., of lattice distortion. Again we can consider an acoustic (or optical) phonon as a periodic modulation of the lattice parameters, which produces, via the piezo-electric effect, a varying electric field which interacts with the electrons and holes. This effect is the so-called piezo (acoustic) coupling.

More details about the polaron concept and on carrier–phonon coupling can be found in [10.9] and references therein.

In the so-called semimagnetic semiconductors, usually II–VI compounds in which the cations are partly replaced by Mn or Fe ions, one finds so-called magnetic polarons, i.e., a spin alignment of the paramagnetic ions in the vicinity of a carrier. For details see [10.10].

10.6 Bandstructures of Real Semiconductors

In this chapter we present bandstructures of real semiconductors.

As already mentioned, in the case of ionic binding, the upper valence bands frequently arise from the highest occupied atomic p-levels of the anions with a more-or-less pronounced mixture of d-levels, or from the bonding state of the sp^3 hybrid orbitals for covalent binding. The lowest conduction bands come from the lowest empty s-levels of the cations or the antibonding sp^3 hybrid, respectively. In Fig. 10.9 we give the essentials of the bandstructure for cubic semiconductors like diamond, Si or Ge (point group O_h) or the III–V, II–VI and I–VII compounds crystallizing in zinc-blende (T_d) structure.

The valence band has its maximum at the Γ point, i.e., at $k=0$. It is six-fold degenerate including spin corresponding to the parent p orbitals. It splits due to spin–orbit coupling at $k=0$ into a two-fold degenerate band (symmetry Γ_7^+ in O_h or Γ_7 in T_d) and a four-fold degenerate band (symmetry Γ_8^+ in O_h or Γ_8 in T_d). Usually the $\Gamma_8^{(+)}$ band is the upper and the $\Gamma_7^{(+)}$ the lower one. The spin–orbit splitting increases in atoms with increasing charge Z of the nucleus and this also applies in the semiconductor. In CuCl the ordering of the valence bands is inverted due to the influence of close-lying d levels.

The $\Gamma_8^{(+)}$ valence band splits for $k \neq 0$ into two bands which have different curvature and are therefore known as heavy- and light-hole bands. All bands

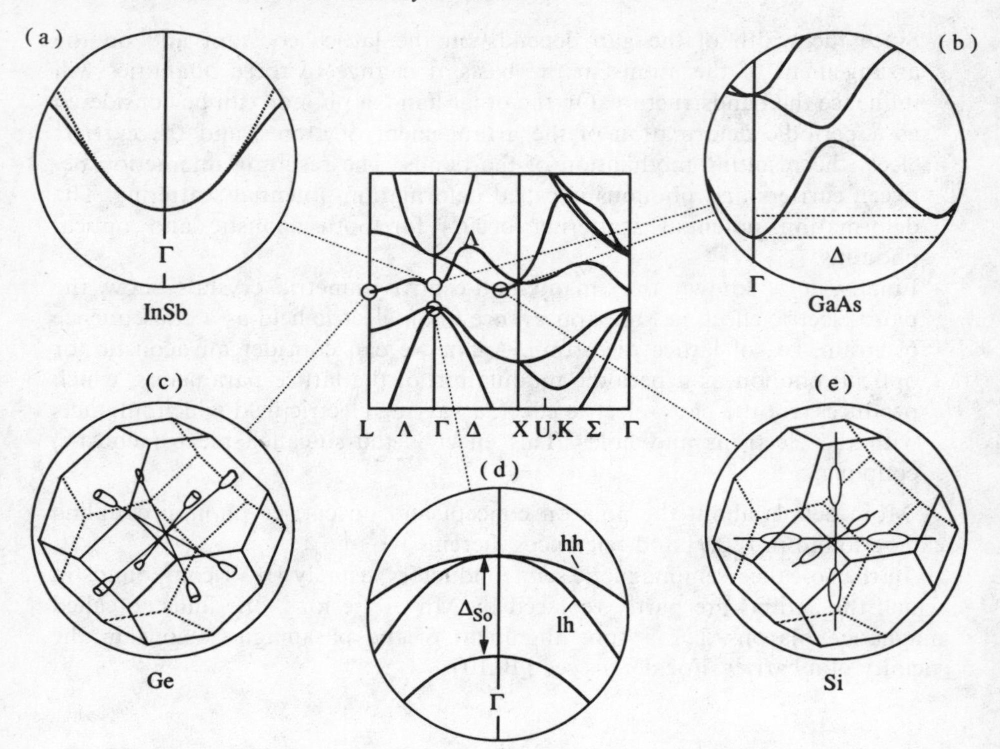

Fig. 10.9a–e. An overview of the band structures of some cubic semiconductors. From [10.11]

have cubic symmetry, which is lower than spherical symmetry. As a consequence the dispersion and thus the hole masses depend on the direction of $\boldsymbol{k}$. This phenomenon is known as band-warping. Often the Γ_8 valence band dispersion is described by the so-called Luttinger parameters γ_1, γ_2 and γ_3 according to

$$E_{1,2} = E_0 + \frac{\hbar^2}{2m_0}\Big[\gamma_1 k^2 \pm [\gamma_2(k_x^4 + k_y^4 + k_z^4) + 3(\gamma_3^2 - \gamma_2^2)(k_x^2 k_y^2 + k_y^2 k_z^2 + k_z^2 k_x^2)]^{1/2}\Big] \quad (10.25)$$

where γ_1^{-1} describes the average effective mass and γ_2 and γ_3 the splitting into heavy- and light-hole bands and the warping.

The conduction band has a minimum at the Γ point and other minima in the direction Δ close to the X points and at the L points.

If the minimum of the Γ point is the deepest one, the semiconductor is said to have a "direct gap" since transitions between the global maximum of the valence band and the global minimum of the conduction band are directly possible with photons, having $k_{\text{photon}} \simeq 0$. In other cases, the semiconductor is called "indirect" since a momentum-conserving phonon is involved in the transitions between the band extrema.

Examples of indirect semiconductors are diamond, Si and Ge, some of the III–V compounds such as AlAs and GaP, and some of the I–VII compounds like AgBr. Direct gap semiconductors include some of the III–V compounds like GaAs or InP, the II–VI compounds ZnS, ZnSe, ZnTe and CdTe, and I–VII materials like CuCl, CuBr and CuI. See also table 10.1.

The conduction band minimum at the Γ point is usually, to a very good approximation, isotropic and parabolic. Only some narrow-gap materials like InSb show significant nonparabolicities in the vicinity of $k=0$ (Fig. 10.9).

The minima of the L points of Ge or in the Δ direction of Si are parabolic but highly anisotropic (Fig. 10.9). The dispersion relation around the minimum at $\boldsymbol{k}_0$ can consequently be expressed by

$$E(\boldsymbol{k}) = E_\mathrm{g} + \frac{\hbar^2}{2}\left(\frac{(k_x - k_{0x})^2}{m_\mathrm{l}} + \frac{(k_y - k_{0y})^2}{m_\mathrm{t}} + \frac{(k_z - k_{0z})^2}{m_\mathrm{t}}\right). \tag{10.26}$$

Here m_l is the effective mass for k-components in the direction from Γ to $\boldsymbol{k}_0$ and m_t that for the two directions perpendicular to it.

In GaAs the minimum at the Γ point is the deepest, but other minima at different points of the Brillouin zone are close in energy giving rise to the Gunn effect, which arises from the transfer of electrons under the influence of a strong electric field from the minimum at $k=0$ with low effective mass to side minima with higher effective masses.

The direct gap semiconductors are also called "single-valley", and the indirect ones "multi-valley" semiconductors because they have several (6 for Si and $8 \cdot 1/2$ for Ge) equivalent conduction-band minima.

Semiconductors with hexagonal wurtzite structure (point group C_{6v}) are usually "direct". They are found preferentially among the II–VI compounds such as ZnO, ZnS, CdS and CdSe, but also among the III–V materials like GaN.

The valence band of the C_{6v} semiconductors is split by spin–orbit coupling and by the hexagonal crystal field into three subbands which are usually labelled from higher to lower energies as A, B and C bands with symmetries Γ_9, Γ_7 and Γ_7 (Fig. 10.10). In ZnO, the symmetries of the two upper bands are inverted as in the case of CuCl. The effective masses of the valence bands are strongly anisotropic in these compounds, $m_\perp$ usually being smaller than $m_\parallel$ where the indices refer to $\boldsymbol{k}$-vectors perpendicular and parallel to the polar crystallographic $\boldsymbol{c}$-axis:

$$m_\perp^\mathrm{vb} \ll m_\parallel^\mathrm{vb} \quad \text{for} \quad C_{6v}\,, \tag{10.27a}$$

$$m_\mathrm{DOS} = (m_\perp^2 m_\parallel)^{1/3}\,. \tag{10.27b}$$

The effective mass that enters in the calculation of the density of states m_DOS is given in (10.27b). A pecularity of the states of Γ_7 symmetry is that they can have a term linear in k in the dispersion relation for $\boldsymbol{k} \perp \boldsymbol{c}$ as shown in

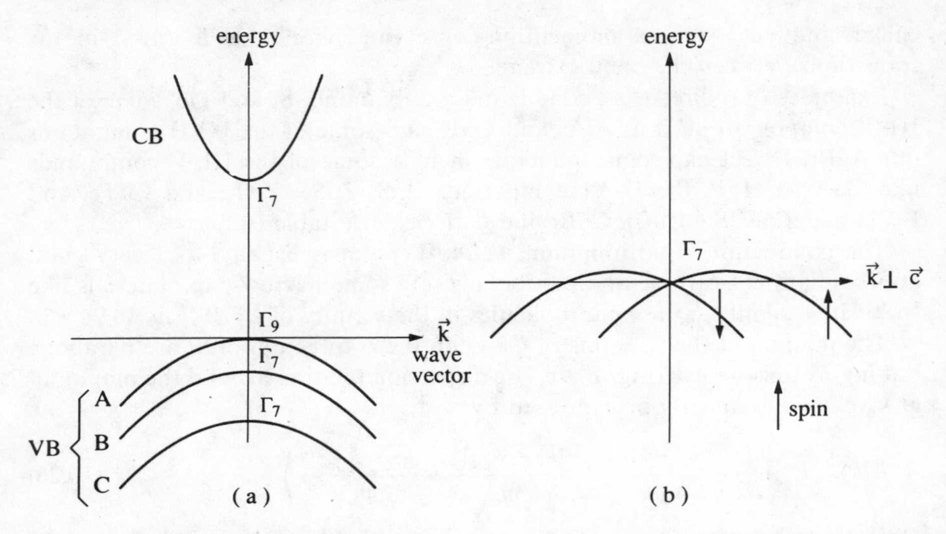

Fig. 10.10a, b. Details of the bandstructure of hexagonal semiconductors around the Γ point. The splitting into three valence bands (**a**) and the influence of a term linear in k (**b**)

Fig. 10.10b. This term has some influence on the optical properties. In principle, it can also occur for the Γ_7 conduction band, but is much smaller there and usually neglected. The same is true for the hexagonal warping which is, in principle, possible for C_{6v} symmetry in the plane normal to $\boldsymbol{c}$.

As already mentioned, the crystal structures of T_d and C_{6v} symmetry are rather similar and differ only in the arrangements of the next-nearest neighbors. The unit cell of C_{6v} symmetry is in one direction twice as long as the primitive unit cell of T_d symmetry. As a consequence the first Brillouin zone is only half as long in one direction. The resulting folding back of the dispersion is shown schematically in Fig. 10.11 neglecting spin.

Due to the *p*- and *s*-type character of valence and conduction bands, respectively, the band-to-band transition is dipole allowed – possibly with some additional selection rules for the hexagonal symmetry.

In Table 10.3 we summarize some band parameters of semiconductors. An exhaustive list is found in [10.7].

It is beyond the scope of this book to review all types of semiconductors. Instead we give only some selected examples beyond those mentioned already above.

The lead salts (PbS, PbSe, PbTe) are narrow-gap semiconductors crystallizing in the NaCl structure. They have a direct gap. In contrast to the above-mentioned direct semiconductors, the band extrema are situated at the L points. Some semiconductors, such as the IV–VI compounds TiO_2 (rutil) and SnO_2 or the I–VI compound Cu_2O have a direct gap. The transitions between the band extrema are, in contrast to the other materials mentioned above, dipole forbidden because of their equal parity. (The symmetry groups of these

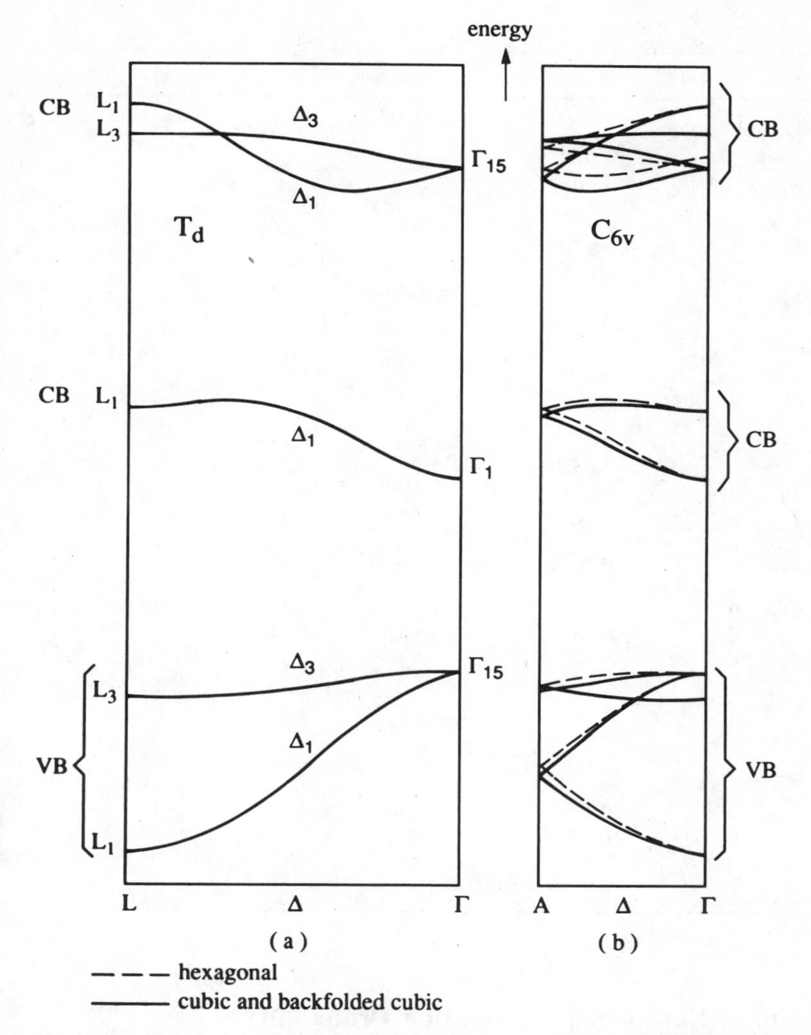

Fig. 10.11. The relation between the bandstructure of semiconductors with symmetry T_d and C_{6v} along the corresponding direction in $\boldsymbol{k}$-space, neglecting spin. After [10.12]

materials contain inversion as an element and parity is therefore a good quantum number). This feature has significant consequences for the optical properties.

A last example is shown in Fig. 10.12 where we demonstrate the transition of the alloy system $Cd_{1-y}Hg_yTe$ from a normal semiconductor ($y = 0$) to a narrow gap material $0.85 \geqslant y \geqslant 1$, and finally to a semimetal for $1 \geqslant y \geqslant 0.85$.

Table 10.3. Effective masses of some selected semiconductors. From [10.7]

SC	Sy	dir/ind	m_e/m_0	m_{hh}/m_0	m_{lh}/m_0	$m_A \perp \parallel /m_0$
C	T_d	i		2.18	0.7	
Si	T_d	i	$\perp$ 0.19 $\parallel$ 0.92	0.54	0.15	
Ge	T_d	i	$\perp$ 0.081 $\parallel$ 1.6	0.3	0.043	
AlAs	T_d	i	$\perp$ 1.56, 5.8 $\parallel$ 0.19	0.76	0.15	
AlSb	T_d	i	$\perp$ 0.26 $\parallel$ 1.0	0.94	0.11	
GaN	C_{6v}	d	0.22			≈ 0.8
GaP	T_d	i	$\perp$ 0.25 $\parallel$ 7.25; 2.2	0.6	0.17	
GaAs	T_d	d	0.066	0.47	0.07	
GaSb	T_d	d	0.042	0.35	0.05	
ZnO	C_{6v}	d	0.28			$\perp$ 0.45 $\parallel$ 0.59
ZnS	C_{6v}	d	0.28			0.5
ZnSe	T_d	d	0.15	0.8	0.145	
ZnTe	T_d	d	0.12	0.6		
CdS	C_{6v}	d	0.2			$\perp$ 0.7 $\parallel$ 2.5
CdSe	C_{6v}	d	0.13			$\perp$ 0.45 $\parallel$ 1.1
CdTe	T_d	d	0.1	0.4		
CuCl	T_d	d	0.4	2.4		
CuBr	T_d	d	0.25	1.4		
CuI	T_d	d	0.3	≈ 2		

10.7 Density of States and Occupation Probability

We start now to consider the density of states $D(E)$ for crystal electrons and holes using simple parabolic bands in the effective mass approximation, i.e.,

$$\text{conduction band (CB)}: E(\boldsymbol{k}_e) = E_g + \frac{\hbar^2 \boldsymbol{k}_e^2}{2m_e}, \tag{10.28a}$$

$$\text{valence band (VB)}: E(\boldsymbol{k}_h) = \frac{\hbar^2 \boldsymbol{k}_h^2}{2m_h}, \tag{10.28b}$$

where we take into account that the energy of a hole increases if it is brought deeper into the valence band.

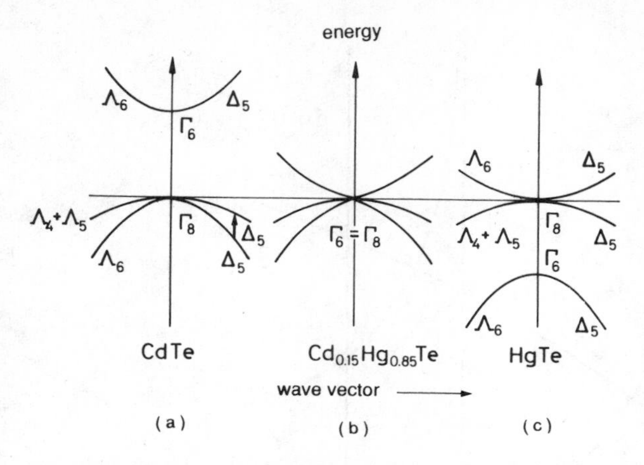

Fig. 10.12a–c. The transition from a semiconductor (**a**) to a semimetal (**c**) for the alloy $Cd_{1-y}Hg_yTe$ (**b**) as a function of the composition y. From [10.11]

With the help of (2.77) we find that the density of states depends on the energy in the conduction and valence bands according to

$$\text{CB}: D(E) = (E - E_g)^{d/2-1}; \quad E > E_g$$

$$\text{VB}: D(E) = E^{d/2-1} \quad E > 0; \quad d = \text{dimensionality} \tag{10.29}$$

for three-, two- and one-dimensional systems. This situation is shown schematically in Fig. 10.13 where we also include a set of δ-functions for the density of states in a quasi-zero-dimensional system.

The appearance of the various subbands for $d < 3$ will be explained in the next section.

The statistics with which we describe the occupation probability are the Fermi–Dirac statistics, as already mentioned. It reads for electrons and holes

$$\text{electrons: } f_{\text{FD}}(E) = (e^{(E-E_F^e)/(k_BT)} + 1)^{-1} \tag{10.30a}$$

$$\text{holes: } f_{\text{FD}}(E) = 1 - (e^{(E-E_F^h)/(k_BT)} + 1)^{-1} = (e^{-(E-E_F^h)/(k_DT)} + 1)^{-1}. \tag{10.30b}$$

The Fermi energies (or chemical potentials) for electrons and holes $E_F^{e,h}$ are non-zero, unlike the situation for phonons or photons, since there is a conservation law for the number of electrons. The Fermi energies depend on the concentrations of electrons (and holes), on the temperature of the electron gas, and on material parameters such as E_g, m_e, or m_h as shown below.

Since electrons can be exchanged between the valence and conduction bands, for example by thermal excitation and recombination, it follows that in thermodynamic equilibrium

$$E_F^e = E_F^h = E_F \tag{10.31a}$$

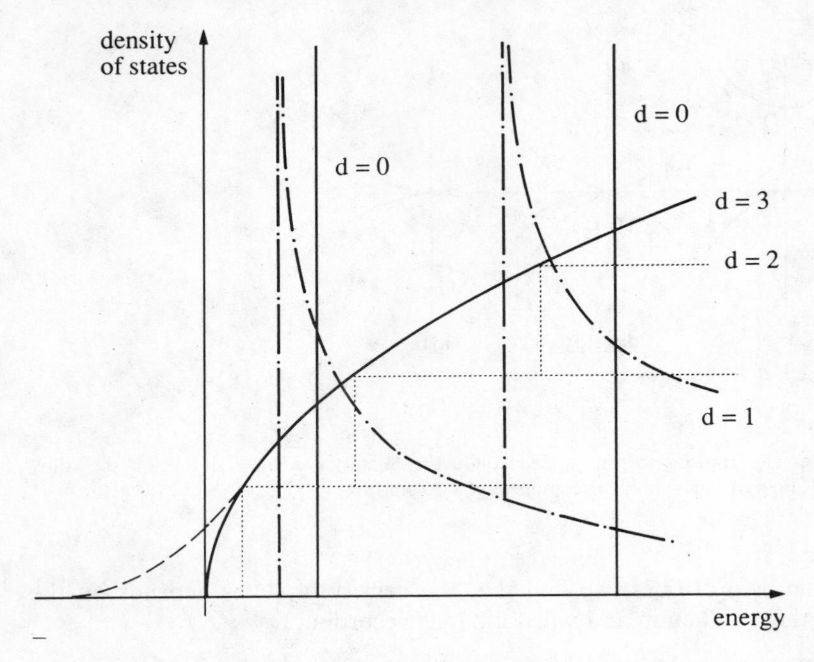

Fig. 10.13. Schematic drawing of the density of states as a function of energy for three-, and quasi-two-, one- and zero-dimensional systems in the effective mass approximation. The dashed line corresponds to localized states (Sect. 10.10)

or, in other words, the chemical potential of the electron–hole pair system μ_{eh} is zero

$$\mu_{eh} = E_F^e - E_F^h = 0 . \tag{10.31b}$$

We shall come back to this point later.

The density of electrons in a certain energy range, say in the conduction band, is then given by the integral over the density of states weighted by the occupation probability. For parabolic bands with a degeneracy g we find for $d = 3$ and for the total density of electrons in the conduction band n:

$$n = \frac{N}{L^3} = \int_{E_g}^{\infty} g_e \frac{1}{2\pi^2}\left(\frac{2m_e}{\hbar^2}\right)^{3/2} (E - E_g)^{1/2}(e^{(E-E_F)/k_BT} + 1)^{-1} dE , \tag{10.32}$$

where we have used our knowledge from Sect. 2.6 and the dispersion relation (10.28). Equation (10.32) leads to the Fermi integral which cannot be solved analytically but can be found in mathematical tables.

Very often E_F lies in the gap. In this case the electron gas is said to be non-degenerate and the part of the Fermi function which overlaps with a finite density of states can be approximated quite well by the Boltzmann function, as can be seen in Fig. 2.5 where we compare Fermi–Dirac, Bose–Einstein and Boltzmann statistics.

With this assumption the Fermi–Dirac statistics (10.32) can be replaced by classical Boltzmann statistics and the integral can be solved analytically giving

$$n = g_e \left(\frac{m_e k_B T}{2\pi\hbar^2} \right)^{3/2} e^{-(E_g - E_F)/k_B T} . \tag{10.33a}$$

The term in front of the exponential is the so-called effective density of states n_{eff} depending on T and on material parameters. So we can write

$$n = n_{eff}(T, m_e) e^{-(E_g - E_F)/k_B T} . \tag{10.33b}$$

If the electron gas is degenerate, i.e., if E_F is situated in the conduction band, we have to use the full (10.32) except for $T \to 0$ where the Fermi function converges to a step function and n is given by

$$n = \frac{2}{6\pi^2} g \left(\frac{2m_e}{\hbar^2} \right)^{3/2} E_F^{3/2} . \tag{10.33c}$$

In analogy we find for a three-dimensional non-degenerate gas of holes

$$p = p_{eff} e^{-E_F/k_B T} \tag{10.33d}$$

with

$$p_{eff} = g_h \left(\frac{m_h k_B T}{2\pi\hbar^2} \right)^{3/2} . \tag{10.33e}$$

We should mention here that in thermodynamic equilibrium, i.e., when (10.31) is valid, one finds that the product np is independent of E_F in the non-degenerate case, i.e.,

$$np = n_{eff} p_{eff} e^{-E_g/k_B T} =: n_i^2(T) . \tag{10.34}$$

10.8 Quantum Wells, Superlattices, Quantum Wires, and Quantum Dots

In this section we want to show how electronic systems of reduced symmetry can be realized.

Quasi-two-dimensional electron and/or hole systems can be created if the motion of the particles is restricted by a (square) potential well in one dimension to a distance comparable to or smaller than the de Broglie length or the average distance between scattering events in the sense of relaxation time approach or the Bohr radius of the exciton. For typical semiconductors this limit is reached for widths of the constraining potential well l_z below a few tens of nm.

In the other two dimensions, the particle can move as a free (Bloch) wave with its effective mass.

Such systems appear either at heterojunctions due to space charges or in single and multiple quantum wells as shown in Fig. 10.14.

Figure 10.14a shows a single heterojunction. Due to the difference of the gap of the two materials and suitable interface space charge layers, which depend on doping and the position of the Fermi levels as outlined in the next section, we get a roughly triangular-shaped potential well for the electrons, which quantizes the motion of the electrons in the z-direction.

The other and presently more widely applied method is to use two heterojunctions with a certain separation l_z fulfilling the above conditions, which results in a potential well of rectangular shape. In both cases a single band in the three-dimensional material yields a (limited) number of subbands with dispersion relation

$$E(\boldsymbol{k}) = E_{n_z} + \frac{\hbar^2(k_x^2 + k_y^2)}{2m_{\mathrm{e,h}}}, \tag{10.35a}$$

where each subband has its constant density of states as shown in Fig. 10.13.

The simplest theoretical approach to Fig. 10.14b is a potential well with infinitely high barriers on both sides, see Fig. 10.14c. In this case the quantization energies E_{n_z} are given by

$$E_{n_z} = \frac{\hbar^2 \pi^2}{2m_{\mathrm{e,h}} l_z^2} \frac{1}{n_z^2}, \quad n_z = 1, 2, 3 \ldots . \tag{10.35b}$$

In contrast to (10.35), in real quantum wells one finds only a limited number of bound states. It is found empirically that they still follow to a good approximation a n_z^{-2} law but the prefactor is smaller since the states can tunnel into the barriers due to their finite height. Compare Figs. 10.14b and c. For $l_z \to 0$, E diverges according to (10.35b). For a real quantum well, the states converge for $l_z \to 0$ to the (Bloch) states of the barrier material.

For a detailed analysis of the calculation of the energy levels and wavefunctions in quantum wells with finite barrier height see for example the solutions presented in [10.13–15].

Quantum wells are usually formed from cubic semiconductors. As a consequence there is a complication for the upper valence band of symmetry Γ_8. It splits for $k \neq 0$ into a heavy- and a light-hole band (Sect. 10.6) and correspondingly with (10.35b) we get two series of quantized hole states as shown schematically in Fig. 10.14b where the indices hh and lh stand for heavy and light-hole, respectively. Due to the fact that the wavefunction of the heavy hole in the z-direction is just the wavefunction of the light hole propagating in the x-and y-directions (and vice versa) and due to the non-crossing rule, the in-plane dispersion looks rather complicated as shown in Fig. 10.15. In many cases however, it is, sufficient to calculate with only the hh and lh masses for the lowest and second-lowest quantized levels.

If we have several quantum wells in a material with barriers so thick that the wavefunctions of adjacent wells do not overlap, we speak about a multiple quantum well (MQW). This situation is depicted in Fig. 10.14d. If the barriers are thin enough to allow overlap, one has a superlattice (Fig. 10.14e). In this

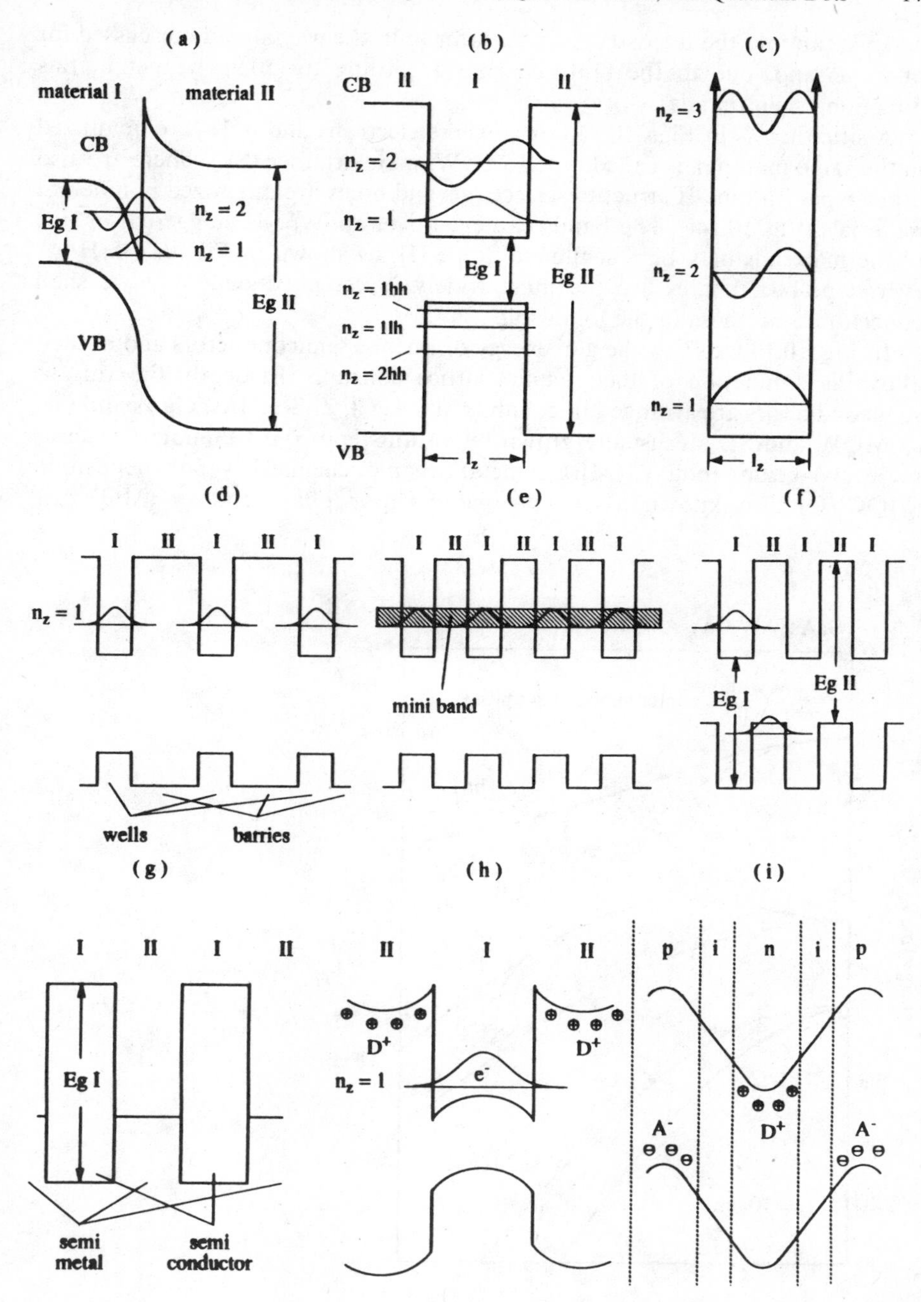

Fig. 10.14a–i. The bandstructures and/or band discontinuities at heterostructures which may lead to quasi-two-dimensional electron (and/or hole) systems. A single heterojunction with space charge layers (**a**), a single quantum well (**b**), an idealized quantum well (**c**), a multiple quantum well (MQW) of type I(**d**), a superlattice of type I(**e**), a superlattice or MQW of type II(**f**), of type III(**g**), a single modulation doped quantum well (**h**) and a doping superlattice (nipi) (**i**)

case we can fold the bandstructure back for k_z in the way already discussed for phonons and, due to the finite overlap according to (10.9), we get in this direction " mini-bands".

A situation as in Figs. 10.14d or e, where electrons and holes are quantized in the same material, is called type-I MQW or superlattice (SL). There are also other types. In type-II structures, electrons and holes are quantized in different materials (Fig. 10.14f). The bands can even overlap (type-II staggered) or one of the materials may be a semimetal (type-III) as shown in Fig. 10.14f. However, type-I structures are the most widely investigated ones and we shall concentrate on these in the following.

In Fig. 10.16 we show the gap energy of various semiconductors and of their alloys as a function of their (cubic) lattice constant. Evidently the various semiconductors are arranged in columns like AgCl, ZnSe, AlAs, GaAs and Ge.

MQW and SL are usually grown by various epitaxial techniques, such as molecular-beam epitaxy (MBE), metal-organic chemical vapor deposition (MOCVD) also known as metal-organic vapor phase epitaxy (MOVPE),

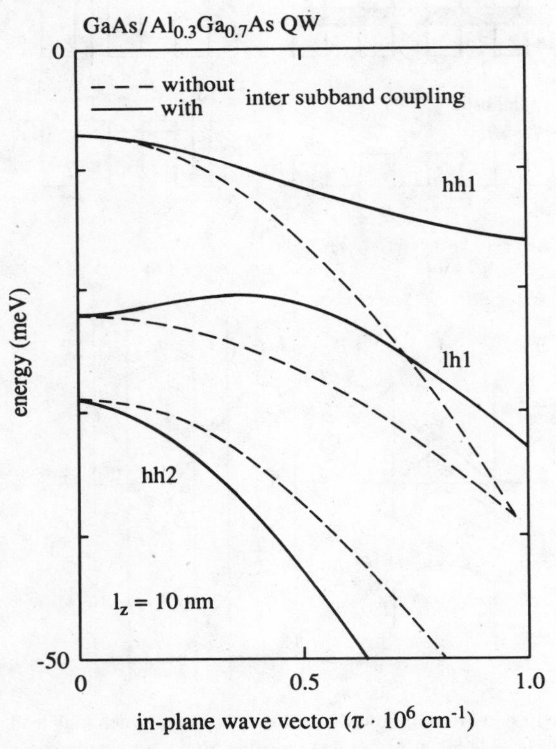

Fig. 10.15. The in-plane dispersion relation of the first three valence subbands in a GaAs/$Al_{1-y}Ga_yAs$ quantum well without (---) and with (—) intersubband interaction. After [10.13]

hot-wall epitaxy (HWE) or atomic-layer epitaxy (ALE). A description of these techniques is beyond the scope of this book, but can be found in [9.9, 10.16–19].

The epitaxial methods work best if the two materials have the same lattice constant. This situation is best achieved for some III–V compounds. By inspecting Fig. 10.16 it is obvious that one of the most suitable systems to produce QW is GaAs/$Al_{1-y}Ga_yAs$ because for all values of y it has almost the same lattice constant. Usually one restricts y to values above 0.5 so that both the GaAs wells and the $Al_{1-y}Ga_yAs$ barriers are direct. The system GaAs/AlAs can form a type-II SL for short superlattice periods in the sense that the first quantized electron state at the Γ point in GaAs can be higher in energy than states at the X points of AlAs [10.19].

Another system of great technical interest is $Ga_{1-y}In_yAs/Al_{1-y}In_yAs$ grown on InP since its bandgap corresponds to the wavelengths around 1.3 μm and 1.5 μm which are important for long-distance signal transmission in glass fibers.

If materials with different lattice constants are grown epitaxially on top of each other, the layers are necessarily strained, leading to "strained layer superlattices". In this case barriers and/or wells cannot be made very thick if one is

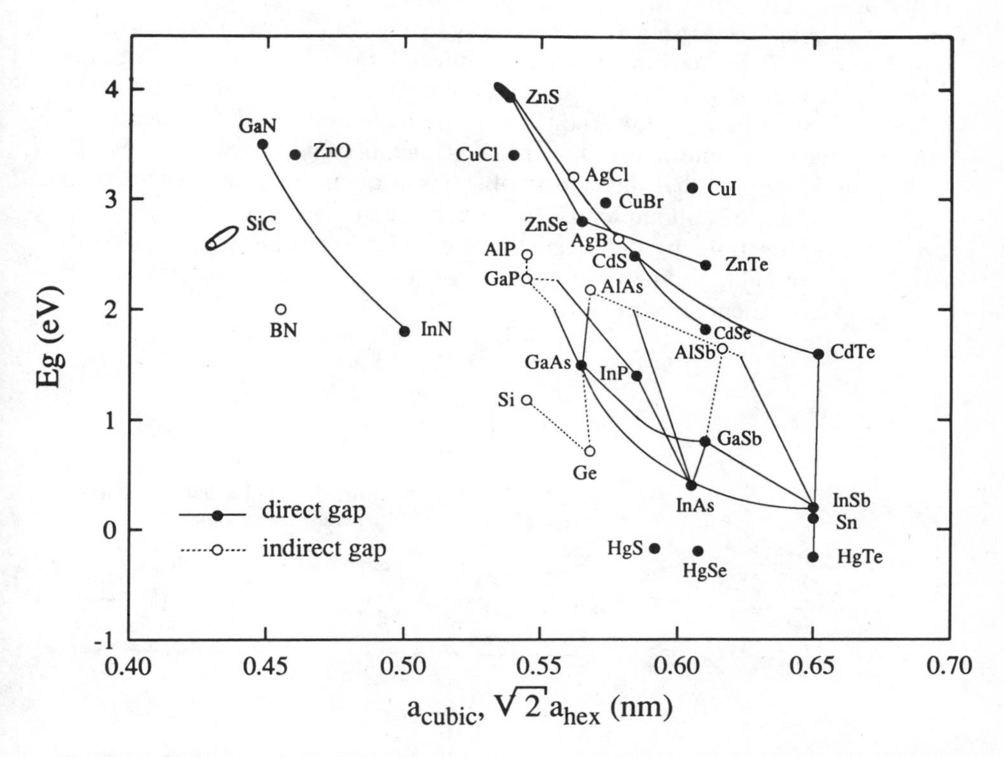

Fig. 10.16. Width of the gap as a function of lattice constants for various semiconductors and their alloys. *Solid lines*: direct gap; *dashed lines*: indirect gap. Data from [10.7]

to avoid the formation of "misfitdislocations" which release the strain but seriously disturb the crystalline quality otherwise obtainable by the modern techniques of epitaxy mentioned above.

Strained-layer SL which are presently under investigation include the system Si/Ge and the II–VI compounds. In the first case it is hoped that the indirect bandstructure can be overcome by the folding back mechanism in the SL which might allow the use of Si/Ge SL as an efficient light-emitting material such as the GaAs- or InP-based (M)QW, despite some drawbacks [10.20]. In the second case one hopes for light-emitting devices in the visible, especially in the green and blue spectral regions for which many applications in display devices can easily be envisaged. First experimental results have been reported recently in [10.21].

Quantum wires involve a quantization of electrons and holes in two directions and result in dispersion relations like

$$E(\boldsymbol{k}) = E_{n_z n_y} + \frac{\hbar^2 k_x^2}{2m_{\mathrm{e,h}}} \tag{10.36}$$

with a density of states shown in Fig. 10.13. The quantum wires are produced either by lateral micro-structuring of quantum wells or by growing the well at an edge or a groove as shown schematically in Fig. 10.17. In Fig. 10.17a the y-dimension of 50 to 200 nm is generally much larger than the z dimension, resulting in correspondingly weaker confinement effects. Generally it can be stated that the technology for producing quantum wires has not yet reached the same state as for quantum wells but that it is making rapid progress [10.19, 22].

It should be noted that the density of states in quantum wires has a certain similarity to that of Landau levels. In the latter case the motion is also quantized in two directions by the magnetic field and free motion is only possible parallel to the field. A difference to quantum wires is that Landau levels are equally spaced in energy according to

$$E(\boldsymbol{k}) = \left(n + \frac{1}{2}\right)\hbar\omega_{\mathrm{c}} + \frac{\hbar^2 k_x^2}{2m}\ , \tag{10.37}$$

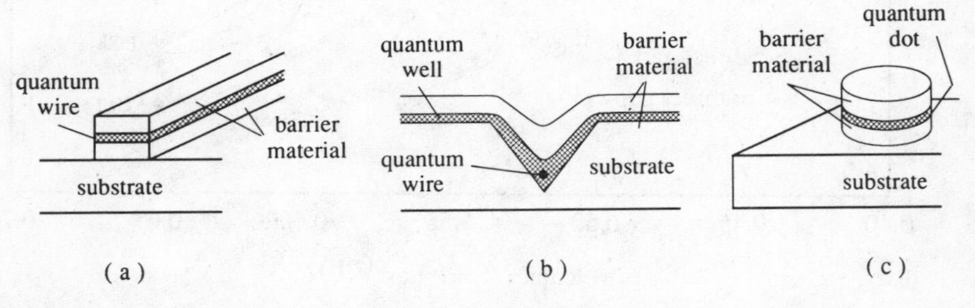

Fig. 10.17a–c. Some possible ways of producing quantum wires (**a, b**) and quantum dots (**c**) by microstructuring of quantum wells (**a, c**) or by special preparation of the substrates (**b**)

with

$$n = 1, 2 \ldots \quad \text{and} \quad \omega_c = \frac{eB}{m_{e,h}},$$

while the $E_{n_z n_y}$ in (10.36) depend on the geometry and tend to a $n_{z,y}^2$ dependence as in (10.35).

An even more complex structure is the type-II quantum-well wire superlattice (QWWSL) which can be grown on highly indexed surfaces [10.23].

Quantum dots (QD) or quantum boxes can be formed by microstructuring of a quantum well in two dimensions [10.19, 22]. In this case one usually gets pancake shaped "dots" since the lateral dimension is often a factor ten wider than the thickness of the dot; see Fig. 10.17c. The density of states is, in principle, δ-function like and is shown schematically in Fig. 10.13. However, there are various effects which lead to a broadening. Some of these are mentioned later. The other possibility is to grow QD by an annealing procedure in glasses doped with the components of a semiconductor. This technique is widely used for II–VI compounds such as $CdS_{1-x}Se_x$ including the cases $x = 0$ and $x = 1$ for CdTe or for the Cu halides. The commercially available edge filters are made in this way. The average diameter of the dots increases with annealing time and temperature. Other techniques involve chemical precipitation of, for example, CdS in (organic) liquids or gels. These techniques have the advantage of giving dots which are almost spherical, but the disadvantage that one always gets a distribution of the dot radii with a certain width which is rarely below 10%. Since the quantization energy depends on the radius this gives one contribution to the inhomogeneous broadening of the ideal δ-function density of states. For recent reviews of these types of QD see [10.24–29].

The properties and applications of a modulation doped structure like in Fig. 10.14 h will be discussed in section 10.9.

An especially tricky system finally are the so-called nipi and related structures. In contrast to MQW they consist of only one semiconductor material which, however, contains alternating layers of n and p doping separated by undoped (intrinsic i) layers. The electrons (holes) from the donors (acceptors) are thermally excited into their bands and recombine, i.e., they annhiliate each other. The remaining ionized donors and acceptors lead to a periodic spatial modulation of the bands (Fig. 10.41i). The remaining electrons and holes accumulate in the corresponding minima and are therefore spatially separated. A nipi structure can thus be considered as a "spatially" indirect semiconductor in contrast to the indirect transitions in $\boldsymbol{k}$-space described in Sect. 10.6.

10.9 Defects, Defect States and Doping

In connection with local phonon modes, we have already mentioned the imperfections which are present in all real crystals. Here we outline the electronic states connected with defects and concentrate again on point defects. These are present, even in good materials, with densities of up to 10^{15}–10^{17} cm^{-3}. Values of 10^{11}–10^{13} cm^{-3} which are given in literature, usually refer only to "electrically active" defects and this means the difference between the concentration of ionized donors and acceptors (see below). The density of dislocations ranges from zero to some 10^4 per cm^2 in high quality materials.

In Fig. 9.21 we classified the point defects according to the way in which they are incorporated in the lattice, e.g., as interstitials or substitutionals. Now we consider their electronic properties and present donors, acceptors, isoelectronic traps and recombination centers in Fig. 10.18. A donor is a shallow center which has an energy level just below the conduction band and can easily give an electron, e.g., by thermal ionization, to this band

$$D^0 \rightleftharpoons D^+ + e\,. \tag{10.38a}$$

Donors are often formed by substitutional atoms situated in the periodic table one column to the right of the atom which they replace, like N or P in Ge or Si, Si on Ga site in GaAs, or Cl on Se sites in ZnSe, etc. Furthermore, donors can be formed by interstitials which have a weakly bound electron such as H, Li or Na, in a II–VI compound.

In analogy, acceptors can easily accommodate an electron from the valence band, i.e., they emit a hole into it

$$A^0 \rightleftharpoons A^- + h\,. \tag{10.38b}$$

Acceptors may be formed by substitutional atoms which have one electron less than the one which they replace. Thus the impurities are often found in the periodic table to the left of the atom which they replace, for example, Ga or B in Si and Ge, Li or Na on the cation site and N on the anion site in II–VI compounds, or Si on the As site in GaAs.

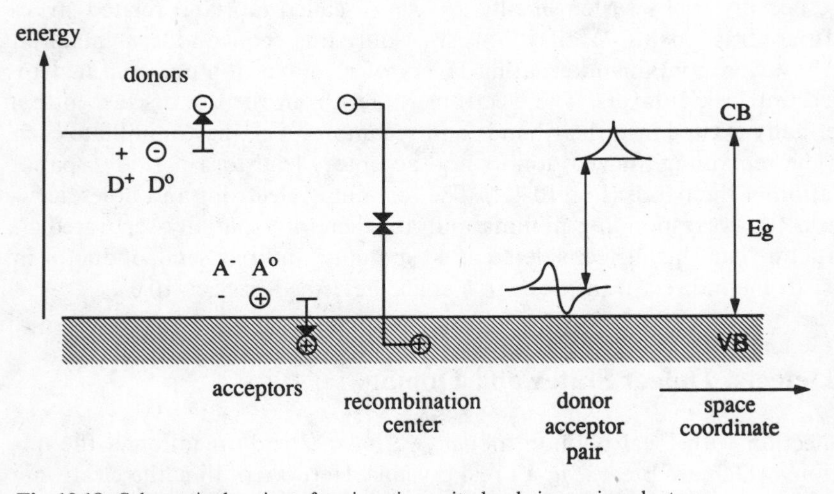

Fig. 10.18. Schematic drawing of various impurity levels in semiconductors

It is clear from some of the above-mentioned examples that the same atom can act as donor or acceptor depending on the way it is introduced into the lattice. This is one possibility for self-compensation. Another arises from the fact that some vacancies or interstitials can act as donors or as acceptors. A material which has a higher concentration of donors (acceptors) is called *n*-type (*p*-type) and we recall that $np = n_i^2(T)$ in thermodynamic equilibrium (10.34). The ability to choose the type and concentration of carriers over a wide range by doping with donors and/or acceptors is the basis for the widespread and important application of semiconductors in electronic devices like diodes, transistors, thyristors, etc. This topic is beyond the scope of this book and we refer the reader to text books on the physics of semiconductor devices, e.g., [10.30, 31].

A shallow donor (acceptor) can be considered as a positively (negatively) charged center to which an electron (hole) is bound by Coulomb interaction. So we are faced with a problem similar to that of a hydrogen atom, leading, in the simplest approximation, to a series of states with binding energy

$$E_b^{D,A} = \mathrm{Ry}\frac{m_{e,h}}{m_0}\frac{1}{\varepsilon^2}\frac{1}{n_B^2} \tag{10.39}$$

where Ry is the Rydberg energy of the H atom ($\mathrm{Ry} = 13.6\,\mathrm{eV}$), n_B the main quantum number and ε a dielectric constant. In Fig. 10.18 we show only the states for $n_B = 1$. Depending on the material parameters, one usually finds values for the donor and acceptor binding energies of

$$5\,\mathrm{meV} \leqslant E_b^D \leqslant 50\,\mathrm{meV}\,, \tag{10.40a}$$

$$20\,\mathrm{meV} \leqslant E_b^A \leqslant 200\,\mathrm{meV}\,. \tag{10.40b}$$

The radius of the $n_B = 1$ state is given by

$$a_{\mathrm{lattice}} < a_{D,A} = a_B\,\varepsilon\frac{m_0}{m_{e,h}}n_B \tag{10.40c}$$

ranging from 1 to 20 nm depending again on the material parameters. The fact that these values are larger than the lattice constant justifies the use of the effective mass approximation. Details concerning which value of ε has to be used are similar to those for excitons, as discussed in Sect. 11.2. In addition, there is some smaller influence of the chemical nature of the atom forming the donor or acceptor which is known as the central-cell correction or chemical shift.

Similar to the localized phonon modes, the wave function of a shallow donor (or acceptor) can be described as a superposition of Bloch states

$$\phi(\boldsymbol{r}) = \sum_{\boldsymbol{k}} a_{\boldsymbol{k}}\phi_{\boldsymbol{k}}^{CB}(\boldsymbol{r})\,. \tag{10.41}$$

The range of $\boldsymbol{k}$ from which significant contributions can be expected increases with decreasing radius in (10.40c).

Pairs of donors and acceptors which are so close in space that their wavefunctions overlap are known as donor–acceptor pairs. As we shall see later in Sect. 15.2, they give rise to a characteristic emission feature.

In connection with (M)QW two special methods of doping should be mentioned:

δ-doping means the introduction of a two-dimensional sheet of doping atoms during epitaxial growth. The concentration in the growth direction then has an almost δ-function-like profile.

Modulation doping means introducing the doping atoms into the barriers of a (M)QW. The electrons (holes) are thermally ionized into the conduction (valence) band of the barrier, reach the well by thermal diffusion, and are captured in it. This allows the production of high two-dimensional carrier densities in the well with high mobility, since the charged impurities are separated spatially from the mobile carriers, thus reducing the scattering with them. The space charges of the ionized doping atoms and of the free carriers lead to a characteristic curvature of the bands which can be calculated by solving the Poisson equation

$$-\Delta\phi = \frac{\rho}{\varepsilon\varepsilon_0}, \tag{10.42}$$

where ϕ is the electrostatic potential and ρ the space charge as shown schematically in Fig. 10.14h. Modulation doping is the basis of devices called high-electron-mobility transistors (HEMT) or modulation-doped field-effect transistors (MODFET). The binding energy of donors and acceptors in SL and (M)QW depends in addition to the parameters (10.39) on the distance from the barrier.

Apart from the shallow donor and acceptors there are deep donors and acceptors and a variety of other deep centers. These are atoms which have one or more energy levels somewhere around the middle of the gap.

For deep centers an approach as in (10.39–41) is not adequate. The wave function is better described by the parent atomic orbitals, modified by the influence of the surrounding atoms, i.e., by the symmetry of the arrangements of the neighbors. Copper, nickel, iron, chromium and other elements can give rise to such deep levels.

Some deep centers can exchange carriers with both the conduction and the valence band (in contrast to the donors and acceptors). In this case they are called recombination centers. The recombination can be radiative or nonradiative and some centers provide fast channels of for de-excitation of electron–hole pairs.

A group of (deep) centers is often formed by so-called isoelectronic traps. These are atoms of the same column of the periodic table as the one which they substitute, i.e., they have the same electron configuration in the outer shell. An example of an isoelectronic trap would be Te replacing S or Se in ZnS or ZnSe.

Some centers have various levels in the forbidden gap and so transitions within the center can be investigated as in Cu, Ni or other ions. Since the chance of having various levels in the gap increases with increasing width of the gap, such internal transitions are best investigated in wide gap semiconductors and in insulators. Carriers in deep centers can sometimes couple strongly to bulk and/or localized phonon modes giving rise to very broad emission (and absorption) features with Huang-Rhys factors S significantly larger than 1. This topic however leads beyond the scope of this book and we refer the reader to [10.34, 35].

To summarize this section we show schematically in Fig. 10.19 the density of states including some impurity centers, the occupation probability in thermodynamic equilibrium and the resulting density of electrons and holes per unit energy.

More information on defects in bulk materials and in MQW structures is given in [10.32, 33] and references therein.

10.10 Disordered Systems and Localization

In this section we will first outline briefly what is new in disordered systems and then we present some examples of how disorder can be realized in

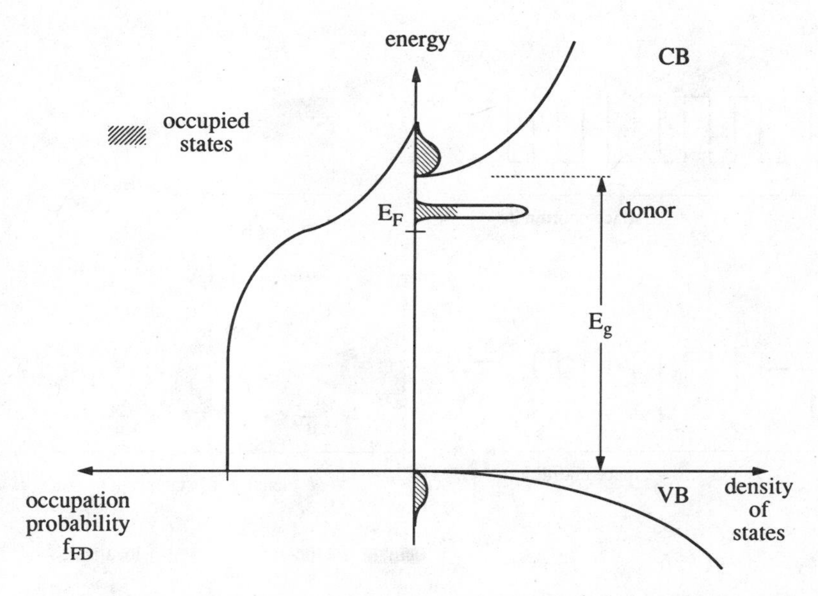

Fig. 10.19. The density of states DOS in the effective-mass approximation of an idealized three-dimensional semiconductor containing some impurities, the occupation probability f_{FD} and the resulting population (*hatched areas*)

semiconductors. More exhaustive treatments of these topics, from which the following facts were largely extracted, are [10.36–41].

In Fig. 10.20 we show schematically a periodic potential. It is known as a Kronig–Penney potential. Due to the tunnelling of the wavefunction into the barriers, we get finite overlap integrals and with (10.9) a band of Bloch states with a certain width B.

We now want to introduce disorder into this system. This can be done by varying the depths of the potentials statistically within a width V_0 ("diagonal disorder") or by varying the widths of the potential wells and barriers and thus the coupling ("off-diagonal disorder"). The first case leads to the so-called Anderson model, the second to the Lifshitz model. In practice, both effects of disorder will occur simultaneously, but from the theoretical point of view it is sufficiently difficult to treat one of them. Figures 10.20 and 10.21 briefly outline the ideas of the Anderson model. If some diagonal disorder is introduced, as in Fig. 10.20c, two things happen – the sharp edges of the density of states are smeared out by exponential tails, and a new type of eigenstate appears, namely localized states. If

$$BV_0^{-1} > 1\,, \tag{10.43a}$$

then there are both localized states at the band tails and extended states in the center, which, however, are different from Bloch states, as we shall see shortly

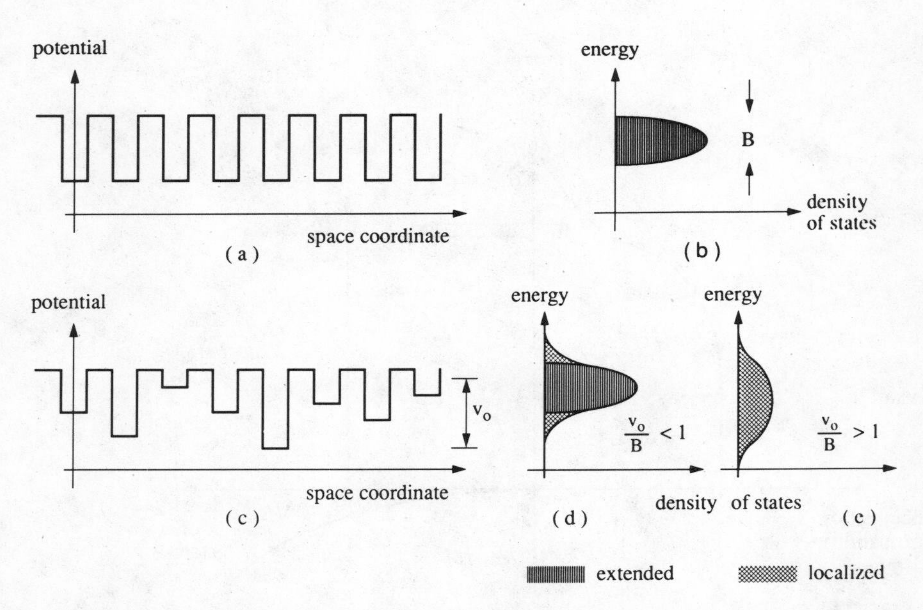

Fig. 10.20a–e. A periodic potential (**a**) and the resulting band, represented by its density of states (**b**), a disordered system (**c**) and bands with localized and delocalized states (**d**) and with localized states only (**e**)

(Figs. 10.20d and 10.21b, c). For

$$BV_0^{-1} < 1\,, \tag{10.43b}$$

there are only localized states in the band (Figs. 10.20c and 10.21c).

To explain the concept of extended and localized states we show in Fig. 10.21a a Bloch wave which is a superposition of atomic orbitals ϕ_n placed at every lattice site $\boldsymbol{R}_i$ with a proper long-range phase correlation (corresponding to the LCAO method mentioned in Sect. 10.1):

$$\phi_k^{\text{Bloch}}(\boldsymbol{r}) = \sum_i \mathrm{e}^{\mathrm{i}\boldsymbol{k}\cdot\boldsymbol{R}_i}\phi_n(\boldsymbol{r}-\boldsymbol{R}_i) = \mathrm{e}^{\mathrm{i}\boldsymbol{k}\cdot\boldsymbol{r}}\sum_i \mathrm{e}^{\mathrm{i}\boldsymbol{k}\cdot(\boldsymbol{r}-\boldsymbol{R}_i)}\phi_n(\boldsymbol{r}-\boldsymbol{R}_i) = \mathrm{e}^{\mathrm{i}\boldsymbol{k}\cdot\boldsymbol{r}}u_k(\boldsymbol{r})\,. \tag{10.44a}$$

In an extended state in a disordered system (Fig. 10.21b) we lose the long-range phase correlation resulting in

$$\phi^{\text{ext}}(\boldsymbol{r}) = \sum_i c_i\phi_n(\boldsymbol{r}-\boldsymbol{R}_i) \tag{10.44b}$$

and for a localized state we get an envelope which decays exponentially with a localization length ξ (Fig. 10.21c)

$$\phi^{\text{loc}}(\boldsymbol{r}) = \sum d_i\phi_n(\boldsymbol{r}-\boldsymbol{R}_i)\mathrm{e}^{-|\boldsymbol{r}-\boldsymbol{r}_0|/\xi}\,. \tag{10.44c}$$

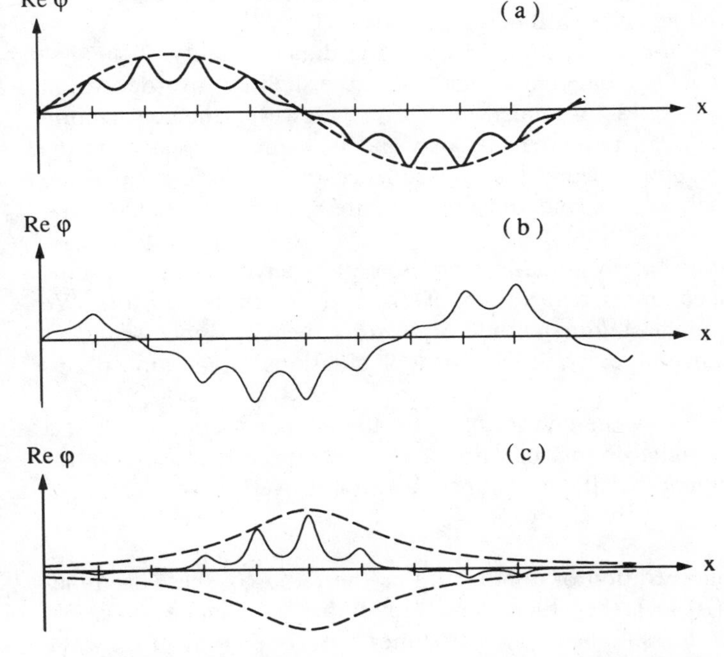

Fig. 10.21a–c. Schematic drawing of the real part of the wave function for a periodic potential **(a)** and for a disordered system an extended state **(b)** and a localized one **(c)**

The energy which separates extended from localized states is called the mobility edge. If the carriers were to have an infinite phase-relaxation time T_2 in their states, it is easy to show that the mobility edge is sharp: if extended and localized states were to coexist in a certain energy range, then the slightest perturbation would mix these states and consequently every localized state would obtain some admixture of an extended wave function (and vice versa) transforming it to an extended state. Since the T_2 times of the carriers in their states are limited, the mobility edge is usually a transition region of a certain width.

Another important consequence of disorder is the following. A disordered system is no longer invariant with respect to translations. As a consequence, the wave vector is no longer a good quantum number and the dispersion relation $E(\boldsymbol{k})$ loses its meaning. A disordered system is in principle characterized only by its density of states $D(E)$.

As usual, there are exceptions to this rule: excitations with a wavelength λ long compared to the typical length scale of the disorder fluctuations "average" over the disorder and can be characterized by a wave vector $\boldsymbol{k}$ in the sense of a continuum approximation. This is fulfilled for example for long wavelength (acoustic) phonons or photons. The latter case is easily checked by looking through a glass window or into water.

Weak disorder, e.g., as defined by (10.43a), will produce some localized states at the band edges (10^{16}–$10^{18}\,\mathrm{cm}^{-3}$) but the extended states in the band (10^{22}–$10^{23}\,\mathrm{cm}^{-3}$) will be close to Bloch-type waves.

A crucial property for localization effects is the dimensionality of the system. It has been shown by general arguments that the slightest disorder will in principle localize all states for dimensions $d \leqslant 2$. However the localization length [ξ in (10.44c)] can be extremely long and in many cases exceeds the dimensions of the sample used. As a consequence these considerations are largely of theoretical interest and only to a limited extent of practical relevance.

After characterizing briefly some properties of disordered systems we now proceed to the inspection of realizations of disorder in semiconductors. We discuss, in roughly increasing magnitude of disorder, heavily doped semiconductors, alloy semiconductors, well-width fluctuations, and amorphous semiconductors.

In Fig. 10.22a, we show the density of states for the conduction band of a semiconductor containing donors of variable concentration. If the concentration of donors is so low that the wavefunctions do not overlap, i.e.,

$$a_{\mathrm{B}}^3 N_{\mathrm{D}} \ll 1\,, \tag{10.45}$$

where N_{D} is the concentration of donors and a_{B} the radius of the wave function according to (10.40c), then their density of states is δ-function-like. We show in Fig. 10.22a, for simplicity, for the donor with one electron only the state with $n_{\mathrm{B}} = 1$. A donor may bind a second electron with opposite spin. As a result of the Coulomb interaction the energy for the two electrons will be

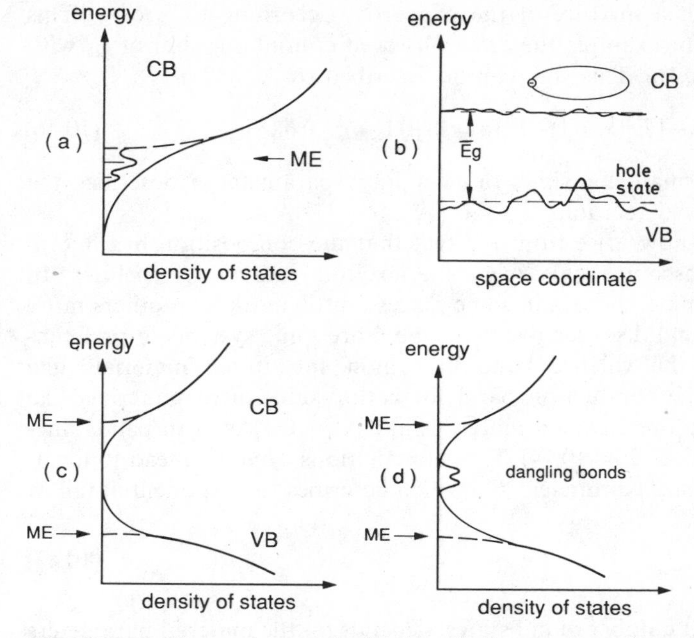

Fig. 10.22a–d. The appearance of tails of localized states due to increasing doping (**a**), potential fluctuations e.g., in an alloy (**b, c**) or in amorphous semiconductors (**d**)

higher. This is indicated by the second peak. The states with be $n_B > 1$, which the electron could also occupy, are again neglected.

If N_D is increased, the wave functions of the donors start to overlap and with (10.9) this again results in the formation of a so-called impurity band. Its width increases with N_D and eventually merges with the conduction band, forming an (exponential) tail of the density of states. Depending on the doping and possibly on compensation through acceptors, the Fermi level can be situated for $T \to 0$ above or below the mobility edge. In the first case, the system is called a metal, since it has finite conductivity at $T \to 0$, in the second it is an insulator. The transition from an insulator to a metal with increasing doping is also known as a Mott transition and there was a long discussion as to whether this transition is continuous or not, i.e., whether there is something like minimum metallic conductivity in the second case. Experiments, e.g., with phosphorous-doped silicon (Si:P), indicate that there is no minimum conductivity, [10.39, 40].

The next case of disorder are alloy semiconductors like $Ga_{1-y}Al_yAs$, $CdS_{1-x}Se_x$, $ZnSe_{1-x}Te_x$ or $Zn_{1-y}Cd_yS$. They can all be grown for every value of x or y between 0 and 1 and they usually have a well-defined crystal structure (e.g., T_d or C_{6v}). The disorder is introduced by the statistically distributed occupation of anion (or cation) sites with the two different atoms.

Though an alloy has no translational invariance one can create it artificially in the so-called virtual crystal approach. One assumes that the unit cell contains atoms with a mixture of the properties according to x or y. This approach explains for example the often observed continuous shift of E_g with composition (Fig. 10.16). This shift can be described by

$$E_g(x) = E_g(x=0) + [E_g(x=1) - E_g(x=0)][(1-b)x + bx^2] , \tag{10.46}$$

where the usually small "bowing" or "bending" parameter b describes the deviation from a linear relation.

The fluctuations now arise from the fact that the composition has fluctuations on a microscopic scale (e.g., the exciton radius, see below). In $CdS_{1-x}Se_x$, for example, there is in some places a little more S, in others more Se. This compositional disorder results in the more ionic systems in preferential fluctuation of the valence band for anion substituted materials like $CdS_{1-x}Se_x$, or of the conduction band for cation-substituted materials like $Zn_{1-y}Cd_yS$. In the more covalent materials like $Al_{1-y}Ga_yAs$ both bands fluctuate with composition (Fig. 10.14). These fluctuations can again lead to localized states. The "tailing parameter" E_0, which describes the exponential tail in

$$D(E) = \frac{N_0}{E_0} e^{-E/E_0} , \tag{10.47}$$

where N_0 is the total number of tail states, depends on the material parameters and on the composition. It is obvious that E_0 disappears for $x, y = 1$ and $x, y = 0$, that it increases for increasing $E_g(x=1) - E_g(x=0)$ and for decreasing effective mass of the particle which is localized. The latter point depends on the fact that in three dimensions a potential well must have a certain width R and depth V to localize or bind a particle of mass in m with

$$VR^2 \geqslant \hbar^2 m_{e,h} \tag{10.48}$$

and explains why holes are more easily localized than electrons.

For the dependence of E_0 on x one finds different but similar formulas in literature. We give here one according to [10.42]. For more recent discussions of this topic concerning also excitons see [10.39–44].

$$E_0 = \frac{1}{178} \beta^4 x^2 (1-x)^2 m_{e,h}^3 (\hbar^6 N^2)^{-1} , \tag{10.49}$$

where N is the density of atoms and β the derivative of the position of the edge of the band with respect to x. Often one finds from experiment a slightly asymmetric behavior of $E_0(x)$ caused by the term β and other effects.

We now consider disorder in systems of reduced dimensionality like MQW. Actually there are two origins of disorder. One is the alloy "broadening" or disorder felt by the carriers if the well or the barrier or both are made from an alloy (see Fig. 10.14). The other is due to well-width fluctuations. Though great progress has been made in growing atomically flat interfaces by modern epitaxial techniques, fluctuations of l_z of at least the order of one atomic (or molecular) layer are hardly avoidable. Due to the dependence of the quantization energy on l_z [see (10.35b)], carriers (and excitons) can be localized or trapped at regions with larger l_z.

The strongest disorder occurs in amorphous semiconductors, the prominent example being α-Si. In these systems only short-range order remains (e.g., a tetrahedral coordination) but no long-range order at all. If we start in a $CdS_{1-x}Se_x$ crystal at a S atom and proceed by a lattice vector $\boldsymbol{R}$ we will hit an atom. We are not sure, however, whether it will be S or Se. In α-Si there is nothing like a lattice "constant" and if we move from a Si atom by a vector $\boldsymbol{R}$ of crystalline Si we are even not sure whether we will find an atom at all. This strong disorder leads to substantial exponential tails of the densities of state both for conduction and valence bands which cover the whole "forbidden gap". Unsaturated covalent bands in Si, so-called "dangling" bonds, form states especially in the center of the gap, since it is undetermined whether they would form a bonding or an antibonding orbital with the next Si. The states of the dangling bonds can be saturated by hydrogen doping, opening interesting technological applications for α-Si:H.

We will end the discussion of the properties of electronic states in semiconductors here and concentrate in the next section on electron–hole pairs, which are actually the relevant complexes for the optical properties, but which cannot be understood without a knowledge of the properties of electrons and holes.

10.11 Problems

1. Try to find the highest and lowest values for effective electron and hole masses in group-IV materials and in binary III–V, II–VI and I–VII semiconductors. (Use a compilation of semiconductor data from the library). What is the trend of the dependence of the effective masses on E_g?

2. Calculate the effective density of states, i.e., the onset of degeneracy, for electrons and holes at 10 K and at room temperature in bulk GaAs and ZnSe.

3. Calculate the effective density of states per unit volume at 300 K for electrons and holes in bulk GaAs and the density of states per unit area in a GaAs QW. Calculate the corresponding density per volume for $l_z = 10\,\mathrm{nm}$.

4. What is the minimum power consumption per unit area required to keep a degenerate electron and hole population at room temperature in ZnSe and in GaAs for a lifetime of 0.3 ns and layers with $l_z = 1\,\mu\mathrm{m}, 0.2\,\mu\mathrm{m}$, and 10 nm thickness? (The first two cases can be considered as bulk material the third one as quasi-two-dimensional).

5. Calculate the binding energy of electrons (holes) to donors (acceptors) for some of the materials mentioned in the first problem. Compare with experimental data such as that given in [10.30]. Deduce the order of magnitude of the

central cell correction. Check the ratio of the binding energy and of the energy of the LO phonon, and try to anticipate the consequence for the choice of ε.

6. Calculate the Bohr radii of acceptors and donors for the extreme values found in the first problem. How many unit cells or atoms are contained in the volume of a donor?

7. Make a sketch of surfaces of constant energy in a two- (or even three-) dimensional $\boldsymbol{k}$-space for spherical, simple cubic, and hexagonal (plane $\boldsymbol{k} \perp \boldsymbol{c}$) symmetries. Is spherical symmetry compatible with cubic and hexagonal symmetry?

11 Excitons

In Chap. 10 we defined the bandstructure for electrons and holes as the solutions to the $(N \pm 1)$-particle problem and later we saw that the number of electrons in a band can be increased or decreased by donors and acceptors, respectively (Sect. 10.9). In contrast, the number of electrons remains constant in the case of optical excitations with photon energies in the eV or band gap region. What we can do, however, is to excite an electron from the valence to the conduction band by absorption of a photon. In this process, we bring the system of N electrons from the ground state to an excited state. What we need for the understanding of the optical properties of the electronic system of a semiconductor is therefore a description of the excited states of the N particle problem. The quanta of these excitations are called "excitons".

We can look at this problem from various points of view.

The ground state of the electronic system of a perfect semiconductor is a completely filled valence band and a completely empty conduction band. We can define this state as the "zero" energy or "vacuum" state. In addition it has total momentum $\boldsymbol{K}=0$, angular momentum $\boldsymbol{L}=0$ and spin $\boldsymbol{S}=0$. From this point $E=0$, $\boldsymbol{K}=0$ we will start later on to consider the dispersion relation of the excitons in connection with Fig. 11.1b.

Another point of view is the following. If we start from the above-defined groundstate and excite one electron to the conduction band, we simultaneously create a hole in the valence band (Fig. 11.1a). In this sense an optical excitation is a two-particle transition. The same is true for the recombination process. An electron in the conduction band can return radiatively or non-radiatively into the valence band only if there is a free place, i.e., a hole. Two quasiparticles are annihilated in the recombination process.

Excitons can be described at various levels of sophistication. We present in the next sections the most simple and intuitive picture using the effective mass approximation. Other approaches are described in [11.1–12] and references therein.

11.1 Wannier and Frenkel Excitons

Using the effective mass approximations, Fig. 11.1a suggests that the Coulomb interaction between electron and hole leads to a hydrogen-like problem with a Coulomb potential term $-e^2/(4\pi\varepsilon_0\varepsilon|\boldsymbol{r}_e-\boldsymbol{r}_h|)^{-1}$.

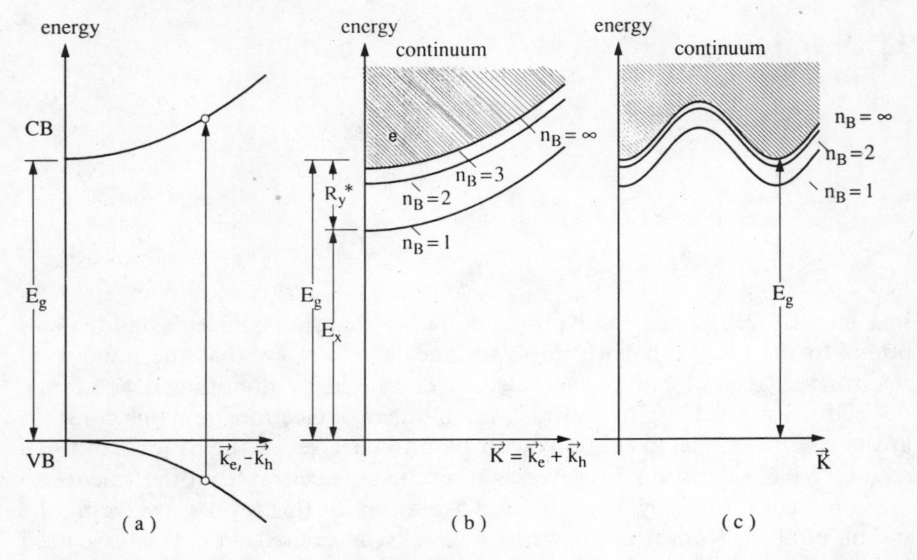

Fig. 11.1a–c. A pair excitation in the scheme of valence and conduction band (**a**) in the exciton picture for a direct (**b**) and for an indirect gap semiconductor (**c**)

Indeed excitons in semiconductors form, to a good approximation, a hydrogen or positronium like series of states below the gap. For simple parabolic bands and a direct-gap semiconductor one can separate the relative motion of electron and hole and the motion of the center of mass. This leads to the dispersion relation of excitons in Fig. 11.1b.

$$E_{\mathrm{ex}}(n_{\mathrm{B}}, \boldsymbol{K}) = E_{\mathrm{g}} - \mathrm{Ry}^* \frac{1}{n_{\mathrm{B}}^2} + \frac{\hbar^2 \boldsymbol{K}^2}{2M} \tag{11.1a}$$

with

$n_{\mathrm{B}} = 1, 2, 3 \ldots$ principal quantum number ,

$$\mathrm{Ry}^* = 13.6\,\mathrm{eV} \frac{\mu}{m_0} \frac{1}{\varepsilon^2} \quad \text{exciton binding energy}, \tag{11.1b}$$

$$M = m_{\mathrm{e}} + m_{\mathrm{h}}, \quad \boldsymbol{K} = \boldsymbol{k}_{\mathrm{e}} + \boldsymbol{k}_{\mathrm{h}} \quad \text{translational mass and wave vector of the exciton} \tag{11.1c}$$

For the moment, we use a capital $\boldsymbol{K}$ for the exciton wave vector to distinguish this two-particle state from the one-particle states. When we are more familiar with the exciton as a new quasi-particle we shall return to $\boldsymbol{k}$.

$$\mu = \frac{m_e m_h}{m_e + m_h} m_0 \quad \text{reduced mass}\,, \tag{11.1d}$$

$$a_B^{ex} = a_B^H \, \varepsilon \frac{m_0}{\mu} \quad \text{excitonic Bohr radius}\,. \tag{11.1e}$$

The series of exciton states in (11.1a) has an effective Rydberg energy Ry* modified by the reduced mass of electron and hole and the dielectric "constant" of the medium in which these particles move; n_B is the principal quantum number. The kinetic energy term in (11.1a) involves the translational mass M and the total wave vector $\boldsymbol{K}$ of the exciton. The radius of the exciton equals the Bohr radius of the H atom again modified by ε and μ. Using the material parameters for typical semiconductors one finds

$$1\,\text{meV} \leqslant \text{Ry}^* \leqslant 200\,\text{meV}\ \text{Ry}^* \ll E_g \tag{11.2a}$$

and

$$50\,\text{nm} \gtrsim a_B \gtrsim 1\,\text{nm} > a_{\text{lattice}}\,. \tag{11.2b}$$

This means that the excitonic Rydberg energy Ry* is usually much smaller than the width of the forbidden gap and the Bohr radius is larger than the lattice constant. This second point is crucial. It says that the "orbits" of electron and hole around their common center of mass average over many unit cells and this in turn justifies the effective mass approximation in a self-consistent way. These excitons are called Wannier excitons [11.10].

It should be mentioned that in insulators like NaCl, or in organic crystals like anthracene, excitons also exist with electron–hole pair wavefunctions confined to one unit cell. These so-called Frenkel excitons [11.11] cannot be described in the effective mass approximation. As a rule of thumb, one can state that in all semi-conductors the inequalities (11.2) hold, so that we always deal with Wannier excitons.

To get an impression of the wavefunction, we form wave packets for electrons and holes $\phi_{e,h}(\boldsymbol{r}_{e,h})$ in the sense of the Wannier function of (10.17) and obtain schematically for the exciton wavefunction

$$\phi(\boldsymbol{K}, n_B, l, m) = \Omega^{-1/2} e^{i\boldsymbol{K}\cdot\boldsymbol{R}} \phi_e(\boldsymbol{r}_e)\, \phi_h(\boldsymbol{r}_h)\, \phi^{env}_{n_B,l,m}(\boldsymbol{r}_e - \boldsymbol{r}_h)\,, \tag{11.3a}$$

with the center of mass $\boldsymbol{R}$

$$\boldsymbol{R} = (m_e \boldsymbol{r}_e + m_h \boldsymbol{r}_h)/(m_e + m_h)\,, \tag{11.3b}$$

where $\Omega^{-1/2}$ is the normalization factor. The plane-wave vector describes the free propagation of a Wannier exciton through the periodic lattice similarly as for the Bloch waves of Sect. 10.1, and the hydrogen-atom-like envelope ϕ^{env} function gives the relative motion of the electrons and holes.

As for the H atom, the exciton states converge for $n_B \to \infty$ to the ionization continuum, the onset of which coincides with E_g.

In indirect gap semiconductors, excitons are also formed with carriers in their respective band minima as shown in Fig. 11.1c. The continuation of the exciton dispersion from the indirect to the direct gap is an oversimplifications, among others because the states are away from the band extrema strongly damped and their binding energy varies.

The discrete and continuum states of the excitons will be the resonances or oscillators which we have to incorporate into the dielectric function of Chaps. 4–7.

For direct semiconductors with dipole allowed band-to-band transitions, one finds an oscillator strength for excitons in discrete states with $S(l=0)$ envelope function proportional to the band-to-band dipole transition matrix element squared and to the probability of finding the electron and hole in the same unit cell. For the derivation of this relation see [11.3]. This latter condition leads to the n_B^{-3} dependence of the oscillator strength for three-dimensional systems.

$$f_{nB} \propto |H_{cv}^{D}|^2 \frac{1}{n_B^3} . \tag{11.4}$$

These f_{nB} result in corresponding longitudinal–transverse splitting as shown in connection with (4.26). Equation (11.4) holds for so-called singlet excitons with antiparallel electron and hole spin. Triplet excitons involve a spin flip, in their creation which significantly reduces their oscillator strength (spin flip forbidden transitions). The oscillator strength of the continuum states is influenced by the so-called Sommerfeld enhancement factor. We come back to this point later in 11.3 and in Chap. 14, when we discuss the optical properties.

In the picture of second quantization, we can define creation operation operators for electrons in the conduction band and for holes in the valence band $\alpha_{k_e}^+$ and $\beta_{k_h}^+$, respectively. The combination of both gives creation operators for electron hole pairs $\alpha_{k_e}^+ \beta_{k_h}^+$. The exciton creation operator B^+ can be constructed via a sum over electron–hole pair operators [11.12]

$$B_k^+ = \sum_{k'_e k_h} \delta[\mathbf{K} - (\mathbf{k}_e + \mathbf{k}_h)] a_{k_e,k_h} \alpha_{k_e}^+ \beta_{k_h}^+ ; \tag{11.5}$$

the expansion coefficients a_{k_e,k_h} correspond, in principle, to those used to form the Wannier states.

It can be shown that the $\boldsymbol{B}_k^+, \boldsymbol{B}_k$, obey Bose commutation relations with a density-dependent correction term which increases with the number of electrons and holes contained in the volume of one exciton $4\pi(a_B^{ex})^3/3$.

This has two consequences: in thermodynamic equilibrium for low densities and not too low temperatures, the excitons can be well described by Boltzmann statistics with a chemical potential ruled by their density and temperature similar to (10.33). For higher densities they deviate more and more from ideal bosons until they end up in an electron–hole plasma made up entirely from fermions (see Chap. 20). This makes the creation of a Bose-condensed state of excitons (or of biexcitons, Chap. 19) a very complicated problem, and indeed no clearcut observation of a spontaneous Bose condensation of excitons has been published so far, though there are some experiments which prove the Bose character of excitons [11.13–15].

11.2 Corrections to the Simple Exciton Model

The simple model outlined in the preceeding section is, as already mentioned, adequate for non-degenerate, parabolic bands. We keep these assumptions for the moment and inspect a first group of corrections which are relevant for the parameters ε and μ entering in (11.1). We already know from Chaps. 4–8 that ε is a function of ω, resulting in the question of which value should be used.

As long as the binding energy of the exciton E_{ex}^{b} is small compared to the optical phonon energies and, consequently, the excitonic Bohr radius (11.1e) larger than the polaron radius (10.23b)

$$E_{ex}^{b} < \hbar\omega_{LO}; \quad a_B > a_{Pol} \Rightarrow \varepsilon = \varepsilon_s , \tag{11.6}$$

we can use for ε the static value ε_s below the phonon resonances and the polaron masses and polaron gap in (10.23). This situation is fulfilled for some semiconductors for all values of n_B, e.g., for GaAs where $Ry^* \simeq 5\,meV$ and $\hbar\omega_{LO} \simeq 36\,meV$.

In many other semiconductors the inequality (11.6) holds only for the higher states $n_B \geqslant 2$, while for the ground state exciton ($n_B = 1$) we get

$$E_{ex}^{b} \gtrsim \hbar\omega_{LO}; \quad a_B \simeq a_{Pol} \longrightarrow \varepsilon_s \geqslant \varepsilon \geqslant \varepsilon_b . \tag{11.7}$$

Examples are CdS, ZnO and CuCl.

In this situation a value for ε between ε_s and ε_b seems appropriate, because the polarization of the lattice can only partly follow the motion of electron and hole. A useful approach is the so-called Haken potential [11.16] which interpolates between ε_s and ε_b depending on the distance between electron and hole, where r_{eh}, r_e^p and r_h^p are the distances between electron and hole, and the polaran radii of electron and hole, respectively:

$$\frac{1}{\varepsilon(\boldsymbol{r}_{e,h})} = \frac{1}{\varepsilon_b} - \left(\frac{1}{\varepsilon_b} - \frac{1}{\varepsilon_s}\right)\left(1 - \frac{\exp(-\boldsymbol{r}_{eh}/\boldsymbol{r}_e^p) + \exp(-\boldsymbol{r}_{eh}/\boldsymbol{r}_h^p)}{2}\right) . \tag{11.8}$$

The next correction concerns the effective masses. The polarization clouds of the polarons (Sect. 10.5) have different signs for electron and hole. If both particles are bound together in an exciton state fulfilling (11.7) the polaron renormalization is partly quenched, with the consequence that values for the effective masses will lie somewhere between the polaron values and the ones for a rigid lattice (10.23). The gap "seen" by the exciton in the 1S state will likewise be situated between the two above extrema. Fortunately, the above effects tend to partly compensate each other. A transition from the polaron gap to the larger rigid lattice gap shifts the exciton energy to larger photon energies. A transition from ε_s to ε_b and a reduction of the effective masses increases the binding energy and shift the 1S exciton to lower photon energies. As a consequence one finds, even for many semiconductors for which inequality (11.7) holds, that the 1S exciton fits together with the higher exciton states reasonably well into the hydrogen-like series of (11.1). We shall use this approach in the future if not stated otherwise and call the experimentally observed energetic distance between the 1S exciton and the polaron gap the exciton binding energy E_{ex}^{b}, in contrast to Ry* in (11.1).

There is a general trend of the material parameters m_{eff} and ε with E_g which results in an increase of the exciton binding energy with increasing E_g as shown in Fig. 11.2.

The next complication comes from the band structure. If the bands are degenerate, as is the Γ_8 valence band in T_d symmetry, it is no longer possible to separate the relative and the center of mass motion—they are coupled together. Similar effects stem from k-linear terms and other sources. We get light- and heavy-hole exciton branches and splittings, e.g., between the 2S and 2P exciton states.

Furthermore it should be mentioned, without going into details, that the splitting between singlet and triplet excitons Δ_{st} and the splitting of the singlet state into a transverse and a longitudinal one Δ_{LT}, are both due to exchange interaction between electron and hole caused by their Coulomb interaction [11.4] if we consider the N-electron problem in the form of Slater's determinant where the ground state consists only of valence-band states and the excited state of a sum of determinants in each of which one valence-band state is replaced by a conduction-band state. This aspect is treated in detail in [11.4, 12].

Usually the following relation holds for Wannier excitons

$$\Delta_{st} \ll \Delta_{LT} \quad \text{with} \quad 0.1\,\text{meV} \lesssim \Delta_{LT} \lesssim 15\,\text{meV}\,. \tag{11.9}$$

Finally we mention that excitons can also be formed with holes in deeper valence bands. These so-called "core-excitons" are situated usually in the VUV

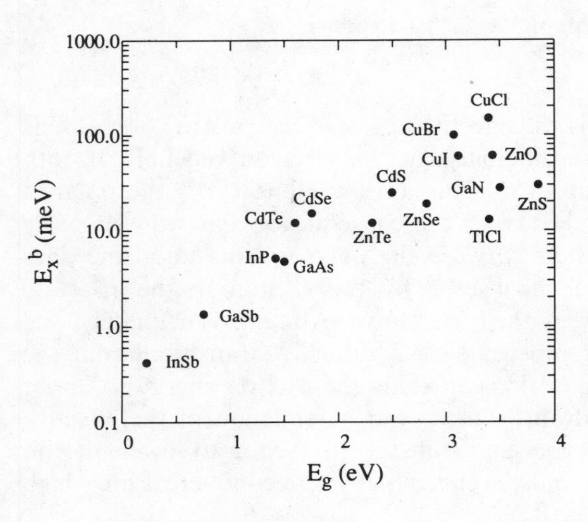

Fig. 11.2. The exciton binding energy E_{ex}^{b} as a function of the band-gap for various direct gap semiconductors. According to [10.7 and 11.12]

or X-ray region of the spectrum and have a rather short lifetime. A detailed discussion of these states is beyond the scope of this book. An exmaple and further references are given in Sect. 14.8 and [11.17].

11.3 The Influence of Dimensionality

If we consider the exciton again as an effective mass particle with parabolic dispersion relations, as given by (11.1), we expect a first influence of the dimensionality on the density of states analogous to the situation shown in Fig. 10.13 for every exciton branch $n_B = 1, 2, 3 \ldots$.

Another effect of the dimensionality manifests itself in the binding energy, the Rydberg series and the oscillator strength. We consider an exciton, for which the motion of electron and hole is restricted to a two-dimensional plane, but the interaction is still a 3d one, i.e., proportional to $e^2/|\boldsymbol{r}_e - \boldsymbol{r}_h|$ and find (11.10b) in comparison to the 3d case of (11.10a):

$$\text{3d: } E(\boldsymbol{K}, n_B) = E_g - \text{Ry}^* \frac{1}{n_B^2} + \frac{\hbar^2(K_x^2 + K_y^2 + K_z^2)}{2M}$$

and for the oscillator strength f and the Bohr radius a_B for the principal quantum number n_B

$$f(n_B) \propto n_B^{-3}; \quad a_B \propto a_B^H n_B; \quad n_B = 1, 2, 3 \ldots \tag{11.10a}$$

$$\text{2d: } E(\boldsymbol{K}, n_B) = E_g + E_Q - \text{Ry}^* \frac{1}{(n_B - \frac{1}{2})^2} + \frac{\hbar^2(K_x^2 + K_y^2)}{2M}$$

with E_Q quantization energy

$$f(n_B) \propto \left(n_B - \frac{1}{2}\right)^{-3}; \quad a_B \propto a_B^H \left(n_B - \frac{1}{2}\right), \; n_B = 1, 2, 3 \ldots \tag{11.10b}$$

Essentially n_B has to be replaced by $n_B - 1/2$ when going from 3d to 2d systems and the quantization energies E_Q [see (10.35)] of electrons and holes must be considered. The excitonic Rydberg Ry* is the same in both cases with the consequence that the binding energy of the 1S exciton is Ry* in three, and 4 Ry* in two dimensions. The oscillator strength increases and the excitonic Bohr radius decreases when going from three- to two-dimensional systems (11.10).

The usual realization of quasi-2d excitons is via (M)QW of type I. In this case the motion in the z-direction is quantized, but the width of the quantum well l_z is non-zero.

In Fig. 11.3 we show the exciton binding energy for GaAs as a function of l_z for infinitely high barriers, where the curve reaches 4Ry* for $l_z = 0$ and for finite barrier height, where the binding energy converges to the value of the barrier material for $l_z = 0$ passing through a maximum of about 2 to 3 times Ry* depending on the material parameters.

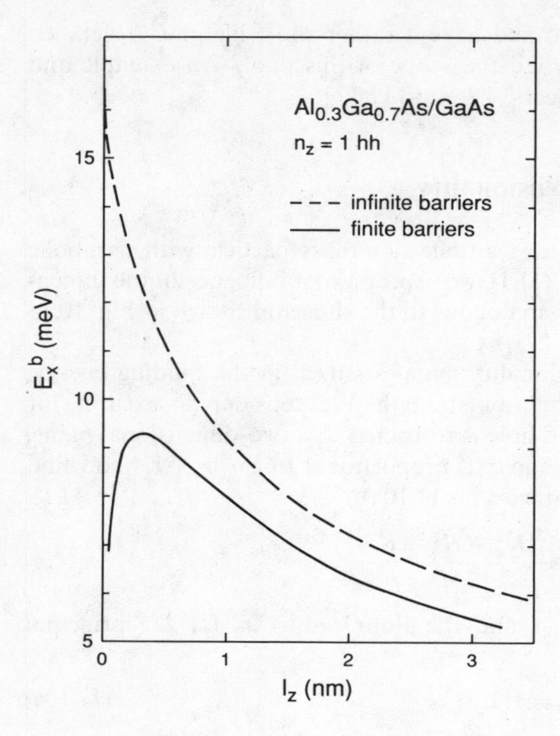

Fig. 11.3. The calculated binding energy of $n_z = 1$ hh excitons in AlGaAs/GaAs quantum wells as a function of the well thickness l_z. According to [11.18]

The increase of the oscillator strength of the 1S exciton comes from the fact that the quantization in the z-direction increases both the overlap between electron and hole and their attraction, which results in turn in a reduction of the two-dimensional Bohr radius.

The Sommerfeld factor F, which describes the enhancement of the oscillator strength of the continuum states and which arises from the residual electron–hole correlation, depends also on the dimensionality [11.19]. It reads

$$\text{3d}: F_{3\text{d}} = \frac{\pi}{W^{1/2}} \frac{\mathrm{e}^{\pi W^{-1/2}}}{\sinh(\pi W^{-1/2})} \quad \text{with} \quad W = (E - E_\mathrm{g})\mathrm{Ry}^{*-1}; \tag{11.11a}$$

$$\text{2d}: F_{2\text{d}} = \frac{\mathrm{e}^{\pi W^{-1/2}}}{\cosh(\pi W^{-1/2})} \quad \text{with} \quad W = (E - E_\mathrm{g} + E_\mathrm{Q})\mathrm{Ry}^{*-1}. \tag{11.11b}$$

In the three-dimensional case it has a square-root singularity at E_g and decreases gradually to unity for $\hbar\omega > E_\mathrm{g}$. In two dimensions it decays only from two to one with increasing energy and in a one-dimensional system it is even below unity just above the gap. We shall see the consequences in Chap. 14.

The corrections which we mentioned in Sect. 11.2 hold partly also in the quasi-2d case, bearing in mind for example the valence-band structure in

Fig. 10.15. However, it should be noted that the most widely investigated (M) QW are based on AlGaAs and InGaAs (Fig. 10.16), which fulfill the inequality (11.6).

The most striking feature of excitons in quasi-two-dimensional systems, however, is the splitting into light- and heavy-hole excitons which results from the corresponding splitting of the valence-band states (Fig. 10.14b). To describe the exciton states we thus need more quantum numbers. Apart from the principal quantum number n_B in (11.10b), we must state which of the quantized conduction- and valence-band states are involved. The simplest optical interband selection rule is $\Delta n_z = 0$, so that we shall see in optical spectra mainly excitons which obey this rule. Finally we must specify whether we are speaking of the light- or the heavy-hole series. Complete information might thus be the $n_z^h = 2\text{hh}$, $n_z^e = 2$, $n_B = 1$ exciton state. Usually one uses the abbreviation $n_z = 2\text{hh}$ exciton involving the above selection rule and the fact that excitons in MQW with $n_B > 1$ are usually difficult to resolve due to broadening effects. For examples see Sect. 14.7.

In strictly one- and zero-dimensional cases the binding energy for the exciton diverges. So it is not possible to give general formulas like (11.10) for these situations. One is always limited to numerical calculations which have to explicitly include the finite dimensions of the quantum wire or quantum dot. Quantum dots made from microcrystallites of semiconductors such as the Cu halides, of $CdS_{1-x}Se_x$ including $x = 0$ and $x = 1$, or of CdTe, represent at this time a very active field of research concerning the investigation of the excitons. Some reviews are given in [11.20].

Three regimes of quantization are usually distinguished in which the crystallite radius R is compared with the Bohr radius of the excitons or related quantities:

$$\text{weak confinement: } R \gtrsim a_B,\ E_Q < \text{Ry}^* \,; \tag{11.12a}$$

$$\text{medium confinement: } a_B^e \geqslant R \geqslant a_B^h;\ E_Q \approx \text{Ry}^* \,, \tag{11.12b}$$

with $a_B^{e,h} = a_B^H \varepsilon \frac{m_0}{m_{e,h}}$;

$$\text{Strong confinement: } R \leqslant a_B^h;\ E_Q > \text{Ry}^* \,. \tag{11.12c}$$

In the first case (11.2a) the quantum dot (QD) is larger than the exciton. As a consequence the center-of-mass motion of the exciton, which is described in (11.3) by the term $e^{i\boldsymbol{K}\cdot\boldsymbol{R}}$, is quantized while the relative motion of electron and hole given by the envelope function $\phi_{n_B,l,m}(\boldsymbol{r}_e - \boldsymbol{r}_h)$ is hardly affected. This situation is found, e.g., for the Cu halides where a_B is small, or for CdSe QD with $R \geqslant 10$ nm. In the second case (11.12b) R has a value between the radii of the electron orbit and the hole orbit around their common center of mass. As a consequence, the electron state is quantized and the hole moves in the potential formed by the dot and the space charge of the quantized electron. This case is the most demanding from the theoretical point of view since Coulomb effects and quantization energies are of the same order of magnitude. However, it is often realized for QD of II–VI semiconductors. The regime given by (11.12c) becomes easier again. The Coulomb energy increases roughly with R^{-1}, and the quantization energy with R^{-2}, so that for sufficiently small values of R one should reach a situation where the Coulomb term can be neglected. However, recent investigation showed that $E_Q \gg Ry^*$ holds only for R values which are comparable with the lattice constant. In this case the applicability of the effective mass approximation becomes questionable, and the dot may be better considered as a huge molecule.

In addition to the above-mentioned difficulties, there are some others which lead to an inhomogeneous broadening as compared to the δ-like density of states in Fig. 10.13 as well as other complications. The dots usually have a certain spread of R values, as already mentioned, which directly influences the quantization and the Coulomb energies. Though the "gap" of the surrounding amorphous glass matrix is usually much larger than that of the semiconductor.

$$E_g^{SC} < E_g^{glass} \tag{11.13}$$

there is no abrupt, infinitely high barrier. Interface states may appear, which depend both on the surrounding matrix and on the growth regime of the QD; image forces have to be considered in quantitative calculations, since the dielectric functions of semiconductor and glass are different; deviations of the QD from an ideal sphere are obvious from TEM investigations but usually neglected; the coupling of excition to phonons is enhanced, especially in the regime of (11.12b), since the different radial distribution of the electron and hole wavefunctions give rise to a dipole layer. Finally a realistic bandstructure has to be taken into account at least as long as the effective mass approximation is still valid. For more details see [10.24–29] and references therein.

11.4 Bound Exciton Complexes

Similar to the way that free carriers can be bound to (point-) defects, it is found that excitons can also be bound to defects.

We discuss first shallow impurities. The binding energy of an exciton (X) is highest for a neutral acceptor (A^0X complex), lower for a neutral donor (D^0X) and lower still for an ionized donor (D^+X). An ionized acceptor does not usually bind an exciton since a neutral acceptor and a free electron are energetically more favorable. The absorption and emission lines of A^0X, D^0X and D^+X are often labelled I_1, I_2 and I_3 lines, respectively. The binding energy of an exciton to a neutral donor (acceptor) is usually much smaller than the binding energy of an electron (hole) to the donor (acceptor). The ratio of the two energies depends only weakly on the material parameters and amounts approximately to 0.1. This fact is known as Heynes rule [11.21].

The binding energy of the exciton to the complex depends also on the chemical nature of the complex (chemical shift) and on the surroundings, leading in high resolution spectroscopy to a splitting of the I_i lines. Furthermore, bound exciton complexes may have a certain manifold of excited states due to the various mutual arrangements and envelope functions of the two electrons (holes) and the hole (electron) in the $D^0X(A^0X)$ complex [11.22]. We shall meet some examples in Chap. 15.

The wavefunctions of excitons bound to shallow centers can be described by a superposition of free exciton wavefunctions in a similar way to that shown in (10.41) for free carriers.

To conclude this section on bound exciton complexes we give first some short statements on excitons bound to point defects other than single shallow donors or acceptors.

Donor–acceptor pairs (Fig. 10.18) can be considered as "polycentric" bound excitons. On the other hand, it has been found that one center can, under certain conditions, bind several excitons. The formation of such multi-exciton complexes is especially favored in indirect semiconductors due to the high degeneracy of the multivalley conduction band and the fourfold degenerate Γ_8^+ valence band [11.23]. Furthermore excitons can be bound to the deep centers, mentioned in Sect. 10.9. A review of bound exciton complexes is found in [11.24].

Finally, it should be mentioned that such complexes also exist in quantum wells. In this case, the energy of the bound exciton depends in addition on the spatial position of the impurity relative to the barriers. The binding energy usually decreases if the impurity is located not in the center of the well but closer to one of the barriers because the wavefunction is pushed away from the impurity [11.25]. This phenomenon results in an additional inhomogeneous broadening of the absorption and emission lines, which then often merge with the tail-states caused by disorder (see below) and/or with the free-exciton line.

11.5 Excitons in Disordered Systems

In our discussion of disordered systems in Sect. 10.10 we saw that disorder leads to the appearance of localized electron and/or hole states.

In a similar way the two-particle complex exciton can be localized in a disordered semiconductor.

If we look to the potential wells and barriers in the valence band e.g., of $CdS_{1-x}Se_x$ (Fig. 10.22) we can envisage two different mechanisms of localization. In very deep potential wells for holes, such a quasiparticle can be localized, and the electron is bound to the localized hole by Coulomb interaction. The other possibility is that we have a wide potential well with dimensions larger than the excitonic Bohr radius. In this case the exciton is localized as a whole.

In Fig. 11.4 we show schematically the density of localized exciton states for $n_B = 1$. At low energies we start with excitons for which one carrier is localized and the other bound to it by Coulomb attraction. With increasing energy there is a continuous transition to excitons which are localized as a whole, which in

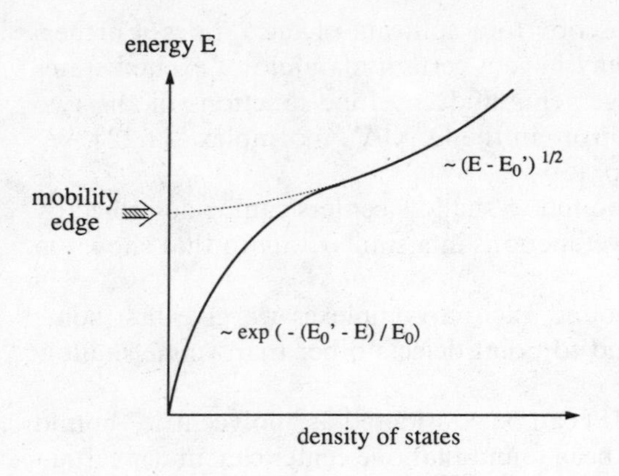

Fig. 11.4. The density of exciton states for $n_B = 1$ in a disordered semiconductor

turn ends at the transition region to extended exciton states known as the mobility "edge". Above this there are the extended exciton states with properties approaching those of free excitons in ordered materials, except for a considerably reduced phase-relaxation time T_2 at low temperatures (Sect. 22.2).

In principle, pictures similar to Fig. 11.4 should hold also for $n_B > 1$. Due to the significant inhomogeneous broadening of the $n_B = 1$ state, however, there is not much chance of identifying higher states of the Rydberg series by optical spectroscopy. Therefore usually only the exciton ground states are considered.

As long as the tailing parameters E_0 (10.47) describing the localized excitons is much smaller than E_g, we can use the effective mass approximation for localized excitons to give the following rules of thumb. Since heavier particles are easier to localize according to (10.47), we will find significant features of localized excitons more frequently in those ionic semiconductors with alloying in the anions that form the valence band like $CdS_{1-x}Se_x$ or $ZnSe_{1-x}Te_x$ than in cation-substituted materials such as $Zn_{1-y}Cd_ySe$. In semiconductors with more covalent binding, localization occurs both in the conduction and valence bands, as in α-Si or in $Ga_{1-y}Al_yAs$. However, especially in the latter example, both the electron and hole masses are relatively low and, as a consequence, it is difficult to localize excitons, i.e., the tailing parameter E_0 and the total number of localized exciton states N_0 in (10.47) are small. Another aspect of the same feature is the following: Due to the low effective masses mentioned above, the exciton Bohr radius is much larger in $Al_{1-y}Ga_yAs$ than in $CdS_{1-x}Se_x$. This means that the exciton averages over a larger volume, thus diminishing the effective fluctuation and reducing the tail of localized exciton states.

As a consequence, in $CdS_{1-x}Se_x$ or $ZnSe_{1-x}Te_x$ for x around 0.1 to 0.4, the tail of localized exciton states contains roughly 10^{18} to 10^{19} states per cm^3, while this number may be one or two orders of magnitude smaller for the

$Al_{1-y}Ga_yAs$ system (or for cation-substituted materials). More information on localized excitons is given in [11.26, 27].

In quantum wells various types of disorder can contribute to the formation of tails of localized exciton states as already discussed for one-particle states:

- alloy disorder if the well and/or the barrier material is an alloy. The first case obviously has a stronger influence since the probability of finding the electron and hole in the well of a type I structure is larger than that of finding it in the barrier (Fig. 10.14b).
- interface roughness, i.e., well-width fluctuations. Usually quantum wells can only be grown with well-width fluctuations of at least one monolayer. These fluctuations of l_z influence the exciton energy via the l_z-dependence of the quantization energy (10.35b), and via the l_z-dependence of the exciton binding energy (Fig. 11.3). Usually the first effect is the dominant one. For some recent reviews of this topic see [11.28] and references therein.

The investigation of excitons in quantum wires is only in its infancy. Therefore not too much is known about localization effects in these structures. For first examples see [11.29].

For systems that are confined in all three dimensions, i.e., quantum dots, the question of localization is irrelevant.

11.6 Problems

1. Calculate the Rydberg energy and the Bohr radius of excitons for some of the semiconductors for which you found the material parameters in the problems of Chap. 10. Compare these with the experimentally determined binding energies and lattice constants, respectively.

2. How many (different) exciton states can be constructed in a semiconductor with zinc-blende (T_d) structure for the principal quantum numbers $n_B = 1, 2$ and 3?

3. Compare the magnitude of the relative splitting between 2s and 2p states in a hydrogen atom (what are the physical reasons?) with the 2s–2p splitting of excitons.

4. Plot the Rydberg series of an idealized three- and two-dimensional exciton and indicate the oscillator strengths.

5. Calculate the (combined) density of states in the continuum of a three- and a two-dimensional exciton in the effective mass approximation. Multiply by the corresponding Sommerfeld enhancement factor.

12 Some Further Elementary Excitations

Here we will briefly address some other collective excitations in semiconductors and the quasi-particles which result from the quantization of these excitations.

12.1 Plasmons and Pair Excitations

The excitons presented in Chap. 11 are the energetically lowest elementary excitations of the electronic system of an ideal semiconductor (or insulator). However, if we produce in a semiconductor a large density of free electrons or holes, e.g., by doping (Sect. 10.9) or by high (photo-) excitation (Chap. 20), other elementary excitations appear in the electronic system.

We consider in the following a semiconductor which contains a large number of electrons (say 10^{17}–$10^{19}\,\mathrm{cm}^{-3}$). Analogous results are found for holes.

The gas of free electrons can perform collective oscillations relative to the positive background of ionized donors. We consider in Fig. 12.1a the three-dimensional situation. A displacement of the electron gas of density n by an amount Δx produces a surface charge density ρ_s

$$\rho_s = ne\Delta x \tag{12.1a}$$

and an electric field, according to (2.1a)

$$E_x = \frac{ne\Delta x}{\varepsilon\varepsilon_0} \tag{12.1b}$$

provided we can neglect boundary effects in the y and z directions.

This electric field acts back on the electrons, leading to an equation of motion

$$eE_x = \frac{e^2 n\Delta x}{\varepsilon\varepsilon_0} = m_e \frac{\partial^2}{\partial t^2}(\Delta x)\,. \tag{12.1c}$$

The solution of (12.1c) is a harmonic oscillation with frequency

$$\omega_{PL}^0 = \left(\frac{e^2 n}{m_e \varepsilon\varepsilon_0}\right)^{1/2} = \omega_L\,, \quad \omega_T = 0\,. \tag{12.1d}$$

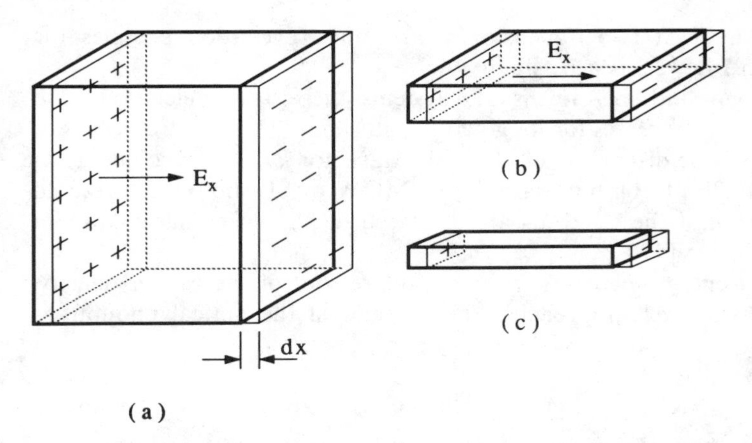

Fig. 12.1a–c. Schematic drawing of a displacement of the electron system in a plasma oscillation in a three- (***a***), two- (***b***) and one-dimensional semiconductor (***c***)

The dielectric "constant" ε which enters (12.1) is ε_s provided $\omega_{PL}^0 \ll \omega_{LO}$. What happens if this condition does not hold will be discussed in a few moments.

The oscillations described by (12.1) are known as plasma oscillations. Their quantization leads to new quasiparticles which obey Bose statistics and which are known as plasmons. In gases, including the electron gas considered here, only longitudinal oscillations can propagate since gases have a non-vanishing compression but no shear stiffness.

Consequently the value of ω_{PL}^0 in (12.1d) gives the longitudinal eigenfrequency. The transverse one is zero. If we go to shorter wavelengths we find a weak parabolic dispersion for the plasmons:

$$\omega_{PL}(\boldsymbol{k}) = \omega_{PL}^0 (1 + \alpha k^2 + \cdots) \quad (12.2)$$

shown in Fig. 12.2a.

For the densities mentioned above, $\hbar\omega_{PI}^0$ is situated in the range 10–100 meV for typical semiconductors, i.e., in the (F)IR. This situation is different in metals, where the plasma frequency is usually situated in the VIS or UV part of the spectrum and causes the high reflectivity of this class of materials which extends from the IR up to ω_{PL}.

For large $\boldsymbol{k}$-vectors, the plasmon modes are strongly damped because they fall in the continuum of one-particle (or, as shown below more precisely, two-particle) intraband excitations. These excitations are shown in Fig. 12.3 where we give the dispersion relation of the conduction band filled up to the Fermi energy E_F by a degenerate electron gas and, for simplicity, $T = 0$. We can produce excitations in this "Fermi sea" of electrons by lifting an electron from a state below E_F into a state above (actually by simultaneously creating a hole in the Fermi sea, but we shall not pursue this aspect). The excitation energies range from zero to values given by the width of the band, i.e., several eV. Small excitation energies can be created for all wave vectors between zero and $2k_F$ if the $\boldsymbol{k}$-space is at least two dimensional. For finite excitation energies the shaded range in Fig. 12.2 is accessible.

For strictly two- and one-dimensional systems, the restoring electric field E is not constant but decreases for long wavelengths as λ^{-1} or λ^{-2}, respectively. As a consequence the dispersion of plasmons starts for $k=0$ at zero energy as shown in Fig. 12.2b [12.1], but plasmons in MQW or SL show rather the 3d dispersion relation, if the electrons are displaced in phase for many adjacent wells. For details see [12.4].

The plasmon energy increases with the square-root of the carrier density. What happens when ω_{Pl}^{0} approaches the energies of the optical phonons is

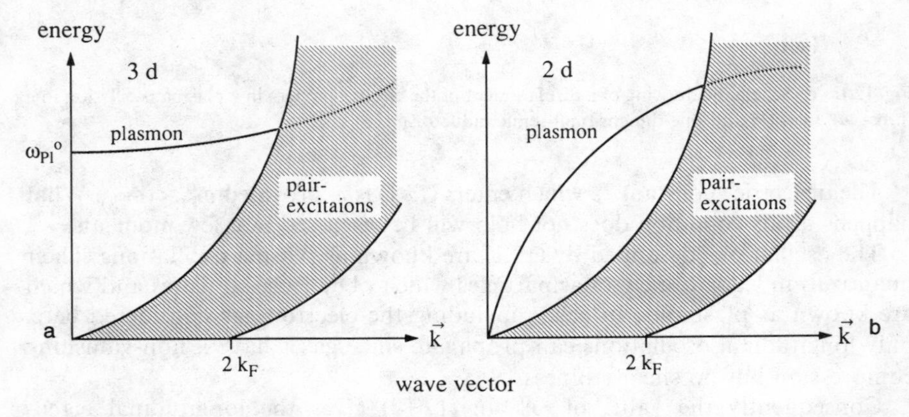

Fig. 12.2a,–b. The plasmon dispersion and the range of two-particle excitations in a three- (**a**) and a two-dimensional system (**b**)

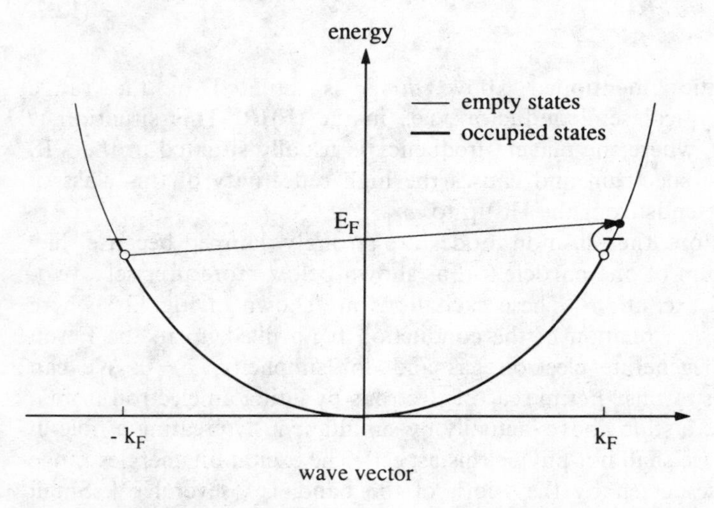

Fig. 12.3. Two-particle excitations within one band with degenerate population

shown in Fig. 12.4. Plasmons and phonons interact with each other due to their electric fields. This interaction results in another example of the quantum-mechanical non-crossing rule (Chap. 5). The plasmon frequency bends over to the transverse optical phonon branch and reappears above the longitudinal one. These two new branches are usually known as ω_- and ω_+ modes of this plasmon–phonon mixed state, respectively. For plasmon frequencies considerably above those of longitudinal optical phonons, the background dielectric constant ε_b has to be used in (12.1d).

To conclude this section we should add two comments. Instead of the plasmon resonance, the optical properties of free carriers can be described by the current term in (2.1d). In this case $\varepsilon(\omega)$ in (4.20) comprises only resonances other than the plasmon mode and we introduce the electrical conductivity σ

$$\boldsymbol{j} = \sigma \boldsymbol{E} \tag{12.3a}$$

with a relaxation-time τ approach for σ, i.e.,

$$\sigma = \sigma_0(1 - \mathrm{i}\omega\tau)^{-1} = \frac{\sigma_0}{1+\omega^2\tau^2} + \frac{\mathrm{i}\sigma_0\omega\tau}{1+\omega^2\tau^2} \quad \text{with} \quad \sigma_0 = \frac{ne^2\tau}{me, h} \tag{12.3b}$$

or a dielectric function for the free electrons

$$\varepsilon(\omega) = 1 - \frac{\omega p^2}{\omega^2 + i\omega/\tau}\,. \tag{12.3c}$$

For an electric field $\boldsymbol{E}$ given by

$$\boldsymbol{E} = \boldsymbol{E}_0 \mathrm{e}^{\mathrm{i}\omega t}\,, \tag{12.4}$$

This leads with $\nabla \times \boldsymbol{H} = \boldsymbol{j} + \dot{\boldsymbol{D}}$ to

$$\nabla \times \boldsymbol{H} = [\sigma + \mathrm{i}\omega\varepsilon(\omega)\varepsilon_0]\boldsymbol{E} =: \varepsilon_{\text{total}}(\omega)\boldsymbol{E}\,. \tag{12.5}$$

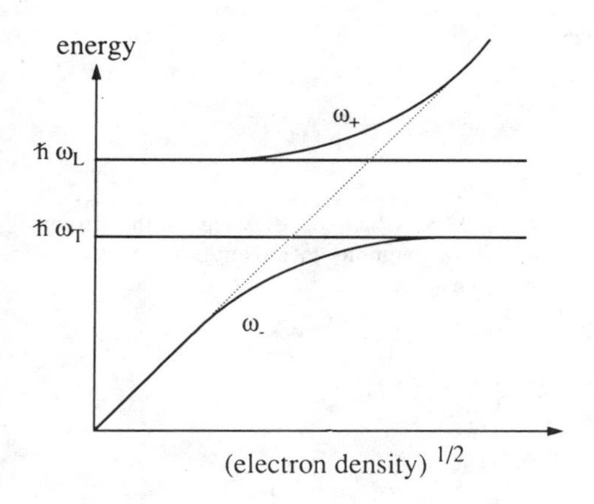

Fig. 12.4. The eigenenergies of the plasmon–phonon mixed state as a function of the square-root of the carrier density n

With this approach we could include a finite σ in the calculations of section 2.3. Evidently the real part of σ contributes then to the imaginary part of ε_{total}, i.e., to the damping, and the imaginary part of σ to the real part of ε_{total}. The optical properties deduced from this concept are the same as those derived from the plasmon concept using the plasma resonance, so that we need not follow this so-called Drude model further.

Finally it should be mentioned that there are also valenceband plasmons connected with collective excitations of the electron system of a filled valence band. Their eigenfrequencies are situated at energies much larger than E_g. Therefore they are not further considered. For details of this aspect see [12.2].

12.2 Magnons

We conclude this chapter by mentioning that in magnetic semiconductors like NiO there are collective excitations of the spin system.

The resulting quanta are known as magnons. In a simple ferromagnet there is only one magnon branch which starts for $\boldsymbol{k} = 0$ at $E = 0$ and shows a dispersion for finite $\boldsymbol{k}$. In antiferromagnetic or ferrimagnetic material there are various branches due to in-phase and antiphase excitation of the spin subsystems. In analogy to the monatomic and diatomic chains, the magnon dispersion curves are called "acoustic" and "optical" magnons. In Fig. 12.5 we give the dispersion relation of magnons in a ferrimagnetic or antiferromagnetic system. Since magnons are of less importance for the optical properties of semiconductors, we refer the reader for details e.g. to [12.3].

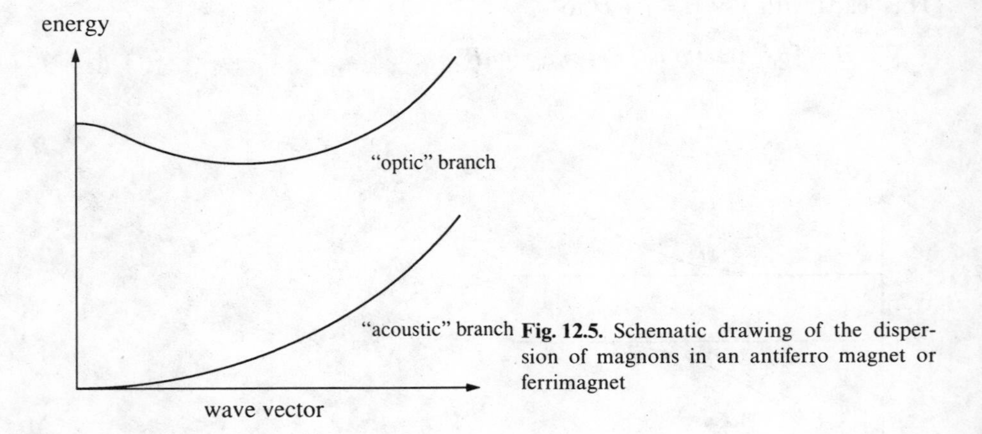

Fig. 12.5. Schematic drawing of the dispersion of magnons in an antiferro magnet or ferrimagnet

12.3 Problems

1. Calculate the plasmon energy $\hbar\omega_{PL}^0$ for a typical three-dimensional semiconductor ($m_e = 0.1\, m_0$) and $n = 10^{16}, 10^{17}$ and $10^{18}\,cm^3$. Compare with the eigenenergies of optical phonons.

2. Calculate $\hbar\omega_{PL}^0$ for a metal ($n \approx 10^{22}$–$10^{23}\,cm^{-3}$). Using the knowledge of Chap. 4, consider which value should be taken for the dielectric "constant" ε?

13 Optical Properties of Phonons and Plasmons

Having presented the optical properties of a system of model oscillators and the elementary excitations in semiconductors, we shall now put these two parts together and investigate as a first example the optical properties of phonons. We start with the properties of bulk material.

13.1 The Phonon Polariton as an Example with Negligible Spatial Dispersion

As already mentioned, optical phonons can couple strongly to the electromagnetic field if the solid has at least partly ionic binding. The $\boldsymbol{k}$ dependence of the eigenfrequency is rather weak, and completely negligible if we concentrate on the region of $\boldsymbol{k}$ vectors reached in the IR or VIS part of the spectrum, i.e., $k < 10^6\,\mathrm{cm}^{-1}$ as seen from Figs. 9.15, 16, bearing in mind that the first Brillouin zone extends up to $\boldsymbol{k}$ values of around $10^8\,\mathrm{cm}^{-1}$.

We can therefore treat the phonon polariton according to Chaps. 4 and 5, neglecting spatial disperison. Figure 13.1a shows the dispersion relation of the phonons in CdS. For the polarisation $\mathrm{E} \perp \mathrm{c}$ the phonon branches with symmetry Γ_5 couple to the radiation field. See also Chap. 25.

13.2 Reflection Spectra

In Fig. 13.1b we show the reflection spectrum of CdS in the IR around the phonon resonance. Spectra of other more or less ionic materials look very similar. We can clearly see the Reststrahlbande between ω_T and ω_L as we predicted in Fig. 4.5. The reflectivity reaches values above 0.9. Figures 13.1c and finally give the spectra of the real and imaginary parts of $\tilde{n}(\omega)$ deduced from the reflection spectrum either by fitting the equation (3.20) with (4.22) and (4.32) to the experiment or by using appropriate Kramers–Kronig relations introduced in Chap. 8. These curves coincide nicely with the results of our model calculations in Figs. 4.4 and 4.5. This agreement allows us to claim good understanding of the optical properties of phonons. It should be remembered here, that optical phonons in semiconductors with purely covalent binding like Si or Ge have, at $k = 0$, zero longitudinal–transverse splitting and zero oscillator strength, as shown in Fig. 9.16.

13.3 Raman and Brillouin Scattering

Raman scattering with phonons is the scattering of exciton polaritons under emission or absorption of a phonon polariton.

In the weak coupling picture one can also say that a photon creates a virtual exciton and is scattered by emission or absorption of an optical phonon, but this model fails in some respects as we shall see below.

In the Raman-scattering process energy and (quasi-)momentum have to be conserved, i.e.,

$$\hbar\omega_{\mathrm{R}} = \hbar\omega_{\mathrm{i}} \pm \hbar\Omega_{\mathrm{Ph}},$$
$$\boldsymbol{k}_{\mathrm{R}} = \boldsymbol{k}_{\mathrm{i}} \pm \boldsymbol{k}_{\mathrm{Ph}}, \qquad (13.1)$$

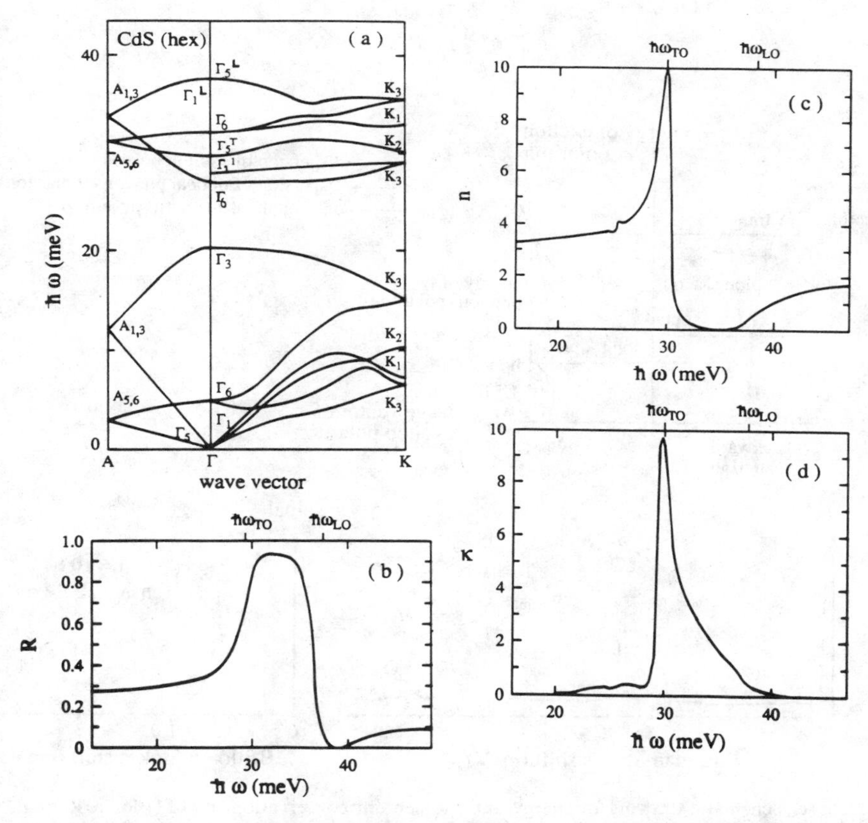

Fig. 13.1a–d. Optical properties of the phonons in CdS: The dispersion of the phonons (**a**), the reflectivity around the Reststrahlbande (stop-band) (**b**) and the real (**c**) and imaginary (**d**) parts of the complex refractive index $\tilde{n}(\omega)$. According to [13.1, 2]

where the index i stands for the initial exciton polariton, R for the Raman signal and Ph for the created or annihilated phonon polariton. Usually one chooses ω_i and ω_R in the transparent spectral region of the semiconductor well above the phonon resonance and below the exciton resonances. Often $\hbar\omega_i$ is determined by a readily available laser such as an Ar^+ laser.

Equation (13.1) results in values around $10^5\,cm^{-1}$ for $\boldsymbol{k}_i$ and $\boldsymbol{k}_R$. In a backward or 90° scattering geometry $\boldsymbol{k}_{Ph}$ is also around $10^5\,cm^{-1}$ and thus clearly lies on the phonon-like part of the dispersion relation (see e.g., Fig. 13.2) or on the longitudinal branch. In Fig. 13.2 we also give typical Raman spectra.

This case is covered by the weak-coupling picture. The selection rules are given in a classical description by the Raman tensor [13.2] and in quantum mechanics by the transition matrix elements. The group theory presented in Chap. 25 gives in this case preliminary information about which transitions are allowed and which are not. We do not want to go into details on this topic which is treated at length in [13.2] but cite here only the easily intelligible formula for the ratio of Stokes to anti-Stokes emission, i.e., scattering under

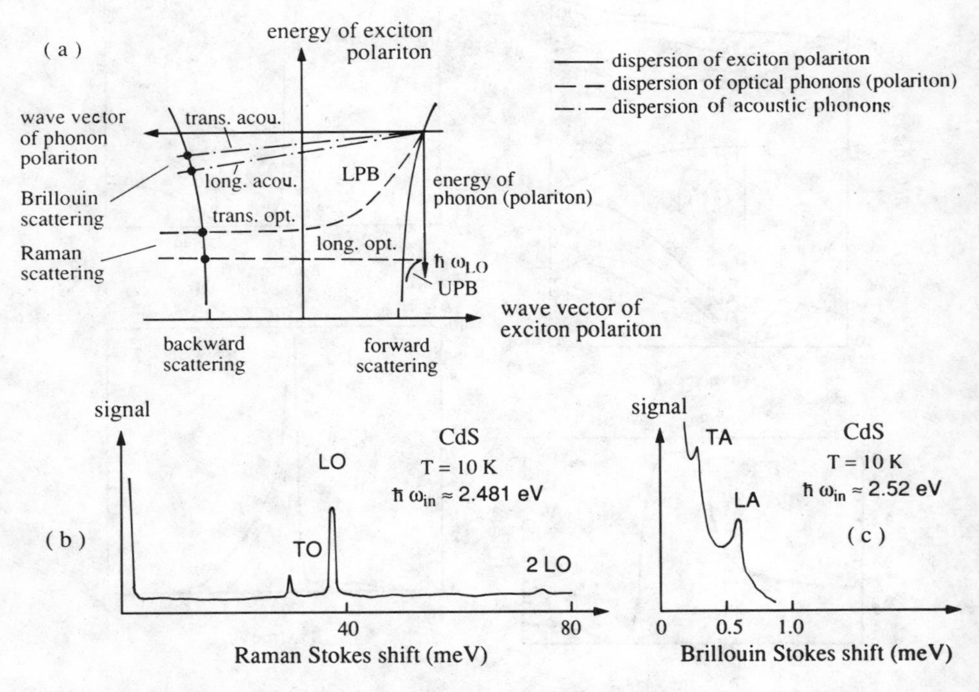

Fig. 13.2a–c. Schematic drawing of energy and momentum conservation in (13.1) for a Raman scattering process (Stokes emission) in a backscattering configuration (**a**), and typical spectra for Raman (**b**) and Brillouin scattering (**c**). Note the different scales on the *x*-axes in (**b**) and (**c**). Spectra according to [13.3,4]

emission or absorption of a phonon:

$$\frac{I_s}{I_a} = \left(\frac{\omega_i + \Omega_{Ph}}{\omega_i - \Omega_{Ph}}\right)^4 \exp\left(\frac{-\hbar\Omega_{Ph}}{k_B T}\right) \tag{13.2}$$

which just reflects the fact that the phonon emission is proportional to $(N_{Ph} + 1)$ and the absorption proportional to N_{Ph}, where N_{Ph} is the phonon occupation number of the respective mode.

If we go now to a forward scattering geometry, we see that $\boldsymbol{k}_{Ph}$ can become small and can fall into the transition region from phonon-like to photon-like polaritons. Measurements in this geometry allow one to measure the dispersion of the phonon polariton rather directly. Beautiful experiments of this type, described e.g., in [13.5], are performed by collecting the scattered light on the entrance slit of a spectrometer so that the height on the slit is a measure of $\boldsymbol{k}_i - \boldsymbol{k}_R$ (Fig. 13.3a). The wavelength dispersion of the spetrometer then gives the Ω axis. We show an example schematically in Fig. 13.3b. Figure 13.4 gives the dispersion relation for the phonon polariton in GaP reconstructed from this type of experiments, which of course, has to take into account the refraction of the beams at the surface of the sample in contrast to the simplifications in Fig. 13.3a. These experiments can obviously only be understood in the strong-coupling or polariton picture.

To conclude our discussion of Raman scattering in bulk semiconductors, we want to mention that Raman processes are also possible in which two phonons are created (or annihilated), e.g., due to the strong anharmonicity of the phonon–phonon interaction mentioned in connection with (9.47). If this is a two-step process, in the sense that the Raman polariton ω_R in (13.1) undergoes a second Raman scattering process, then both phonons have small wavevector compared to the size of the Brillouin zone. If, on the other hand, both phonons are emitted simultaneously, only the sum of the phonon wavevectors has to fulfill (13.1b). The individual phonon can come from any part of the Brillouin zone. Consequently the Raman spectrum then reflects to a certain extent the density of states of the phonons. Details of such processes are given in [13.7].

For the direct observation of the THz emission of optical phonons see [13.15] and references therein and for the influence of isotopes [13.16]

Brillouin scattering is the analog of Raman scattering for acoustic phonons. Because of the rather flat dispersion relation of acoustic phonons, with a slope given by the velocity of sound (instead of the c/n), one finds even in a back scattering configuration only much smaller (anti-) Stokes shifts, which are usually $\leqslant 1$ meV (Fig. 13.2c). Since the coupling of acoustic phonons to photon-like exciton polaritons is also much weaker than for optical phonons, some high-resolution techniques and an efficient suppression of stray light are necessary to detect Brillouin scattering, as can be seen by comparing the abscissa of Figs. 13.2b and c. Since the dispersion relation of acoustic phonons is linear in the range of interest, (13.1) can be rewritten for Brillouin scattering in a backscattering configuration as

$$\frac{\Omega_{\mathrm{Ph}}^{\mathrm{LA,TA}}}{v_{\mathrm{s}}^{\mathrm{LA,TA}}}=\frac{\omega_{\mathrm{i}}-\omega_{\mathrm{R}}^{\mathrm{LA,TA}}}{v_{\mathrm{s}}^{\mathrm{LA,TA}}}=2\frac{\omega_{\mathrm{i}}}{c}n(\omega_{\mathrm{i}}) \tag{13.3}$$

if we assume that we are so far away from the exciton resonance that $n(\omega)$ does not change significantly over the Brillouin shift. In this case either $n(\omega)$ or v_{s} can be determined from Brillouin scattering if the other quantity is known. If ω_{i} approaches the resonance of the exciton polariton, Brillouin scattering is an

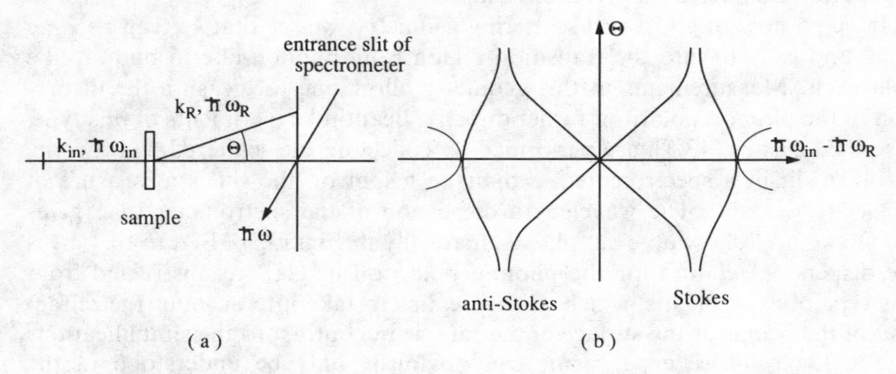

Fig. 13.3a, b. Schematics of a Raman scattering experiment in forward direction with a spectrometer (**a**) and the visualization of the relation between Θ and $\hbar\Omega$ of (**a**) in the output plane of the spectrometer (**b**)

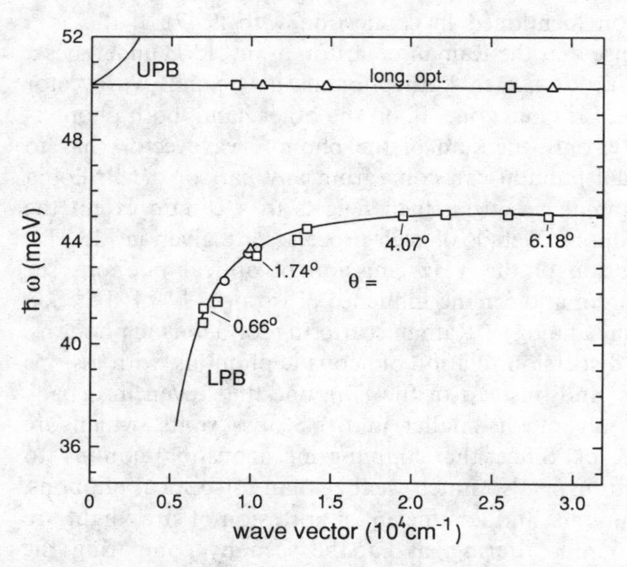

Fig. 13.4. The dispersion of the phonon polariton in GaP, measured by the method shown in Fig. 13.3. According to [13.6]

efficient means of $\boldsymbol{k}$-space spectroscopy of these resonances. We come back to this aspect in Sect. 14.4.

As a final aspect in this section we treat the optical properties of phonons in superlattices (see also Sect. 9.8). The folding back of the acoustic (and optical) phonon branches imposed by the period d of the superlattice compared to the lattice constant a gives rise to new states at $k=0$ with finite energy as shown in Fig. 9.20. These states can couple in a Raman-type process to the radiation field. In Fig. 13.5a we show a part of the Raman spectrum. The peaks $A_1^{(1)}$ and $B_1^{(1)}$ are attributed to the first folded-back $k=0$ states of the longitudinal acoustic phonon branch of the GaAs/AlAs superlattice (Fig. 13.5b). The scattering with TA phonons is forbidden in the geometry used in this example. The splitting is due to the different sound velocities in GaAs and AlAs. A detail of the dispersion relation is shown in Fig. 13.5c where the states observed in (a) are indicated, thus giving support to the model developed in Sect. 9.8. For a more recent discussion of this topic see e.g. [13.17] and references given therein.

13.4 Surface Phonon Polaritons

We already mentioned in connection with Figs. 7.1 and 7.2 that surface polariton modes exist in the range between the transverse and longitudinal eigenmodes. These modes can be observed by attenuated total reflection (ATR), i.e., by coupling the evanescent wave of Fig. 3.3 to the material under investigation. More details of this technique are given in Chap. 14.

Here we show in Fig. 13.6a ATR spectra and in b the dispersion relation of the surface phonon polariton in GaP. The agreement with the schematic drawing of Fig. 7.2 is obvious. For more details see [13.18] and references given therein.

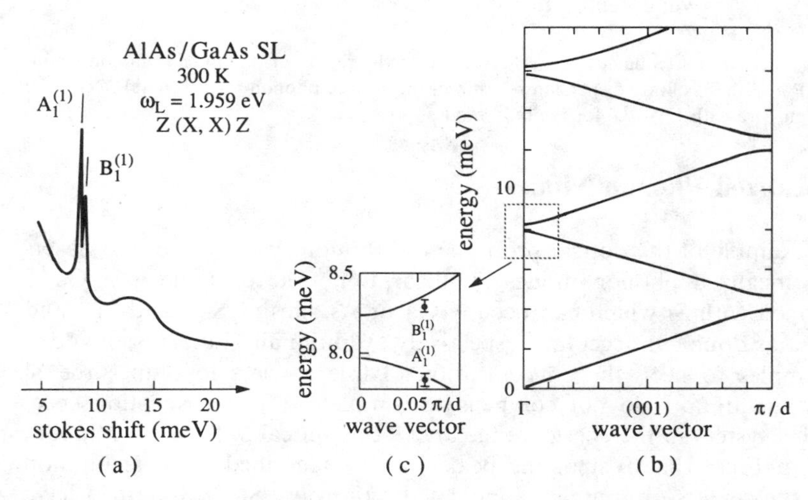

Fig. 13.5a–c. Raman spectrum of a GaAs/AlAs superlattice (**a**), the dispersion of the folded-back acoustic phonon branch (**b**) and a detail of (**b**) around $k \approx 0$ (**c**). According to [13.8]

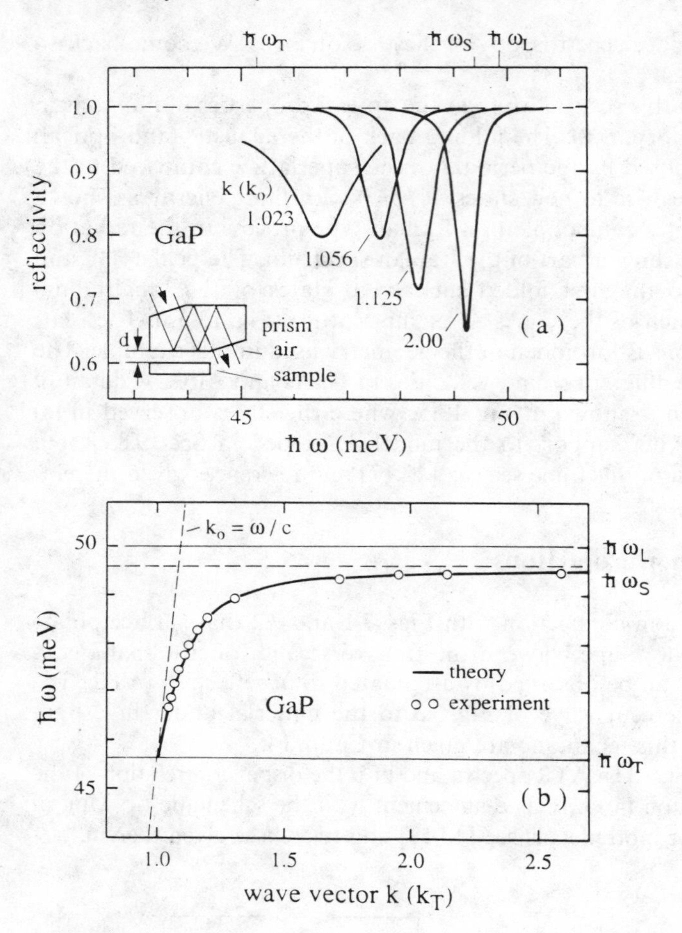

Fig. 13.6a, b. Spectra of attenuated total reflection (**a**) in the region of the optical phonon resonances of GaP and the resulting dispersion relation of the surface phonon polariton. (**b**) According to [13.9]. Compare with Fig. 13.4 for the bulk modes

13.5 Localized Phonon Modes

As a last example of the optical properties of phonons we consider the observation of localized phonon modes in absorption. The example is a GaAs crystal doped with Si which can occupy Ga or As sites as Si_{Ga} and Si_{As} and thereby act as donor or acceptor, respectively. Additionally Li has been added to the samples to keep the concentration of free carriers low since the Si acceptors and donors do not compensate completely. The absorption spectrum of this system in the energy range above the optical bulk phonon modes is shown in Fig. 13.7. Most of the peaks can be identified as local phonon modes of the centers or complexes indicated by arrows. Since Si and Li have

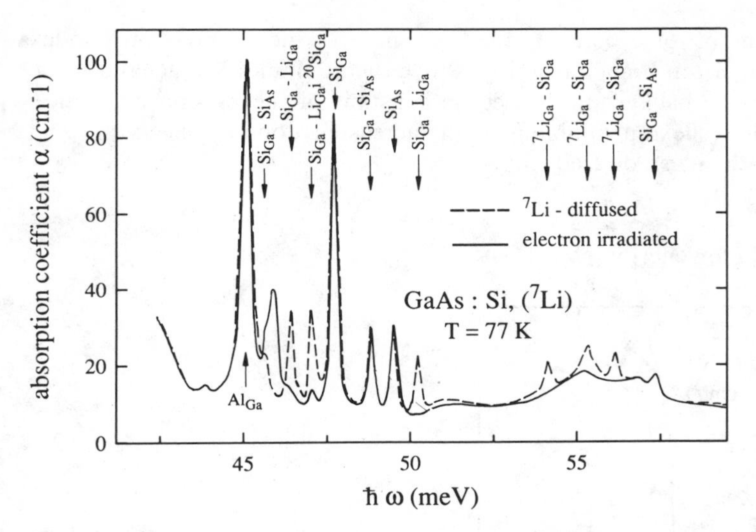

Fig. 13.7. Absorption spectrum of a GaAs: Si, Li, doped sample, showing localized phonon modes. According to [13.10]

considerably lower atomic masses than Ga or As and since the force constants are not too much lower, it is not surprising that the localized phonon modes appear at frequencies above the bulk modes. The isotope shift introduced by ^{7}Li and ^{8}Li is clearly visible in further data of [13.10]. More information on this topic may be found in [13.19] and references given therein.

13.6 Plasmons

Plasmons as collective excitations of the carriers in a partly filled band do not usually exist in pure semiconductors under thermodynamic equilibrium, since the probability of thermal excitation across the forbidden gap is negligible.

There are, however, two conditions under which plasmons can be observed, and these are either highly doped semiconductors or highly excited ones. Under high doping and in thermodynamic equilibrium one has (see Chap. 10)

$$np = n_i^2(T) = N_{\text{eff}}^{\text{e}} N_{\text{eff}}^{\text{h}} \, \mathrm{e}^{-E_g/k_B T}, \tag{13.4}$$

i.e., one has a high density either of electrons or of holes, and the data of the majority carriers have to be used in the calculation of ω_{Pl} according to (12.1).

Under high excitation an electron–hole plasma can be formed, which consists of electrons and holes (for details see Chap. 20). In this case the reduced mass of electron and hole enters in (12.1).

In Fig. 13.8 we show the IR reflection spectra of InSb samples with different n-doping. In agreement with our statements that the transverse eigenfrequency of plasmons is zero, we see reflectivity close to one from zero up to the plasma frequency, which is, as we remember, the longitudinal eigenfrequency. The reflection minimum corresponds to the frequency at which the refractive index of the upper polariton branch is unity. Above the minimum R reaches a value determined by the background dielectric constant of the plasma resonance. The shift of the reflexion minimum with increasing doping reflects the $n^{1/2}$ dependence of the plasmon frequency.

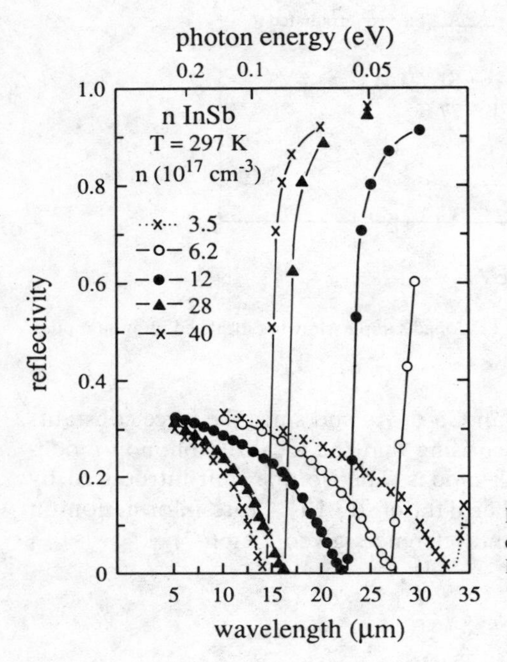

Fig. 13.8. The reflection spectrum in the vicinity of the plasmon resonance in doped InSb samples. According to [13.11]

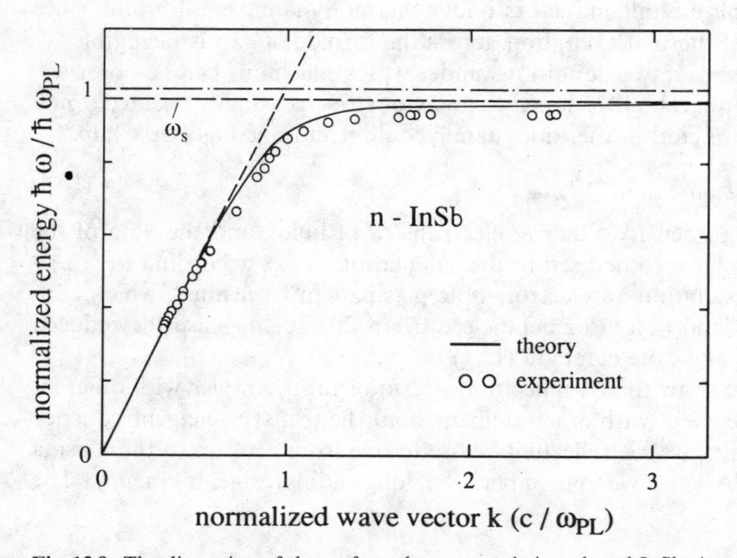

Fig. 13.9. The dispersion of the surface plasmon mode in n-doped InSb. According to [13.12]

Adapting Fig. 7.2 to the situation $\omega_T = 0$ we expect, for the case of plasmons, that surface plasmon modes exist in the whole frequency range between zero and ω_{PL} (Fig. 13.9). These surface plasmons can be investigated either by attenuated total reflection (see Sect. 24.1 or 13.4) or by another method which we will now outline briefly, since it has been used to obtain the data in Fig. 13.9. If a grating with spacing Λ is engraved in the surface, then the parallel component of the wave vector is conserved only modulo reciprocal vectors of this surface grating (see Sect. 3.1.3), i.e., modulo

$$G_{\parallel} = \frac{2\pi}{\Lambda} m; \quad m = 0, \pm 1, \pm 2 \ldots . \tag{13.5}$$

Adding such $G_{\parallel}$ values to the wave vectors of the incident light beam allows coupling to the surface polariton modes for $m \neq 0$. By varying the frequency and the angle of the incident light beam and thus

$$\omega, \boldsymbol{k}_{\parallel} + \boldsymbol{G}_{\parallel} \tag{13.6}$$

independently, it is possible to measure the dispersion of the surface plasmon polariton.

By changing the doping concentration (or the pump power in the case of an electron–hole plasma) it is possible to deliberately vary the carrier concentration in a semiconductor.

Since optical phonons and plasmons can be observed around $\boldsymbol{k} = 0$ in Raman scattering (the second case is an electronic Raman process), by this technique it is

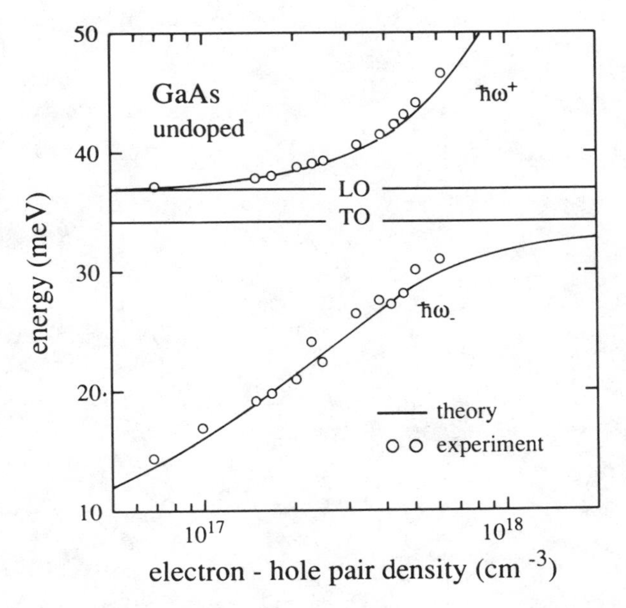

Fig. 13.10. The dependence of the plasmon–phonon mixed mode on the electron–hole pair density in highly photon-excited GaAs. According to [13.13]

also possible to investigate the phonon–plasmon mixed states introduced in Sect. 12.1 and Fig. 12.4. In Fig. 13.10 we show the dependence of this state in GaAs under high excitation on the carrier concentration. By changing the carrier density, it is possible to observe the phonon, the plasmon, and the plasmon–phonon mixed states, as is obvious by comparing fig. 13.10 with Fig. 12.4.

With this example we close the discussion of the optical properties of plasmons in semiconductors. Indeed, plasmons play only a minor role in the understanding of the optical properties of semiconductors, although they are essential for the optical properties of metals. More information about this topic is found in [13.14] and, especially for plasmons in semiconductor structures of reduced dimensionality in [13.20].

13.7 Problems

1. Try to find more reflection spectra like that in Fig. 13.1b in the literature. Deduce from these spectra data for $\hbar\omega_T$, Δ_{LT}, ε_s and ε_b and compare with values in the literature.

2. Calculate the plasmon energy for *n*- (or *p*-) doped GaAs and carrier concentrations of 10^{10}, 10^{15} and $10^{20}\,\text{cm}^{-3}$.

3. Calculate the density of electrons at which $\hbar\omega_{PL} = \hbar\Omega_{LO}$ for ZnO and InAs. Up to which temperatures are the electron gases degenerate.

4. Show that a treatment of the optical properties of free carriers like the Drude model outlined in (12.3–5) results in similar optical properties to those discussed above.

14 Optical Properties of Intrinsic Excitons

Having treated the optical properties of phonons and plasmons, we come in this and the following chapters to the essence of semiconductor optics, namely the optical properties of excitons.

Phonons are necessary to describe the optical properties of semiconductors and of insulators in the IR; plasmons determine the optical properties of metals from the IR through the visible to the near UV, and in semiconductors, if present at all, they contribute along with the phonons to the IR spectra. Excitons, on the other hand, determine the optical properties around the band gap, i.e., in the visible including the near UV and IR in the case of semiconductors and in the (V)UV for insulators. Although inorganic insulators like the alkali halides and organic ones such as anthracene have specific optical properties, many of the aspects presented in the following for excitons in semiconductors also apply to them.

We concentrate in this chapter on the intrinsic linear optical properties, starting from semiconductors with a dipole-allowed direct band-to-band transition, proceed to those with a dipole-forbidden direct band-to-band transition, and end up with indirect gap materials.

14.1 Exciton–Photon Coupling

In semiconductors with dipole-allowed direct band-to-band transitions, the excitons couple strongly to the radiation field. As a consequence many optical properties can be understood quantitatively only in the strong-coupling or polariton picture. Thus we use this occasion to elucidate once more for this particular case the concept of weak and of strong coupling to the radiation field, heeding the classical dogma "repetitio est mater studiorum".

In Sects. 2.1–4 we introduced the electromagnetic radiation field, and in Sect. 2.5 the photons as its quanta. In Chaps. 9–13 we presented the properties of various elementary excitations. The interaction between the two can be treated in perturbation theory. This is the so-called weak coupling approach. The one-photon absorption coefficient $\alpha(\omega)$ is then proportional to the dipole matrix element squared in (14.1a), i.e., by first order-perturbation theory with the initial and final state:

$$\alpha(\omega) \propto |\langle f|H_D|i\rangle|^2 \, \delta(E_i - E_f + \hbar\omega). \tag{14.1a}$$

The refractive index is obtained at this level of approximation either by a Kramers–Kronig transformation of $\alpha(\omega)$ or, away from the resonance, by second-order perturbation theory according to

$$n(\omega) - 1 \propto \left| \sum_{z} \frac{\langle i|H_{\mathrm{D}}|z\rangle \langle z|H_{\mathrm{D}}|i\rangle}{E_z - E_{\mathrm{i}} - \hbar\omega} \right| \tag{14.1b}$$

A photon $\hbar\omega$ creates virtually an excited intermediate state $|z\rangle$ under momentum conservation, which, after a time Δt limited by

$$\Delta E\, \Delta t = (E_z - E_{\mathrm{i}} - \hbar\omega)\, \Delta t \lesssim \hbar\,, \tag{14.2}$$

emits again a photon which is identical to the incident one, while the electronic system returns to the initial state $|i\rangle$. The time Δt during which the energy is "stored" in the virtual excited state reduces the phase velocity of the light and thus evidently describes an $n(\omega)$ which diverges when ω approaches the resonance energy $E_z - E_{\mathrm{i}}$, since ΔE goes to zero and Δt can be very long, in agreement with Fig. 4.4.

In the polariton concept, on the other hand, one quantizes the mixed state of the electromagnetic radiation and the excitation of the medium, i.e., the polarization wave. We already introduced this concept in Chap. 5. Since it is a very important one we want to demonstrate it here again for the exciton polariton.

For readers who are not satisfied with the statement that the polaritons are the quanta of the mixed states of electromagnetic radiation and excitation (or polarization), we give two other approaches. The first is just a diagrammatic representation of what was said before.

In Fig. 14.1 an incident photon creates an electron–hole pair, which recombines again to give a photon, and so on. The Coulomb interaction between electron and hole, which is responsible for the formation of the exciton, is represented by a virtual exchange of photons between electron and hole, i.e., by the vertical lines. Consequently the whole diagram of Fig. 14.1 can be considered as a representation of the exciton polariton.

In the other approach, which follows [14.1], we start with the electron and hole operators, construct from them the exciton and finally the exciton polariton. (see also Sect. 11.1)

We start with the creation and annihilation operators for excitons:

$$B^{+}_{\nu,\boldsymbol{k}};\; B_{\nu,\boldsymbol{k}}. \tag{14.3}$$

The index ν stands for the quantum numbers n_{B}, l, m in (11.3).

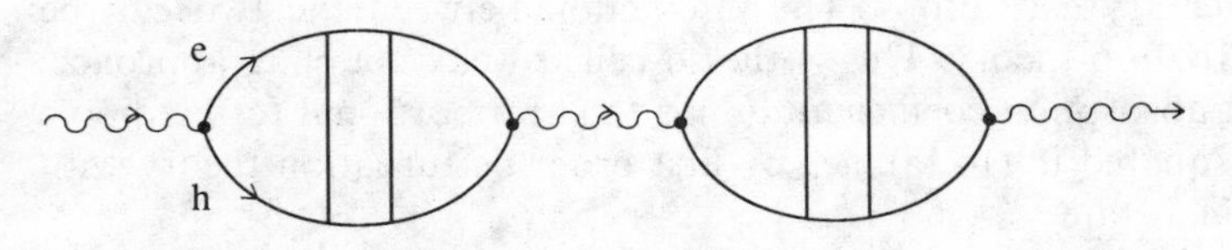

Fig. 14.1. Diagrammatic representation of an exciton polariton

It can be shown that the $B_{v,k}$ deviate from the commutator relations of ideal bosons by a term proportional to the mean number of electron–hole pairs, n, contained in the volume of an exciton a_B^d

$$\langle [B_{0,0}, B_{0,0}^+]^- \rangle = 1 - O(na_B^d) , \tag{14.4}$$

where d is the dimensionality of the system. The Hamiltonian of a non-interacting exciton gas is then

$$H = \sum_{v,k} E(v,\boldsymbol{k}) B_{v,k}^+ B_{vk} . \tag{14.5}$$

Using an analogous expression for the photons with the number operator $c_k^+ c_k$, for the interacting system of excitons and photons considering the leading, i.e., resonant terms around a specific resonance only, we obtain

$$H = \sum_k \left[\sum_v E_{vk} B_{vk}^+ B_{vk} + \hbar\omega_k c_k^+ c_k - \mathrm{i}\hbar \sum_v g_{vk} (B_{vk}^+ c_k - \mathrm{H.c.}) \right] . \tag{14.6}$$

The coupling coefficients g_{vk} contain the transition matrix elements as in (14.1).

If we consider the third term on the right-hand side of (14.6) as a perturbation, we are back once more to the weak coupling limit.

The polariton concept is obtained if we diagonalize the whole Hamiltonian (14.6) by a suitable linear combination of the $\boldsymbol{B}_{vk}$ and the c_k, leading to the polariton operator P_k

$$P_k = u_{vk} B_{v,k} - v_k c_k \tag{14.7}$$

$$\text{with} \quad |u_{vk}|^2 + |v_k|^2 = 1 . \tag{14.8}$$

The u_{vk} give the exciton-like character of the polaritons. They are close to one around the exciton energy $E_{v,0}$ and decrease on the upper and lower polariton branches with decreasing energetic distance from the resonance. The v_k give the photon-like part and, according to (14.8), show the opposite behavior. In Fig. 14.2 we give an example for the upper polariton branch. As a rule of thumb we can state that the polariton wavefunction contains considerable exciton-like parts over energies

$$|\hbar\omega - E_{\mathrm{ex}}| \lesssim 10\Delta_{\mathrm{LT}} . \tag{14.9}$$

It is interesting to note that the dispersion relation that we obtain from this approach is identical to the one obtained from the set of classical coupled oscillators treated in Chaps. 4–6 and shown in Figs. 5.1 and 6.2.

14.2 Consequences of Spatial Dispersion

In contrast to that of optical phonons, the $\boldsymbol{k}$ dependence of the exciton energy (11.1a) is significant. For $\boldsymbol{k}$ vectors in the transition region from the photon-like to the exciton-like part of the dispersion relation, the so-called bottle-neck, the kinetic energy term in (11.1a) becomes comparable to the longitudinal transverse splitting Δ_{LT}. So we have to consider Fig. 6.2 instead of Fig. 5.1. The consequences of spatial dispersion have already been outlined in connection with Figs. 6.3–5, so that we can restrict ourselves to just recalling them here. For all frequencies there is at least one propagating mode. This fact reduces the reflectivity in the reststrahlbande to values below 1, even in the case of negligible damping. For frequencies above ω_L there are several propagating modes, and below it there is at least one propagating mode and one or more evanescent ones. This situation is not covered by the boundary conditions deduced from Maxwell's equations and additional boundary conditions (abc) have to be introduced containing the information about what fraction of the energy transmitted through the interface travels on which polariton branch. Since this "branching" ratio is ω-dependent and since the imaginary parts of the various branches differ, the decay of the intensity into the depth of the sample can be nonexponential. This means the "effective" absorption coefficient can be thickness dependent. Furthermore it looks more complex (Fig. 6.5) than Fig. 4.4.

The abc which have been introduced by Pekar and by Hopfield [14.2] assume that the excitonic part of the polarization at the surface vanishes, or its derivative normal to the surface, or a linear combinations of both. Furthermore one can assume that excitons do not "leak out" of the semiconductor

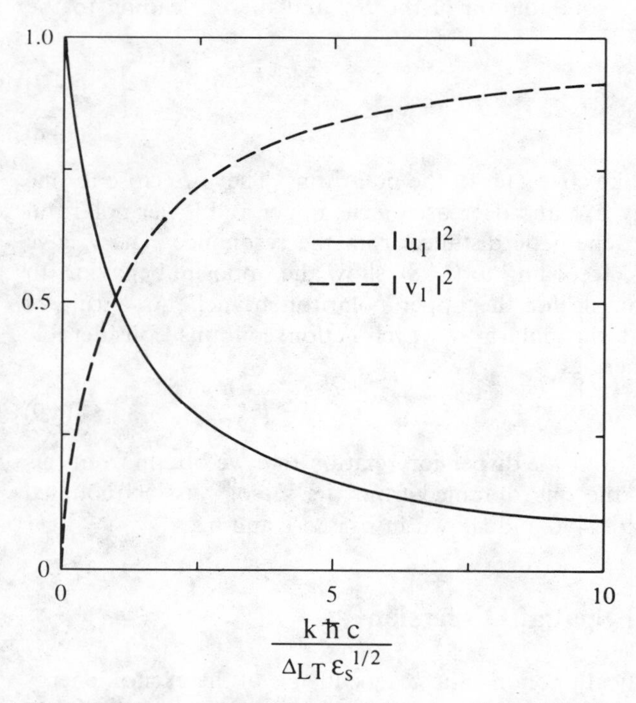

Fig. 14.2. The energy dependence of the $u_{v,k}$ and v_k of (14.8) for the upper polariton branch. According to [14.1]

into vacuum and that there should consequently be an exciton-free surface layer (dead layer), the optical properties of which are described by ε_b and which has a minimum thickness of the excitonic Bohr radius of (11.1e). Electric fields, which occur often at surfaces, can ionize, i.e., destroy the exciton and lead to an increase in the thickness of the dead layer. The problem which thus has to be solved to calculate a reflection spectrum is shown in Fig. 14.3. An incident beam passes first the dead layer in which multiple reflection occurs and then enters the semiconductor in which several modes can be excited. Sometimes scientists apply even more complex models assuming, e.g., that the damping and/or the eigenfrequencies are depth dependent [14.3, 4].

The formulas to calculate the spectra are rather complex and we do not give them here but refer the reader to [14.2–9]. Instead we give in the next section examples of reflection, transmission, and luminescence spectra of the exciton polariton in bulk semiconductors with direct, dipole-allowed band-to-band transitions.

14.3 Spectra of Reflection, Transmission and Luminescence

In Fig. 14.4a we show the bandstructure of CdS (compare Fig. 10.10) in part b the dispersion of the exciton polariton for the orientation $\boldsymbol{k} \perp \boldsymbol{c}$, $\boldsymbol{E} \perp \boldsymbol{c}$ and

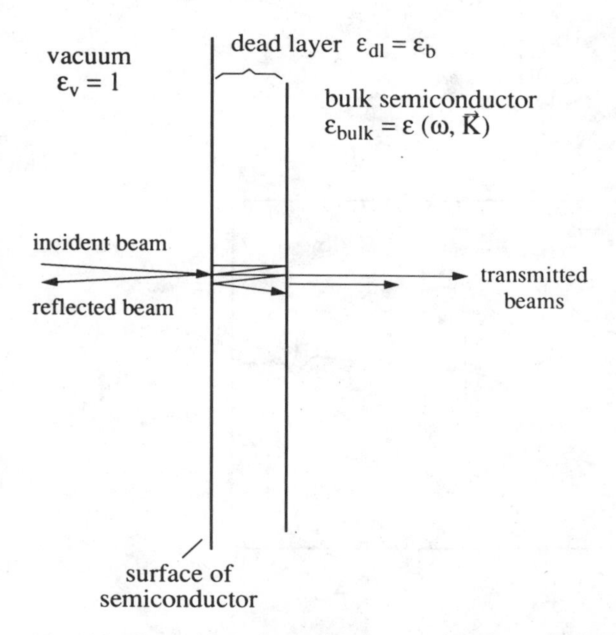

Fig. 14.3. The problem of reflection for a semiconductor in the vicinity of an exciton resonance for normal incidence including multiple reflection in a dead layer and two propagating modes due to spatial dispersion

including the $n_B = 1$ excitons involving a hole either in the A or in the B valence band. The A exciton is a rather simple resonance for this orientation, comparable to our model system in Fig. 6.2. The k-linear term of the B valence band (Fig. 10.10b) mixes the singlet and triplet states for $\boldsymbol{k}_\perp \neq 0$ and gives rise to an additional polariton branch. Figure 14.4c finally gives the reflection spectra of the two resonances for $\boldsymbol{E} \perp \boldsymbol{c}$ and $\boldsymbol{E} \parallel \boldsymbol{c}$ and of some higher states ($n_B \geqslant 2$). The combination of the Γ_1(S-) envelope function for $n_B = 1$ with the symmetries of the electron (Γ_7) and the holes (AΓ_9, BΓ_7) gives excitons of symmetries AΓ_5 and AΓ_6 and BΓ_1, BΓ_2, BΓ_5 as explained in more detail in

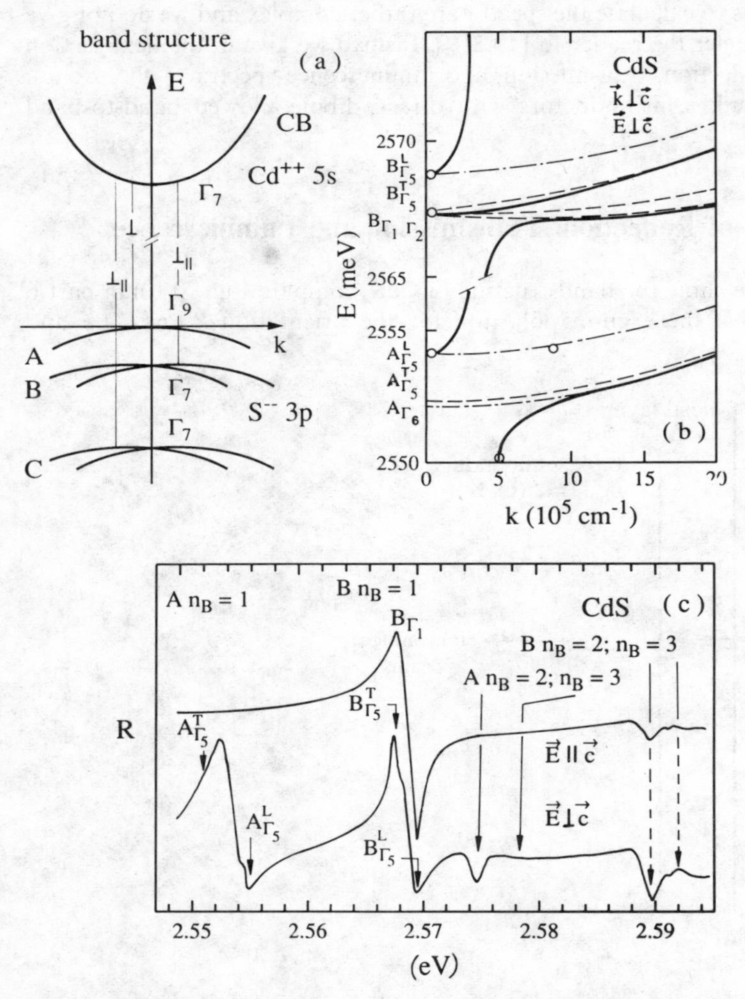

Fig. 14.4a–c. The bandstructure of CdS around the Γ-point (**a**); the dispersion of the $n_B = 1$ A and B exciton polariton resonances (**b**); and the reflection spectra for the polarizations $\boldsymbol{E} \perp \boldsymbol{c}$ and $\boldsymbol{E} \parallel \boldsymbol{c}$ (*c*). According to [14.5, 9]

Chap. 25. The Γ_6 and Γ_2 states are triplets which couple only weakly to the radiation field since they are spin-flip and dipole-forbidden and do not show up in reflection. Γ_5 and Γ_1 states couple to the radiation field for the orientations $\boldsymbol{E} \perp \boldsymbol{c}$ and $\boldsymbol{E} \parallel \boldsymbol{c}$, respectively. These selection rules show up clearly in the reflection spectra, the $n_B = 1$ A excitons being seen only in $\boldsymbol{E} \perp \boldsymbol{c}$.

A fit to the reflection spectra (not shown here) using spatial dispersion, an exciton-free layer and some abc coincides with experiments within a few percent.

The $A\Gamma_5$ resonance is quite simple, as already mentioned. R remains significantly below 1 due to spatial dispersion as predicted above. For smaller oscillator strength or longitudinal–transverse splitting, the maximum almost disappears and only a narrow spike close to the longitudinal eigenfrequency remains, as shown in Fig. 14.5 for ZnTe.

The small spike around $A\Gamma_5^L$ in Fig. 14.4c is partly caused by the onset of the UPB but mainly by the dead layer. An increase of its thickness increases the importance of the spike due to multiple reflections (Fig. 14.3) and may even lead to an "inversion" of the usual reflection spectrum, i.e., to a dip at low energies and a maximum above. A set of calculated spectra showing this phenomenon is given in Fig. 14.6.

The $B\Gamma_1$ exciton resonance is again a simple one, but the $B\Gamma_5$ has a small dip stemming from the additional polariton branch shown in Fig. 14.4b, which at this energy reaches exactly $n = 1$; see (4.34).

At higher energies we see $n_B \geqslant 2$ exciton states which split into several sublevels due to the various L and/or L_z values of the envelope function. The reflection signal of these higher states decreases due to the n_B^{-3} dependence of the oscillator strength (11.4). In the band-to-band transition region the

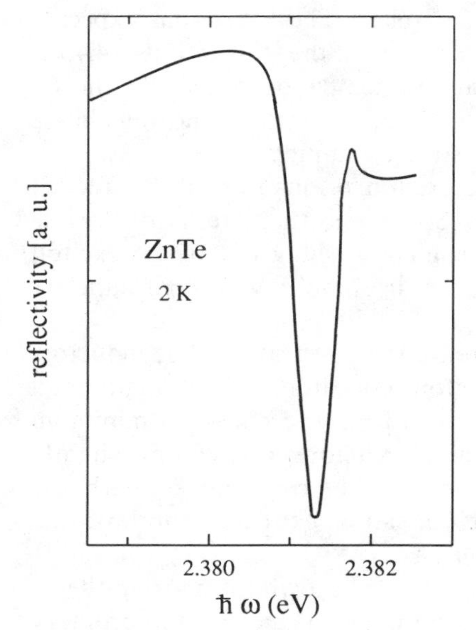

Fig. 14.5. A reflection spectrum for ZnTe. According to [14.10]

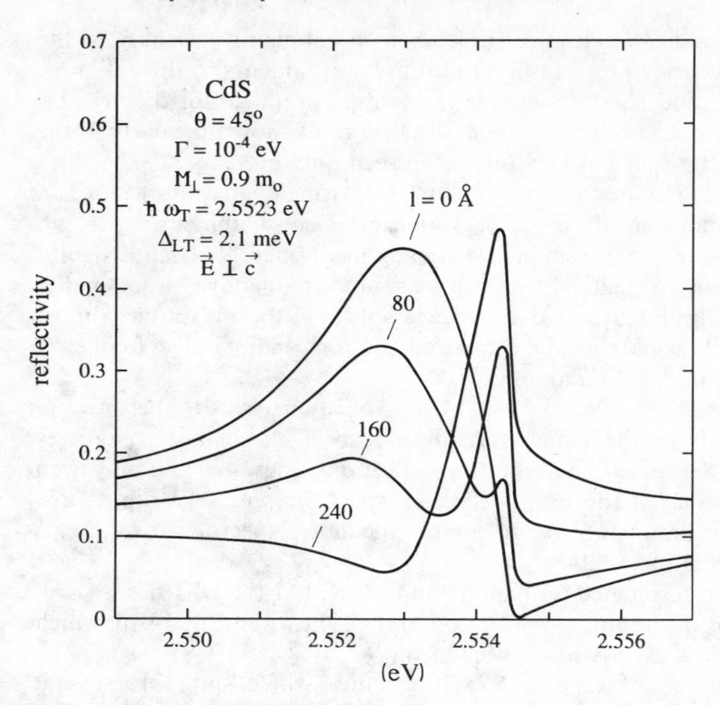

Fig. 14.6. A set of calculated exciton reflection spectra of CdS for 45° incidence and for various thicknesses of the exciton-free layer with all other parameters kept constant. According to [14.11]

reflection spectra are usually flat and structureless. The C-excitons expected from Fig. 14.4a are situated around 2.61 eV and are off the scale of Figs. 14.4b, c. These resonances are washed out even at low temperature because the C-exciton is situated in the continuum of the A- and B-excitons and thus has a rather short phase relaxation time T_2, i.e., strong damping.

With increasing lattice temperature the exciton resonances are broadened due to increasing scattering with phonons. Sometimes they are hardly visible at RT as shown in Fig. 14.7 for CdS. A similar washing out of the exciton resonance can occur even at low temperatures in samples with high impurity content and/or lower crystalline quality.

To summarize, we can state that the reflection spectra of semiconductors are determined around the gap by exciton polaritons. The longitudinal eigenenergy can be reasonably well determined from the reflection minimum which corresponds to $n = 1$ on the UPB and is therefore situated only slightly (< 1 meV) above $\hbar\omega_L$. The transverse eigenenergy, the oscillator strength, the damping, and the effective mass of the exciton can be extracted only with the help of a rather complicated line-shape analysis.

For uniaxial materials like CdS or ZnO it can be shown from group-theoretical considerations that the Γ_5-excitons are the resonances for the ordinary

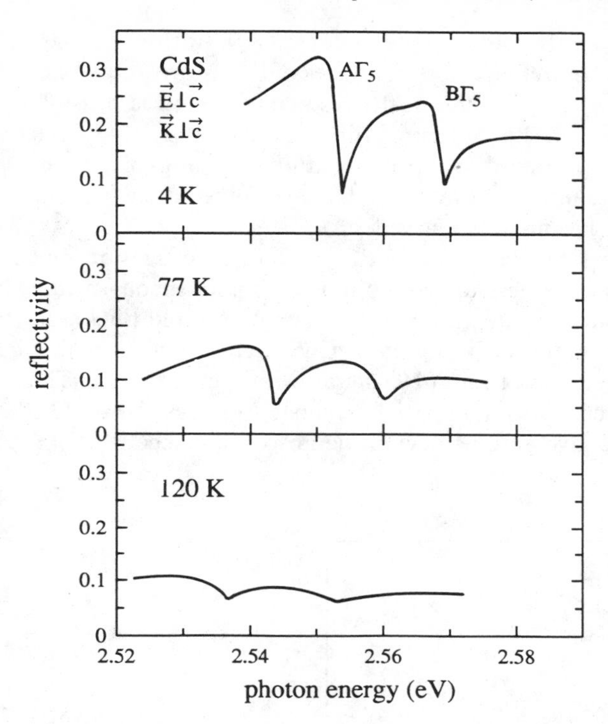

Fig. 14.7. A set of reflection spectra of the A- and B-exciton resonances of CdS for various lattice temperatures. According to [14.12]

beam (Sect. 3.1.7) because the polarization $\boldsymbol{E} \perp \boldsymbol{c}$ can be realized for all angles between $\boldsymbol{k}$ and $\boldsymbol{c}$. The Γ_1 resonances have dipoles oriented parallel to $\boldsymbol{c}$. As a consequence they have maximum coupling to the radiation field for $\boldsymbol{E} \parallel \boldsymbol{c}$ and $\boldsymbol{k} \perp \boldsymbol{c}$, but develop as extraordinary or mixed-mode polaritons to the longitudinal state if the angle $\angle$ $(\boldsymbol{k}, \boldsymbol{c})$ is continuously changed from $\boldsymbol{k} \perp \boldsymbol{c}$ to $\boldsymbol{k} \parallel \boldsymbol{c}$. In Fig. 14.8 we show the dispersion of the $\mathrm{C}\Gamma_1$ exciton polariton in ZnO, which has an oscillator strength varying according to

$$\Delta_{\mathrm{LT}} = \Delta_{\mathrm{LT}}^{0} \sin^2(\angle \boldsymbol{k}, \boldsymbol{c}) \,. \tag{14.10}$$

Experimental data for mixed mode polaritons can be found in [14.6, 14.15].

The absorption spectra of $n_{\mathrm{B}} = 1$ polariton resonances are usually difficult to measure quantitatively, since in the resonance region the effective absorption, or rather extinction, coefficient reaches values in the range

$$10^4\,\mathrm{cm}^{-1} \lesssim \alpha_{\mathrm{eff}} \lesssim 10^6\,\mathrm{cm}^{-1} \,. \tag{14.11}$$

As a consequence for samples with $d \gtrsim 1\,\mu\mathrm{m}$, the transmitted light intensity goes to "zero", i.e., to values comparable to or smaller than the stray light of the spectrometer. We show such a situation in Fig. 14.9a for $n_{\mathrm{B}} = 1$ for the $\mathrm{A}\Gamma_5$

resonance of CdS. There are Fabry–Perot modes (Figs. 3.8, 9) in the transparent region due to the natural reflectivity of the surfaces of the as-grown platelet type sample. A detailed analysis of $\alpha_{eff}(\omega)$ in the resonance region is possible only for sample thicknesses of the order of 1μm, such that $\alpha d \lesssim 4$. Such samples cannot usually be produced by grinding and polishing of thicker samples, since these processes introduce so much damage to the lattice that the damping γ of the resonance becomes too large (i.e., $\hbar\gamma > \Delta_{LT}$).

Sometimes epitaxial layers can be used for this type of investigation, but care has to be taken that these layers do not contain (inhomogeneous) strain since transmission measurements integrate over the whole sample thickness. Fortunately some semiconductors, such as CdS or CdSe, tend to grow as thin platelets. The absorption spectrum of Fig. 14.9a stems from such a sample. The deviations from the simple case without spatial dispersion of Figs. 4.4 or 13.1 are evident. In Fig. 14.9c we give an overview of the absorption spectrum of a

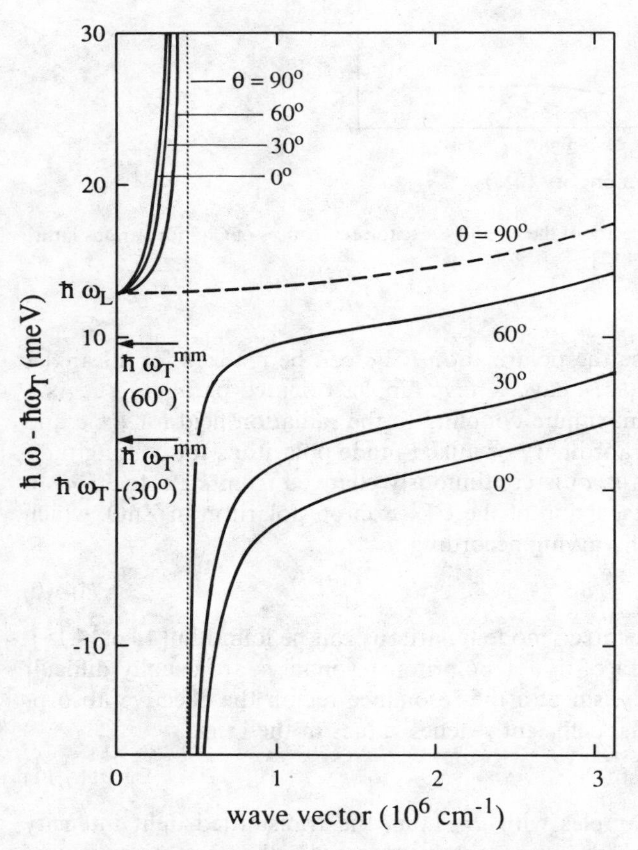

Fig. 14.8. Dispersion of the exciton polariton in a uniaxial material for various angles between $\boldsymbol{k}$ and $\boldsymbol{c}$. Data for the C1 exciton polariton in ZnO. According to [14.13]

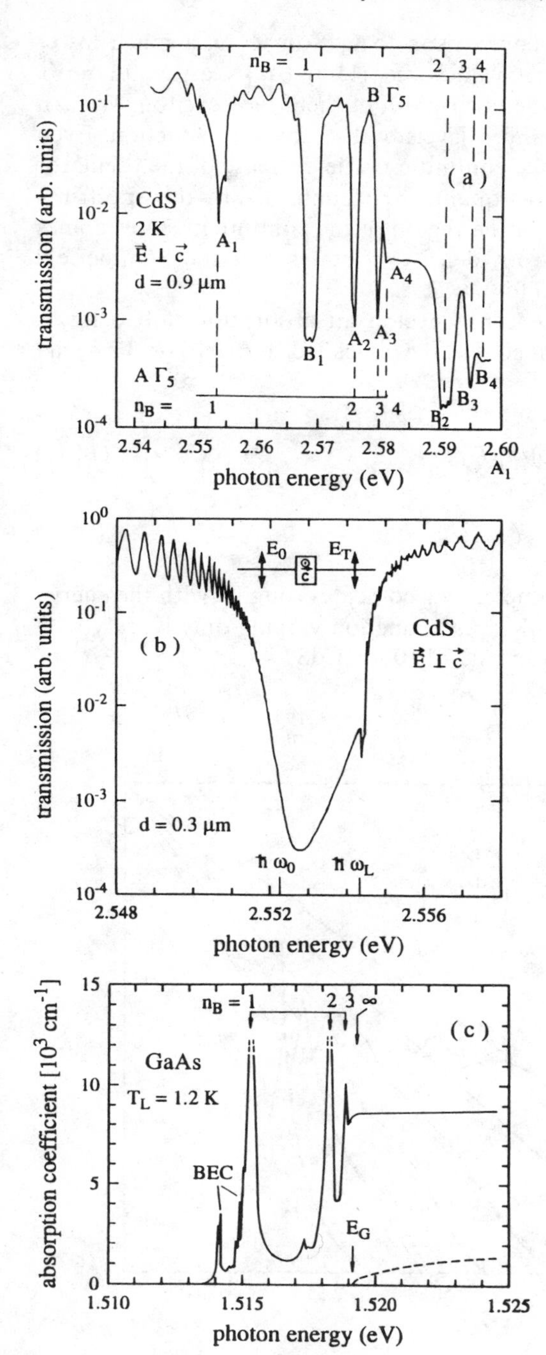

Fig. 14.9a–c. An experimental transmission spectrum of CdS for the orientation $\boldsymbol{E} \perp \boldsymbol{c}$ in the region of the A and B exciton resonances (**a**); a calculated one for the AΓ_5 resonance for $n_B = 1$ (**b**); and an absorption spectrum for a thin GaAs sample (**c**). According to [14.16], [14.11] and [14.17], respectively

thin GaAs sample. GaAs has, in comparison to CdS or ZnO, a much lower oscillator strength due to the larger value of a_B (11.4); (Δ_{LT} GaAs ≈ 0.1 meV). Therefore it is easier to measure the absorption of the $n_B = 1$ exciton. We can see in this figure also the $n_B = 2$ and 3 levels with S-envelope function. Even higher states ($n_B > 3$) merge with the continuum. The decrease of the oscillator strength with n_B^{-3} (11.4, 10) is at least qualitatively confirmed by this spectrum. The rather constant value of $\alpha(\omega)$ in the region of the continuum states comes from the product of the square-root density of states and the Sommerfeld factor, already discussed in (11.11) for $d = 3$.

At higher temperatures, the excitons develop an absorption tail to lower photon energies, which is described by the so-called Urbach or Urbach–Martienssen rule [14.18–25]

$$\alpha(\hbar\omega) = \alpha_0 \exp[-\sigma(T)(E_0 - \hbar\omega)/k_B T], \tag{14.12}$$

$$\hbar\omega < E_0$$

where α_0 and E_0 are material parameters. E_0 coincides roughly with the energy of the lowest free exciton at $T_L = 0$K. σ is a function varying only weakly with temperature. An example is given in Fig. 14.10 for CdS.

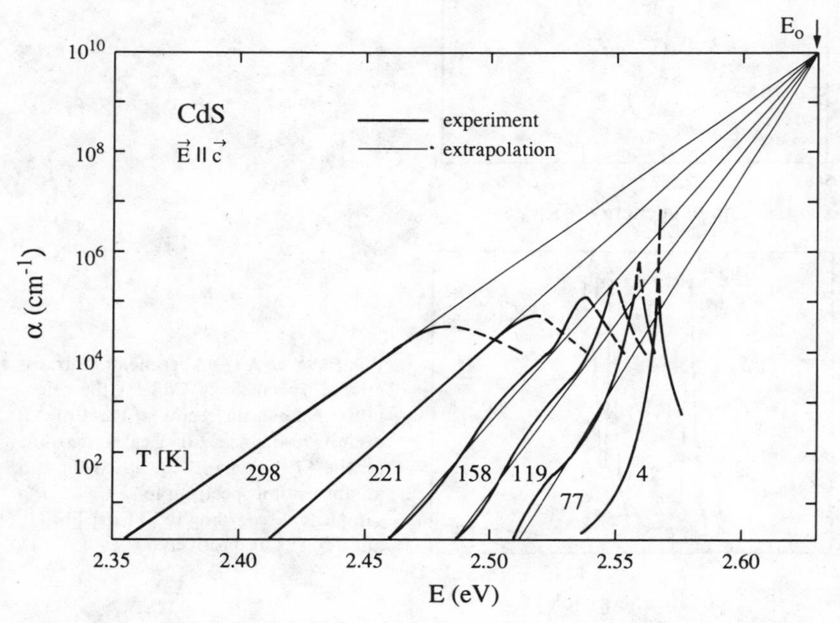

Fig. 14.10. The absorption edge of CdS for various temperatures and the polarization $\boldsymbol{E} \parallel \boldsymbol{c}$. According to [14.22]

The reason for the behavior described by (14.12), which is of rather universal nature in semiconductors and insulators, is the interaction of excitons with optical phonons. Two effects are usually discussed in theory: a momentary localization of the excitons in the randomly fluctuating field of optical phonons, or an ionization in their electric field. These two effects seem to contribute with a weight that depends on the material parameters. Details about the theory can be found in [14.23–25] and references therein.

The absorption of direct excitons in indirect materials will be addressed in Sect. 14.6.

The investigation of the luminescence from exciton polaritons is a rather difficult task since the emission from these states is very weak even at low temperatures. This has various reasons: the total luminescence yield η_{lum} of semiconductors is usually very low. Often one finds, even for direct gap materials,

$$10^{-1} \gtrsim \eta_{\mathrm{lum}} \gtrsim 10^{-3}, \tag{14.13}$$

i.e., the main recombination channel is non-radiative involving defect centers. A large part of the emission stems from bound-exciton complexes and donor–acceptor pairs, or other defect centers which are considered in Chap. 15. Furthermore, the emission is limited by various effects: One is the internal reflection, another the small escape depth. These points will be further clarified below.

If we excite an electron–hole pair, e.g., in the continuum states, it will relax to lower energies and thermalize by emission of phonons, as described in more detail Chap. 22. At very low temperatures ($k_{\mathrm{B}}T < \Delta_{\mathrm{LT}}$) the excitons end up on the LPB where they further relax by acoustic phonon emission. Since the scattering matrix element and the density of final states both decrease in the transition region between exciton- and photon-like dispersion, the excitons accumulate there. This is the reason why this region is called a bottle-neck. At higher temperatures the excitons reach essentially a Boltzmann-like distribution on the exciton-like part of the LPB, on the longitudinal branch, and on the UPB. In the polariton picture the luminescence from these states cannot be described as a "recombination" of the exciton polariton with emission of a photon, since the photon will be immediately reabsorbed to form an exciton, or in other words, since we are considering the quanta of the mixed state of exciton and photon.

The proper description is the following: The exciton polariton moves with its group velocity through the sample. It can be scattered by impurities or phonons or be trapped. Eventually it reaches the surface. In most cases it will be reflected back into the sample. The limiting angle for total internal reflections α_{TR} is, e.g., for $n = 5$ – a typical value on the LPB in the bottle neck region as shown in Sect. 14.4, only about 13°. Of the excitons impinging under an angle smaller than α_{TR}, a considerable fraction are also reflected back into the sample, as becomes clear if one combines (3.20) or (4.34) for normal incidence and Fig. 3.6b, or if one averages over the corresponding formula (3.19).

Furthermore, the luminescence yield of free-exciton polaritons is limited by the small escape depth l_{esc}, i.e., the depth from which they can reach the surface. If one excites in the band-to-band transition region, the exciting light penetrates about 0.1–1 μm into the sample corresponding to α-values of 10^4–$10^5\,cm^{-1}$ in this spectral range. By diffusion, the excitons spread out over a region of 1–2 μm. The depth from which they can reach the surface is much less than this. A rough estimate can be obtained either from the inverse effective absorption coefficient in the exciton resonance or from the product of phase-relaxation time T_2 (Sect. 22.1) and the group velocity in the exciton resonance (Sect. 14.4):

$$\alpha_{eff}^{-1} = (10^4\text{–}10^6\,cm^{-1})^{-1} = 0.01\text{–}1\,\mu m\,; \tag{14.14a}$$

$$l_{esc} = v_g\,T_2 = (10^{-3}\text{–}10^{-5}\cdot c)(10\text{–}40\,ps) = 10\text{–}0.03\,\mu m\,. \tag{14.14b}$$

In spite of all these difficulties it was possible to observe the emission from the exciton polariton and we give an example for ZnO in Fig. 14.11. On the

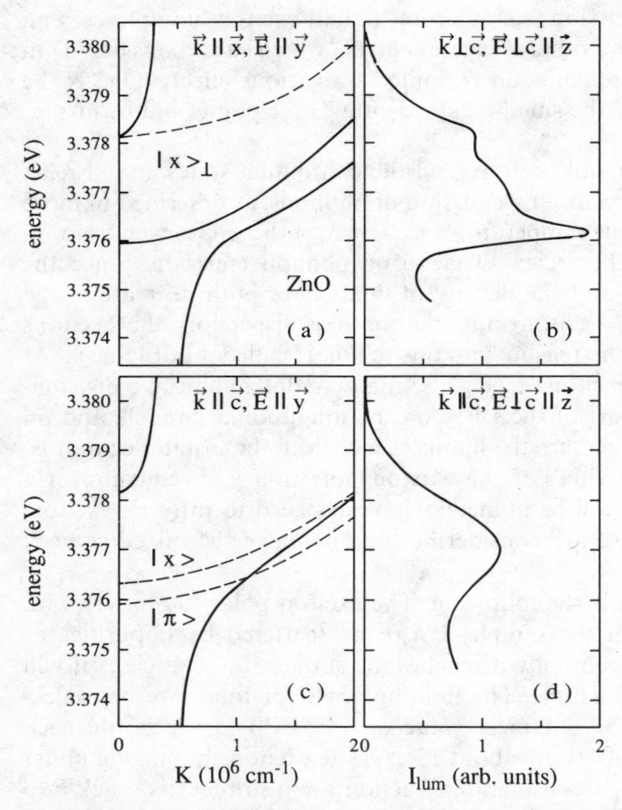

Fig. 14.11a–d. The dispersion of the $n_B = 1$ A-exciton resonance in ZnO for the polarizations $\boldsymbol{E} \perp \boldsymbol{c}\,(\Gamma_5)$ and $\boldsymbol{E} \parallel \boldsymbol{c}\,(\Gamma_1)$ (**a, c**) and the observed polariton luminescence (**b, d**). According to [14.13, 26a]

left the dispersion relation of the A-exciton is shown for $\boldsymbol{k} \perp \boldsymbol{c}$ and $\boldsymbol{k} \parallel \boldsymbol{c}$. Since the Γ_7 and Γ_9 valence bands are inverted in ZnO as compared to CdS. See section 10.6 and (Fig. 10.10), the k-linear term appears in the A-exciton for the orientation $\boldsymbol{k} \perp \boldsymbol{c}$. The influences of the resulting additional polariton branch, of the longitudinal branch, and of the UPB are seen by comparison with the orientation $\boldsymbol{k} \parallel \boldsymbol{c}$ where the longitudinal branch and the k-linear term are missing. Some other examples for the polariton luminescence are found e.g. in [14.26 b,c].

A luminescence channel of the exciton polaritons with higher luminescence yield are the LO-phonon replicas. In this case a polariton on the exciton-like part of the dispersion relation or in the bottleneck is scattered onto the photon-like branch by emission of one or more longitudinal optical phonons. The coupling with this type of phonon is stronger than with transverse optical or acoustic phonons since the lattice distortion of the polaron (Sect. 10.5) can be described largely as a superposition of longitudinal optical phonons. Once the polariton is on the photon-like branch, it travels over long distances with almost negligible damping and is transmitted through the surface into vacuum with quite high probability, since $n \lesssim 2$ in the corresponding spectral range.

In Fig. 14.12 we show the appearance of the LO-phonon satellites schematically. If we neglect the bottleneck region for the moment, we can deduce with the Boltzmann occupation probability the distribution of the excitons as a function of their kinetic energy E_{kin}

$$N(E_{\mathrm{kin}}) \propto \begin{cases} E_{\mathrm{kin}}^{1/2} \exp\{-E_{\mathrm{kin}}/k_{\mathrm{B}}T\} & \text{for } E_{\mathrm{kin}} \geqslant 0 \\ 0 \text{ otherwise} & \end{cases} \quad \text{with } E_{\mathrm{kin}} = \frac{\hbar^2 k^2}{2M} . \tag{14.15a}$$

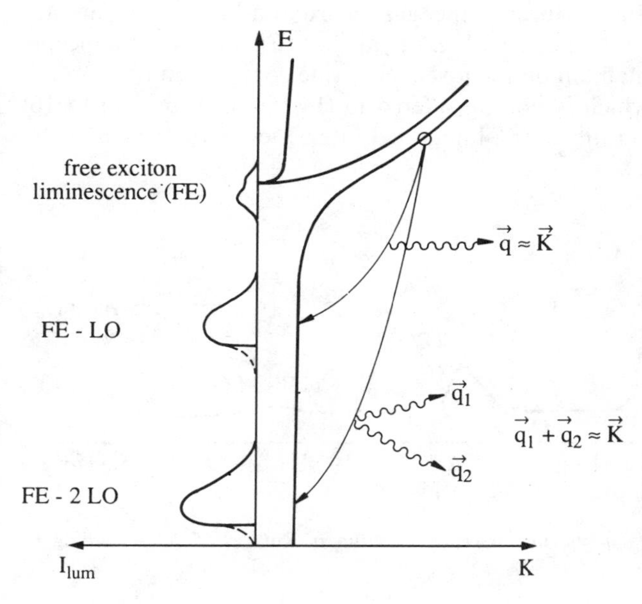

Fig. 14.12. Schematic drawing of the decay mechanisms of the exciton-mLO phonon emission processes

The lineshape of the luminescence of the m-th LO-phonon replica is then given by [14.27]:

$$I_m^{\mathrm{lum}}(\hbar\omega) \propto \begin{cases} E_{\mathrm{kin}}^{1/2} \exp(-E_{\mathrm{kin}}/k_{\mathrm{B}}T) W_m(E_{\mathrm{kin}}) & \text{for} \quad E_{\mathrm{kin}} \geqslant 0 \\ 0 \text{ otherwise} & \end{cases}$$

$$\text{with} \quad \hbar\omega = E_0 - m\hbar\omega_{\mathrm{LO}} + E_{\mathrm{kin}} \,. \tag{14.15b}$$

where E_0 is the energy of the dipole allowed exciton at $\boldsymbol{k}=0$.

The transition probability $W_m(E_{\mathrm{kin}})$ can be often expressed by a power law, i.e.,

$$W_m(E_{\mathrm{kin}}) \propto E_{\mathrm{kin}}^{lm} \,. \tag{14.16}$$

For $m=1$ one finds $l_1=1$ since the density of final states for the LO phonons increases with $E_{\mathrm{kin}} \propto k^2$ assuming that the wave vector of the photon-like exciton polariton in the final state is negligible.

For $m=2$ many different combinations of the two-phonon wave vectors are possible for a given $\boldsymbol{k}$ of the exciton-like polariton. As a consequence l_2 is zero and the lineshape of the second LO phonon replica directly reflects the distribution of exciton polaritons in the initial state. In Fig. 14.13 we show the emission of ZnO at 55 K. The free exciton polariton is not seen in emission for the reasons given above. There is a little bound exciton emission (Sect. 15.1) around 3.34 eV and the LO phonon replicas for $m=1,2,3$. The theroetical curves are calculated according to (14.15, 16) assuming that the lattice temperature and the temperature of the gas of exciton-like polaritons are equal. The fit coincides very nicely with experiment, thus confirming concepts developed above. The small tail on the low energy side comes from the population in the bottleneck which is not considered in (14.15, 16). From (14.15, 16) one can deduce that the ratio of the integrated intensities of the first and the

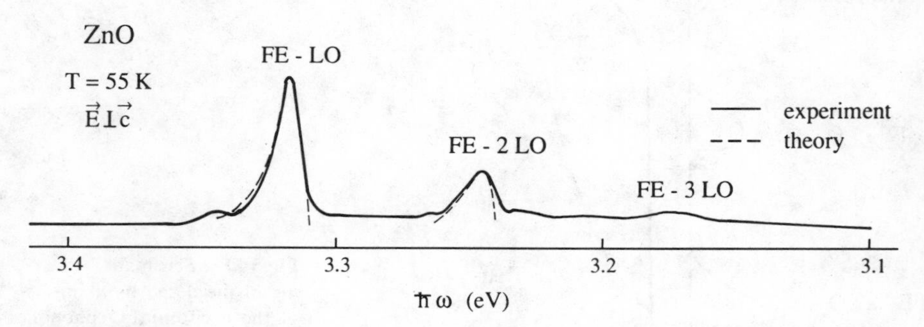

Fig. 14.13. LO-phonon replica in the luminescence spectrum of ZnO at 55 K. According to [14.28]

second LO-phonon replica is proportional to T

$$Q_{1,2} = \frac{\int I_1^{\text{lum}}(\omega)\mathrm{d}\omega}{\int I_2^{\text{lum}}(\omega)\mathrm{d}\omega} \propto T\,. \tag{14.17}$$

In Fig. 14.14 we give experimental data for $Q_{1,2}$ in ZnO for volume excitation and surface excitation. The first case has been realized by two-photon excitation with a ruby laser, which allows relatively homogeneous excitation of samples up to thicknesses in the mm range, and the second by UV excitation in the continuum states where the excitation depth is limited mainly by diffusion to values of the order of μm as discussed above.

Up to temperatures of 100 K the points follow nicely the predictions of (14.17), then they drop. This deviation is due to reabsorption effects caused by the absorption tail described by (14.12), which starts to influence the escape depth of the polaritons also in the $m = 1$ range at higher temperatures. This effect is evidently more pronounced for volume excitation than for surface excitation.

In semiconductors with less polar coupling, such as GaAs, the LO-phonon replicas are less pronounced. For those who work on luminescent ions in insulators and are therefore familiar with the concept of the Huang–Rhys factor S, it should be mentioned that for free excitons in most semiconductors

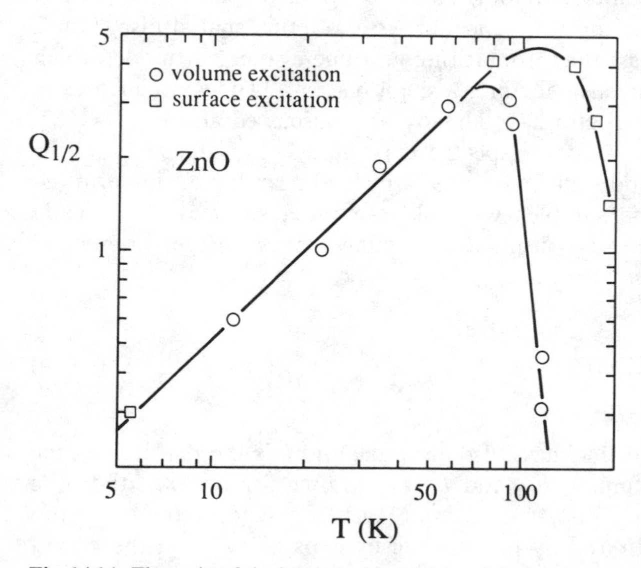

Fig. 14.14. The ratio of the integrated intensities of first and second LO-phonon replica in ZnO as a function of temperature for two different excitation conditions. According to [14.28]

S is below one. The large ratio of first to zero phonon intensity seen in Fig 14.13 is not connected with a large value of S but with other processes that reduce the zero phonon emission, as outlined above (See e.g. [14.26 b]).

14.4 Spectroscopy in Momentum Space

It is clear from the above discussion that the concept of exciton polaritons allows one to understand the spectra of reflection, transmission, and luminescence, but that a quantitative interpretation of the data usually involves a rather elaborate theoretical fit. Therefore various techniques have been developed which allow more or less directly the spectroscopy of exciton polaritons in momentum space, i.e., they provide the possibility of measuring the dispersion relation $E(\boldsymbol{k})$ more directly.

In this section we therefore recall briefly the consequences of "spatial dispersion", i.e., of the dependence of ω_0 on $\boldsymbol{k}$, and then present various methods of $\boldsymbol{k}$- or momentum-space spectroscopy.

As mentioned earlier in Chap. 5, the combination of the dielectric function $\varepsilon(\omega, \boldsymbol{k})$ (6.1) and the polariton equation (2.33) forms an implicit representation of the polariton dispersion as given in Figs. 6.2 and 14.4. For $\omega > \omega_{0,\mathrm{L}}$ we get two propagating modes in the sample or even more if the dispersion relation is complex (Fig. 14.4b or Fig. 14.11) or if the longitudinal branch couples to the radiation field as may occur for $\boldsymbol{k} \neq 0$ or for oblique incidence in uniaxial crystals. We showed such a situation in Figs. 6.4 and 14.4. Since the $\boldsymbol{k}$ vectors of the various modes in the sample are different, the diffracted beams propagate in different directions, thus giving some meaning to the term "spatial dispersion". Below $\omega_{0,\mathrm{L}}$ we have at least one propagating and one evanescent mode, which however, for finite damping also acquires a small real part. The consequences of this fact for the reflection spectra have already been discussed above.

The first method of $\boldsymbol{k}$-space spectroscopy uses the analysis of the Fabry–Perot modes introduced in Fig. 3.9, which appear for example in as-grown, thin, platelet type samples with plane-parallel surfaces. As already pointed out in (3.41) transmission maxima occur when an integer number of half-waves fit into the resonator, i.e.,

$$k_m(\omega_m) = m\frac{\pi}{d}\,, \quad m = 1, 2, 3 \ldots, \tag{14.18}$$

where d is the geometrical thickness of the sample and k is the real part of the wave vector in the medium. If ω_m and k_m are known for one m_1, then the dispersion relation can be reconstructed from (14.18) by reading the $\hbar\omega_m$ from the transmission spectrum and by progressing in steps of πd^{-1} on the k-axis. In Fig. 14.15 we show the dispersion of the $n_\mathrm{B} = 1$ A-Γ_5-exciton polariton in CdSe, together with a measured and a calculated reflection spectrum. The

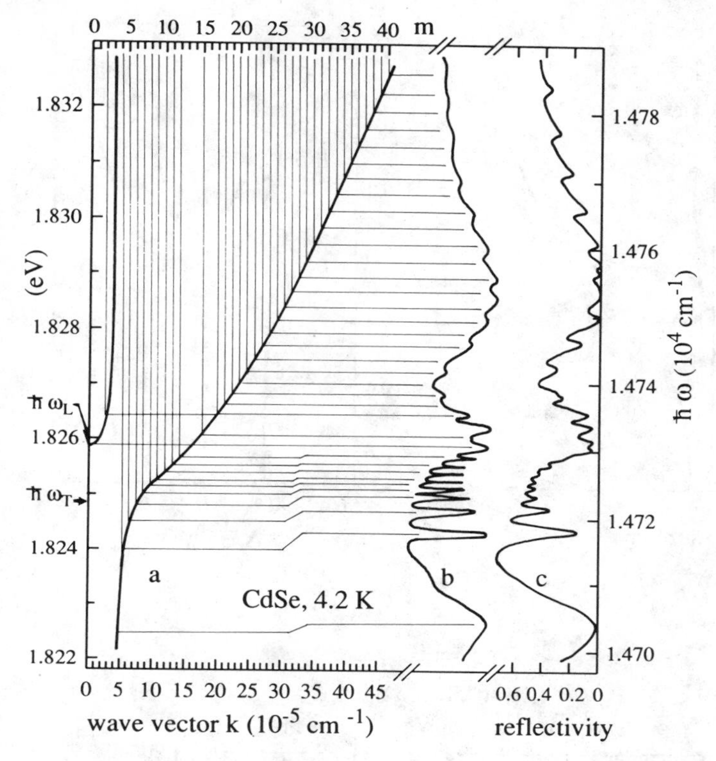

Fig. 14.15. The polariton dispersion in CdSe with an equidistant ruling of the ***k***-axis (**a**) and the observed (**b**) and calculated structure (**c**) of the resulting Fabry–Perot modes. According to [14.29]

condition for the reflection minima coincides with that for the transmission maxima in (14.18).

The equidistance on the ***k***-axis is clearly visible and the good agreement of theory and experiment is obvious. Above ω_{0L} one can clearly see closely spaced modes of the LPB and superimposed widely spaced ones from the part of the light travelling on the UPB. The fact that the small modulation decreases with increasing photon energy indicates that the fraction of light travelling on the LPB through the sample decreases with increasing energy above ω_{0L}.

Similar data have also been found in CdS and CuCl [14.30, 31].

The next method is resonant Brillouin scattering. We introduced this scattering process with acoustic phonons in Sect. 13.3 (Fig. 13.2) and know that the Brillouin shift is directly proportional to the transfer of momentum due to the linear dispersion relation of acoustic phonons around the Γ point.

Figure 14.16 shows schematically the Brillouin scattering in the resonance region of an exciton polariton in a backward scattering configuration. The

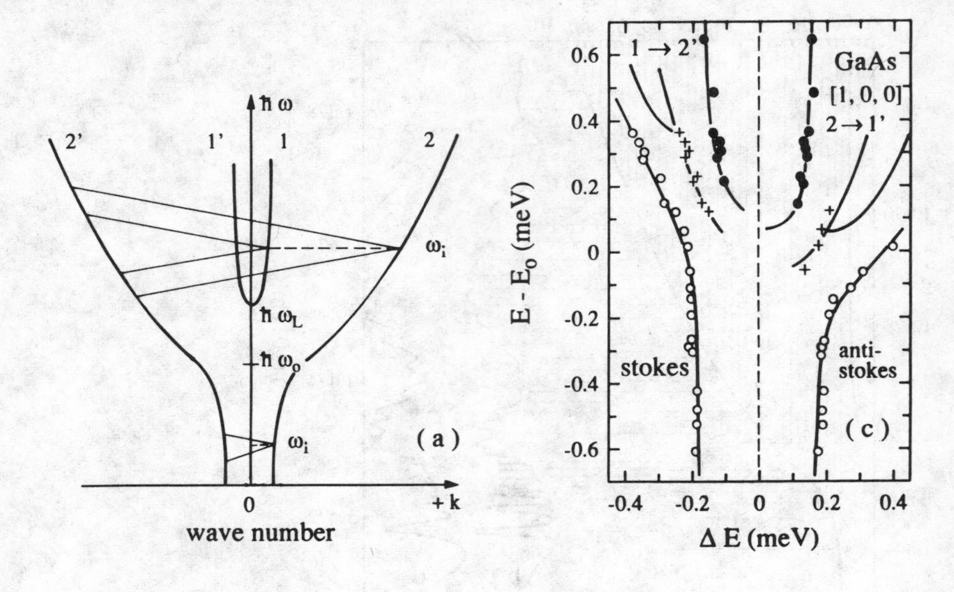

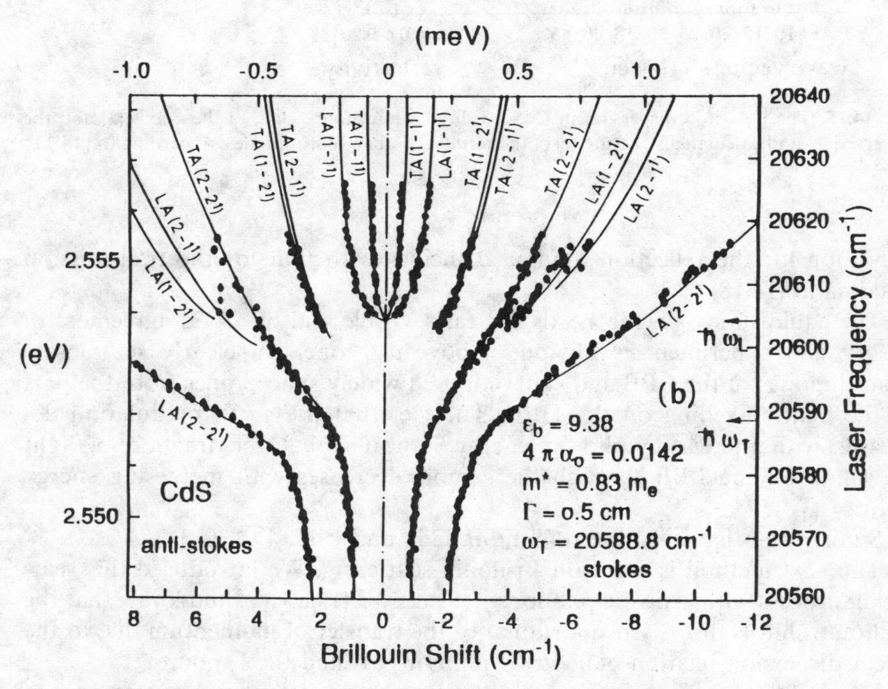

Fig. 14.16a–c. Schematic drawing of Brillouin scattering in the resonance of an exciton polariton (**a**) and experimental data for GaAs (**c**) and for CdS (**b**). According to [14.32, 33], respectively

effect has been observed in GaAs, CdS and ZnSe [14.32–34]. In Fig. 14.16b and c we give data for CdS and GaAs partly involving TA and LA phonons. The soild lines have been calculated in a fit procedure with the parameters in the inset; these of course, also allow the calculation of the polariton dispersion itself.

It is well known that refraction from a prism can be used to determine the real part of $\tilde{n}$ and thus of $\boldsymbol{k}$ (cf. Fig. 5.3). If sufficiently thin ($d \leqslant 1\,\mu\mathrm{m}$), prism-shaped samples are available, it is also possible to extend this technique to the resonance region of the exciton polariton. Fortunately some CdS platelets grow in the desired form, presumably involving a small-angle grain boundary.

Figure 14.17 gives two examples for CdS, where n is given and not k, but where both quantities are simply connected with each other according to $\mathrm{Re}\{k\} = \omega/c\,\mathrm{Re}\{\tilde{n}\}$, see (2.36). In Fig. 14.17a one sees again the birefringence and dichroism for the polarizations $\boldsymbol{E} \perp \boldsymbol{c}$ (i.e., Γ_5) and $\boldsymbol{E} \parallel \boldsymbol{c}$ (i.e., Γ_1) of the $n_\mathrm{B} = 1$ A-exciton resonance known already from the reflection spectra of Fig. 14.4. In Fig. 14.17b the dispersion of the LPB can be followed up to $n \approx 20$ corresponding to $\varepsilon_1 = 400$ or $k = 2.5 \cdot 10^6\,\mathrm{cm}^{-1}$.

If one compares the time of flight of a picosecond pulse through a sample with its propagation through vacuum one can deduce the group velocity v_g and, with (2.13b), the slope of the dispersion relation. In Fig. 14.18b experimental values are given for v_g at the lowest exciton polariton resonance in CuCl, together with a curve deduced from the dispersion relation of Fig. 14.18a. The excellent agreement between experiment and theory again proves the validity of the polariton concept. One can see from Fig. 14.18b that in CuCl v_g can be as low as $5 \cdot 10^{-5} c$. In the region above ω_0L two pulses are created in the medium from one incident pulse due to the different group velocities on the LPB and the UPB. The spatial distance between them increases with sample thickness and this is another reason to call the $\boldsymbol{k}$-dependence of ω_0 "spatial dispersion". Similar experiments have been performed in GaAs, CdSe and CdS [14.38].

We conclude this section with two nonlinear methods of $\boldsymbol{k}$-space spectroscopy, namely two-photon spectroscopy or absorption (TPA) and two-photon (or hyper-) Raman scattering (TPRS, HRS), anticipating some of the results of Sect. 19.3. In TPA two laser beams are directed onto the sample, sometimes with a finite angle between them. They have energies $\hbar\omega_{1,2}$ and momenta $\hbar\boldsymbol{k}_{1,2}$. A TPA signal, i.e., a signal which occurs only if both beams are on, is observed only if the sum of energies and momenta coincide in the sample with those of an excited state, i.e., (see Fig 14.19)

$$\hbar\omega_1 + \hbar\omega_2 = E_\mathrm{f}\,,$$
$$\hbar\boldsymbol{k}_1 + \hbar\boldsymbol{k}_2 = \hbar\boldsymbol{k}_\mathrm{f}\,. \tag{14.19}$$

Due to the curvature of the LPB it is only possible with TPA to reach states on the longitudinal exciton branch or on the UPB. The selection rules are usually different from those of one-photon absorption and depend in addition

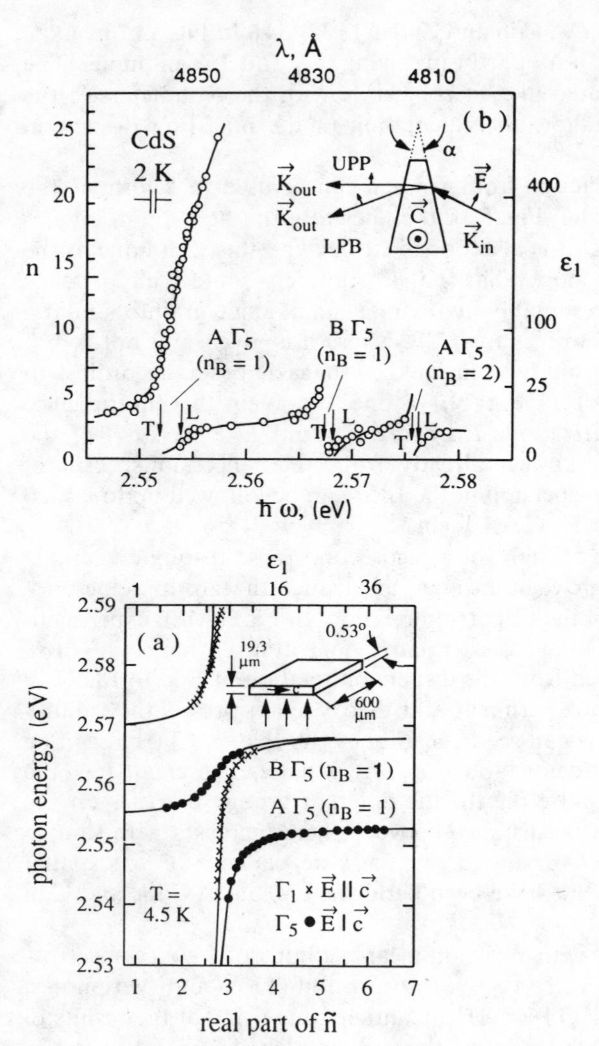

Fig. 14.17a, b. Dispersion relations of the A- and B-exciton resonances determined with the thin-prism method. According to [14.35, 36], respectively

on the polarizations of the beams relative to each other and to the crystallographic axes. Examples will be given in Sect. 25.5. More recently, even three-photon spectroscopy has been used to determine the exciton polariton branches in various semiconductors and insulators [14.15, 39].

In HRS one or two incident laser beams create in the sample two (generally photon-like) polaritons. This two polariton state decays under energy and momentum conservation into a photon-like polariton $\hbar\omega_R$, which is observed as a Raman-like emission, and another final state particle $\hbar\omega_f$ which can be photon- or exciton-like. If a single incident beam is used to deliver both

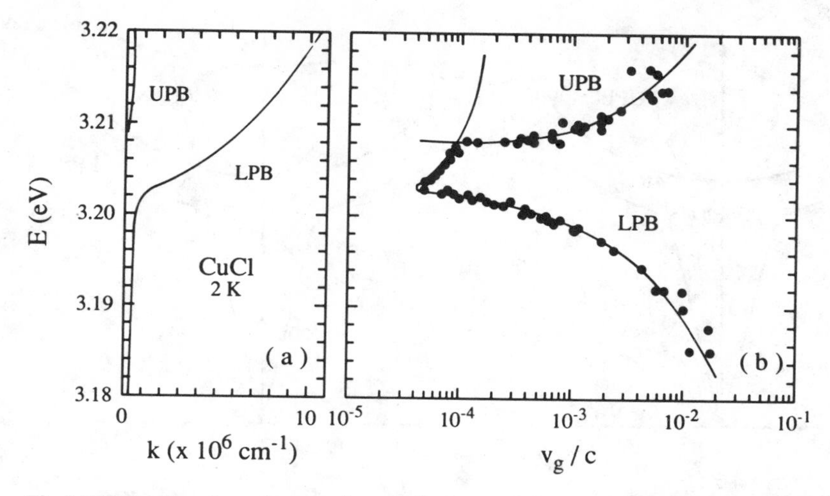

Fig. 14.18a, b. The dispersion relation of the lowest free exciton resonance in CuCl (**a**) and the group velocity determined from the time-of-flight of picosecond laser pulses and from the dispersion relation (**b**). According to [14.37]

incident quanta, i.e., $\hbar\omega_1 = \hbar\omega_2$, this process reads

$$2\hbar\omega_1 = \hbar\omega_R + \hbar\omega_f$$
$$2\hbar \boldsymbol{k}_1 = \hbar \boldsymbol{k}_R + \hbar \boldsymbol{k}_f . \tag{14.20}$$

Usually one aims to have one or both of the incident quanta almost in resonance with the exciton and/or the biexciton state (see below). This choice enhances the transition probability due to the small resonance denominators appearing in perturbation theory. In Fig. 14.19 we show schematically TPA and the HRS processes where in both cases the longitudinal state has been reached. Figure 14.20 shows the polariton dispersion in CuCl around the lowest $n_B = 1\,\Gamma_5$-resonance and the states reached by two photon absorption and by HRS.

A large number of semiconductors have been investigated by TPA and HRS. TPA has also been used to detect higher states ($n_B > 1$). For recent reviews see, e.g. [14.41, 42]. Three-photon absorption has proved the validity of the polariton concept also in insulators like KCl [14.39, 43].

14.5 Surface-Exciton Polaritons

As already mentioned connection with Fig. 7.2, one also finds surface-exciton polaritons in the spectral region between the LPB and the longitudinal exciton. Their dispersion can be measured by attenuated (or frustrated) total internal reflection (ATR). A light beam is sent into a prism in such a way that

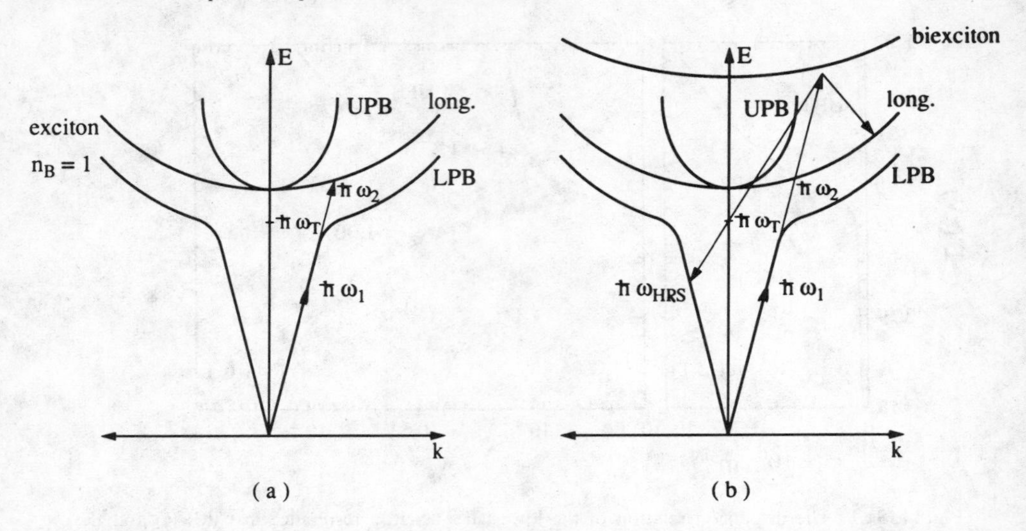

Fig. 14.19a, b. Two-photon transition from the crystal ground state to the longitudinal exciton branch (**a**) and the hyper-Raman process which is almost resonant with the exciton and the biexciton state (**b**)

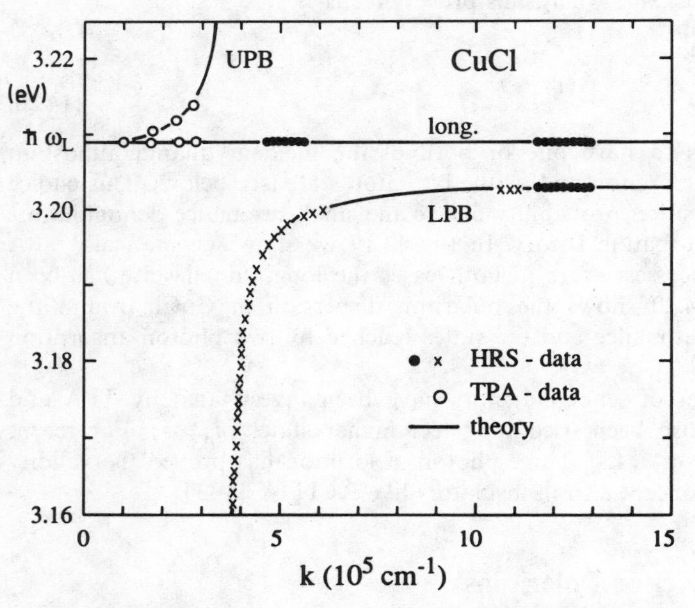

Fig. 14.20. The polariton dispersion of the lowest Γ_5-exciton resonance in CuCl measured by two-photon absorption and by hyper-Raman scattering. According to [14.40–42], respectively

total internal reflection occurs as its base (Fig. 14.21a). Under these conditions an evanescent wave propagates into the vacuum as is indicated schematically (see also Fig. 3.3). The frequency of this wave can be tuned trivally, and its wave vector $\boldsymbol{k}_{\parallel}$ by varying the angle of incidence. If a semiconductor is brought close enough to the base of the prism, the evanescent wave can couple to the surface-exciton polariton modes in the semiconductors if both $\hbar\omega$ and $\boldsymbol{k}_{\parallel}$ coincide. "Close enough" means distances of the order of λ, i.e., fractions of a μm and the realization of this condition involves some experimental skill. The coupling to the surface-exciton polariton mode attenuates the total reflection. In Fig. 14.21b measured and calculated ATR spectra are shown for the $n_B = 1$ $C\Gamma_1$-exciton resonance in ZnO, which has a rather large Δ_{LT} splitting of about 12 meV. The calculated dispersion relation of the surface-exciton polariton is given in Fig. 14.21c together with the states reached by ATR. Good agreement between experiment and theory can be claimed in both Figs. 14.21b and c. More details about surface-exciton polaritons can be found in [14.44, 45].

14.6 Forbidden Transitions

In the preceding section, we treated the optical properties of exciton resonances with large oscillator strength, requiring the polariton concept for a quantitative description. There are also intrinsic excitons which, for various reasons, couple only weakly to the radiation field. In this case the weak coupling approach is usually but not always sufficient for their description.

Reasons why excitons can have low oscillator strength include spin-flips, dipole-forbidden transitions, or the necessity to involve a third particle, e.g., a phonon in their creation. We give some examples in the following.

Even in direct gap semiconductors with dipole-allowed band-to-band transitions there are exciton states with very small oscillator strength, so-called forbidden transitions. Among them there are triplet states, which involve a spin-flip, longitudinal excitons, and states whose symmetry forbids the transition in the dipole approximation (Sect. 25.5). The latter situation can occur for $n_B \geqslant 2$ and L or $L_z = 1, 2 \ldots$. Such exciton states can sometimes be observed in one-photon transitions in higher-order perturbation theory, e.g., electric quadrupole or magnetic dipole transitions. This, however, is only possible if no strong one-photon transition occurs in the same spectral range. An example is again CdS. We have already stated that the $n_B = 1$ A-excitons are dipole allowed only for the polarization $\boldsymbol{E} \perp \boldsymbol{c}$, i.e., Γ_5. For $\boldsymbol{E} \parallel \boldsymbol{c}$ one can weakly see the $A\Gamma_6$ triplet or the $A\Gamma_5^L$. In the latter case the finite angle of aperture of every real light beam plays a role because small deviations from $\boldsymbol{k} \perp \boldsymbol{c}$ lead to a small oscillator strength of mixed-mode states according to (14.10). In Fig. 14.22 we show a transmission spectrum of CdS around the $n_B = 1$ A-exciton states for the orientation $\boldsymbol{E} \parallel \boldsymbol{c}$; $\boldsymbol{k} \perp \boldsymbol{c}$, where the $A\Gamma_6$ triplet and the $A\Gamma_5^L$ states are seen.

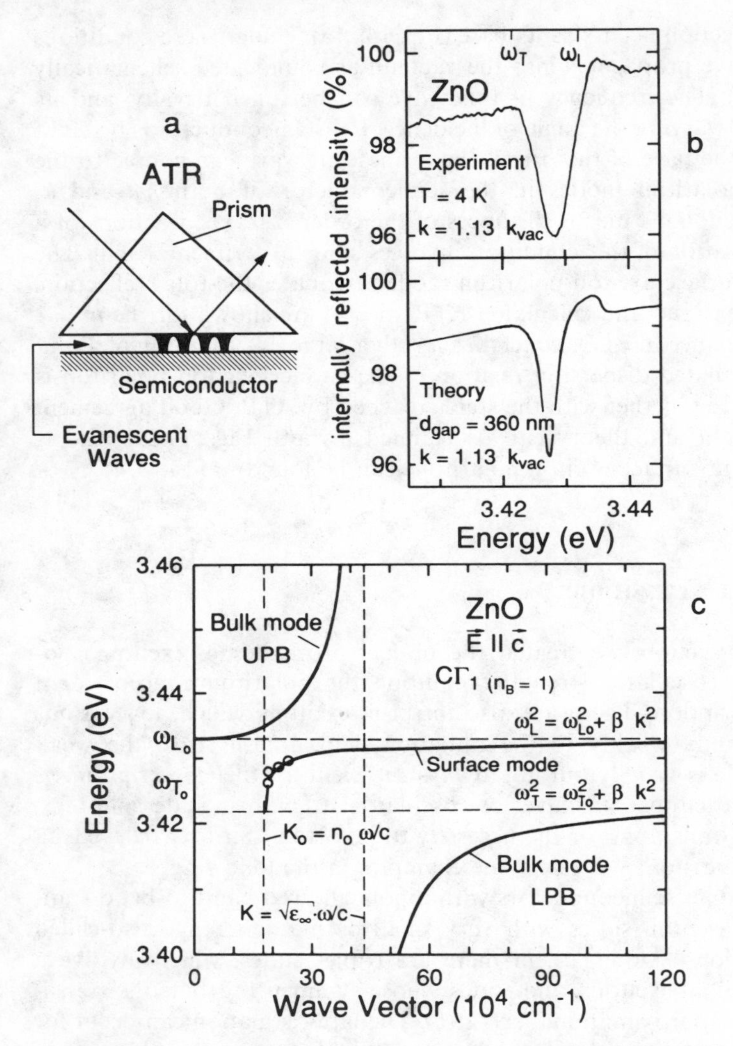

Fig. 14.21a–c. The principle of the experimental technique of attenuated total reflection (ATR) (**a**) experimental and calculated ATR spectra for the $n_B = 1$ $C\Gamma_1$-exciton in ZnO (**b**) and the resulting dispersion of the surface and volume polaritons (**c**). According to [14.44]

The other way of reaching some of these states is via two-photon absorption (TPA). The selection rules for TPA differ from those for one-photon absorption as will be outlined in Sect. 25.5.

A first example has already been presented in Fig. 14.20, where the longitudinal branch of the Γ_5 exciton is seen in TPA. Another example for states with $n_B \geq 2$ will be given in Sect. 16.1, where we discuss the influence of magnetic fields on excitons.

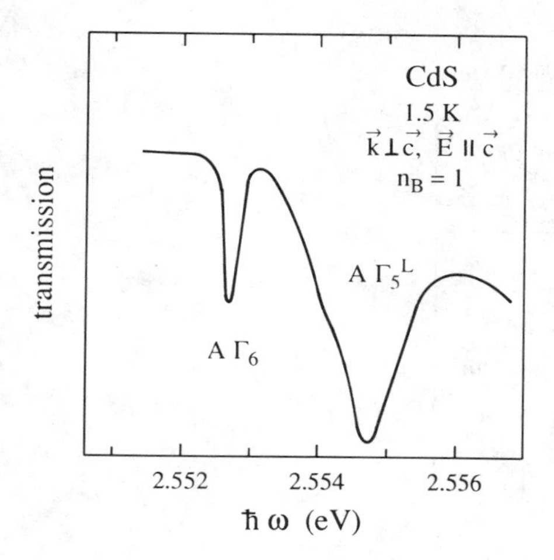

Fig. 14.22. A transmission spectrum of CdS in the resonance region of the $n_B = 1$ AΓ_5-exciton for the orientation $\boldsymbol{E} \parallel \boldsymbol{c}$ showing the dipole-forbidden transitions AΓ_6 and AΓ_5^L. According to [14.46]

There is a group of semiconductors that have a direct gap at the Γ point. But the transition between the uppermost valence band and the lowest conduction band is dipole forbidden because both bands have the same parity. The best investigated material of this group is Cu_2O, but there are many others such as SnO_2, TiO_2, and GeO_2. In these materials exciton states with an S envelope (i.e., $L = 0$) are all dipole forbidden.

For $n_B \geqslant 2$ there are also envelope functions with $L = 1$, which have odd parity. Via their envelope, these states acquire a weak oscillator strength and can be seen in absorption.

Figure 14.23 gives an example of the "yellow" series in Cu_2O, where the one- and two-photon absorption spectra are compared. The $n_B = 1$ state has been found by TPA. Its position is indicated by an arrow. This state does not fit into the n_B^{-2} series, for the reasons discussed already in Sect. 11.2. However, even for the tiny oscillator strength of this $n_B = 1$ S-exciton, it has been shown that the polariton effect exists and can be measured [14.48].

Many semiconductors, among them some of great technological importance such as Si or Ge, have an indirect bandstructure, as explained in connection with Fig. 10.9. Consequently the lowest free excitons occur at $\boldsymbol{k} \neq 0$, as outlined in Fig. 11.1c. Due to momentum conservation, these states cannot couple directly to the radiation field. In both absorption and emission processes a third, momentum conserving, particle has to be involved, usually a phonon. At low temperature only photon absorption with creation (i.e., emission) of one or more phonons is possible. At higher temperatures, when the relevant phonon states are populated with finite probability, absorption of light quanta is possible with both phonon emission and absorption.

Figure 14.24a shows schematically the process of absorption accompanied by the creation or emission of a phonon, and in Fig. 14.24b, as an example, the absorption spectrum of GaP at low temperature. The exciton states do not

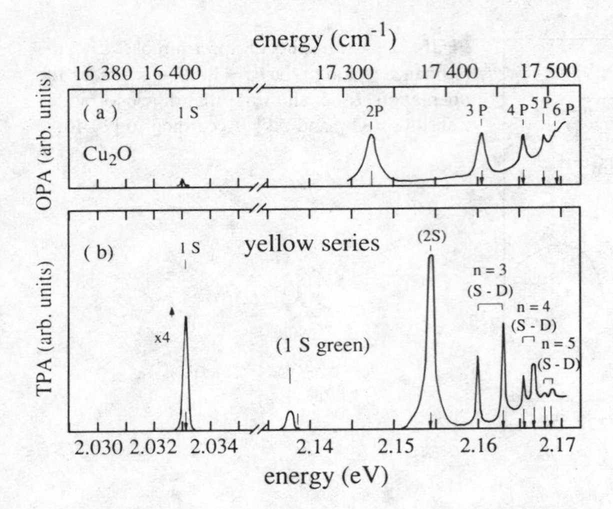

Fig. 14.23a, b. The absorption of the forbidden "yellow" exciton series in CuO_2. According to [14.47]

appear as peaks as in direct gap materials (Fig. 14.23) but instead as the onset energies, since the participation of phonons allows one to reach the whole density of states. Thus the absorption spectra are given in the simplest approximation by sums of expressions like

$$\alpha_{\mathrm{em}}^{(\hbar\omega)} \propto \alpha_0 [\hbar\omega - E_{\mathrm{ex}}(\boldsymbol{k}_0) - \hbar\omega_{\mathrm{ph}}]^{1/2}(1 + N_{\mathrm{ph}}) , \tag{14.21a}$$

$$\alpha_{\mathrm{abs}}^{(\hbar\omega)} \propto \alpha_0 [\hbar\omega - E_{\mathrm{ex}}(\boldsymbol{k}_0) + \hbar\omega_{\mathrm{ph}}]^{1/2} N_{\mathrm{ph}} , \tag{14.21b}$$

where α_0 contains the transition matrix element squared, which is assumed not (or only weakly) to depend on the momentum of the phonon. $E_{\mathrm{ex}}(\boldsymbol{k}_0)$ is the exciton energy at the indirect minimum, $\hbar\omega_{\mathrm{ph}}$ the phonon energy for $\boldsymbol{k} \approx \boldsymbol{k}_0$, and N_{ph} is the number for the phonons in a given mode.

The photon energy in the square-root term which describes the density of states of the excitons above $E_{\mathrm{ex}}(\boldsymbol{k}_0)$ has to be chosen so that the argument is positive. For other values of $\hbar\omega$, α_{em} and α_{abs} are zero.

The participation of the phonon makes the transition probability for absorption in indirect-gap semiconductors several orders of magnitude smaller than that for direct, dipole-allowed transitions. Typical values of the "indirect" absorption coefficient are in the range

$$1\mathrm{cm}^{-1} \leqslant \alpha_{\mathrm{ind}} \leqslant 10^2\mathrm{cm}^{-1} . \tag{14.22}$$

These values are so low that no significant structures appear in the reflection spectra.

Above the indirect exciton, there can also be direct excitons which show absorption structures of the type already discussed in the preceding section.

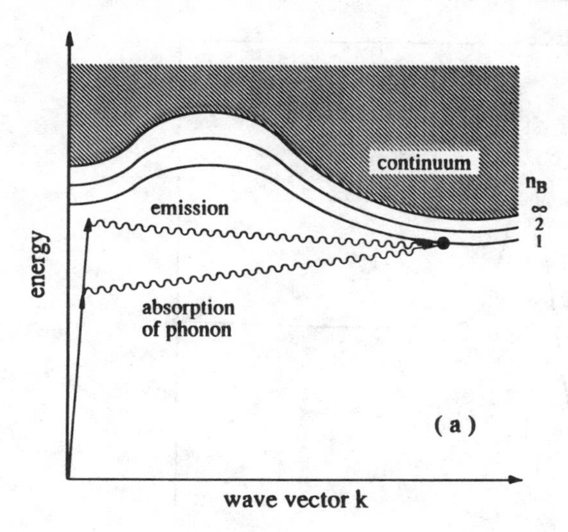

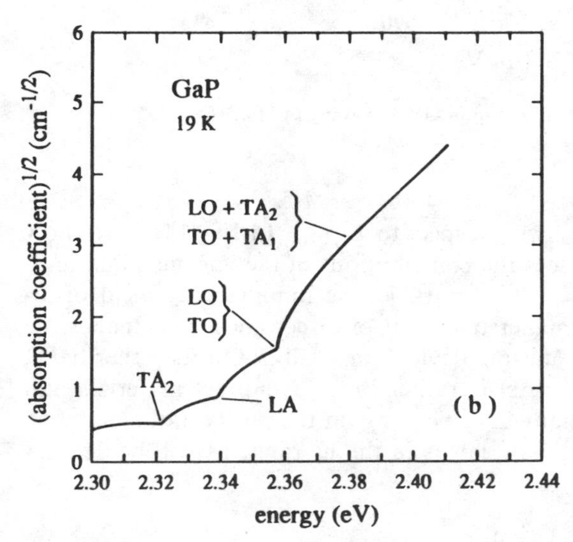

Fig. 14.24a, b. The creation of an exciton in an indirect gap semiconductor accompanied by creation or annihilation of a momentum-conserving phonon (**a**). The absorption spectrum of GaP (**b**). According to [14.49]

We give an example in Fig. 14.25 for Ge. Since the direct exciton can decay rapidly into lower states, it is strongly damped, preventing the observation of finestructure such as states with $n_B > 1$.

The luminescence of free excitons in indirect-gap materials involves – as does the absorption – one or more phonons (Fig. 14.24a). Consequently the

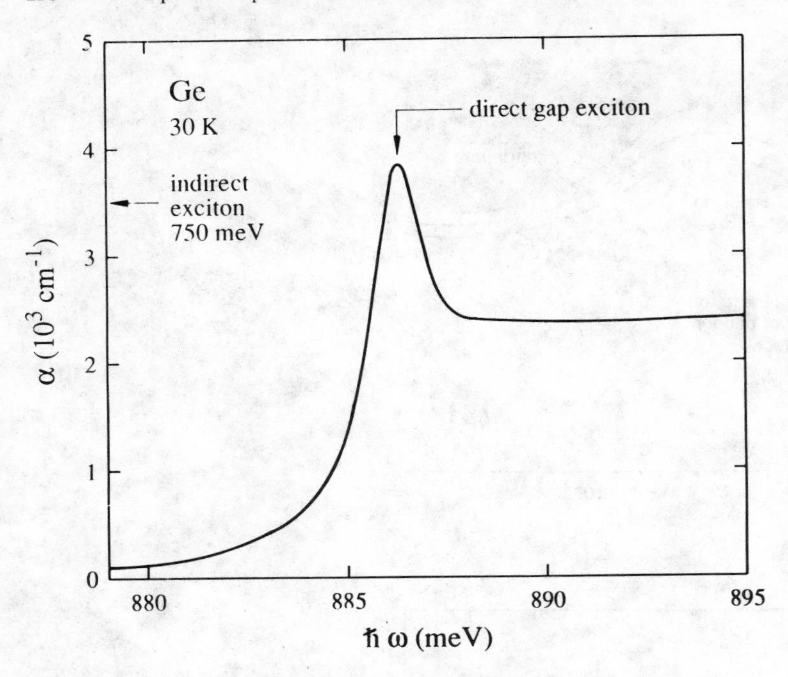

Fig. 14.25. Absorption around the direct exciton in Ge. According to [14.50]

excitonic emission is red-shifted with respect to $E_{ex}(\boldsymbol{k}_0)$. In Fig. 14.26 we show an example for Ge where one sees the contributions of the various phonons.

Due to the participation of a third particle, the luminescence yield of indirect-gap materials is much smaller than that of direct ones. This makes the application of bulk Si in light emitting devices impossible. On the other hand, the lifetime of excitons (or carriers) is rather long in indirect materials and ranges from less than 1 μs to many μs depending on the purity of the material, while the lifetime in direct-gap materials is in the ns range as will be demonstrated later.

14.7 Excitons in Systems of Reduced Dimensionality

As in the presentation of elementary excitations in Chaps. 9–12, we now proceed from bulk materials to the optical properties of excitons in systems of reduced dimensionality. Ample results are available for quantum wells, superlattices, and quantum dots, whereas results for quantum wires are presently still rather limited. For some recent reviews see [14.52–54] or [10.22].

In Sects. 10.8, 11.3 we outlined the influence of the dimensionality on the eigenstates of carriers and of excitons. Now we briefly describe the optical

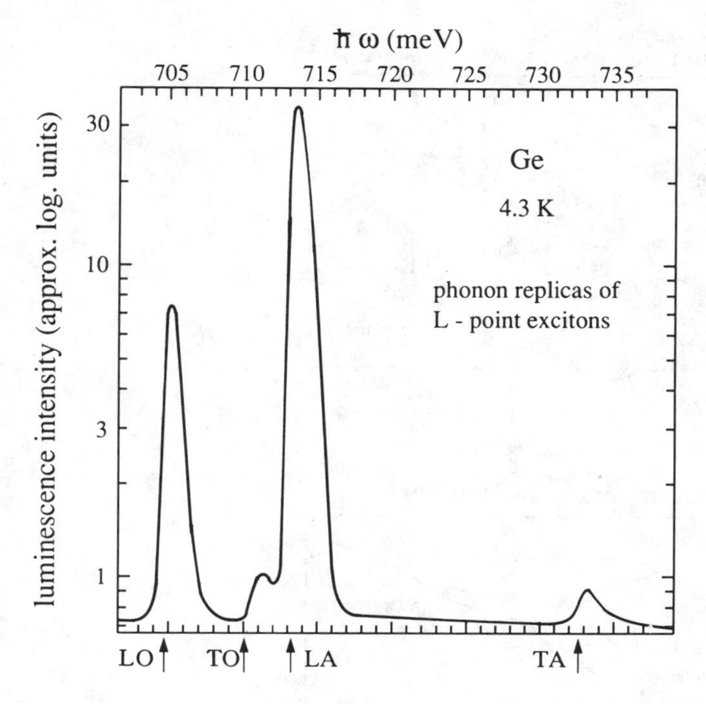

Fig. 14.26 A luminescence spectrum of Ge showing various phonon satellites. According to [14.51]

properties of excitons in quantum wells. We start with type-I structures, where the electron and hole are quantized in the same material.

The description will remain almost exclusively in the weak coupling limit, mainly because the quality of presently available samples in relation to interface roughness or alloy disorder is not sufficient for the fine details stemming from the polariton concept to be observable. The most widely investigated groups of MQW are based on GaAs/$Al_{1-y}Ga_yAs$ and on InP/InAlP/GaAlP (Fig. 10.16). In Fig. 14.27 we show absorption spectra of both systems.

We see the exciton and the continuum transitions between the first quantized heavy- and light-hole levels and the first quantized electron state, $n_z = 1$ hh and $n_z = 1$ lh, respectively. The peaks are due to the $n_B = 1$ state. The higher states $n_B \geqslant 2$ are usually not visible, due to their relatively small binding energies and oscillator strengths [see (11.10)] and due to the broadening mechanisms mentioned above. The absorption coefficient in the continuum states simply reflects the constant density of states for two-dimensional effective-mass particles. It is not influenced much by the Sommerfeld enhancement unlike the three-dimensional systems mentioned in relation to (11.11). The next prominent structures are the $n_z = 2$ hh and lh excitons. Higher states are usually less

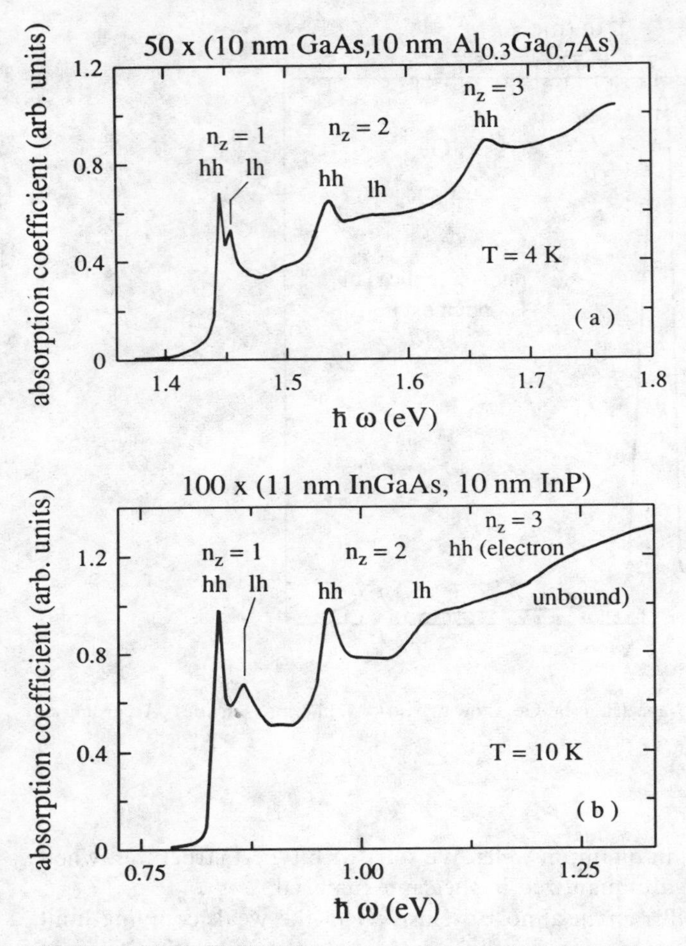

Fig. 14.27a, b. The low temperature absorption spectra of multiple quantum well samples of GaAs/$Al_{1-y}Ga_yAs$ (**a**) and of $In_{1-y}Ga_yAs$/InP (**b**). According to [14.52] and [14.55], respectively

clearly seen, among other reasons because the electron states are often no longer quantized for $n_z > 3$, depending on l_z, m_e and the band offset.

In contrast to the InP system, the GaAs substrate of the AlGaAs MQW is opaque for the exciton resonances. So it has to be removed by selective etching [14.56] for the observation of absorption spectra. Before this technique was known, scientists often measured the luminescence–excitation spectra, i.e., they detected the exciton luminescence as a function of the photon energy of excitation $\hbar\omega_{exc}$ for constant incident intensity (or rather constant photon flux density). Similarly, the dependence of the photo current on $\hbar\omega_{exc}$ can be used. The quantities

$$\left.\begin{aligned} I^{\text{lum}} &= f(\hbar\omega_{\text{exc}}) \\ j^{\text{photo}} &= f(\hbar\omega_{\text{exc}}) \end{aligned}\right\} I_{\text{exc}} = \text{const} . \tag{14.23}$$

are related to, but not identical with, the absorption spectrum $\alpha(\hbar\omega)$.

The above-mentioned techniques are still useful if a single QW is investigated. The absorption coefficients of two-dimensional excitons in the above-mentioned systems are of the order of $10^4\,\text{cm}^{-1}$. A SQW with $l_z = 10\,\text{nm}$ therefore gives an optical density of only

$$l_z \cdot \alpha \approx 10\,\text{nm} \cdot 10^4\,\text{cm}^{-1} = 10^{-2} , \tag{14.24}$$

which is hardly detectable in a simple transmission experiment. The observation of such a small variation needs either highly developed modulation techniques or photo-luminescence excitation spectroscopy.

The advantage of studying only a SQW is that fluctuations of l_z from one QW to the next can be avoided. Furthermore some (opto-) electronic devices contain only a SQW, e.g., some field-effect transitors (MOD-FET, HEMT). The additional optical selection rules which arise when the excitons are confined to a QW are rather simple:

$$\Delta n_z = 0 . \tag{14.25}$$

Transitions with odd Δn_z are forbidden by parity and those for even non-zero Δn_z are forbidden for rectangular shaped QW in the limit of infinetly high barriers. Transitions which violate (14.25) can be observed if there are perturbations such as external or internal electric fields. Examples will be given in Sect. 16.2.

The luminescence spectra of excitons in QW are often rather broad and partly Stokes-shifted with respect to the absorption. We give two-rather different examples in Fig. 14.28. The effects can be explained by assuming that the luminescence is a superposition of the recombination of excitons in extended states at $\boldsymbol{k}_\parallel \approx 0$ (excitons with larger $\boldsymbol{k}_\parallel$ cannot decay radiatively because of momentum conservation), of excitons localized in tail states due to well-width fluctuations and/or alloy disorder, and finally of excitons bound to impurities such as neutral acceptors or donors. As we shall see in Chap. 15, the latter complexes lead to spectrally very narrow emission bands in 3d systems, but in QW, their energies are strongly inhomogeneously broadened because the energy of the complex depends on its positions relative to the barriers.

In samples of very high quality the mechanisms mentioned above are less pronounced, resulting simultaneously in a reduction of the spectral widths of absorption and emission bands and of the Stokes shift. An example is given in Fig. 14.28. The shoulders in the emission on the low energy side can have various origins: fluctuations of l_z by one monolayer, recombination of a bound exciton complex, or recombination of a biexciton (see Sects. 15.1 and 19.3 for the latter two processes). Due to the weaker coupling of carriers and excitons

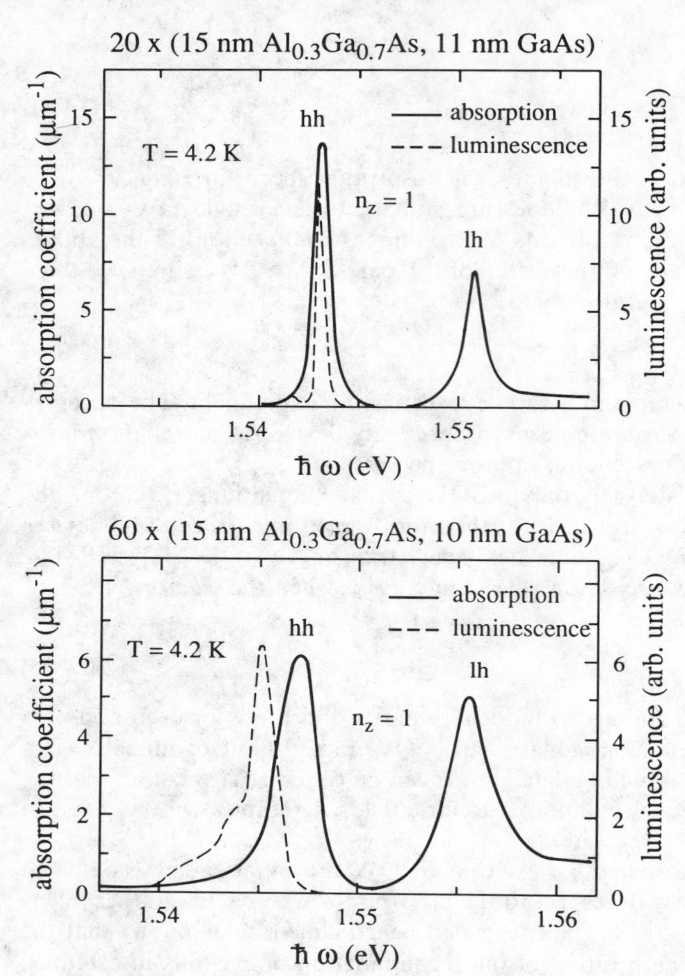

Fig. 14.28. A comparison of absorption and luminescence spectra in two samples of different quality. According to [14.57]

to optical phonons in the III–V compounds compared to the more ionic II–VI semiconductors, LO-phonon replica are hardly detectable in the former materials.

Reflection spectra of excitons in QW have not been investigated very thoroughly so far. Usually the Fabry–Perot modes of some layers in the QW sample, such as buffer layers or thin films of adhesive (with which the QW are attached to a transparent substrate) are much more pronounced than the exciton resonance itself (see Fig. 14.29).

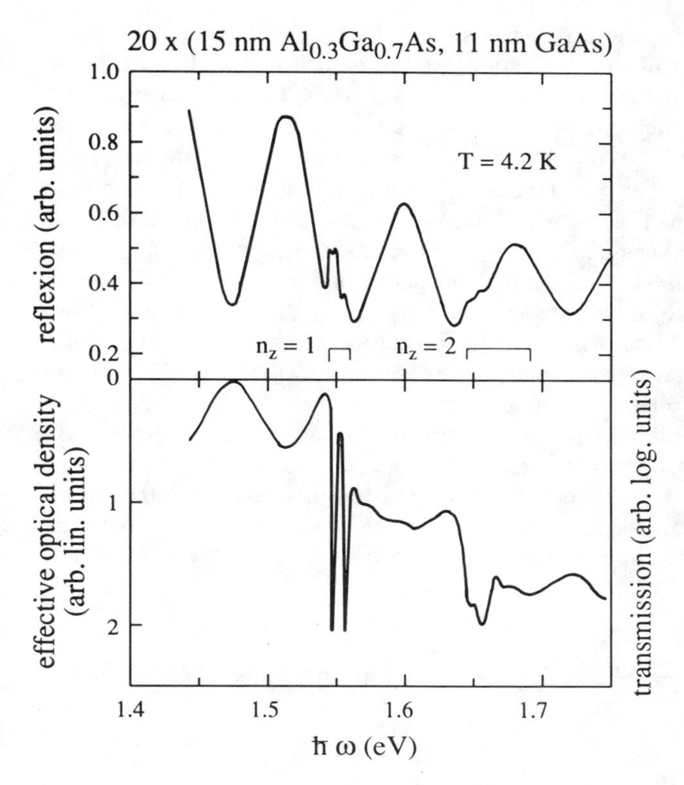

Fig. 14.29. The reflection (*top*) and transmission (*bottom*) spectra of an $Al_{1-y}Ga_yAs/GaAs$ MQW sample. According to [14.57]

In bulk GaAs and InP the exciton resonance shows up in reflection only as a small dip (see also Fig. 14.5). The increase of oscillator strength caused by the transition from three to two dimensions is usually more than compensated by the broadening mechanisms mentioned above.

This is also the reason why the polariton concept has only recently been elaborated for excitons in QW. For first results see [14.58]. The techniques of momentum-space spectroscopy outlined in Sect. 14.4 for bulk materials have not yet been applied to systems of reduced dimensionality. When the $\boldsymbol{k}$-vector of the light is perpendicular to the layer, the situation should be relatively simple. In the wells one has a resonance with infinite mass, i.e., without spatial dispersion because of the quantization in this direction, but for MQW one has to consider multiple reflection and interference caused by the various QW. When the $\boldsymbol{k}$-vector of the incident light is parallel to the layers, a new situation will arise because the light beam covers both regions containing a resonance (the wells) and others without (the barriers). Since l_z is much smaller than the wavelength of the light, these two parts cannot be treated independently, so that interesting new results in the field of exciton polaritons can be expected in the future, especially if high quality samples are used. The optical properties will depend, among other things, on the orientation of $\boldsymbol{E}$ relative to the layers.

The optical properties of type-I superlattices (SL) do not deviate very much from those of MQW except for some broadening due to the formation of mini-bands in the direction of periodicity of the SL and the increase of the importance of monolayer fluctuations with decreasing l_z; see (10.35).

The II–VI compounds usually have a rather strong lattice mismatch, as can be seen from Fig. 10.16. As a consequence only rather thin films (a few monolayers) can be grown if one is to avoid the formation of misfit dislocations. The results are so-called strained-layer superlattices where the two different materials forming the wells and the barriers adapt their lattice constants. In Fig. 14.30a we show the absorption spectra of a series of ZnS/ZnSe SL with decreasing well-width. The quantized $n_z = 1$ hh and 1h excitons are the main features. Figure 14.30b gives the reflection, the luminescence and the photoluminescence excitation spectrum for one sample. It is obvious that the structures are much broader than for the GaAs- and InP-based MQW in Fig. 14.27,

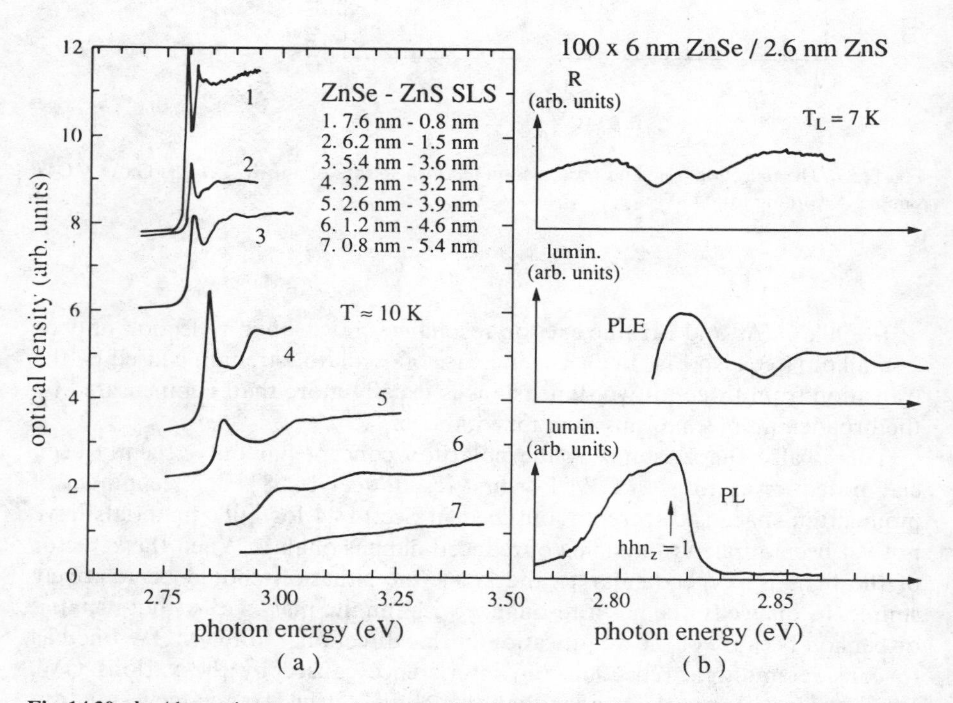

Fig. 14.30a, b. Absorption spectra of a series of ZnS/ZnSe super-lattices with various values of the barrier and well width (**a**) and a luminescence (PL), a photoluminescence excitation (PLE), and a reflection (R) spectrum of one sample (**b**). According to [14.59] and [14.60], respectively

mainly since the methods of growing II–VI SL are not so advanced as for the III–V compounds. Some exceptions to this statement are discussed in [14.61]. A good review of the present state of the art of II–VI epitaxy and of their optical properties is found in the proceedings of recent conferences on this topic [14.62].

Since the luminescence tail of excitons localized by well-width fluctuations in II–VI superlattices is sometimes spectrally broader than the energy of the optical phonons, and since the coupling of the excitons to the phonons is stronger than in the III–V system AlGaAs, one can observe interesting phenomena related to the relaxation or "freezing" of excitons in these tail states. More information about this topic and about the magnetic polaron effects which appear for II–VI SL containing Mn can be found in [14.63, 64].

An other interesting topic is the piezo-field in strained-layer SL, its influence on the bandstructure, and its screening by photocarriers, which results in a blue-shift of the emission of as much as 0.4 eV in hexagonal CdS/CdSe SL [14.65].

The transition from a type-I to type-II band alignment can be observed in short period GaAs/AlAs SL. GaAs is a direct gap material and AlAs an indirect one with a larger gap. If the width of the GaAs barrier is made smaller and smaller, then the $\boldsymbol{k} = 0$ $n_z = 1$ conduction band state shifts to higher and higher energies until it is situated above the indirect conduction band state in AlAs. This situation is shown schematically in Fig. 14.31.

In contrast to the type-II GaSb/AlAs SL, the short period GaAs/AlAs SL are indirect in real space and in $\boldsymbol{k}$-space. The peculiarities of absorption in the well, relaxation of the electrons into the "barrier", and subsequent recombination are discussed in detail in [14.53]. An example is shown in Fig. 14.31, where the photoluminescence excitation spectra and the luminescence spectra are given for a type-I and a type-II GaAs/AlAs SL.

Finally we mention that efforts are also being made to grow strained Ge/Si SL, though there are considerable technical problems in growing good quality samples due to the different lattice constants. The idea is that a similar folding-back mechanism as outlined in connection with Fig. 9.20 for phonons, should yield a direct electronic bandstructure which would allow the highly developed Si technology to be used also for light-emitting devices. For a recent review see [14.66] and references therein.

There are various ways of producing quantum wires from QW. One is by masking and etching, possibly followed by regrowth with the barrier material to avoid surface states at the etched lateral surfaces of the well. Another possibility is the focussed-ion-beam (FIB)-induced intermixing of the well and barrier material keeping narrow undisturbed stripes (the wires). An alternate method, which is presently receiving increasing attention, is the growth of wires on high index surfaces or on *V*-grooves. For details of these techniques see [14.53, 54, 67].

The problem encountered in lateral structuring is that the quantization length l_z in the growth direction of the QW is usually much smaller than the

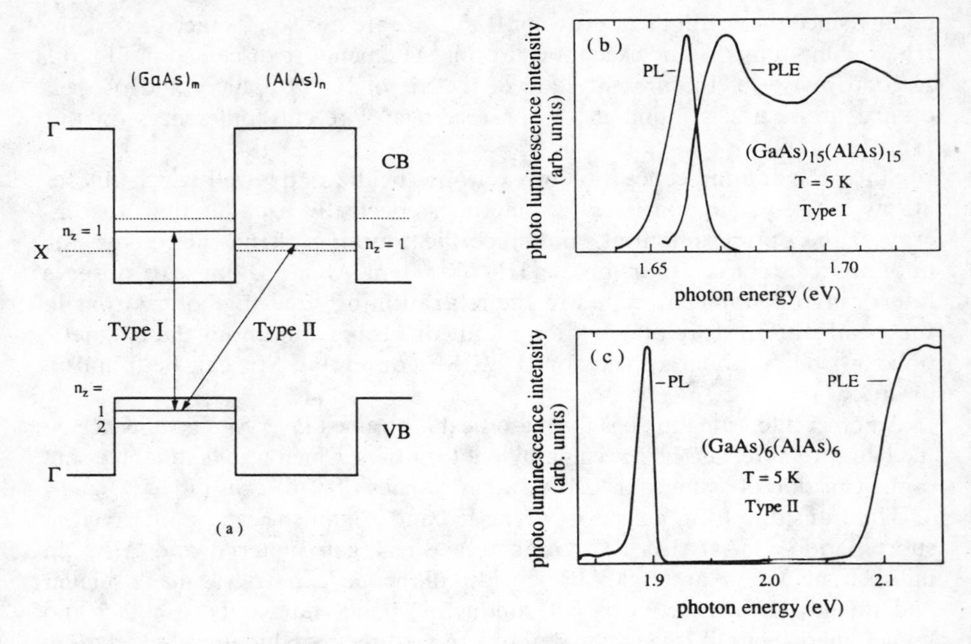

Fig. 14.31a–c. The band alignment of the direct and indirect transitions in a short period AlAs/GaAs type-II superlattice (**a**). Luminescence (PL) and photoluminescence excitation (PLE) spectra of a type-I (**b**) and a type-II (**c**) AlGaAs superlattice. The indices give the number of monolayers. According to [14.53]

length l_y of the subsequent lateral structuring. Typical values are

$$l_z \approx 10\,\text{nm},$$
$$l_y \approx 50-100\,\text{nm}\,. \tag{14.26}$$

Consequently, the confinement energies resulting from confinement in the two directions are very different. Usually one observes only a narrow modulation of the luminescence or photoluminescence excitation spectra due to quantization in the y-direction. An example is given in Fig. 14.32.

At present there are two main possibilities for producing semiconductor quantum dots (QD) to achieve three-dimensional confinement. One involves lateral structuring of QW in two dimensions in the ways mentioned above for quantum wires or the use of dot-like electrodes on the sample which produce a confining potential [14.69]. The results are disk-shaped "dots". Such structures are partly used for the investigation of cyclotron resonances and of transport properties not covered in this book. References to this topic are found in [14.69]. The other possibility is to grow QD during an annealing process in semiconductor-doped glasses or by some chemical precipitation in a liquid or

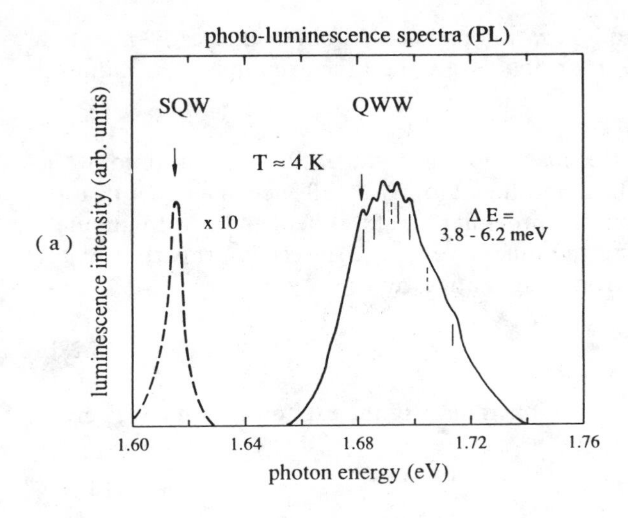

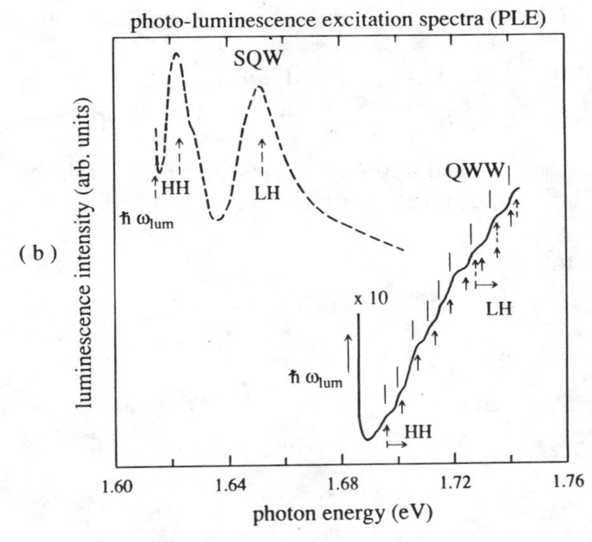

Fig. 14.32a, b. Luminescence (**a**) and photoluminescence excitation spectra (**b**) from quantum well wires (QWW) compared to the corresponding single quantum well. According to [14.68]

solid organic matrix [14, 70]. In this way microcrystallites (sometimes also called nano-crystals) with radii of 1–30 nm can be grown.

The dots produced by the latter technique are more-or-less spherical but inherently have a certain spread in the size distribution of the crystallite radii R which is given by general thermodynamic considerations as

$$\frac{\Delta R}{\bar{R}} \gtrsim 0.1\,. \tag{14.27}$$

One distinguishes three quantization regimes. As outlined already in section 11.3 the regime I is obtained for dots larger than the excitonic Bohr radius;

$$\text{regime I: } /\bar{R} \gtrsim a_B \, . \tag{14.28a}$$

In this case, only the center of mass motion of the exciton is quantized, while the relative motion of electron and hole are hardly affected. The quantization energy is given, in the simplest approximation of a spherical dot with infinitely high barriers and neglecting the difference in the dielectric properties of the microcrystallite and the surrounding matrix, by

$$\Delta E_Q^I = \frac{\hbar^2 \pi^2}{2M\bar{R}^2} \, , \tag{14.28b}$$

where $M = m_e + m_h$ is the translationed mass of the exciton. Regime II is given by

$$\text{regime II: } a_B^e > \bar{R} > a_B^h \, , \tag{14.28c}$$

where $a_B^{e,h}$ are the respective radii of the electron and hole orbits in the exciton around their common center of mass.

In this case the motion of the electron is quantized and the hole moves in the potential of the QD and of the space-charge of the electron for the usual situation $m_e \ll m_h$.

Continuing to use the above-mentioned approximations, and neglecting the Coulomb energy between electron and hole, one finds

$$\Delta E_Q^{II} = C\,\mathrm{Ry}^* \left(\frac{a_B \pi}{\bar{R}} \right)^2 ; \quad \text{with} \quad C = 0.67 \tag{14.28d}$$

Regime III is defined by

$$\text{regime III: } \bar{R} \ll a_B^{e,h} . \tag{14.28e}$$

Since the quantization energy scales with $\bar{R}^{-2}$ and the Coulomb energy with $\bar{R}^{-1}$, for this regime one can expect

$$\frac{\hbar^2 \pi^2}{2\mu\bar{R}^2} = E_Q^{III} \gg E_{\text{Coulomb}}^{III} \tag{14.28f}$$

where μ is the reduced mass of electron and hole. Thus the influence of the Coulomb energy is negligible as compared to the quantization energy. However, more recent calculations have shown that the Coulomb energy is necessary for a quantitative understanding of the eigenstates and the optical spectra also in regime III. In other words, the values of $\bar{R}$ where (14.28e, f) hold are comparable to the lattice constant. In this situation, the whole effective mass concept breaks down and a description of the dot in terms of molecular orbitals becomes more appropriate.

Recently it became obvious that the size of the dots is not the only relevant quantity affecting their linear and especially nonlinear optical properties. The three different growth regimes of the dots which occur during the annealing process of the glasses, namely nucleation, normal growth of the dots and growth through coalescence, result is significantly different properties of, among other things, the interface states between glass matrix and semiconductor dot. For recent reviews see [14.71] or [10.24–29] and references therein.

From our considerations of the density of states in Fig. 10.13 we would expect QD to have an absorption spectrum consisting of a series of δ-functions. Such behavior is not observed in reality since various processes contribute to a broadening of the resonances, e.g., the size distribution of the crystallites considered in Fig. 14.33, the coupling to optical phonons, which is enhanced in the quantum dots as compared to the bulk material, and the influence of impurity, surface, or interface states; see [14.70–72] and again [10.24–29] and references therein.

Regime I is usually realized for QD made from the copper halides CuCl, CuBr and CuI since the values of a_B for these materials are in the range of 1 nm. Figure 14.34a shows an example for CuBr which clearly illustrates the blue-shift of the $n_B = 1$ exciton resonances from the two uppermost valence bands with decreasing $\bar{R}$.

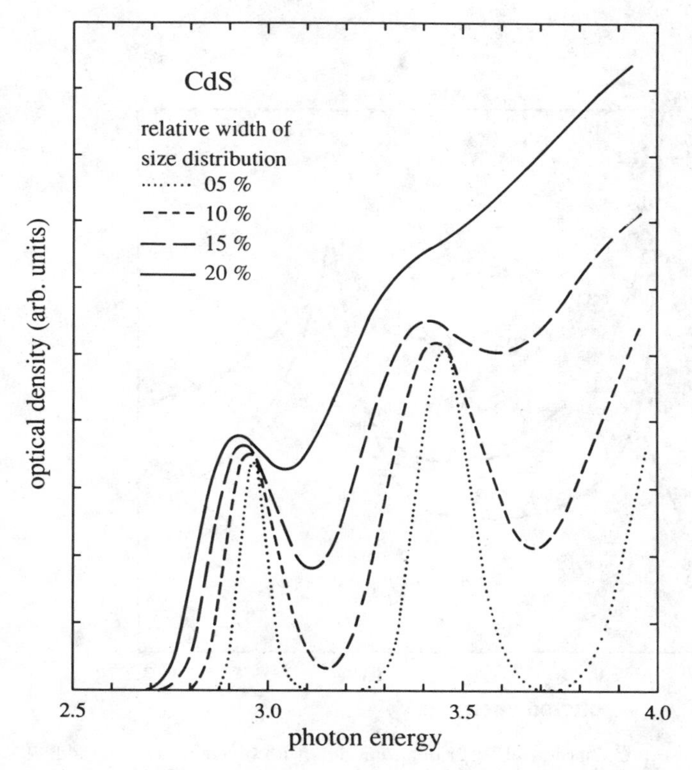

Fig. 14.33. Calculated spectra of the optical density of quantum dots for various relative widths of thier size distribution. According to [14.72]

In Fig. 14.34b we show the blue-shift of the absorption edge of glasses containing CdSe QD with decreasing values of $\bar{R}$. The calculated transition energies and their oscillator strengths are given by the positions and heights of the vertical lines. For more data see also [14.74b].

The luminescence is usually considerably Stokes-shifted with respect to the absorption, in Fig. 14.35 by roughly 200 meV. This shift can be partly attributed to a relaxation of the carriers into defect or surface states, and partly to stronger phonon coupling resulting in Huang-Rhys factors S^{QD} in the QD of around 1, which is larger than for the corresponding bulk materials. For $CdS_{1-x}Se_x$ QD, for example, one finds

$$0.2S^{bulk} < S^{QD} \lesssim 1. \tag{14.29}$$

It should be mentioned, that semiconductor-doped glasses have been used for several decades as edge filters, transmitting long wavelength. It is known that they obtain their color only during an annealing process and that the absorption edge shifts to longer wavelength with increasing annealing time and annealing temperature. But only recently have scientists become aware of the fact that this is the physical demonstration of a simple problem of

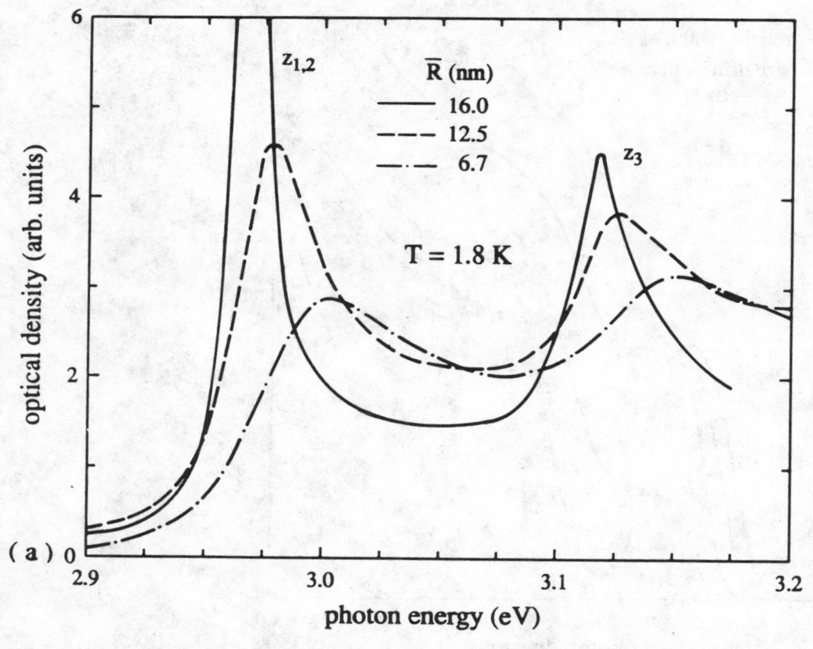

Fig. 14.34a, b. A set of absorption spectra of CuBr quantum dots with different average raddi $\bar{R}$ (**a**) and a set of measured absorption spectra together with the calculated positions of the optical transitions in CdSe QD (**b**). According to [14.73] and [14.74a] respectively

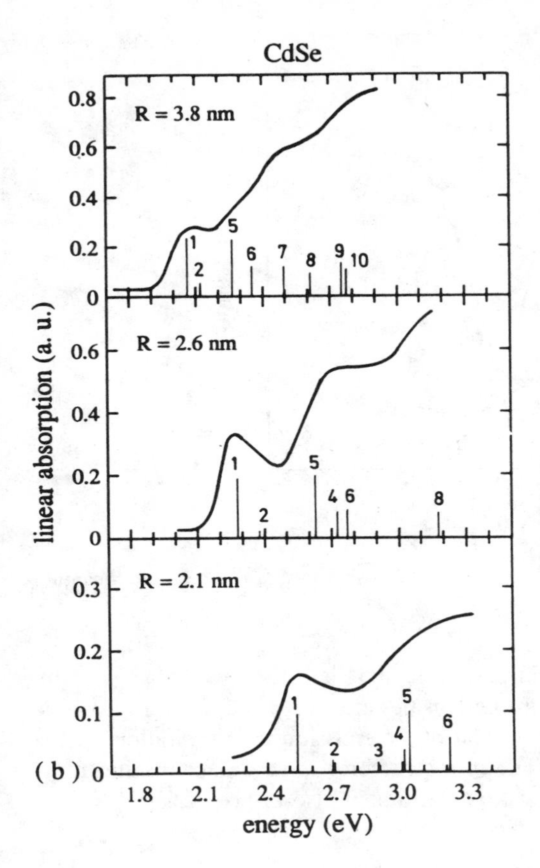

Fig. 14.34b

quantum mechanics, which can be found in many textbooks, namely the quantization of a particle in a box. Many questions are still open in our understanding of the linear and nonlinear optical properties of QD so that exciting results can be expected in the future.

To conclude this section we briefly return to the concept of nipi structures in connection with Fig. 10.14i. The probability of transitions between electron states in the conduction-band minima and holes in the valence band maxima is low because of the "spatially indirect" bandstructure. Consequently strong absorption sets in only around the energy of the "direct" gap, i.e., at an energy corresponding to that between the bands at the same place, modified by the Coulomb interaction. The carriers are spatially separated after their creation resulting in a very long lifetime. The radiative recombination which can nevertheless occur due to the small overlap between electron and hole wavefunctions in their respective minima is strongly Stokes-shifted with respect to the

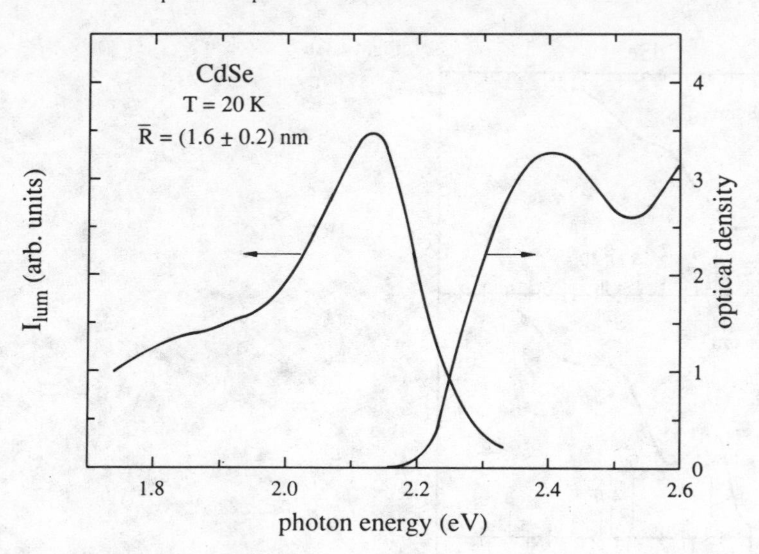

Fig. 14.35. The Stokes-shift between the absorption and emission spectra of $CdS_{1-x}Se_x$ quantum dots. According to [14.75]

direct absorption edge. We give an example in Fig. 14.36, which shows the emission spectra for two different pump intensities and the position of the absorption edge of bulk GaAs. An analogous shift of the emission with excitation is also seen in piezo-superlattices [14.65] The nipi structures have quite amazing nonlinear optical properties [14.76] as will be mentioned also in Sect. 23.5.

14.8 Optical Transitions Above the Fundamental Gap

So far, this chapter has concentrated on electronic transitions in semiconductors close to the fundamental gap. But transitions from deeper valence bands and/or into higher conduction bands are also possible. The structures connected with these transitions are obviously found at $\hbar\omega > E_g$, i.e., in the (V)UV part of the spectrum.

One of the main features discussed in the following are the so-called critical points or van Hove singularities in the combined density of states. To elucidate this term, we first consider the density of states of a single band. We assume that there is a critical point in the dispersion relation at $\boldsymbol{k}_0$, i.e., a point with zero slope $\mathrm{grad}_k\ E(\boldsymbol{k}_0) = 0$; see also (2.77). This can be a maximum, a minimum, or a saddle point at an energy $E(\boldsymbol{k}_0)$. It is usually possible to use a parabolic approximation in the vicinity of this point and we can thus write

$$E(\boldsymbol{k}) = E(\boldsymbol{k}_0) + \sum_i a_i (k_i - k_{0i})^2 \tag{14.30}$$

with $i = x, y, z$.

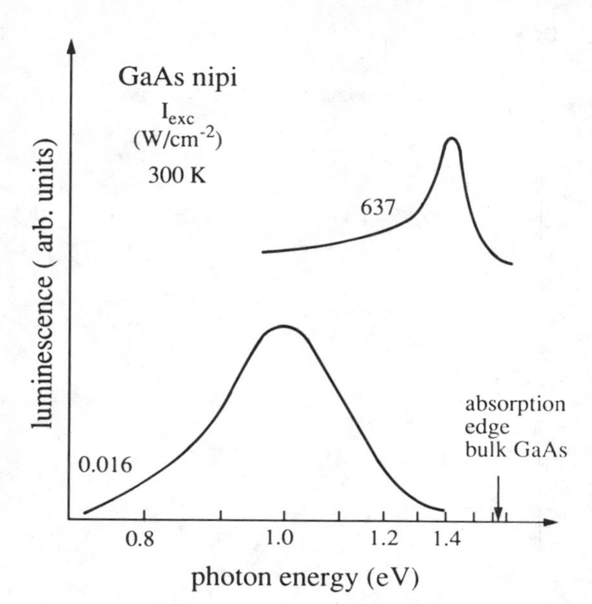

Fig. 14.36. Luminescence spectra of a nipi structure compared to the RT absorption edge of bulk GaAs. According to [14.76]. Compare with Fig. 14.31c

If all the a_i are positive, we have a band minimum at $\boldsymbol{k}_0$ or a so-called M_0 critical point. If all a_i are negative there is a local maximum or an M_3 critical point at $\boldsymbol{k}_0$. Situations where one a_i is positive and the other two negative or vice versa correspond to saddle points in the dispersion, and are called M_1 and M_2 critical points, respectively. Critical points are generally found at points or in directions of high symmetry in $\boldsymbol{k}$-space. This is one of the reasons why one generally gives $E(\boldsymbol{k})$ for these directions; see Fig. 10.11.

In Fig. 14.37 we give the density of states in the vicinity of the four types of critical points mentioned above; it is additionally assumed that the density from parts of the Brillouin zone other than the neighborhood of the critical point is relatively constant. The energy $E(\boldsymbol{k}_0)$ is indicated by E_0. For the M_0 and M_3 critical points we observe the well-known square-root dependence of the density of states while at the saddle points we get a transition from a two-dimensional to a one-dimensional behavior.

Now we proceed to band-to-band transitions with $\boldsymbol{k}$-conservation, but neglecting for the moment the Coulomb interaction between electron and hole.

In Fig. 14.38 we show transitions between a filled valence band and an empty conduction band for a one-dimensional $\boldsymbol{k}$-space. At an arbitrary photon energy (indicated by arrow 'a') there is only one possibility for a given photon energy, i.e., only one point in $\boldsymbol{k}$-space for which transitions between the bands are possible. If the two bands are parallel it is possible for transitions with the same energy to occur over a small interval in $\boldsymbol{k}$-space and we can expect a

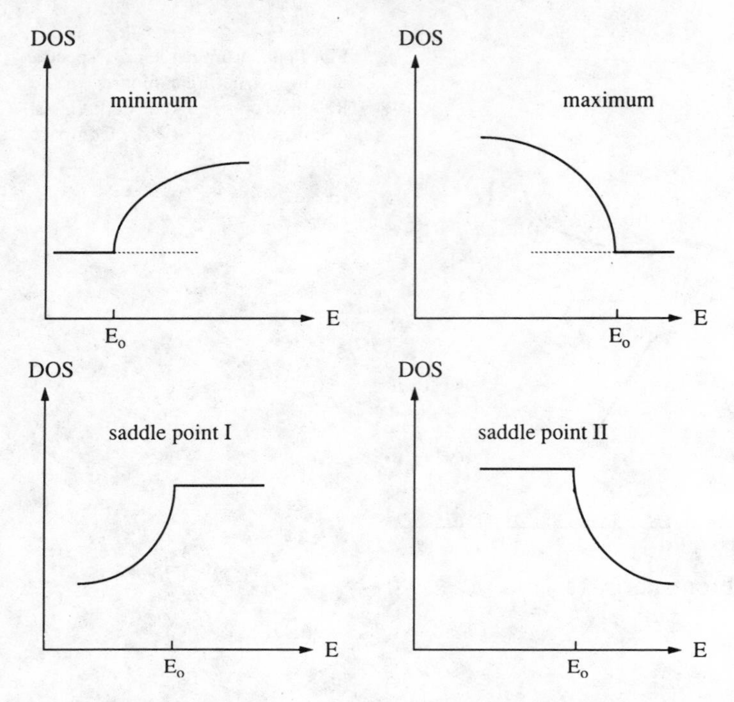

Fig. 14.37. The density of states in the vicinity of various critical points

distinct structure in the absorption spectrum at this photon energy. Such a situation occurs if either the $\mathrm{grad}_k E_j(\boldsymbol{k})$ of both bands is zero as shown in Fig. 14.38 for the Γ-point or if the distance between the bands is constant, i.e., they are parallel, having the same value of $\mathrm{grad}_k E_j(\boldsymbol{k})$.

If we go from the one-dimensional $\boldsymbol{k}$-space of Fig. 14.38 to the usual three-dimensional one, the situation remains qualitatively similar. This means, that we can expect structures, whenever

$$\mathrm{grad}_k [E_i(\boldsymbol{k}) - E_j(\boldsymbol{k})] = 0, \tag{14.31}$$

where the indices i and j stand for valence and conduction bands, respectively. If we insert (14.31) into (2.77) we get the so-called combined density of states, which is the one relevant for the description of optical band-to-band transitions. In the absorption spectrum one can therefore expect structures like those in Fig. 14.37 in the vicinity of the critical points.

In experiments, one usually measures the reflection spectrum from $\hbar\omega \approx E_g$ up to higher energies and applies Kramers–Kronig relations (Chap. 8) and formulas like (2.35), (3.20) or (4.32) to deduce from $R(\omega)$ either $\varepsilon_1(\omega)$ and $\varepsilon_2(\omega)$ or $n(\omega)$ and $\kappa(\omega)$.

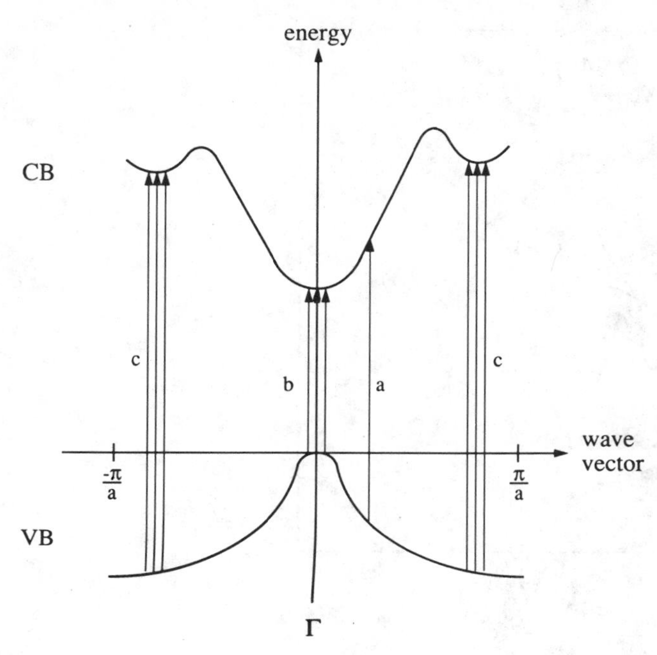

Fig. 14.38. Band-to-band transitions at an arbitrary point of the Brillouin zone (**a**) and at critical points of the combined density of states where the two bands are parallel (**b**, **c**)

In Fig. 14.39 we show an example for the group-IV semiconductor Ge. Other examples may be found in [1.4, 14.78] and references given therein. Fig. 14.39a gives the measured spectrum of $\varepsilon_2(\omega)$ and the calculated one, using the bandstructure shown in Fig. 14.39b. In this part, the transitions contributing to the various structures are identified and the type of the respective critical point is given. Apart from a small overall renormalization of the energy axis, the calculated and the measured spectra of $\varepsilon_2(\omega)$ coincide nicely. The ε_2 values connected with the indirect transition at the fundamental gap of Ge are so small that they do not appear on the scale used in Fig. 14.39a. The fact that the structures in Fig. 14.39 look generally more like peaks and less like Fig. 14.37 is to a large extent due to the neglect of the electron–hole Coulomb interaction in the latter. The resulting excitons, which are responsible for the rich variety of phenomena discussed in the preceding part of this chapter, also modify the optical properties of the higher energy transitions (core excitons, saddle-point excitons). However, these excitons are often not resolved as individual structures, even at low temperatures, because the damping of these states is too high (or the phase-relaxation time too short) since they can decay rapidly into lower energy states. More information about core excitons and related topics can be found in [14.46, 77, 78] and references given therein.

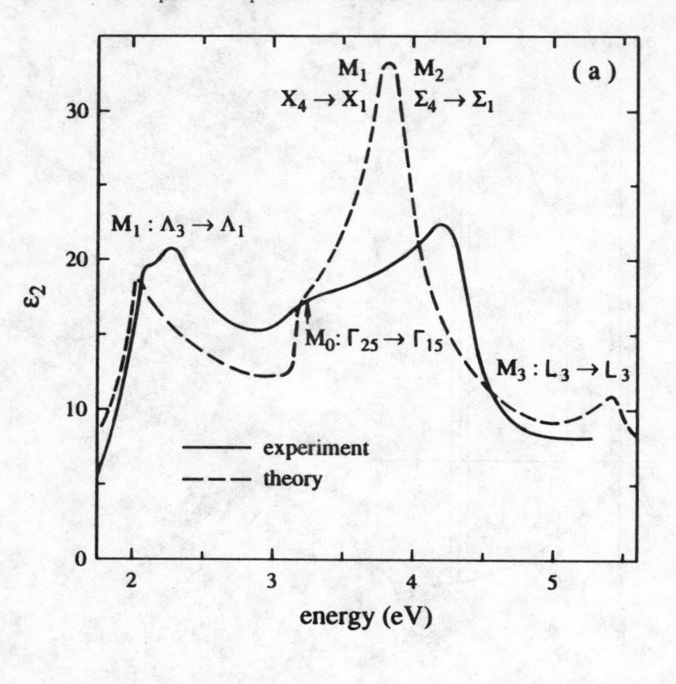

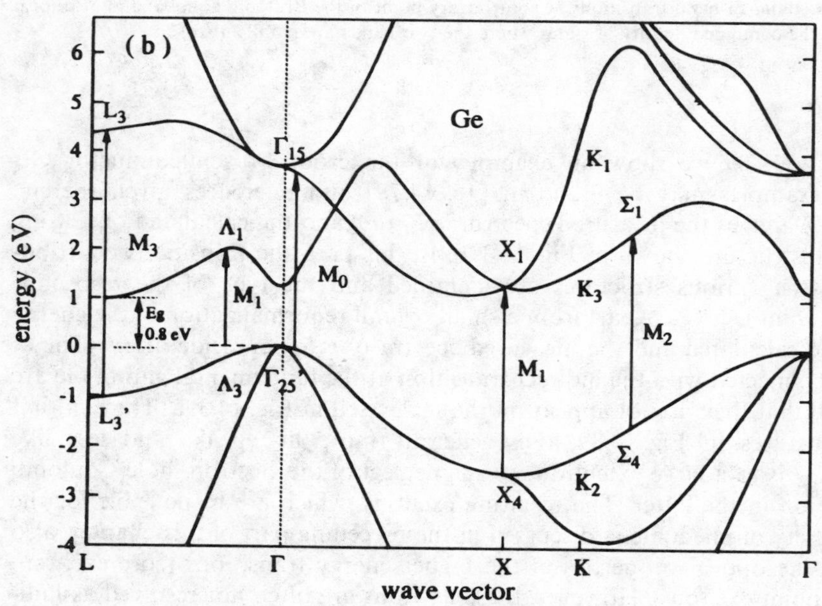

Fig. 14.39a, b. The imaginary part of the dielectric function $\varepsilon_2(\omega)$, measured and calculated **(a)**, using the bandstructure of Ge **(b)**. According to [14.77]

14.9 Problems

1. Try to make a simple estimate of the radius of an ideal quantum dot at which the quantization energy and the Coulomb energy are equal.

2. Make a plot of the longitudinal–transverse splitting of $n_B = 1$ exciton resonances as a function of the exciton binding energy. Include only semiconductors with direct, dipole-allowed band-to-band transitions. Compare with Fig. 11.2.

3. Consider a band-to-band transition in a direct-gap semiconductor neglecting the Coulomb interaction between electron and hole and calculate the absorption spectrum for a dipole-allowed and a dipole-forbidden transition.

4. Consider the $n_B = 1$ $A\Gamma_5$-polariton resonances in CdS (Fig. 14.4b) and determine for a light beam incident at 45° to the surface ($\boldsymbol{E} \perp \boldsymbol{c}, \boldsymbol{k} \perp \boldsymbol{c}$) the length and direction of the wave vectors of the propagating modes in the sample and their phase and group velocities. Select a few characteristic photon energies. Explain the term "spatial dispersion".

5. Explain the difference between the concepts of polaritons and of spatial dispersion.

15 Optical Properties of Bound and Localized Excitons and of Defect States

In the previous chapter we discussed mainly the optical properties of intrinsic excitons.

Here we consider optical properties of defect and localized states.

15.1 Bound-Exciton and Multi-exciton Complexes

We have already introduced point defects in connection with Fig. 10.18. Some of these defects can bind an exciton resulting in a bound exciton complex (BEC), see also Sect. 11.4. In Fig. 15.1 we visualize excitons bound to an ionized donor (D^+X), a neutral donor (D^0X), and a neutral acceptor (A^0X). Excitons are usually not bound to ionized acceptors, as explained in Sect. 11.4. The binding energy E^b of the exciton to the complex usually increases according to

$$E^b_{D^+X} < E^b_{D^0X} < E^b_{A^0X} \,. \tag{15.1}$$

The binding energy is defined as the energetic distance from the lowest free exciton state at $\boldsymbol{k} = 0$ to the energy of the complex.

There is a rule of thumb, known as Hayne's rule, which relates the binding energy of the exciton to the neutral complex with the binding of the additional carrier to the point defect. For the D^0X complex, for example, this says that the ratio of $E^b_{D^0X}$ to the binding energy of the electron to the donor E^b_D is a constant, depending only on material parameters such as the effective masses:

$$E^b_{D^0X}/E^b_D = \text{const} \,. \tag{15.2}$$

The constant is often found to have a value around 0.1 (Hayne's rule, see [15.1]). Isoelectronic traps (such as a Te on a Se site in ZnSe) sometimes form deep centers, with binding energies for excitons exceeding those of neutral acceptors.

The BEC do not have any degree of freedom for translational motion. As a consequence BEC often show up in luminescence and absorption spectra as extremely narrow peaks. In Fig. 15.2 we give as an example a luminescence spectrum of ZnO.

The emission and absorption lines of the BEC are sometimes labelled I_1, I_2 and I_3 for A^0X, D^0X and D^+X complexes, respectively. The influence of the chemical nature of the impurity, e.g., whether a Li^+ or a Na^+ ion is on the

Zn^{2+} site as an acceptor, results in energetic splittings of the energies in the same group of complexes as demonstrated in Fig. 15.2 for the case of the A^0X (or I_1) lines. This splitting is known as the chemical shift or central-cell correction. Similar splittings can be caused by different surroundings of the BEC, e.g., by the presence of other defects in the vicinity.

Since $\boldsymbol{k}$-conservation is relaxed for BEC due to the lack of translational invariance, BEC can also couple to acoustic phonons, for example via the

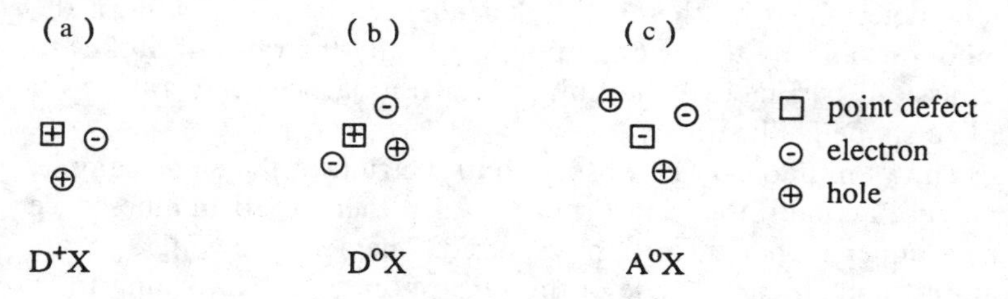

Fig. 15.1 a–c. Visualization of an exciton bound to an ionized donor (**a**), a neutral donor (**b**), and a neutral acceptor (**c**)

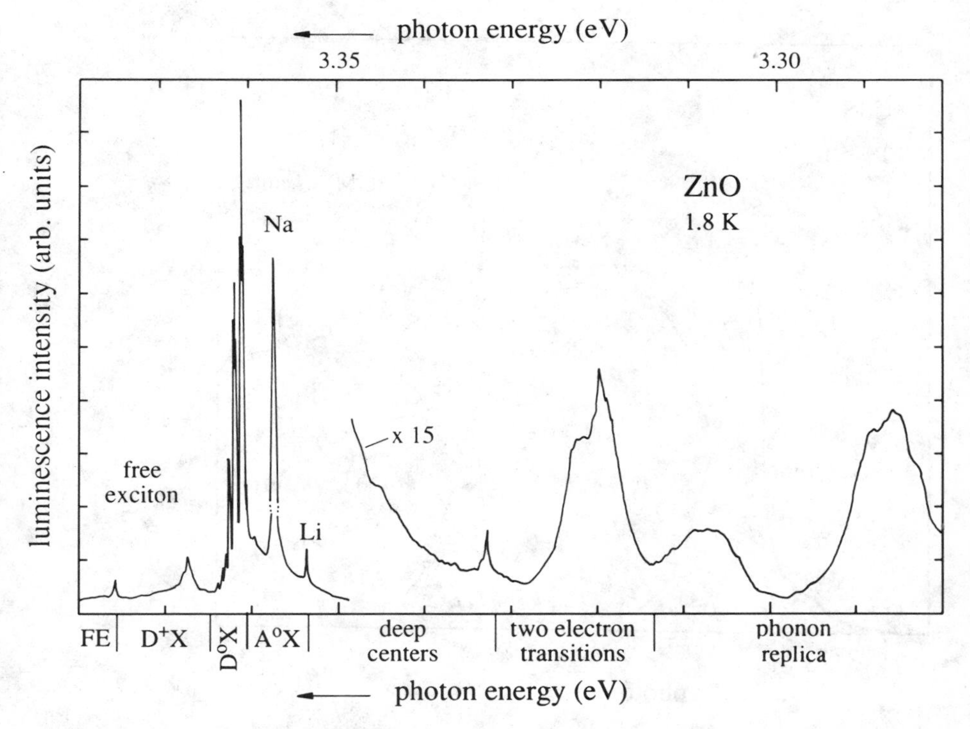

Fig. 15.2. A luminescence spectrum of ZnO taken at low temperature and showing predominantly bound-exciton emission features. According to [15.2]. Compare with Fig. 14.11

deformation potential. This coupling leads to a so-called acoustic wing in the emission or absorption spectra of a BEC. Figure 15.3 shows an example for the I_1 line in CdS. Since only few phonon states are thermally populated at low temperature, the emission manifests essentially a wing on the low energy side according to

$$E_{\mathrm{A^0X}} \longrightarrow E_{\mathrm{A}}^0 + \hbar\omega_{\mathrm{lum}} + \hbar\Omega_{\mathrm{A}} \,, \tag{15.3a}$$

where $\hbar\Omega_{\mathrm{A}}$ is the energy of the emitted acoustic phonon. The shape of the phonon wing is determined by the energy dependence of the coupling of the BEC to the phonons and by their population, especially the ratio of Stokes to anti-Stokes emission. In Fig. 15.4 the influence of the lattice temperature on the emission line shape is illustrated.

Figure 15.3 shows in addition an absorption spectrum of the same sample. In this special case actually the reabsorption of a rather broad luminescence band appearing under higher excitation was used. The acoustic wing appears in absorption on the high energy side of the zero-phonon line according to

$$E_{\mathrm{A^0}} + \hbar\omega_{\mathrm{abs}} \longrightarrow E_{\mathrm{A^0X}} + \hbar\Omega_{\mathrm{A}} \,. \tag{15.3b}$$

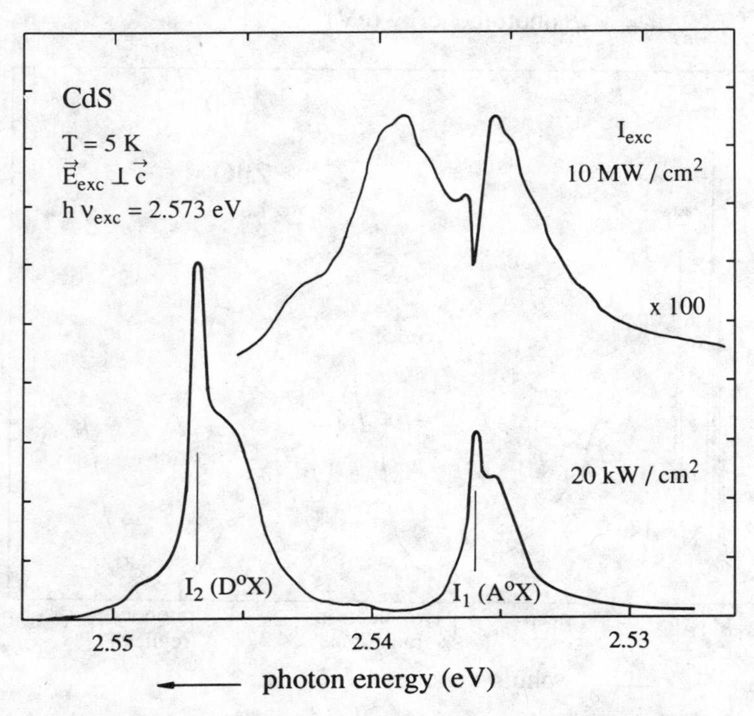

Fig. 15.3. The emission and (re-)absorption spectrum of a CdS sample around the I_1 BEC, showing the acoustic wing. According to [15.3]

Another recombination process leaves the point defect in an excited state, this means that the donor electron of a D^0X complex is transferred in the recombination process from its ground state into an excited state. The energy conservation for these "two-electron transitions" reads in the simplest approximation for the D^0X complex, and assuming that the binding energy of the donor forms a hydrogen-like series of states described by the main quantum number n_B,

$$\hbar\omega_{\text{lum}} = E_{D^0X} - E_{D^0}(1 - n_B^{-2}) \tag{15.4}$$

with $E_{D^0} = \text{Ry}\frac{1}{\varepsilon^2}\frac{m_0}{m_e}$.

In Fig. 15.5 we show an example for ZnO.

The BEC have a very rich and complex spectrum of excited states. Apart from the quantum numbers known from free excitons n_B, l, m, [see (11.3)], excited states can be created by the relative orientations of the spins of identical carriers (with corresponding parities of the envelope function), or in the A^0X complex, for example, by the participation of holes from deeper valence bands. As an example we give in Fig. 15.6 the photoluminescence excitation spectrum of the A^0X complex in ZnO. The peaks labelled R_2 and R_5 are thought to be due to complexes which involve one or two holes from the B valence band instead of the A valence band. The former is separated from the uppermost A valence band by about 5.4 meV. A more elaborate treatment of this topic is described in [15.7].

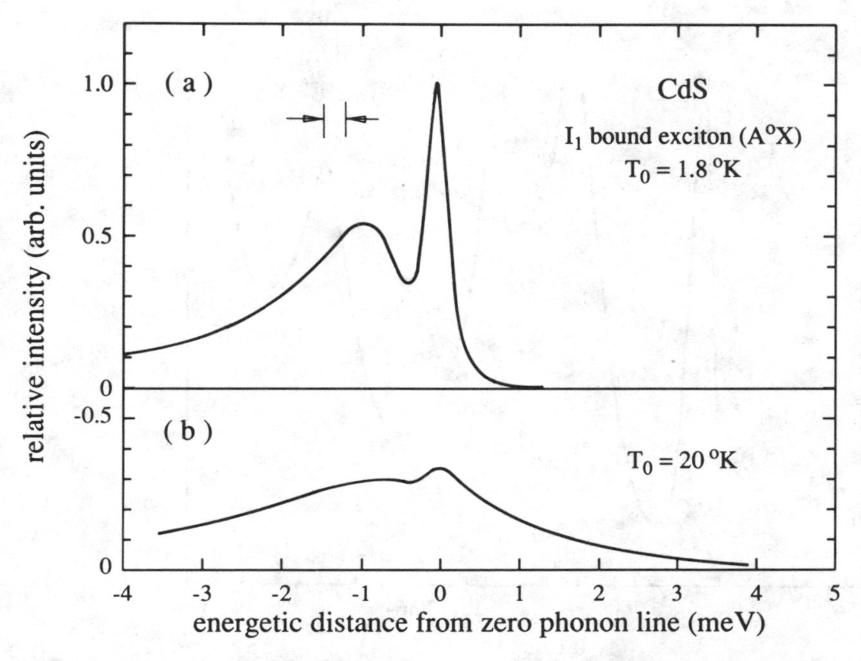

Fig. 15.4. The Stokes and anti-Stokes emission in the vicinity of the I_1 BEC in CdS for two lattice temperatures. According to [15.4]

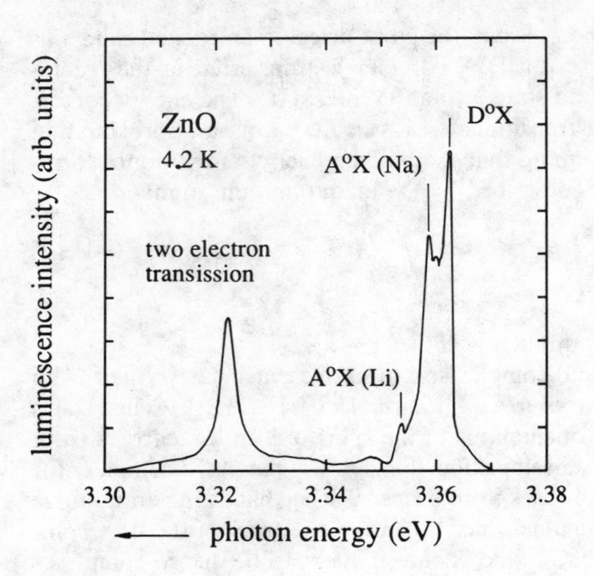

Fig. 15.5. An emission spectrum of a ZnO sample with an especially pronounced two-electron transition. According to [15.5]

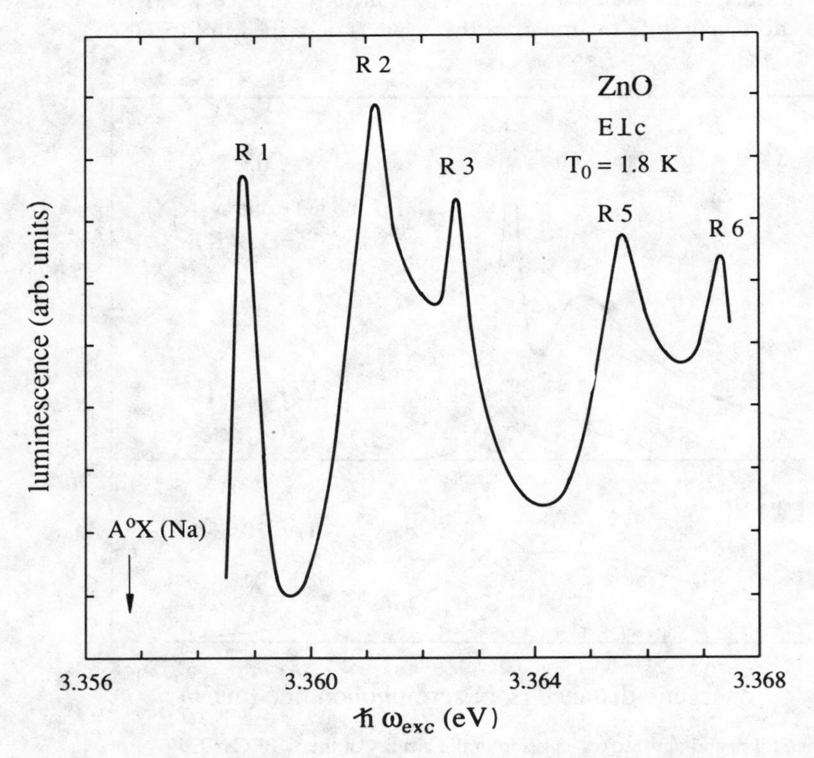

Fig. 15.6. Photoluminescence excitation spectrum of an A^0X complex in ZnO. According to [15.6]

The bandstructure of some indirect semiconductors allows multiple occupation of the conduction band due to the many-valley structure, and of the valence band due to the four-fold degeneracy of the Γ_8 valence-band maximum. In these materials multiple bound-exciton complexes can be formed, i.e., BEC which contain one, two, three or even more electron–hole pairs. The emission lines are due to transitions of a complex containing m electron–hole pairs to one with m-1 pairs. In Fig. 15.7 we give as an example a luminescence spectrum of Si:P. The usual (m = 1) emission line of the phosphorous D^0X complex dominates. It is followed at lower energies by a series of lines with indices m up to 6. The decay of the higher members of the series is obviously faster than of the lower ones, as expected already from simple reaction kinetics. These multiple BEC can be considered as a precursors or nucleation centers for the electron–hole plasma droplet discussed in Chap. 20. Another noteworthy point is that Si is an indirect semiconductor and optical transitions usually involve a momentum conserving phonon. Since the k-selection rule is relaxed for BEC as mentioned above, we can see these lines even without phonon participation.

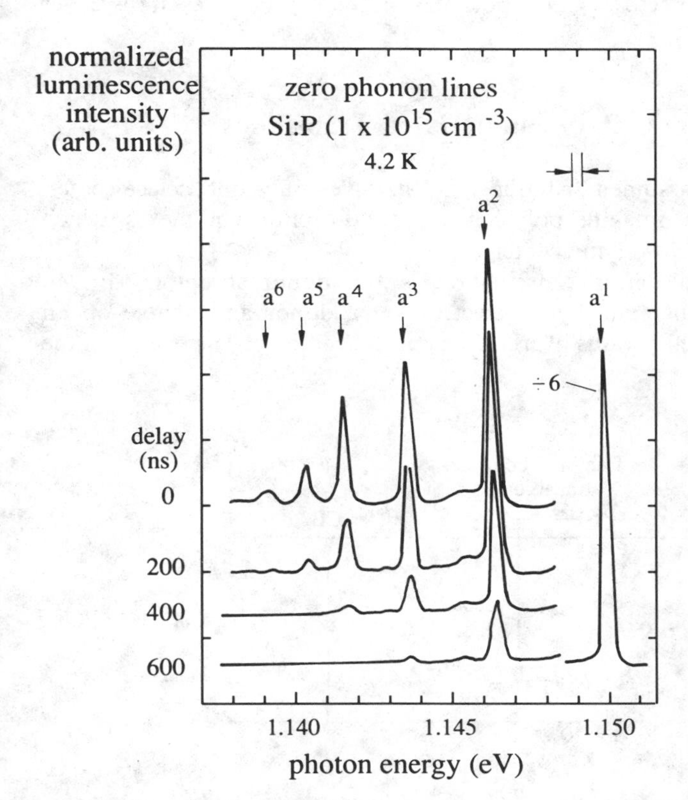

Fig. 15.7. The zero-phonon luminescence spectra of a multiple-bound-exciton complex in Si:P after pulsed excitation, normalized to the a_1 line. According to [15.8]

BECs can be best observed at low lattice temperatures. With increasing T_L the BEC disappear (depending on the material, at temperatures between 20 and 70 K) due to thermal dissociation of the exciton from the complex. For an example see Fig. 15.4. Below this temperature the halfwidth of the emission lines increase partly linearly and partly quadratically with T_L.

For optimal observation conditions the concentration of the point defects should be low enough ($\lesssim 10^{16}\,\mathrm{cm}^{-3}$) to avoid broadening due to interaction between the BEC.

Due to this low concentration BEC are usually best observed in emission and only partly in absorption, and only very rarely do they give rise to reflection structures, simply because the modulation of n and/or κ is too small to result in a significant signal in $\boldsymbol{R}(\omega)$. For an exception to this rule see [15.9].

To conclude this section it should be mentioned that BEC are sometimes said to have a "giant" oscillator strength. This expression is partly the result of a misconception because the maximum absorption coefficient $\alpha^{\max}$ for the free excitons per unit cell was compared with the corresponding quantity of the BEC per defect center. However, in both cases the (Wannier-) excitons cover many unit cells.

15.2 Donor–Acceptor Pairs and Related Transitions

Until now we have assumed that the BEC involves only one defect center. Actually one can also imagine poly-centric bound-exciton complexes, which involve two or more close-lying centers.

The simplest such defect combination is the donor–acceptor pair. In Fig. 15.8 we show schematically an electron on a donor and a hole on an acceptor. If their wavefunctions overlap, they can recombine. The energy of the

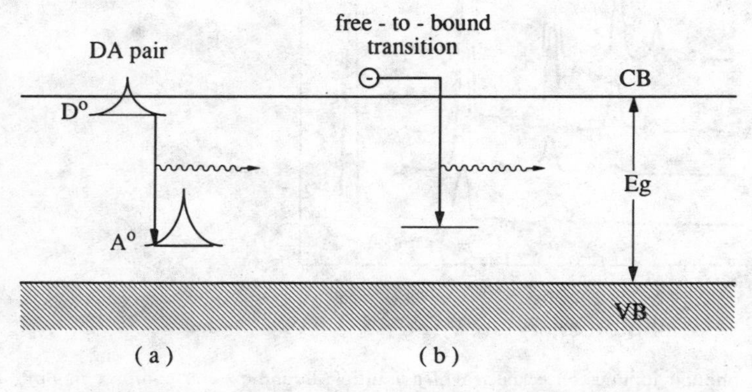

Fig. 15.8a, b. A donor–acceptor pair (**a**) and a free-to-bound recombination process (**b**)

photon resulting from radiative recombination is given by

$$\hbar\omega_{DA} = E_g - E^b_{D^0} - E^b_{A^0} + \frac{e^2}{4\pi\varepsilon\varepsilon_0 r_{DA}} - m\hbar\omega_{LO} \tag{15.5}$$

for singly charged centers if we neglect the energetic shift caused by the overlap of the wavefunctions of the neutral donor and acceptor. $E^b_{D^0A^0}$ are the binding energies of electron and hole to their respective centers.

The forth term on the rhs of (15.5), which depends on the distance r_{DA} of the centers, reflects the Coulomb energy of the ionized centers after the recombination. The last term describes LO phonon replica. We give an example in Fig. 15.9.

If the donors and acceptors are introduced substitutionally in a rather simple (e.g., cubic) lattice, then only discrete, well-defined values for r_{DA} are possible, to nearest neighbors and next-nearest neighbors, etc. Consequently under high resolution one sometimes observes that the zero-phonon band consists of discrete lines corresponding to the discrete values of r_{DA} [15.11].

With increasing pump intensity, the number of occupied donor and acceptor centers increases and their average distance r_{DA} necessarily decreases. As a consequence one finds that the emission maximum of the pair-band shifts to

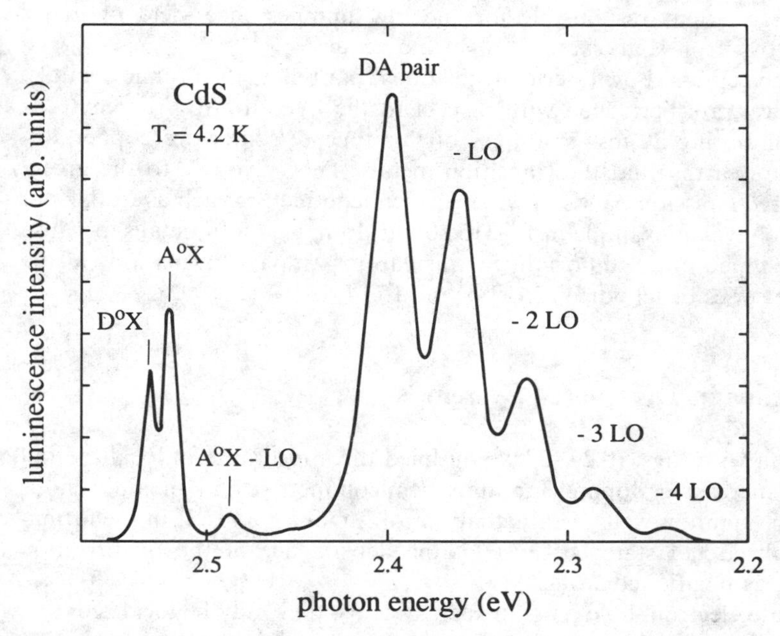

Fig. 15.9. Donor–acceptor pair luminescence in CdS including some phonon replica. According to [15.10]

the blue with increasing excitation due to the Coulomb term in (15.5). This is a very characteristic feature of donor–acceptor pair recombination.

Another recombination process connected with neutral donors or acceptors is the so-called free-to-bound transition. In this case a free electron or hole recombines with a netural acceptor or donor, respectively, as shown schematically in Fig. 15.8. The corresponding emission peak is at

$$\hbar\omega_{FB} = E_g - E^{b}_{D^0/A^0} - m\hbar\omega_{LO} \,. \tag{15.6}$$

Obviously $\hbar\omega_{FB}$ is blue-shifted as compared to $\hbar\omega_{DA}$ and often both processes overlap together with their phonon replica in one complex or broadened luminescence spectrum. More details on BEC and related topics are given in [15.11] and references therein.

15.3 Internal Transitions and Deep Centers

Some defects have not only one (or a few) levels close to one band, but they have several of them, partly around the middle of the gap. The chance of encountering such a situation evidently increases with increasing E_g.

Such deep centers can sometimes interact with both the conduction and valence band and serve then as recombination centers. If this recombination is fast and non-radiative, these centers are known as "luminescence killers". They are to a large extent responsible for the low luminescence yield of many semiconductors. Iron is an example of such a center.

Other deep centers show internal transitions that reflect the atomic orbitals. These, however, are perturbed with respect to the free atom by the environment of neighboring atoms. Examples of this type of center are copper, sodium, the rare earths, and the transition metals. They give rise to the green, orange and red emission bands of wide-gap semiconductors such as CdS, ZnO and ZnS. We give an example in Fig. 15.10 but do not go into details, because this topic is much more important for insulators. Ample information can be found in the series of schools [15.13] or in [15.14–16].

15.4 Excitons in Disordered Systems

In connection with Figs. 10.20–22 we outlined the appearance of localized tail states with increasing doping, for alloy semiconductros and in amorphous materials. In quantum wells the fluctuations of l_z may also result in the formation of localized tail states because of the dependence of the quantization energy on l_z as mentioned earlier.

If we create electron–hole pairs in such systems, they may be localized, too, in the energetic regime below the mobility "edge" which is usually a transition region of finite width.

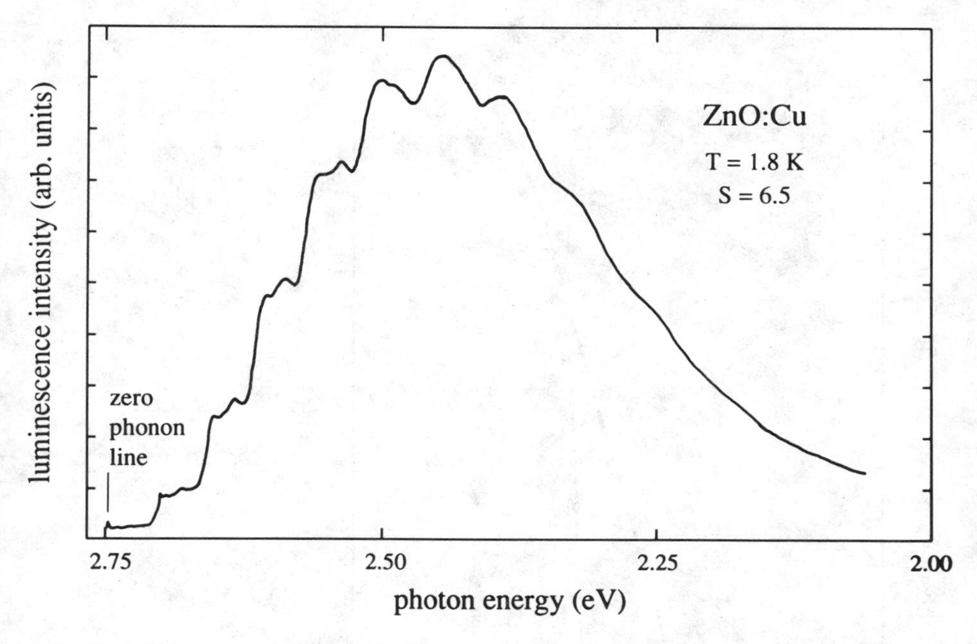

Fig. 15.10. Deep center emission of ZnO:Cu. According to [15.12]

Whole excitons can be localized in potential fluctuations of sufficient depth and with diameters larger than their Bohr radius. In other cases, only one of the carriers is localized—usually the hole, because of its heavier mass—and binds the other carrier to it by Coulomb attraction.

The typical shape of the absorption spectrum of a strongly disordered (e.g., amorphous) system is shown in Fig. 15.11.

In regime I the absorption is weak and is caused by impurities in the alloy or amorphous semiconductor, i.e., atoms of a different chemical nature.

Then region II follows where the absorption coefficient increases exponentially with photon energy. This regime is also known as the "Urbach tail" in analogy to Fig. 14.10 and (14.12). Its origin in disordered semiconductors is, however, not the interaction with phonons, but simply the exponential tail of the density of states of localized excitons which can be reached without *k*-selection. Above follows region III comprising the absorption via transitions into extended states. An extrapolation of this part to lower energies, e.g., with a $(E' - E_0)^{1/2}$ law, allows one to define an optical or mobility gap. The latter quantity can also be found by excitation-spectroscopy of the photocurrent. Absorption spectra corresponding to the schematic drawing of Fig. 15.11 have been found in amorphous Si, in chalcogenide glasses, and in alloy semiconductors such as $CdS_{1-x}Se_x$ and $ZnSe_{1-x}Te_x$. Examples are presented in [15.17, 18].

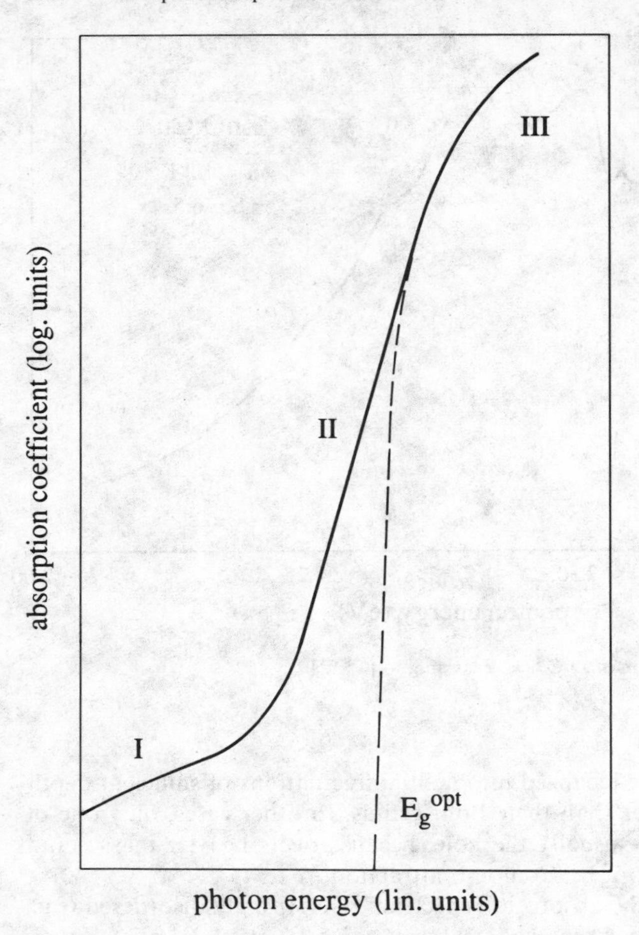

Fig. 15.11. The various regimes of the absorption coefficient in disordered semiconductors. According to [15.17]

The gradual increase of the absorption, and via Kramers–Kronig analysis also of the real part of the reflective index, in many cases prevents the observation of discrete exciton resonances in reflection.

The analysis of the "Urbach" part of the absorption spectrum gives the tailing parameter ε_{loc} describing the density of localized states below the mobility edge.

$$N(E) = \frac{N_0}{\varepsilon_{loc}} e^{-E/\varepsilon_{loc}} . \tag{15.7}$$

In Fig. 15.12 we show ε_{loc} for $CdS_{1-x}Se_x$ as a function of x. This quantity necessarily vanishes for the ordered binary compounds $x = 0$ and $x = 1$, see also (10.50). In between it goes through a maximum with a shape that is not

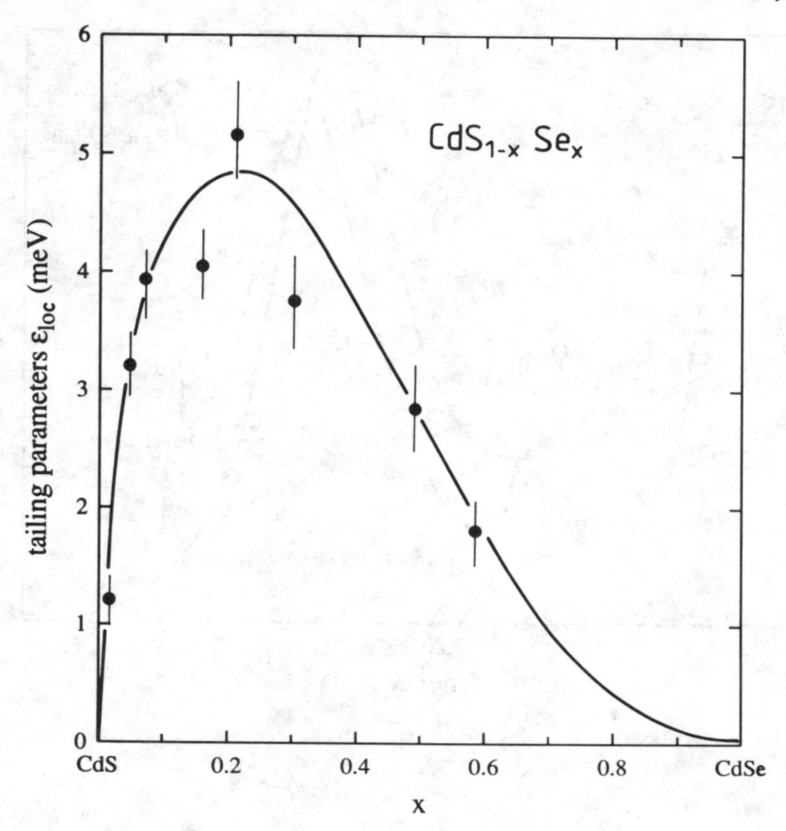

Fig. 15.12. The dependence of the tailing parameter ε_{loc} on the composition x in $CdS_{1-x}Se_x$. According to [15.19]

symmetric with respect to $x=0.5$, partly since the quantity dE_g/dx which enters (10.50) changes with x.

Further information on the localized states can be deduced from the luminescence spectra. In Fig. 15.13 we show the luminescence of a $CdS_{1-x}Se_x$ crystal under excitation of the excited states and under resonant excitation of the localized ones; see also the spectra in [15.19]. One sees here a rather broad zero-phonon line peaking at 2.21 eV and the first LO-phonon replica around 2.18 eV. The phonon replicas are resolved under resonant excitation into the vibrations of CdS and of CdSe, since in this alloy the phonons are of the "persistent-mode" type, i.e., they keep approximately the energies corresponding to $x=0$ and $x=1$ and contribute to the emission with a weight changing with x, while the gap in this case varies continuously with x. See Fig. 10.16.

The high energy edge of the luminescence band can be identified with the transition region from extended to localized states [15.19]. This statement can

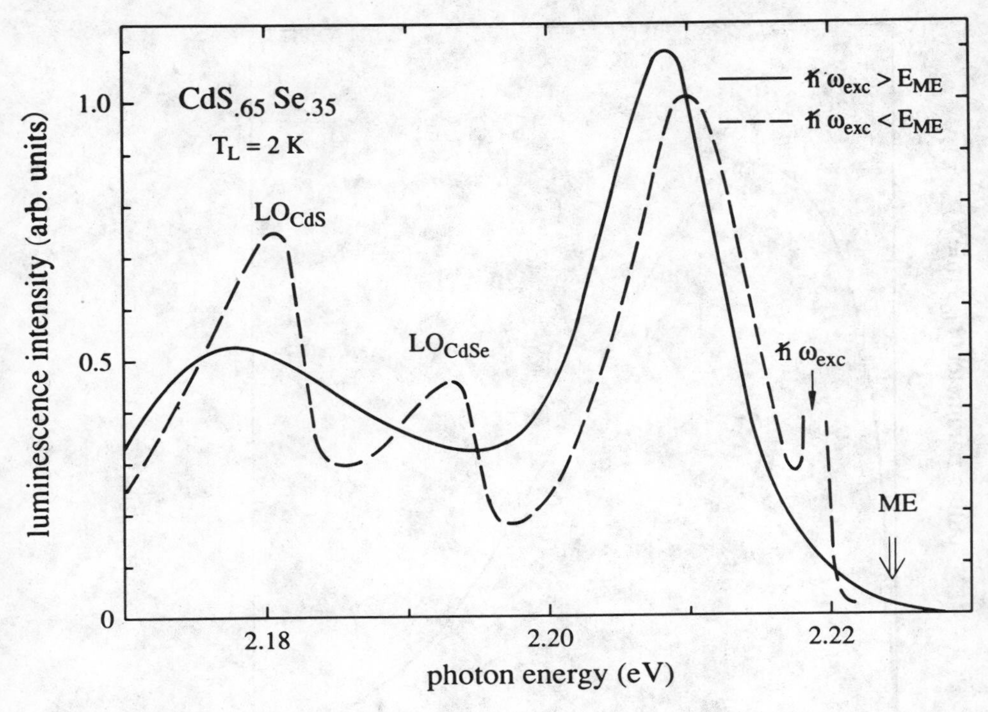

Fig. 15.13. Luminescence spectrum of $CdS_{1-x}Se_x$ under resonant and non-resonant excitation. According to [15.20]

be supported by the following argument: As long as excitons are in extended states, they "scan" the sample and have a good chance of hitting one of the fast nonradiative recombination centers mentioned above in Sect. 15.3. Once it is sitting in a localized state which does not have such a center within the localization length, the exciton can either recombine radiatively or reach a deeper localized state, e.g., by phonon-assisted tunneling. The probability for the latter process decreases with decreasing energy and density of the localized states, resulting in increasing luminescence yield. This idea has been confirmed by investigations of the phase-relaxation times and of the polarization memory as will be outlined in Chap. 22.

At higher lattice temperatures defined by

$$k_B T_L \gtrsim \varepsilon_{loc} \tag{15.8}$$

the majority of the excitons are thermally (re-) excited from the localized into extended states where they behave similarly to free excitons in ordered crystals except for a rather short phase-relaxation time due to the alloy scattering.

Similar results as presented for $CdS_{1-x}Se_x$ have also been observed in $ZnSe_{1-x}Te_x$ [15.19] and in $Ga_{1-y}Al_yAs$ [15.21]. In the first case the localisation effects are even more pronounced than in $CdS_{1-x}Se_x$. Furthermore, self-trapping of excitons seems to occur for $x \approx 0.01$ at single Te atoms or at small Te clusters. In the second case, ε_{loc} is much smaller, partly due to the smaller translational exciton mass which makes localization less probable, and partly because of the larger Bohr radius of the exciton which averages over a larger area of the lattice.

For recent reviews on localized excitons in disordered semiconductors see [15.18–26] and references therein.

15.5 Problems

1. For some standard semiconductors, such as Si, Ge, or GaAs, calculate the binding energy of electrons and holes to donors and acceptors, respectively. Find some data for the binding energies of excitons to these complexes and compare the results with Haynes' rule.

2. Calculate the shift in energy for the donor–acceptor pair recombination when donor and acceptor are nearest possible neighbors in a zinc-blende lattice, and when they have a separation of three, and of ten lattice constants. Can one observe emission from pairs with a separation $d \gg a_B \varepsilon \frac{m_0}{m_e}$

16 Excitons Under the Influence of External Fields

A technique which reveals fascinating new phenomena as well as providing a powerful tool to detect and probe the properties of excitons is the application of external fields. For a rather early treatment see [16.1]

An exhaustive review of this field including the cases of temporally constant fields [16.2] and of modulation techniques [16.3] would itself fill a whole book. Therefore, we restrict ourselves here to the presentation of general features and of some selected topics and examples.

We consider the influence of magnetic, electric and strain fields on the optical properties of excitons.

16.1 Magnetic Fields

First we consider magneto-optics, i.e., the influence of a magnetic field $\boldsymbol{B}$ on excitons. There are two "natural" energy units which can be compared. One is the excitonic Rydberg energy Ry* the other the cyclotron energy $\hbar\omega_c = \hbar(eB/\mu)$, where μ is the reduced mass of the exciton.

The regime

$$\mathrm{Ry}^* \gg \hbar\omega_c; \Rightarrow \quad \gamma = \hbar \frac{eB}{\mu \mathrm{Ry}^*} \ll 1 \tag{16.1a}$$

characterizes the weak field limit. The Coulomb energy dominates and the magnetic field can be treated as a perturbation.

In the strong-field limit

$$\mathrm{Ry}^* \ll \hbar\omega_c; \Rightarrow \quad \gamma \gg 1\,, \tag{16.1b}$$

we have to consider first the Landau levels resulting from the free particle states by the quantization of the motion in the plane perpendicular to $\boldsymbol{B}$, and then the Coulomb energy. The intermediate regime $\gamma \approx 1$, which pertains in many semiconductor in typical dc-fields of superconducting or Bitter-type magnets ($\boldsymbol{B}_{max} \approx 30\,\mathrm{T}$) is more complicated to describe quantitatively.

In accordance with the philosophy of this book, we present here the main effects, namely the diamagnetic shift, the Zeeman splitting, and the appearance of Landau levels, in a very elementary way and give some examples. References leading deeper into this field are [16.2–4].

If we apply a magnetic field to an exciton, the relative motion of electron and hole is deformed by Lorentz forces. In perturbation theory, this deformation can be described by a weak admixture of other states. If we consider the ground-state $n_B = 1$, $l = 0$, which has an S-like envelope function without angular momentum, we can describe this deformation as a small admixture of $l = 1$ (or P-) envelope states. The angular momentum resulting from this admixture is proportional to $\boldsymbol{B}$ and is oriented according to Lenz' rule (i.e., the minus sign in (2.1c)) antiparallel to $\boldsymbol{B}$. Since the energy of a magnetic dipole in a magnetic field also increases linearly with $\boldsymbol{B}$ we get in total a quadratic, so-called diamagnetic shift ΔE_{dia} to higher energies

$$\Delta E_{dia} = a\boldsymbol{B}^2. \tag{16.2}$$

The constant a is a material parameter which is proportional to the square of the Bohr radius of the exciton. The dependence on a_B or n_B and on the material parameters explains that the diamagnetic shift increases with n_B. A typical value for the A-excitons in CdS ($a_B \approx 2.8$nm) is $a(n_B = 1) \approx 2.10^{-6}$ eVT^{-2}.

Exciton states that at $B = 0$ already have a non-vanishing magnetic moment, which can be aligned parallel or antiparallel to $\boldsymbol{B}$, exhibit in addition the linear Zeeman splitting.

This magnetic moment can come from the spins of electron and hole. For singlet and triplet excitons with S envelope the difference or the sum of electron and hole g-factors enters, respectively,

$$\Delta E_z = \pm \frac{1}{2} |g_e \pm g_h| \mu_B B \tag{16.3}$$

due to the relative alignment of electron and hole spin.

For states with $n_B \geqslant 2$ there is an additional contribution from the magnetic moment of the envelope function for $l \geqslant 1$, depending on the orientational quantum number m in (11.3).

For a quantitative calculation, perturbation theory for (almost) degenerate states has to be used, e.g., in the eight-fold space of $n_B = 1$ excitons which can be constructed from the four-fold degenerate Γ_8 valence band and the two-fold degenerate Γ_6 conduction band in T_d symmetry. The terms describing the diamagnetic shift appear together with the kinetic energy terms, the singlet–triplet and the longitudinal–transverse splitting in the main diagonal of the resulting 8×8 matrix, while the Zeeman terms, k-linear terms and others contribute to the off-diagonal elements. A detailed description is – as already mentioned – beyond the scope of this book; see e.g., [16.2]. Instead we give in the following some results.

Figure 16.1 shows the behavior of the $n_B = 1$ $A\Gamma_5$-exciton resonance of CdS with increasing magnetic field. The splitting is obviously connected with the Zeeman effect.

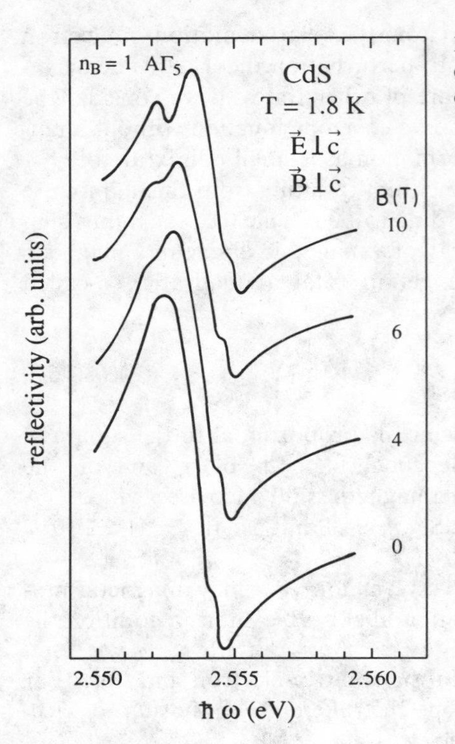

Fig. 16.1. Reflection spectra of the $n_B = 1 A\Gamma_5$-exciton resonance for various magnetic fields. According to [16.5]

In Fig. 16.2 we give transmission spectra of ZnO in the spectral region of the $n_B = 1$ A-exciton resonances, for the polarization $\boldsymbol{E} \parallel \boldsymbol{c}$. In this orientation all transitions have only very weak oscillator strength. Again we observe the Zeeman splitting of the $A\Gamma_{1,2}$ triplet state with increasing $\boldsymbol{B}$ and the $A\Gamma_5^L$ state. The narrow spike S is due to an isotropic point, i.e., a photon energy for which the dispersion relations for the polarizations $\boldsymbol{E} \parallel \boldsymbol{c}$ and $\boldsymbol{E} \perp \boldsymbol{c}$ cross or in other words where the ordinary and extraordinary refractive indices are equal. In this situation energy can be transferred fom one polarization (here $\boldsymbol{E} \parallel \boldsymbol{c}$) to the other under energy and momentum conservation, leading to the dip in transmission.

Figure 16.3 depicts the splitting pattern of the $n_B = 1$A- and B-exciton resonances in CdS as a function of the magnetic field for the orientation $\boldsymbol{B} \perp \boldsymbol{c}$, $\boldsymbol{E} \perp \boldsymbol{c}$, showing nicely the combined influences of terms linear and quadratic in $\boldsymbol{B}$.

Figure 16.4a gives the dispersion relation of the $n_B = 1$ Γ_5-exciton-polariton resonance in cubic (T_d) ZnTe. The influence of the light and heavy holes on the dispersion is clearly visible. Figure 16.4b shows the situation for finite $\boldsymbol{B}$ in a fairly arbitrary direction. In this situation all degeneracies are lifted and all exciton branches acquire some admixture of oscillator strength resulting in a total of eight different exciton and nine polariton branches. This is a rather frightening example for the dispersion of so-called magneto-polaritons. From

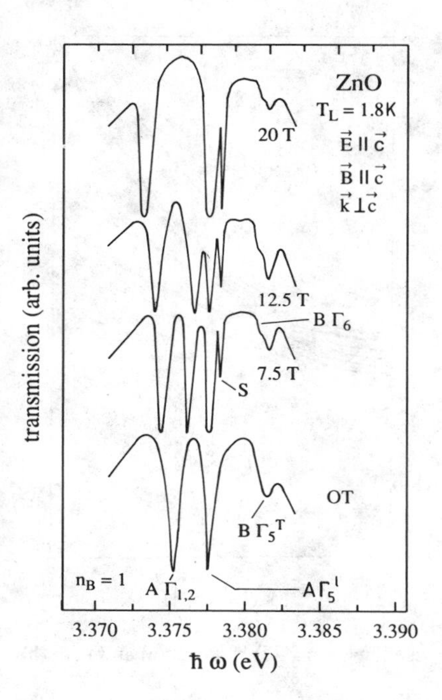

Fig. 16.2. Transmission spectra in the region of the $n_B = 1$A and B exciton resonances of ZnO for the orientation $\boldsymbol{E} \parallel \boldsymbol{c}$. According to [16.5]

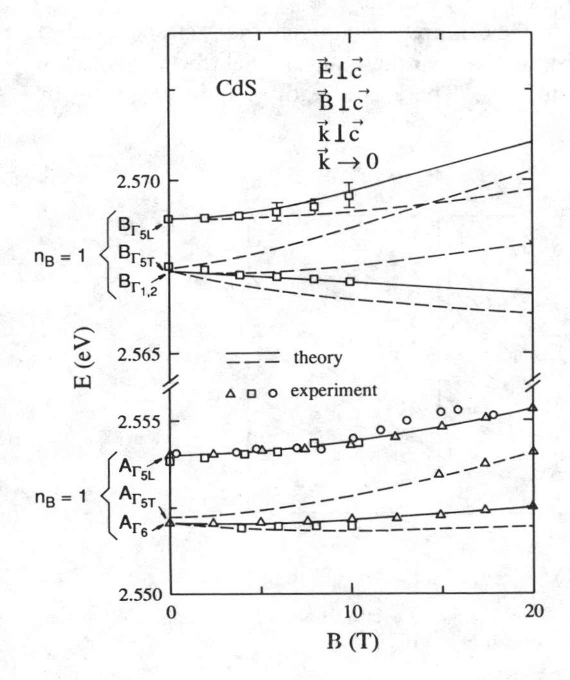

Fig. 16.3. The splitting pattern of the $n_B = 1$ A- and B-exciton resonances in CdS with increasing magnetic field. According to [16.2]

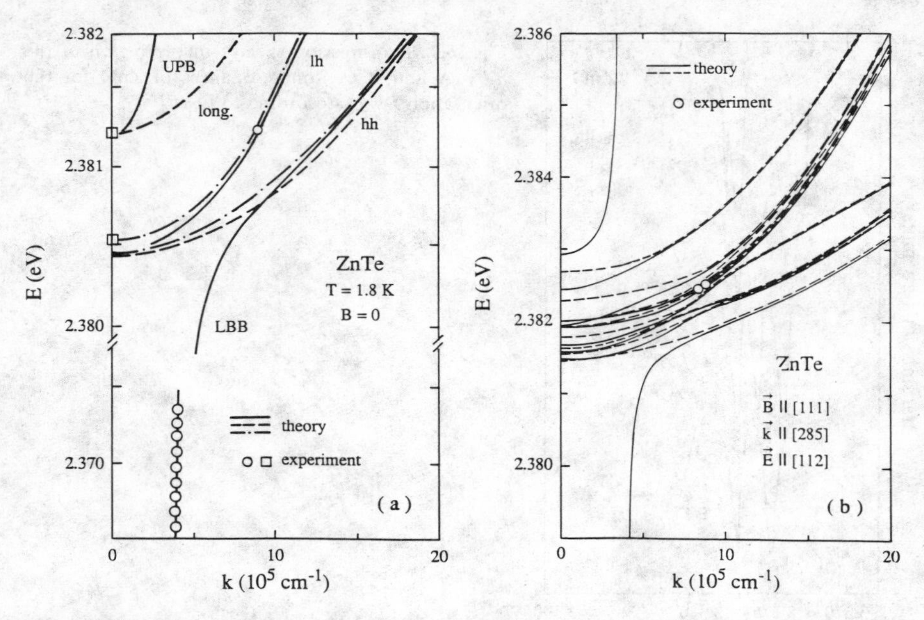

Fig. 16.4a, b. Dispersion relation of the $n_B = 1$ exciton resonances in ZnTe without (**a**) and with (**b**) an applied magnetic field. According to [16.2b]

Fig. 16.4b it is clear that it is wise to use "simple" geometries for the measurements, so that the effects remain relatively simple.

Until now we have considered in the examples exclusively the excitonic ground state $n_B = 1$. Therefore we now give in Fig. 16.5 the magnetic field behavior of the $n_B = 2$, $l = 1$ exciton resonances in ZnSe. There is already a finite splitting for

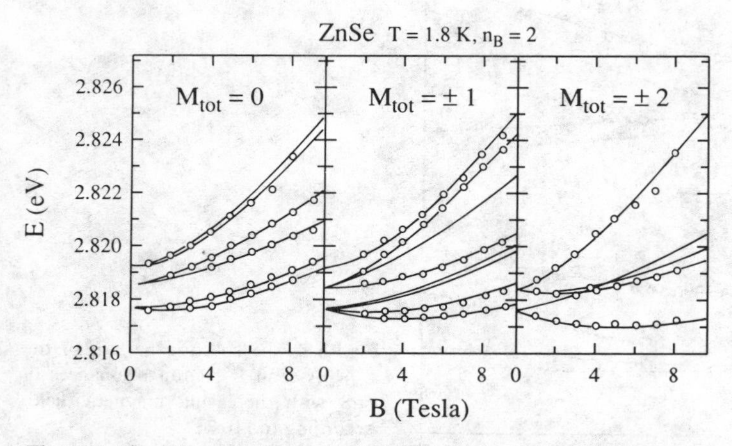

Fig. 16.5. The influence of a magnetic field on the $n_B = 2$ exciton resonances in ZnSe, determined in two-photon absorption spectroscopy. According to [16.6]

$B=0$. For higher fields this splitting increases due to Zeeman terms. In addition, there is a pronounced diamagnetic shift due to the large Bohr radius. The total angular momentum M, consisting of spin and envelope contributions, could be chosen in the two-photon absorption experiments of Fig. 16.5 by using circular and/or linear polarizations of the two beams (see also Sects. 19.1 and 25.5).

Similar splitting patterns as for free excitons can also be observed for BEC, where one has to take into account that the $\boldsymbol{B}$-field influences both the initial and the final state, e.g., in a recombination process from A^0X to A^0. Examples of the magnetic field behavior of BEC are given in [16.7–9].

At the beginning of this section we noted that the observation of Landau levels might be difficult, since for most semiconductors the limit $\gamma \gg 1$ can be reached only for fields $\boldsymbol{B} \gtrsim 10^2$ T, which, if available at all, are usually only in pulsed form.

There is, however, a way to overcome this difficulty by investigating the continuum states, where the Coulomb interaction still influences the oscillator strength but where the motion of electron and hole is almost free. To observe the Landau levels, their broadening must be smaller than the cyclotron energy as in micro wave experiments, in other words

$$\omega_c T_2 \gtrsim 1 \tag{16.4}$$

where T_2 is the phase-relaxation time. Bulk samples with direct, dipole-allowed band-to-band transitions are usually opaque in this region so that reflection spectroscopy is the appropriate tool.

In Fig. 16.6 we give as an example the reflectivity of ZnO at the beginning of the continuum states of the C-exciton series. The Landau-level structure becomes obvious for the highest B-fields. The observation of the Landau levels allows the sum of electron and hole masses to be determined. If the electron mass is known, e.g., from cyclotron absorption by n-type materials, the hole mass can then be determined.

In quasi two-dimensional QW systems transmission spectra in the exciton and in the continuum regime are readily measured. We give an example in Fig. 16.7. With increasing $\boldsymbol{B}$ we can see the diamagnetic shift of the $n_z=1$, $n_B=1$ hh and lh exciton states and above this the evolution of the Landau-level fan. This type of measurement gives detailed information about the bandstructure. See [16.11–13] and references therein. Some further references, concerning especially the influence of magnetic fields on systems of reduced dimensionality will be given in section 22.5.

16.2 Electric Fields

In the same way that we know the quadratic (diamagnetic) and linear (Zeeman) effects of a $\boldsymbol{B}$-field from atomic physics, we can guess what will happen if we apply a static electric field $\boldsymbol{E}_s$ to the exciton resonances in semiconductors. There will be a Stark effect, i.e., a shift (and splitting) of the exciton resonances, usually quadratic in $\boldsymbol{E}_s$. The influence of $\boldsymbol{E}_s$ on the band-to-band transitions, i.e., on the continuum states is also known as the Franz–Keldysh effect. The modifications of the absorption spectra due to $\boldsymbol{E}_s$ also influence the refractive index in the transparent spectral region below the exciton resonances according to the Kramers–Kronig relations (Chap. 8), resulting in Pockels- or Kerr-effect-like phenomena. We shall concentrate here on the influence of $\boldsymbol{E}_s$ on the eigenfrequencies.

The observation of the Stark effect for excitons in three-dimensional (or bulk) samples is not easy, for the following reason: To get an observable shift of the eigenstates, the "electric field energy" $e\boldsymbol{E}_s a_B$ should be comparable with the spectral width of the absorption bands, i.e.,

$$e\boldsymbol{E}_s a_B \gtrsim \Delta_{LT} . \tag{16.5}$$

Equation (16.5) requires fields of the order of 10^6 Vm^{-1} depending on the material parameters. On the other hand, such fields broaden or even destroy the exciton resonances due to two effects. One is field ionization of the exciton, by tunneling through the finite Coulomb barrier at finite fields, as illustrated in Fig. 16.8. The other is impact ionization; this means that carriers which are always present in a semiconductor at finite temperature can gain such high

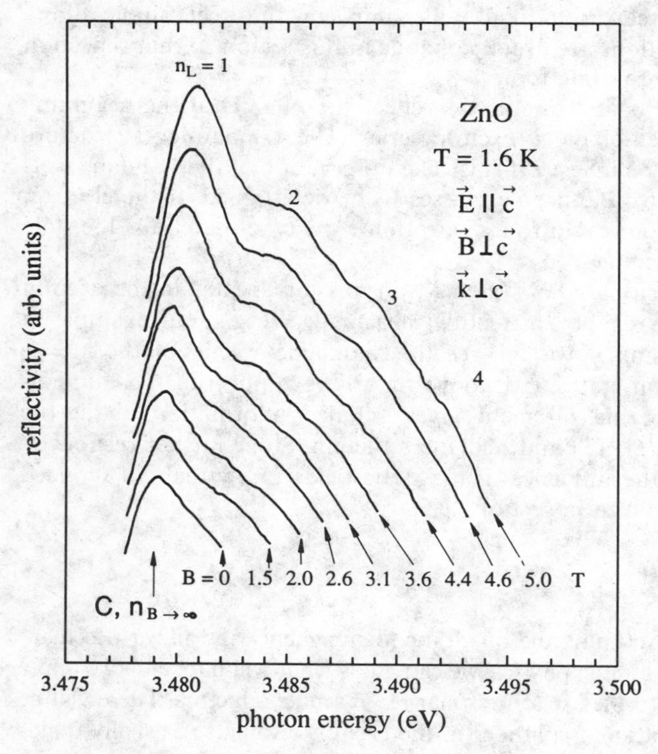

Fig. 16.6. The reflectivity in the exciton continuum in ZnO for various magnetic fields. According to [16.10]

energies in the electric field, that they can ionize another exciton if they hit it, resulting in two more carriers and a collision-broadening of the exciton resonance. For the above reason, not many successful attempts have been reported to observe the Stark effect for excitons in bulk materials directly.

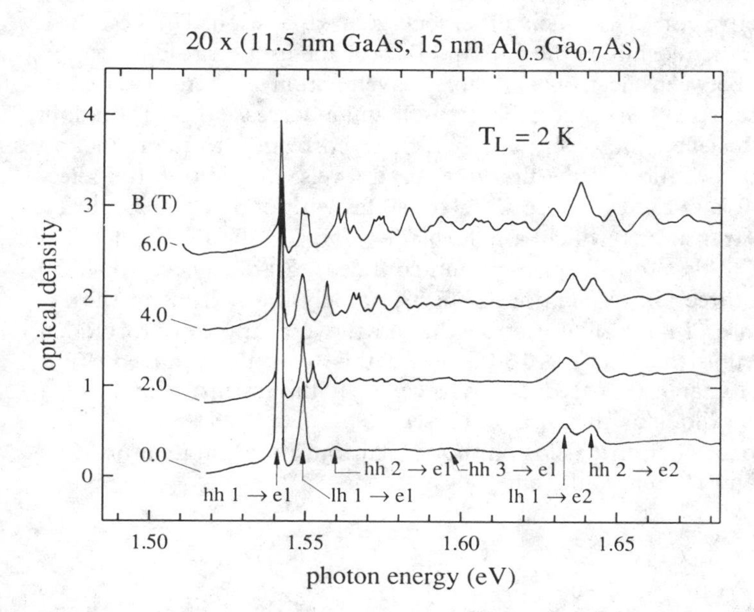

Fig. 16.7. The absorption spectrum of excitons in an AlGaAs MQW sample for various magnetic fields. According to [16.11]

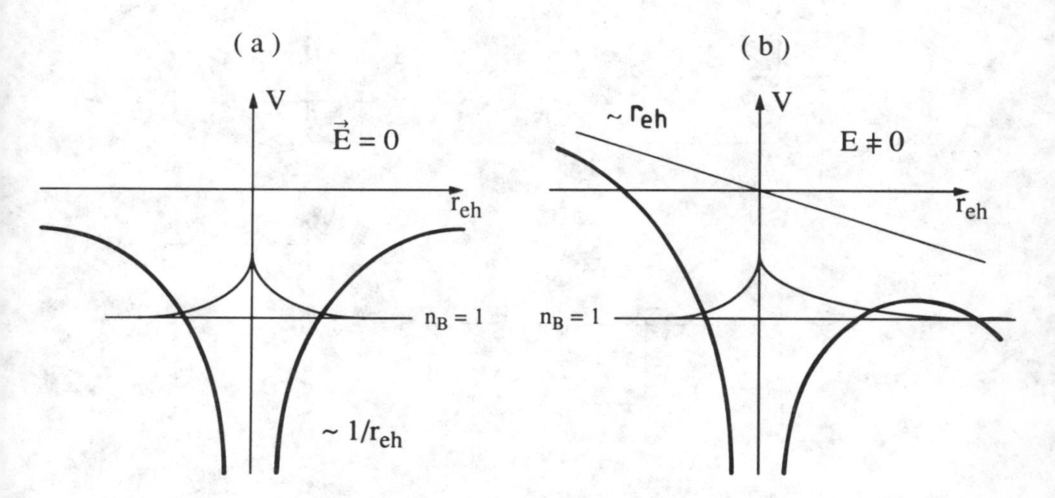

Fig. 16.8a, b. The Coulomb potential of a three-dimensional exciton without (**a**) and with (**b**) an applied constant electric field

Instead, small modifications occurring at low fields have usually been used in modulation spectroscopy [16.3, 14–16].

This problem can be overcome by confining the electron and hole between barriers, e.g., in quantum wells, as shown schematically in Fig. 16.9. With increasing $\boldsymbol{E}_s$ oriented perpendicular to the layers the electron and hole shift into their respective corners, reducing their energetic separation. This results in a roughly quadratic redshift of the gap and thus of the exciton resonance.

The overlap between electron and hole wavefunction is thereby reduced, resulting in a decrease of the oscillator strength and a decrease of their binding energy. This latter effect, however, is only a small correction to the reduction of the energetic separation of the first quantized levels. In addition the selection rule $\Delta n_z = 0$ is relaxed because $\boldsymbol{E}_s$ mixes states with odd and even parity, inducing some transitions which are forbidden for $\boldsymbol{E}_s = 0$. For a detailed elaboration of this so-called quantum-confined Stark effect (QCSE) see [16.17]. All three above-mentioned effects, namely the redshift of the exciton, the decrease of its oscillator strength, and the appearance of forbidden transitions, are illustrated in Fig. 16.10, where this quantum-confined Stark effect is shown for an InP based MQW located in the intrinsic region of a reverse-biased pin diode as shown in the inset.

For the influence of electric fields on the optical transitions of quantum dots see [16.19–22] and references therein.

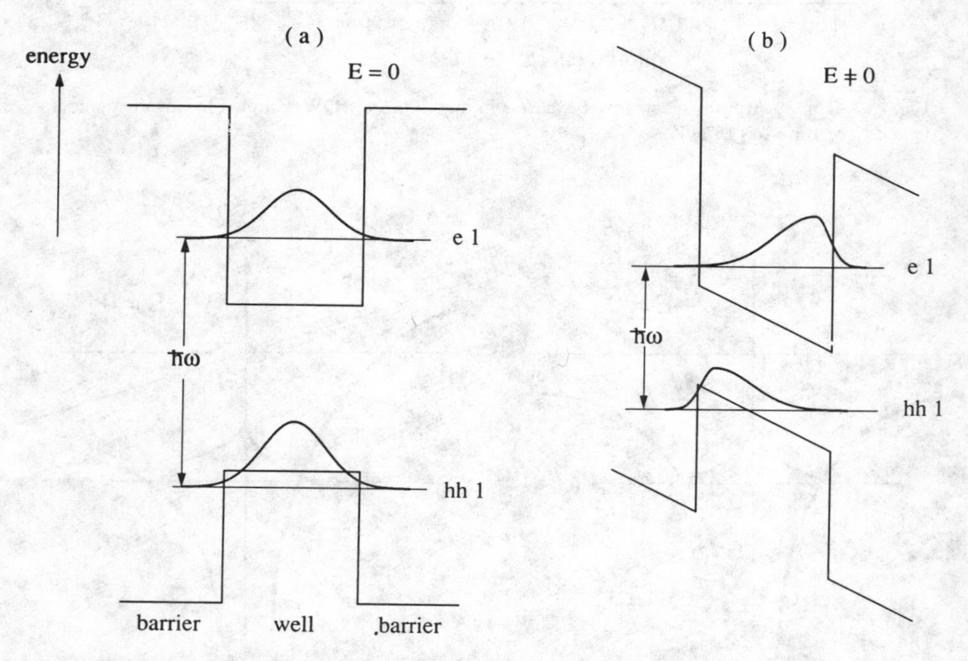

Fig. 16.9a, b. The bandstructure of a quantum well without (**a**) and with (**b**) an applied constant electric field perpendicular to the wells, but neglecting the Coulomb interaction

In section 22.5 we present the influence of an electric field on the optical properties of a supelattice in combination with methods of nonlinear and ultrafast spectroscopy.

Phenomena like that of Fig. 16.10 lead into the regime of opto-electronics or electro-optics, to which we shall return briefly in Sect. 23.5.

16.3 Strain Fields

The third external perturbation which we shall consider here are mechanical strain fields. The crucial quantity which enters here is the dependence of the energy of the band extrema on the strain, the so-called deformation potential. It is defined for conduction and valance band by

$$V_{\mathrm{def}} = a \frac{\mathrm{d}E_{\mathrm{c,v}}}{\mathrm{d}x} \tag{16.6}$$

where a is the lattice constant.

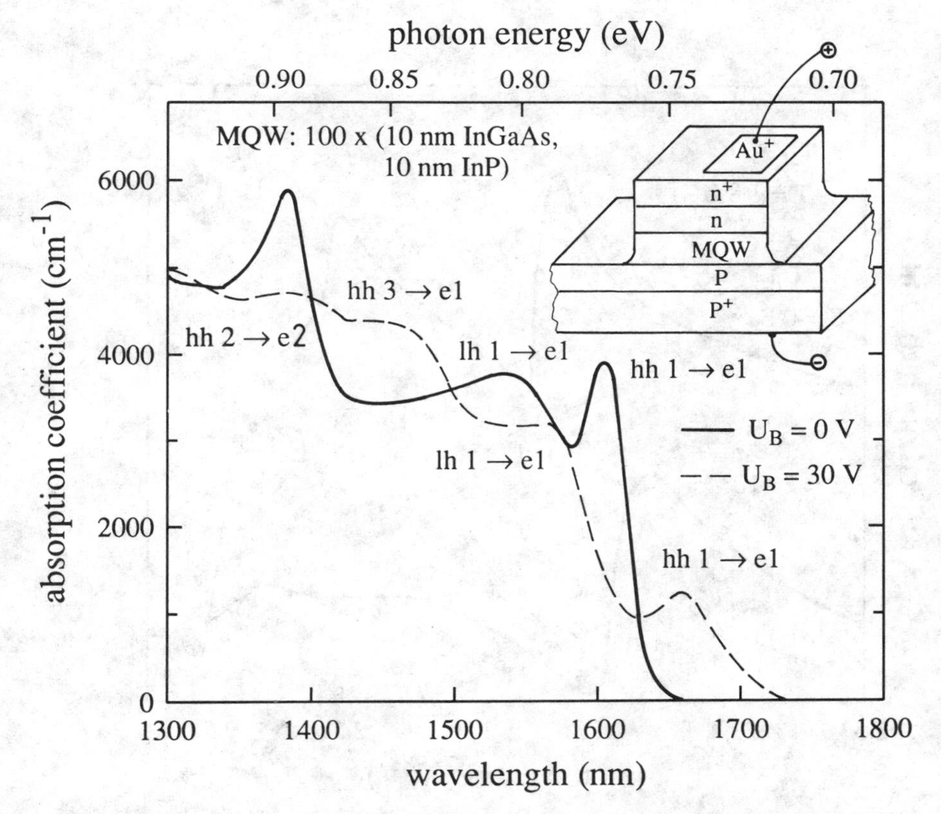

Fig. 16.10. The quantum-confined Stark effect in InGaAs MQW samples. According to [16.18]

For excitons the relevant quantity is the sum of the deformation potentials of conduction and valence bands. As in the case of magnetic and electric fields, an applied stress changes the eigenenergies of the exciton states and may also lead to a splitting of degenerate states, if the strain is oriented such that it reduces the symmetry of the lattice. The latter situation is e.g. realized when stress is applied perpendicular to the crystallographic ***c*** axis of a uniaxial material (e.g., the Wurtzite structure) where it lifts the degeneracy in the plane perpendicular to ***c***, but not if the stress acts in the direction of ***c***. This orientation leads only to a shift of eigenenergies but not to a splitting since the symmetry is not changed. For details see Chap. 25 on group theory.

We now give various examples: Figure 16.11 shows reflection spectra of CdS under increasing strain. The shift and the splitting are obvious. Measurements of this type can be used to determine the deformation potentials. For materials with zinc-blende structure see [16.24].

Since BEC often have very narrow absorption and emission bands (Sect. 15.1) one can easily study the influence of magnetic or strain fields, which

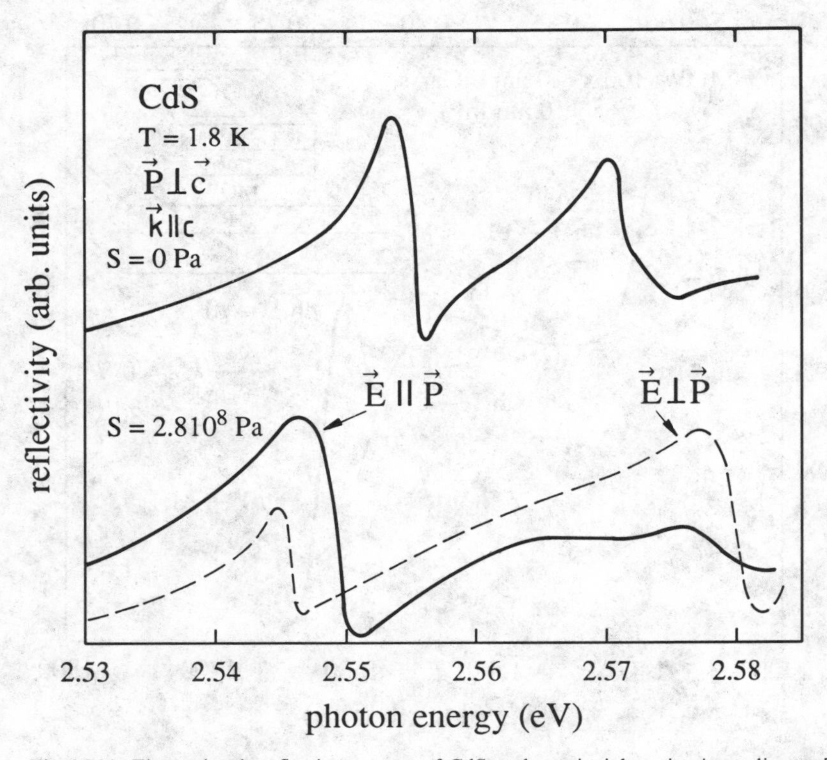

Fig. 16.11. The excitonic reflection spectra of CdS under uniaxial strain. According to [16.23]

affect both the ground and the excited state as already mentioned above. In Fig. 16.12 we give the shift of the luminescence of the A^0X BEC in ZnO which shows clearly the influence of the strain. According to the statement given above, a shift but no splitting occurs for the orientation $\boldsymbol{S} \parallel \boldsymbol{c}$. For more details see [16.25].

Recently the influence of strain acquired new importance in the field of epitaxial growth of heterolayers.

There are two contributions to the "biaxial" strain which occurs in the plane of epitaxial growth, and which can be considered as a superposition of a hydrostatlic and a uniaxial strain. One contribution to the strain already arises during the growth process if the lattice constants of substrate and layer do not coincide at growth temperature. The first epitaxial layers will often grow with some strain, trying to match the lattice constant of the substrate. With increasing layer thickness it is energetically more favorable for the layer to create a dislocation network which relaxes the strain [16.26, 27]. Sometimes a small residual strain remains, independent of the layer thickness [16.28, 29]. The critical thickness l_c for the onset of dislocation formation depends on the

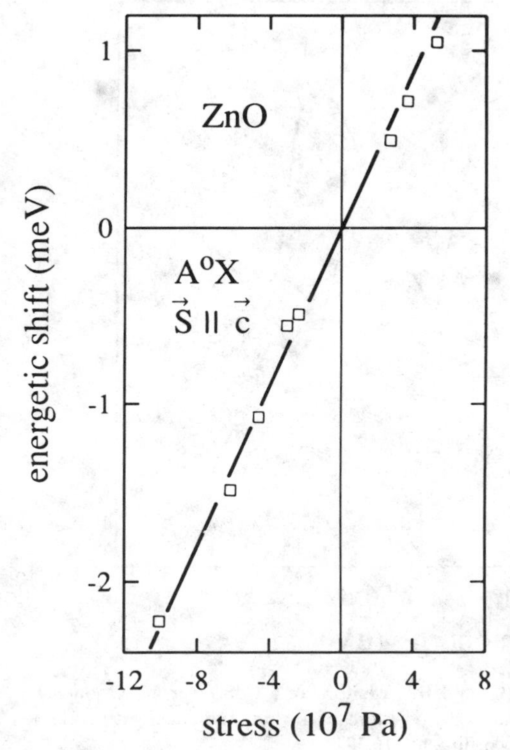

Fig. 16.12. The shift of a A^0X BEC luminescence line in ZnO under the influence of strain. According to [16.25]

lattice misfit and on the energy necessary to produce the dislocations. It can be as small as a few atomic layers only. The dislocation network explains the function of a buffer layer which is often used if two materials of different lattice constant are grown on top of each other. this case the crystalline quality of the interface is considerably poorer than that of the surface of the layer after a thickness of about 0.1 μm.

The other part of the biaxial strain arises during the cooling from the growth temperature to the temperature at which the (optical) measurements are performed if the coefficients of thermal expansion are different for substrate and layer, as is usually the case. A part of this thermal strain can sometimes, be accommodated by modifications in the dislocation network. In Fig. 16.13 we show reflection spectra in the region of the $n_B = 1 A\Gamma_5$ and $B\Gamma_5$ excitons in CdS grown on SrF_2.

Since the cubic SrF_2 substrate was oriented (111), the hexagonal CdS grows with the $\boldsymbol{c}$ axis perpendicular to the interface. Consequently the C_{6v} symmetry

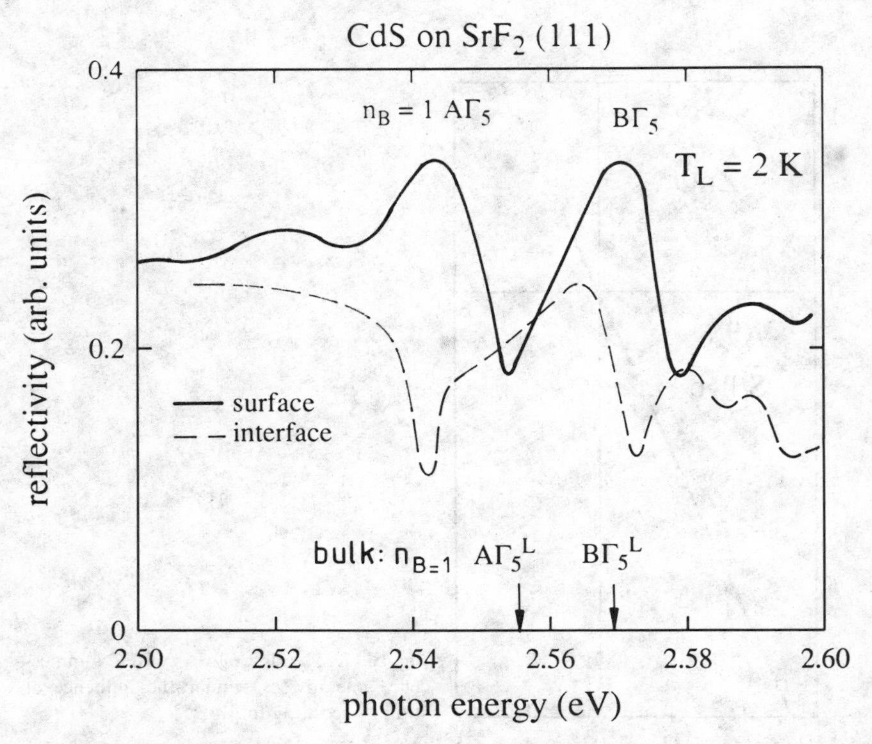

Fig. 16.13. Reflection spectra of the $n_B = 1\,A\Gamma_5$ and $B\Gamma_5$ excitons of a CdS layer grown on SrF_2 and measured at the surface and at the interface. The positions of the longitudinal eigenenergies in the bulk are indicated by vertical arrows. According to [16.30]

is not affected by the biaxial strain, but the eigenenergies are shifted. The bulk values for the longitudinal eigenenergies (or the reflection minima) are given by the vertical lines. In agreement with the discussion above, we see that the excitons are shifted and that this shift is different at the interface and the surface. Furthermore, the damping is larger at the interface than at the surface due to the dislocations in the former region.

In contrast to Fig. 16.13, we show in Fig. 16.14 a luminescence spectrum of the cubic material ZnTe grown on GaAs(001). In this case, the biaxial strain reduces the cubic T_d symmetry of bulk ZnTe and the Γ_5 exciton of Fig. 14.5 splits into two components.

Splittings as in Fig. 16.14 also play a role in strained-layer superlattices in addition to the mass-dependent quantization energies.

With this example we finish the analysis of the linear optical properties of semiconductors. We give in the next chapter a short review and proceed then to the second major topic of this book, namely, to nonlinear optics.

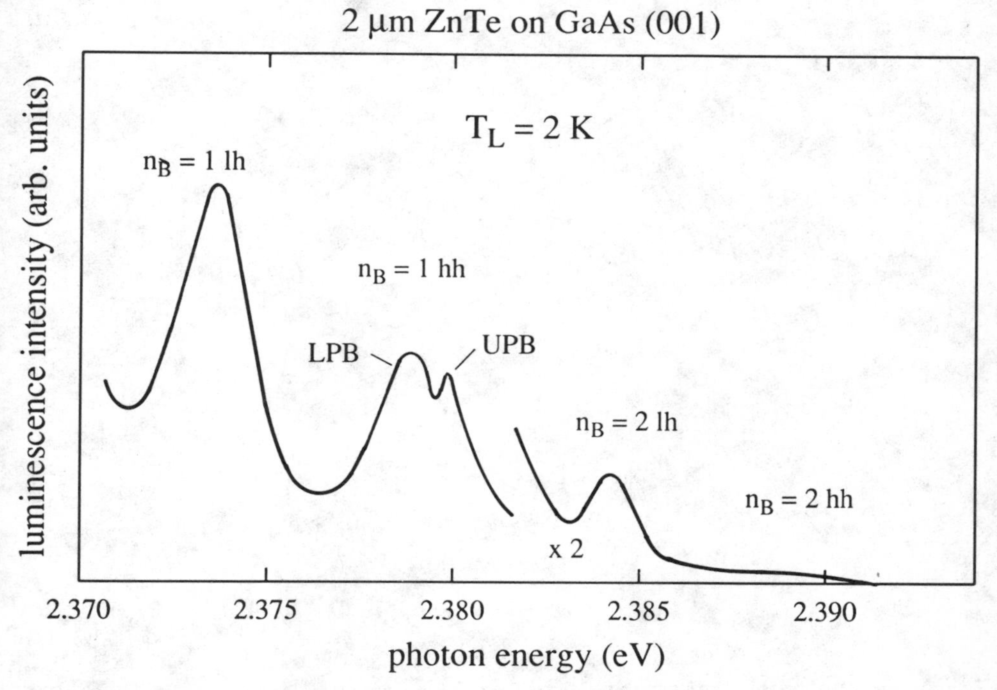

Fig. 16.14. Luminescence spectrum of a ZnTe layer grown on GaAs (001), showing the strain-induced splitting of the light and heavy hole $n_B = 1$ exciton resonances. According to [16.28]. Compare with Figs. 14.5 and 16.4

16.4 Problems

1. Typical values of the deformation potential are around 10 eV. What is the shift of the band edges for $\Delta a/a = 10^{-3}$?

2. Do you expect the band-gap shifts due to the lattice deformation by acoustic and by optical phonons to be identical or not? Why?

3. Calculate the diamagnetic shift at $\boldsymbol{B} = 10$ T for excitons characterized by $E_{ex}^{b} = 5$ meV, $\varepsilon = 15$ and by $E_{ex}^{b} = 100$ meV, $\varepsilon = 6$, respectively.

4. Calculate the Zeeman splitting for a singlet and a triplet exciton for $n_B = 1$, $g_e = 1.6$ and $g_h = 2.2$. What is the difference if at $B = 0$ the two states already show a finite energy splitting δ?

5. Do you expect that the linear Stark effect can occur for $n_B = 1$ and/or for $n_B = 2$ excitons? Why?

6. Calculate in the simplest approximation the Stark shift for excitons with the data given in connection with problem 3 and for an electric field strength E of 10^3 Vm^{-1}, 10^6 Vm^{-1}, and $10^{-2} E_{ex}^{b}(e\alpha_B)^{-1}$.

17 Review of the Linear Optical Properties

In this brief chapter, we shall review and summarize some of the aspects of the linear optical properties of semiconductors that were presented in the preceeding chapters in some detail.

In Fig. 17.1 we give a schematic overview of the spectra of the complex dielectric function, of the complex index of refraction and of the reflectivity for a typical direct-gap semiconductor over the whole spectral range from the IR to the UV. For simplicity, we consider one optically active phonon mode, one exciton resonance, and one further resonance which represents all continuum and band-to-band transitions. We include some damping for every resonance but neglect details including spatial dispersion. The additive structure of the resonances in the dielectric function [see, e.g.(4.20)] is clearly visible. The back ground dielectric constant ε_b of one resonance is simultaneously the "static" one ε_s for the next higher resonance.

Figure 17.1a includes contributions which might come from two other effects. The dashed-dotted line gives the modifications introduced by a plasma. [This plasma can be caused either by high doping levels (Sects. 12.1, 13.6) or by high excitation (Chap. 20)]. In the first case it consists of either electrons or of holes, in the second case it is a bipolar plasma. If we consider Fig. 4.3 or Fig. 13.9 and remember that the transverse eigenfrequency of a plasma is zero, we immediately obtain the dashed-dotted line in Fig. 17.1. We should mention that the presence of a plasma also influences the other resonances, e.g., the phonon resonance, due to plasmon–phonon mixed states (Fig. 12.4), or the exciton resonances by increasing damping, by screening of the Coulomb attraction of electron and hole and by band renormalization and band filling (Chap. 20). These effects are not shown in Fig. 17.1 for sake of clarity.

The other contribution to the dielectric function is the so-called orientational polarization shown by the dashed line. It describes the contribution of freely rotatable permanent electric dipoles to $\varepsilon(\omega)$, i.e., a para-electric contribution. Freely rotatable means that the dipoles do not have a restoring force, but that an orientation of the dipoles which has been established, e.g., by an external field, decays after switching off of this field with a relaxation time τ by (thermal) collisions with the surroundings. The contribution of the orientational polarization to ε_1 is constant for low frequencies as long as the dipoles can follow the driving field and decays for $\omega\tau \geqslant 1$ as is usual for relaxation phenomena. The imaginary part has its maximum at $\omega\tau = 1$. This is reasonable

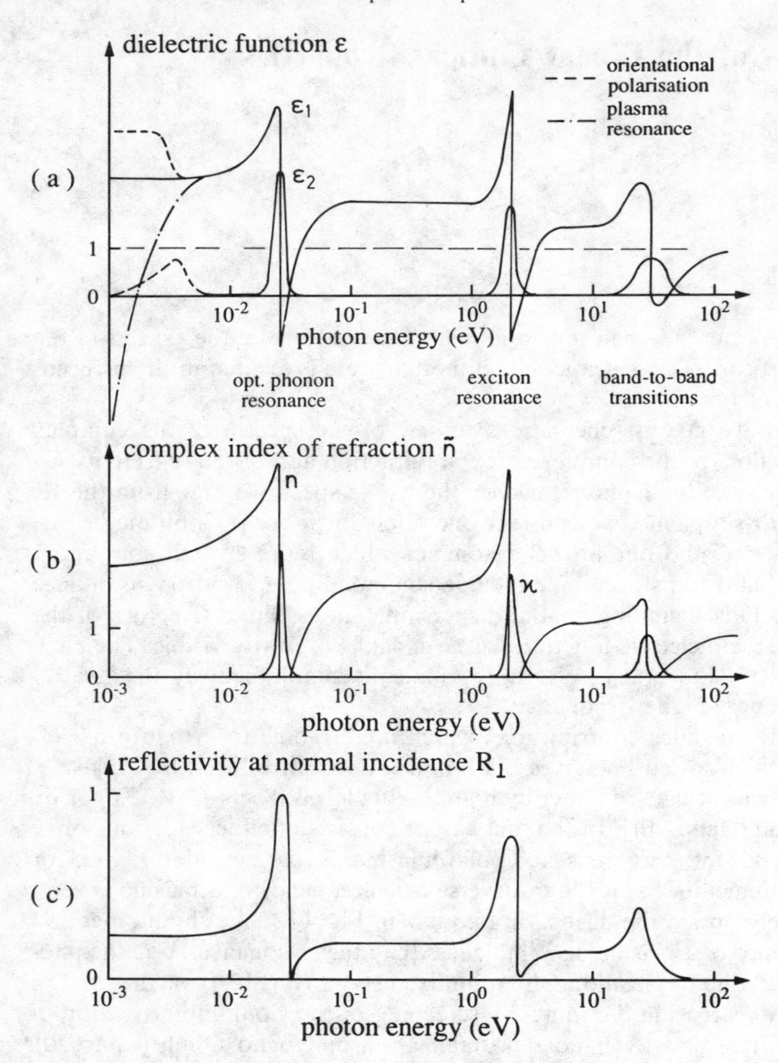

Fig. 17.1 a–c. A schematic overview from the IR to the UV of the spectra of the real and imaginary parts of $\varepsilon(\omega)$ (**a**), of $\tilde{n}(\omega)$ (**b**), and of the reflectivity (**c**) for a semiconductor

if we assume that the "friction losses" of the rotating dipoles are proportional to the amplitude and the speed ω of the rotation. The orientational polarization is a very important effect for polar liquids (e.g., water) and organic solids containing some rotatable radicals. In semiconductors this effect is less common but is mentioned here for sake of completeness.

Another "summary" of what we have learned up to now is the dispersion relation in Fig. 17.2, where we again include one phononic and two electronic

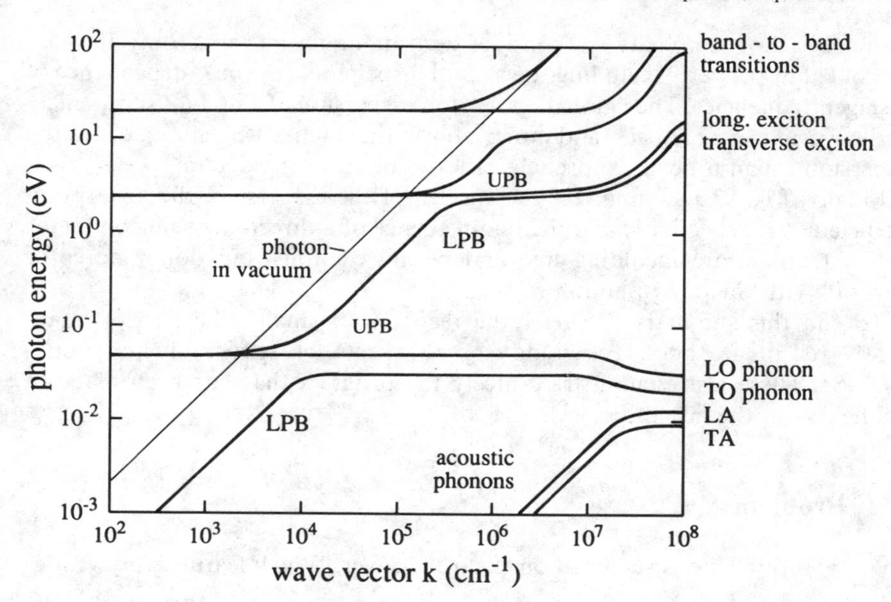

Fig. 17.2. A schematic overview of the real part of the dispersion relation of light in a semiconductor from the IR to the UV, neglecting damping

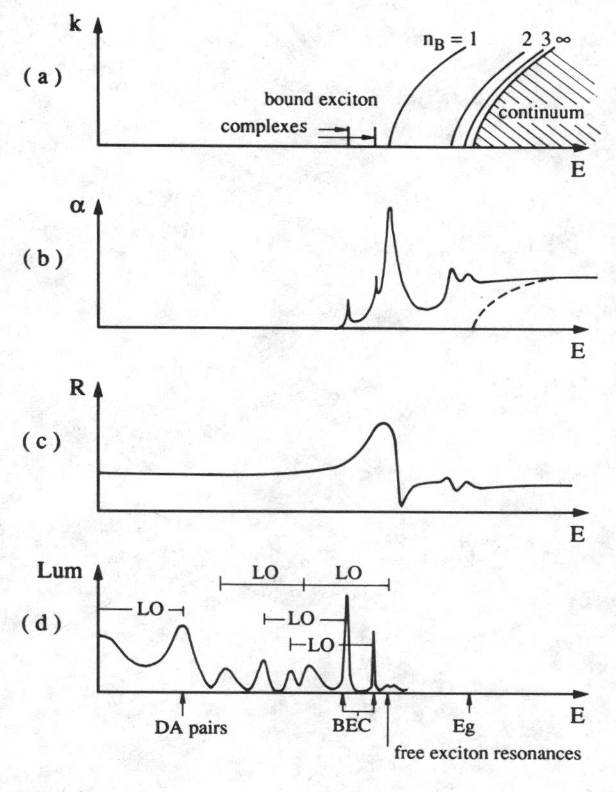

Fig. 17.3 a–d. A schematic drawing of the dispersion (**a**), and the spectra of the absorption (**b**), reflection (**c**) and luminescence (**d**) of a high-quality, direct-gap semiconductor at low temperatures in the region of the exciton resonances

resonances. For simplicity we consider zero damping and give only the real part of $\boldsymbol{k}$, but properly including the spatial dispersion, i.e., the $\boldsymbol{k}$-dependence of the eigenfrequencies. The "global" polariton dispersion of Fig. 17.2 shows nicely the various resonances and how, above the highest eigenfrequency, the dispersion asymptotically approaches that of photons in vacuum.

Finally, Fig. 17.3 summarizes the spectra of the dispersion, the absorption coefficient, the reflectivity, and the luminescence of a direct-gap semiconductor at low temperature including free and bound excitons and donor–acceptor pairs, all with their LO-phonon replica.

To end this summary, we refer the reader to a small collection of more-or-less recent textbooks covering some aspects of semiconductor optics [17.1–8]. These also demonstrate nicely the progress that has been made in this field over the last three decades.

17.1 Problem

Try to identify what is sketched on the front cover. Which features are wrong?

18 High Excitation Effects and Nonlinear Optics

In this and in some of the following chapters we shall leave the regime of linear optics introduced with the linear response equation (2.26) and proceed to the field of nonlinear optics. Nonlinear optics including high excitation phenomena and electro-optics, together with the investigation of semiconductors of reduced dimensionality, are presently the most active fields in semiconductor science.

In the next two sections we give the definition and the general scenario. Then we continue in Chaps. 19–12 with discussions of the most important effects and phenomena contributing to nonlinear optics in semiconductors.

For a rather early work on nonlinear optics see [18.1] and for more recent reviews [18.2–8] and references therein.

18.1 Introduction and Definition

In linear optics we learned how the optical properties of matter depend on the frequency of the incident radiation field and on the direction of polarization or propagation relative to the crystallographic axes. But we explicitly assumed that the optical properties do not depend on the field amplitude(s) $\boldsymbol{E}_{\mathrm{i}}$ (or the intensities, I) of the incident light beam(s). As a consequence, the polarization of matter oscillates with the same frequency as the incident field(s) and two light beams can cross each other in matter without mutual interaction.

The regime of nonlinear optics comprises all effects for which the above assumptions are no longer valid. As a definition of the term "nonlinear optics" we can thus say that these are all phenomena in which the optical properties like $\varepsilon(\omega)$ or $\tilde{n}(\omega)$ depend in a reversible way on the illumination. We stress here the term "reversible". This means that the system returns to its initial state, after the illumination has been switched off, possibly after some time delay. We therefore exclude from our considerations phenomena like the photographic process or simply drilling a hole into a sample by intense laser excitation, a process known as "laser ablation". An example from every-day life of an optical nonlinearity are phototropic sun-glasses, which are transparent under weak illumination, become dark under the influence of UV radiation, e.g., from sunlight, and become transparent again after a few seconds if the illumination is reduced.

There are two approaches to describing nonlinear optical phenomena. In the first we assume that the response of the medium to the incident field(s) depends only on the instantaneous field amplitudes. This condition is fulfilled if the electronic excitations, on which we shall concentrate in the following, are created only virtually using the language of weak coupling or, in the language of strong coupling, introduced in Chap. 5 and Sec. 14.1, if we treat only interactions of coherent polaritons.

In fact, the weak coupling approach is used in most cases to describe optical nonlinearities, and we follow this trend here. But we also give some examples and hints for the proper description of the phenomena in the polariton picture.

As already stated, we now assume that the dielectric susceptibility depends on the instantaneous field amplitudes, i.e.,

$$\chi(\omega_\mathrm{i}, \boldsymbol{E}_\mathrm{i}) = \varepsilon(\omega_\mathrm{i}, \boldsymbol{E}_\mathrm{i}) - 1 \ . \tag{18.1}$$

Since the dependence on $\boldsymbol{E}_\mathrm{i}$ is usually not known explicitly we expand $\chi(\omega_\mathrm{i}, \boldsymbol{E}_\mathrm{i})$ into a power series of the incident field amplitudes [18.2, 3]:

$$\frac{1}{\varepsilon_0}\boldsymbol{P}_\mathrm{i} = \sum_j \chi_{ij}^{(1)} \boldsymbol{E}_j + \sum_{j,k} \chi_{ijk}^{(2)} \boldsymbol{E}_j \boldsymbol{E}_k + \sum_{i,k,l} \chi_{ijkl}^{(3)} \boldsymbol{E}_j \boldsymbol{E}_k \boldsymbol{E}_l + \dots \ . \tag{18.2}$$

If all frequencies are equal, a similar expansion can be formulated for the refractive index using intensities, i.e.,

$$\tilde{n}(\omega, I) = \tilde{n}_0(\omega) + \tilde{n}_2 I + \dots \tag{18.3}$$

where $\tilde{n}_0$ is again the linear refractive index and $\tilde{n}_2$ describes changes induced by I.

The first term on the right-hand side of (18.2, 3) describes the linear optical properties discussed in Chaps. 2–17. The second term of (18.2) containing the phenomenologically introduced parameter $\chi^{(2)}$ describes effects like second-harmonic, sum- and difference-frequency generation or the dc effect, i.e., rectification of the electric field of a light beam.

It immediately becomes clear what is meant by these terms if we direct a (laser) light beam with frequency ω onto the sample. The second term then reads

$$\chi^{(2)} \boldsymbol{E}_0^2 \sin^2 \omega t = \frac{1}{2}\chi^{(2)} \boldsymbol{E}_0^2 (1 - \cos 2\omega t) \ . \tag{18.4}$$

The $\cos 2\omega t$ term in (18.4) tells us that a contribution to the polarization is created, which oscillates at 2ω, and is radiated by the sample, the so-called second harmonic generation. The first term in the bracket describes a temporally constant polarization, which results in a voltage across the sample as shown schematically in Fig. 18.1. This corresponds to a partial rectification of the ac field of the light beam. The corresponding effect is therefore also called the "dc effect". If two fields with different frequencies ω_1 and ω_2 interact in the sample via $\chi^{(2)}$ effects, a similar approach as in (18.4) gives contributions to the polarization which oscillate with frequencies

$$\text{sum-frequency generation: } (\omega_1 + \omega_2) ,$$

$$\text{difference-frequency generation: } (\omega_1 - \omega_2) , \tag{18.5}$$

which are also radiated.

$\chi^{(2)}$ and all other even terms in the expansion (18.2) vanish for crystals whose symmetry elements contain the inversion. This is immediately clear by letting $\boldsymbol{E} \rightarrow -\boldsymbol{E}$ and consequently $\boldsymbol{P} \rightarrow -\boldsymbol{P}$.

For non-centrosymmetric crystals, the $\chi^{(2)}$ tensor contains one or more non-zero elements. Which elements vanish and which are equal depends on the point group of the crystal. Details are given in [18.2].

Another aspect of the $\chi^{(2)}$ phenomena concerns interference. When the fundamental beam propagates in the sample, it creates everywhere the second harmonic with the same relative phase between fundamental wave and second harmonic. On the other hand, the second harmonic, created at one place propagates and interferes with the second harmonic generated deeper in the sample. In order to get a maximum output of the second harmonic, this interference should be always constructive. The requires that

$$n(\omega) = n(2\omega) \tag{18.6a}$$

or

$$\boldsymbol{k}(2\omega) = 2\boldsymbol{k}(\omega) . \tag{18.6b}$$

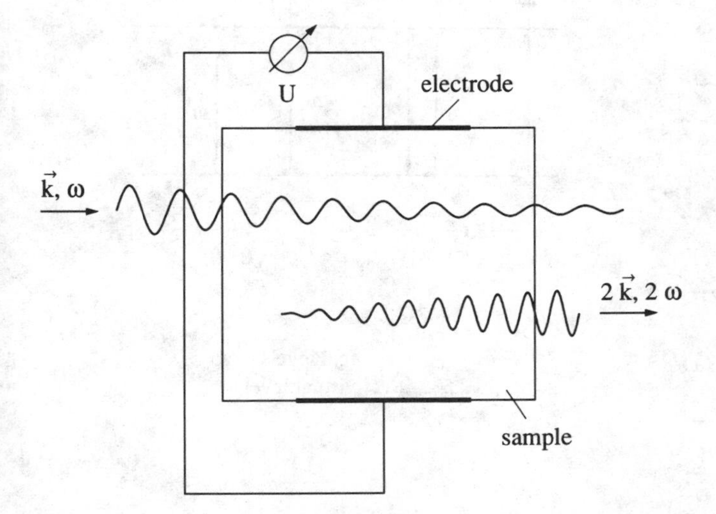

Fig. 18.1. Schematic drawing of an experimental situation in which second harmonic generation and the dc effect can be observed

The second version is nothing but momentum conservation i.e., two light quanta of frequency $\hbar\omega$ and (quasi-) momentum $\hbar\boldsymbol{k}_\omega$ are annihilated to give one with $2\hbar\omega$ and $\hbar\boldsymbol{k}_{2\omega}$. The condition (18.6) is also known as phase-matching and can usually be fulfilled only in some birefringent materials. Additionally one needs a special orientation for a given frequency. Similar arguments to (18.6b) hold also for sum- and difference-frequency generation.

The $\chi^{(3)}$ effects describe four-wave mixing (FWM) and hyper-Raman scattering (HRS), coherent anti-Stokes Raman scattering (CARS) etc. Some of these effects are described with the examples below.

The linear and nonlinear optical effects discussed here can be described in perturbation theory of increasing order either with the dipole operator $H_1 \Rightarrow H^{\mathrm{D}}$ or the second-order term $H_2^{(2)}$ of (3.55).

The linear one-photon absorption is then given by [see also Fig. 18.2a and (3.56)]:

$$\alpha(\omega)\boldsymbol{A}^2 \propto w_{\mathrm{if}} \propto |\langle f|H^{(1)}(\omega)|i\rangle|^2 \tag{18.7a}$$

and the corresponding linear refractive index by a process involving one virtually excited intermediate state $|z\rangle$ (Fig. 18.2b) which after some phase delay spontaneously emits a photon, which is otherwise identical to the

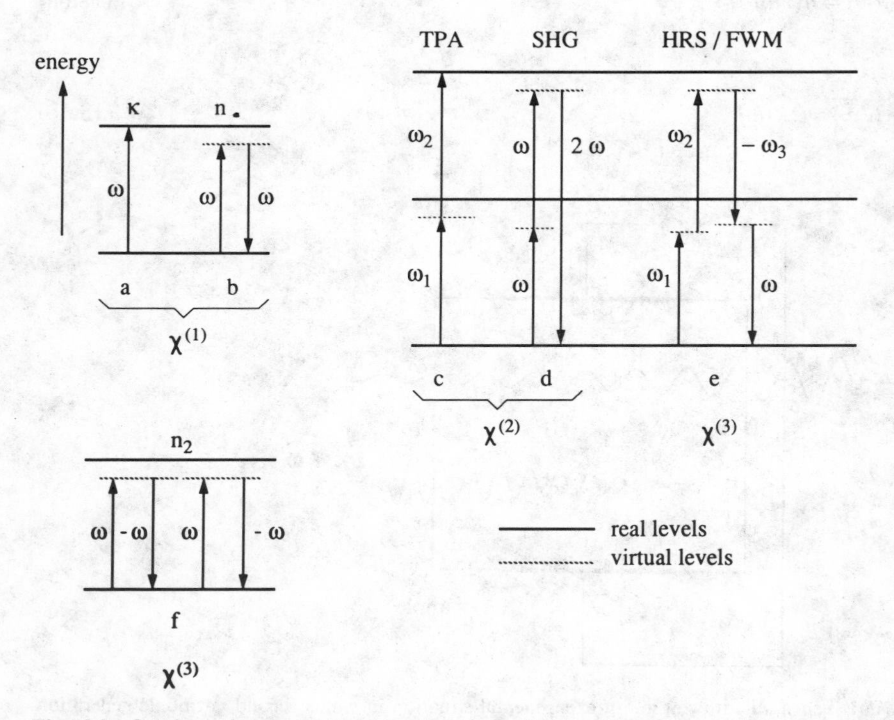

Fig. 18.2a–f. Schematic presentation of various linear (**a, b**) and nonlinear optical processes (**c–f**)

incident one. We use the following notation: Absorption processes leading to real or virtual excited states are described by positive frequencies and stimulated or spontaneous emission processes induced by an incident field or by the zero-point fluctuations, respectively, by negative frequencies.

We recall the dipole operator H^{D} from (3.68) and write for small imaginary parts $\kappa(\omega)$ of $\tilde{n}(\omega)$, i.e., away from the resonance energy E_{zj}

$$n(\omega)-1 \propto \left| \sum_{z_j} \frac{\langle i|H^{\mathrm{D}}(-\omega)|z_j\rangle\langle z_j|H^{\mathrm{D}}(\omega)|i\rangle}{\hbar\omega-(E_{z_j}-E_{\mathrm{i}})} \right|^2 . \tag{18.7b}$$

Alternatively $n(\omega)$ can be obtained from $\kappa(\omega)$ via the Kramers–Kronig relations (Chap. 8) or vice versa.

The two-photon absorption coefficient β introduced phenomenologically for a one-beam experiment in (18.8) and Fig. 18.2c,

$$-\frac{\mathrm{d}I}{\mathrm{d}x}=\beta I^2, \tag{18.8}$$

is then given by

$$\beta \propto \left| \sum_{z_j} \frac{\langle f|H^{\mathrm{D}}(\omega_2)|z_j\rangle\langle z_j|H^{\mathrm{D}}(\omega_1)|i\rangle}{\hbar\omega_1-(E_{z_j}-E_{\mathrm{i}})} + \mathrm{c.p.} \right|^2 , \tag{18.9}$$

where we allow for two different frequencies ω_1 and ω_2, represented by their respective dipole operators, and where c.p. stands for cyclic permutations.

Energy conservation has to be fulfilled between initial state $|i\rangle$ and final state $|f\rangle$, momentum conservation in every step, i.e.,

$$\left.\begin{aligned} &E_{\mathrm{f}}=E_{\mathrm{i}}+\hbar\omega_1+\hbar\omega_2, \\ &\boldsymbol{k}_{\mathrm{i}}+\boldsymbol{k}(\omega_1)=\boldsymbol{k}_{z_j}, \quad \text{and} \\ &\boldsymbol{k}_{z_j}+\boldsymbol{k}(\omega_2)=\boldsymbol{k}_{\mathrm{f}}. \end{aligned}\right\} \tag{18.10}$$

The dipole approximation actually means $\boldsymbol{k}(\omega_{1,2})=0$; see (3.65).

After the short discussion of n, κ and β we turn now to the $\chi^{(n)}$. For the linear susceptibility we get

$$\chi^{(1)}=\sum_{z_j} \frac{\langle i|H^{\mathrm{D}}|z_j\rangle\langle z_j|H^{\mathrm{D}}|i\rangle}{\omega-(E_{\mathrm{i}}-E_{z_j})+\mathrm{i}\gamma}, \tag{18.11}$$

where we have introduced a small damping γ to avoid a singularity.

From the imaginary part of $\chi^{(1)}$ we get $\kappa(\omega)$ and $\alpha(\omega)$ and from the real part $n(\omega)-1$ for weak absorption as indicated above.

Second-harmonic and sum-frequency generation involve two virtually excited intermediate states $|z_i\rangle$ and $|z_j\rangle$ and are thus described by a contribution to $\chi^{(2)}$ which reads, according to Fig. 18.2d:

$$\chi^{(2)} \propto \sum_{z_j, z_k} \frac{\langle i|H^{\mathrm{D}}[-(\omega_1+\omega_2)]|z_k\rangle\langle z_k|H^{\mathrm{D}}(\omega_2)|z_j\rangle\langle z_j|H^{\mathrm{D}}(\omega_1)|i\rangle}{[\hbar(\omega_1+\omega_2)-(E_{z_k}-E_{\mathrm{i}})][\hbar\omega_1-(E_{z_j}-E_{\mathrm{i}})]} + \mathrm{c.p.} \tag{18.12}$$

Another contribution to the second-harmonic generation comes from first-order perturbation theory, using the term $H^{(2)}$ in (3.55) which was considered to be small of second order.

$$\chi^{(2)} \propto \sum_{z_k} \frac{\langle i|H^{\mathrm{D}}(-2\omega)|z_k\rangle\langle z_k|H^{(2)}|i\rangle}{2\hbar\omega - E_{z_k}} . \tag{18.13}$$

If the decay of the second virtually excited intermediate state is stimulated by a third photon field, this induces a $\chi^{(3)}$ effect of the type shown in Figs. 18.2e, f. We denote the frequencies in these contributions to $\chi^{(3)}$ in the following way:

$$\chi^{(3)}(\omega\colon \pm\omega_j, \pm\omega_k, \pm\omega_l), \tag{18.14}$$

where the incident, absorbing, and stimulating frequencies are given with their respective sign after the colon, and the resulting frequency of the polarization that is radiated in the process under consideration before the colon.

One of the many contributions to $\chi^{(3)}$ (Fig. 18.2e) is then given by

$$\chi^{(3)}(\omega\colon -\omega_3, \omega_2, \omega_1)$$

$$= \sum_{z_j, z_k, z_l} \frac{\langle i|H^{\mathrm{D}}(\omega)|z_l\rangle\langle z_l|H^{\mathrm{D}}(-\omega_3)|z_k\rangle\langle z_k|H^{\mathrm{D}}(\omega_2)|z_j\rangle\langle z_j|H^{\mathrm{D}}(\omega_1)|i\rangle}{[\hbar(\omega_1+\omega_2-\omega_3)-(E_{z_l}-E_{\mathrm{i}})][\hbar(\omega_1+\omega_2)-(E_{z_k}-E_{\mathrm{i}})][\hbar\omega_1-(E_{z_j}-E_{\mathrm{i}})]} + \mathrm{c.p.} \tag{18.15}$$

If we inspect the terms describing one- and two-photon absorption (18.7a) and (18.9), or for three photon absorption not given here, or those which describe processes in which a photon is emitted like (18.7b) (18.10) and (18.13), the rules become intuitively clear. For every next higher order of perturbation theory there is one more dipole matrix element and one more resonance denominator. A product of two dipole matrix elements and of two resonance denominators can be replaced by one $H^{(2)}$ term as can be seen by comparing (18.10) and (18.11).

With increasing order of the perturbation there is an increasing number of cyclic permutations and of $\pm$ as in (18.14). An exhaustive presentation of all phenomena up to order three is given in [18.2] and an investigation of higher orders in [18.4].

As a last example, we give in (18.16) a combination which contributes to an excitation-induced change of the real part of the refractive index, i.e., to $\mathrm{Re}\{n_2\}$

in (18.3) in the way sketched in Fig. 18.2f:

$$\chi^{(3)}(\omega:\omega,-\omega,\omega)$$

$$=\sum_{z_l,z_k,z_j}\frac{\langle i|H^{\mathrm{D}}(-\omega)|z_l\rangle\langle z_l|H^{\mathrm{D}}(\omega)|z_k\rangle\langle z_k|H^{\mathrm{D}}(-\omega)|z_j\rangle\langle z_j|H^{\mathrm{D}}(\omega)|i\rangle}{[\hbar\omega-(E_{z_l}-E_{\mathrm{i}})][\hbar(\omega-\omega)-(E_{z_k}-E_{\mathrm{i}})][\hbar\omega-(E_{z_j}-E_{\mathrm{i}})]}$$

$$+\,\mathrm{c.p.} \tag{18.16}$$

The square of the sum of all analogous $\chi^{(3)}(\omega:\pm\omega,\pm\omega,\pm\omega)$ is then proportional to $\mathrm{Re}\{\tilde{n}_2(\omega)\}$. These effects lead e.g. to a self-focussing or defocussing of a light beam with a Gaussian beam profile, depending on the sign of $\mathrm{Re}\{\tilde{n}_2(\omega)\}$.

In all of the above processes, momentum has to be conserved in every step and energy must be conserved between initial and final states, as already mentioned. More details are found in [18.2–8].

The second group of nonlinear optical phenomena involves modifications of the optical properties by really or incoherently excited species, e.g., electron–hole pairs, excitons, or phonons. These species have finite lifetimes T_1 which can be of the order of ns to ms. Due to this finite lifetime, their density N does not instantaneously follow the incident light field, but depends on the generation rate in the past weighted by some decay function such as an exponential.

In this case we have:

$$\frac{1}{\varepsilon_0}\boldsymbol{P}=\chi(\omega,N)\boldsymbol{E} \tag{18.17}$$

with

$$N(t)=\int_{-\infty}^{t}G(t')\exp[-(t-t')/T_1]\,\mathrm{d}t'. \tag{18.18}$$

The generation rate $G(t')$ is connected with the intensity at time t' e.g., in the presence of one- and two-photon excitation by

$$G(t')=\alpha I(t')+\beta I^2(t'), \tag{18.19}$$

where α is the linear absorption coefficient treated in (18.7a), while β (see (18.8, 9)) is proportional to $\mathrm{Im}\{\tilde{n}_2\}$ in (18.3).

A further complication is introduced if we assume that the parameters T_1, α, β in (18.18, 19) are not constants but depend themselves on N or I. In this case we are left with rather complex systems of coupled integro-differential equations, which generally can be solved only numerically, often using some further approximations.

It is obvious that a power expansion as in (18.2, 3) is not adequate to describe optical nonlinearities which depend on an incoherent population $N(t)$. This population is created by a laser pulse which ends e.g. at $t=0$. For $t>0$, the electric fields in (18.2) are zero; nevertheless there are changes of the optical properties for as long as the incoherent population lives.

In order to be able to compare the magnitude of the coherent and non-coherent optical nonlinearities, some authors prefer to describe, e.g., the diffraction efficiency of a laser-induced grating (Sect. 24.3) as a $\chi^{(3)}$ process also in the case of incoherent population gratings. Though this approach is basically wrong, it may be of some practical use for the above-mentioned purpose especially if quasi-stationary conditions are used. We shall see later an example for such an effective $\chi^{(3)}_{\text{eff}}$.

At the end of this section, we want to introduce some terms that are closely, related to nonlinear optics and which may even be used as a synonym.

At the beginning of the investigation of optical nonlinearities in semiconductors in the 1960s, these physical effects were and sometimes still are known as high-excitation, high-density or many-particle effects.

The reason is that optical nonlinearities can often be observed under high excitation, usually with laser light. Typical values of the light intensities in the field of nonlinear optics range from 10^3 to more than 10^6 W/cm^2. Under these conditions many particles, i.e., electron–hole pairs, are created, really or virtually. This means that they exist in high density. The nonlinearities are due to the interactions between these many particles.

Changes of the optical properties under strong illumination are also known as renormalization effects. When they are induced by a laser of frequency ω_{exc} at this very frequency one thus speaks of self-renormalization of the optical properties.

To conclude this excursion into semantics we may state that optical nonlinearities are the consequence of the many-particle or renormalization effects that occur under high excitation.

18.2 General Scenario for High Excitation Effects

In the following chapters we shall discuss optical nonlinearities in detail. We concentrate on those due to many-particle effects in the electron–hole pair system of semiconductors. This means that we again concentrate on the spectral region around the exciton resonances and the band gap. We later consider interaction processes between excitons and phonons, but we shall not treat nonlinearities that occur in the spectral region of the (optical) phonon resonance, e.g., due to anharmonic phonon–phonon interaction.

A general scenario for the many-particle effects and the resulting optical nonlinearities has been developed over the last three decades. It is shown schematically in Fig. 18.3. At low light levels (i.e., in the low-density regime) the optical properties are determined by single electron–hole pairs, either in the exciton states or in the continuum. At low temperatures these excitons may also be bound to some defects to form BEC.

With increasing excitation intensity we reach the so-called intermediate density regime. In this regime, excitons are still good quasi-particles, but their density is so high that they start to interact with each other. There are elastic

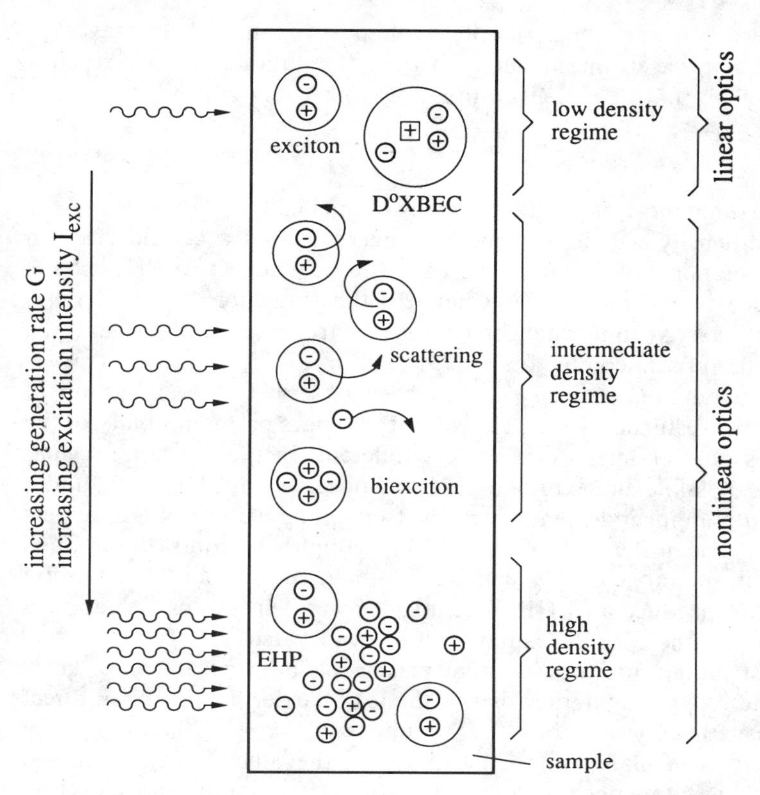

Fig. 18.3. The general scenario for many-particle effects in semiconductors

and inelastic scattering processes between excitons and (at higher temperatures) between excitons and free carriers. These scattering processes may lead to a collision-broadening of the exciton resonances and to the appearance of new luminescence bands, to an excitation-induced increase of absorption, to bleaching or to optical amplification, i.e., to gain or negative absorption depending on the excitation conditions. Another group of coherent and incoherent interaction processes in this intermediate density regime involves transitions to the excitonic molecule or biexciton. The biexciton, which was briefly mentioned in Sect. 14.4, is a new quasiparticle which consists of two electrons and two holes. If the exciton can be seen in analogy to the hydrogen or rather the positronium atom, then the biexciton corresponds to the H_2 or the positronium molecule. As we shall see in Sect. 19.3, transitions involving biexcitons give rise to a large variety of optical nonlinearities.

If we pump the sample even harder, we leave the intermediate and arrive at the high density regime, where the excitons lose their identity as individual quasiparticles and where a new collective phase is formed which is known as the electron–hole plasma (EHP). In this regime, the density of electron–hole pairs n_p is at least in parts of the excited volume so high that their average distance is comparable to or smaller than their Bohr radius, i.e., we reach a "critical density" n_p^c in an EHP, given to a first approximation by

$$a_B^3 n_p^c \approx 1 . \tag{18.20}$$

We shall meet some more elaborate formulas than (18.20) in Chap. 20.

In this high-density regime, we can no longer say that a certain electron is bound to a certain hole; instead, we have the new collective EHP phase. As we shall see in more detail in Chap. 20, the transition to an EHP is connected with very strong changes of the electronic excitations and the optical properties of semiconductors.

The scenario outlined in Fig. 18.3 has been observed in every group of semiconductors investigated so far. However, the ranges of observability of the various groups of phenomena can be very different. In the indirect gap material Ge it is possible in favorable cases to observe an EHP even under illumination with an incandescent lamp, due to the long lifetime of the carriers which can reach ms in the bulk of high quality samples. In contrast, for CuCl pump powers in the range of GW/cm^2 have to be used to fulfill the condition of (18.20) and to produce an EHP. Such high power densities can be applied only for a few ps, otherwise the sample will be evaporated.

As a rule of thumb, one can say that plasma phenomena are most easily observed in indirect gap materials due to the long carrier lifetime or in direct gap materials with large excitonic Bohr radius, so that (18.20) can also be readily satisfied as in GaAs. The Cu halides, on the other hand, are model substances for biexciton phenomena, due to the large exciton and biexciton binding energies. The II–VI compounds have an intermediate position allowing observation of all of the effects shown in Fig. 18.3 (and some more). This is very satisfying but poses the challenge of separating the various contributions, which partly overlap spectrally. Recent reviews which cover high excitation effects are [18.5–8] and references therein.

In the following chapters we describe the various effects in some detail, always beginning with three-dimensional materials but also giving examples of systems of reduced dimensionality. We start in Chap. 19 with the intermediate density regime and continue in Chap. 20 with the electron–hole plasma.

18.3 Problems

1. Try to make a rough guess of the electric field strength due to the positively charged nucleus of Si at the distance of the outer electron shell. At what light intensities would you expect to observe coherent optical nonlinearities assuming that the electric field in the light beam is about 0.1 of the electric field from the nucleus. How do the field strength and the light intensity change if you consider excitons instead of atoms?

2. Calculate the light intensity necessary to create in a semiconductor with $E_g \approx 1.5$ eV a stationary density n_p of $10^{18}\,\mathrm{cm}^{-3}$ electron–hole pairs, if their lifetime is 0.1 ns (direct gap semiconductor) or 1 μs (indirect gap semiconductor).

3. Give a schematic drawing as in Fig. 18.2 for a three-photon absorption process and for third-harmonic (3ω) generation. Give the formula for the three-photon absorption coefficient β_3 and for the third-harmonic generation in perturbation theory. Can third-harmonic generation occur in centrosymmetric crystals?

4. Try to find some more processes which contribute to Re $\{\tilde{n}_2(\omega)\}$.

5. Calculate the critical densities n_ρ^c for some semiconductors using the formula (18.20).

19 The Intermediate Density Regime

In the following sections we present selected examples from the intermediate density regime where excitons and biexcitons are still good quasiparticles.

19.1 Two-Photon Absorption by Excitons

We have mentioned already the two-photon absorption (TPA) to the exciton level in connection with $\boldsymbol{k}$-space spectroscopy of polaritons (Sect. 14.4) and with the magnetic properties of excitons (Sect. 16.1). The process is described in the weak coupling limit by (18.9), i.e., a first photon virtually excites an intermediate state which is converted by a second photon to the exciton in the final state. In the strong coupling limit, one would say that two photon-like polaritons merge to form one exciton-like polariton.

The momentum conservation

$$\hbar \boldsymbol{k}_1 + \hbar \boldsymbol{k}_2 = \hbar \boldsymbol{k}_{\mathrm{exc}} \tag{19.1}$$

usually allow only the longitudinal and the UPB to be reached. The two-photon selection rules are different from those for one photon. Consequently it is sometimes possible by TPA to reach states that are forbidden in one-photon absorption and vice versa. Examples are given in Chap. 25 on group theory. An experimental result for the two-photon spectroscopy of exciton has already been given in Figs. 14.20, 23 and 16.5. More details, including three-photon absorption, are found in [19.1].

19.2 Elastic and Inelastic Scattering Processes

If we increase the density of really excited excitons, it can happen that two excitons meet during their lifetimes on their diffusive motion through the sample and scatter via their mutual dipole–dipole interaction. All of these scattering processes will disturb the phase of the exciton (exciton-like polaritons). We will meet an example of this phenomenon in Chap. 22.

The scattering processes themselves can be categorized into elastic and inelastic. For elastic scattering the sums of kinetic energies before and after the collision are equal, in addition to momentum conservation, i.e.,

$$\boldsymbol{k}_{i,1} + \boldsymbol{k}_{i,2} = \boldsymbol{k}_{f,1} + \boldsymbol{k}_{f,2} \tag{19.2a}$$

$$\frac{\hbar^2}{2M}(\boldsymbol{k}_{i,1}^2 + \boldsymbol{k}_{i,2}^2) = \frac{\hbar^2}{2M}(\boldsymbol{k}_{f,1}^2 + \boldsymbol{k}_{f,2}^2) \tag{19.2b}$$

in analogy to the definition in classical mechanics. These processes show up mainly in the reduction of the phase relaxation time T_2. In the inelastic processes on which we shall concentrate now, an exciton is scattered into a higher excited state with principal quantum number $n_{B,f} \geqslant 2$, while another is scattered on the photon-like part of the polariton dispersion and leaves the sample with high probability as a luminescence photon, when this photon-like particle hits the surface of the sample. This process is shown schematically in Fig. 19.1. The momentum conservation law is given by (19.2a) with one of the $\boldsymbol{k}_f$ being ≈ 0. If we assume that both excitons are initially in states $n_{B,i} = 1$ and that the momentum of the photon-like polariton is zero, then energy conservation reads

$$\hbar\omega_{P_{n_{B,f}}} = E_{exc}(n_B = 1,\ \boldsymbol{k} = 0) - E_{exc}^{b}\left(1 - \frac{1}{n_{B,f}^2}\right) - \frac{\hbar^2}{M}\boldsymbol{k}_{i,1}\boldsymbol{k}_{i,2}\ . \tag{19.3}$$

The resulting emission bands are usually called P-bands with an index given by $n_{B,f}$. The bands are broadened by averaging over the last term on the right of (19.3) and by the fine structure of the exciton states, e.g., the splitting of the $n_B = 2$ exciton into states with S, P_0 and P_1 envelope functions.

The transitions into the continuum (P_∞) decay rather fast with increasing excess energy due to a decreasing transition probability. A summary of the calculations for the transition probabilities and further references are given in [19.2].

In the simplest approximation, one expects that the luminescence intensity of these scattering processes increases quadratically with the density of excitons. Indeed one finds a superlinear increase of these bands with increasing pump power with exponents ranging from 1.5 to 2.

In Fig. 19.2 we give an example for ZnO showing the P_2 and P_∞ bands. References to further experimental results are compiled in [19.2].

At higher lattice (and exciton) temperatures a fraction of the excitons will be thermally ionized. In this situation a similar inelastic scattering process becomes possible, in which an $n_B = 1$ exciton-like polariton is again scattered onto the photon-like branch of the dispersion curve, while a free carrier (electron or hole) is scattered under energy and momentum conservation into a higher state in the respective band. A characteristic feature of the resulting rather broad and unstructured emission band is that its maximum shifts with increasing temperature considerably faster to lower energies than the band gap does. Examples are given in Fig. 19.3, for CdS and ZnO. While the basis for exciton–exciton scattering is the dipole–dipole interaction, we have to consider dipole–monopole interaction for the scattering between excitons and free carriers.

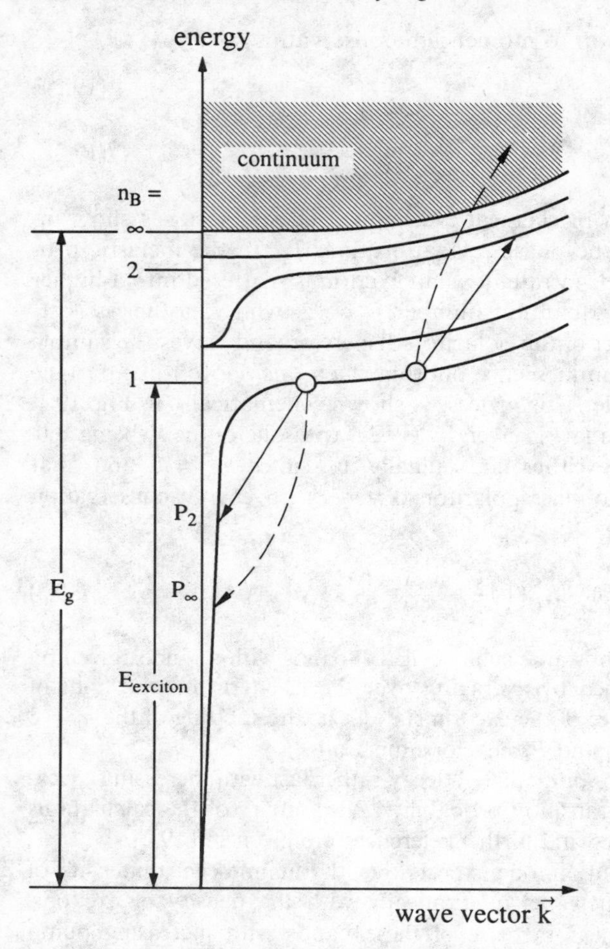

Fig. 19.1. Schematic representation of the inelastic exciton–exciton scattering processes P_2 and P_∞

In the literature one can find many other inelastic scattering processes not considered here in detail involving both free and localized states, for example, biexciton–biexciton scattering, scattering between a bound-exciton complex and a free carrier, and scattering processes involving phonons. A review is given in [19.2].

Most of these inelastic scattering processes give rise to the appearance of new emission bands (which usually grow more than linearly with the generation rate), to induced absorption and, eventually, to optical gain. The latter point will be addressed specifically in Chap. 21.

The inelastic scattering processes have been studied in great detail in the hexagonal and cubic II–VI semiconductors, but they have also been observed in III–V and I–VII compounds [19.2–8].

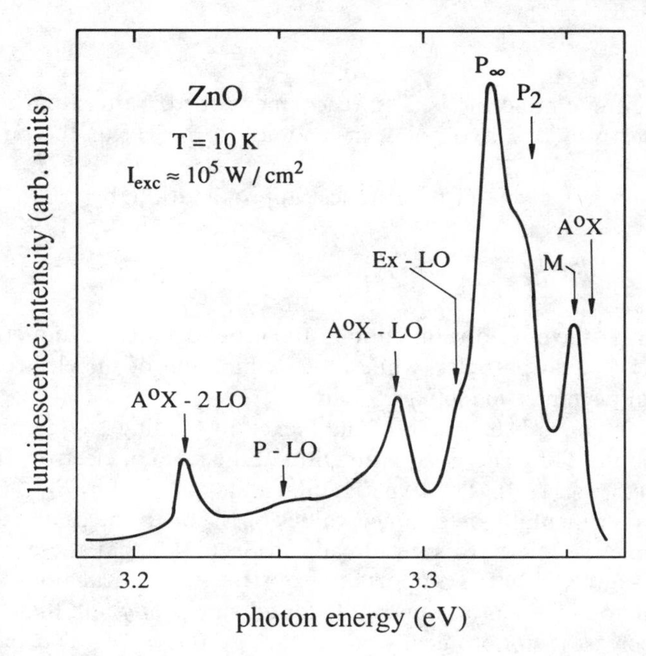

Fig. 19.2. The P_2 and P_∞ bands in the luminescence spectra of ZnO. According to [19.3]

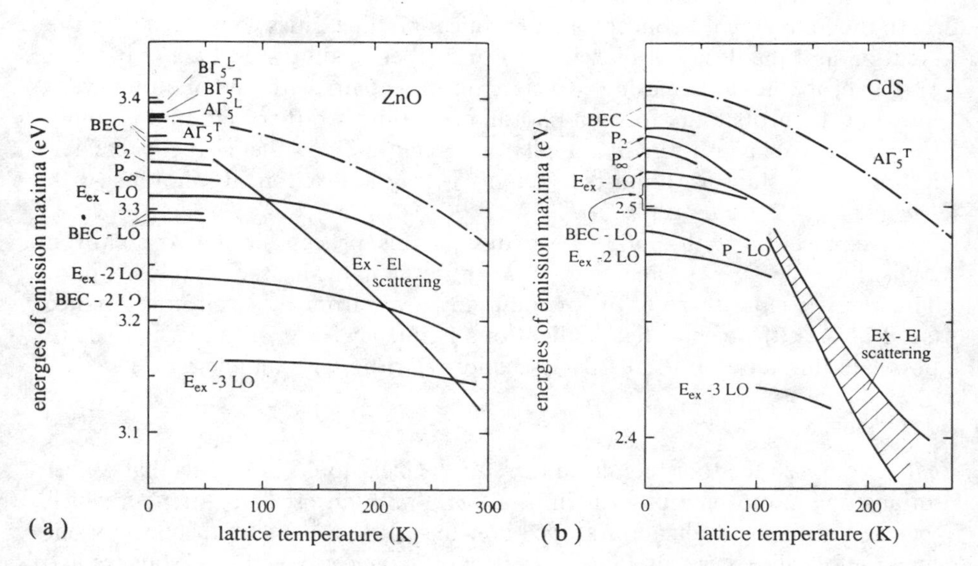

Fig. 19.3a, b. The temperature dependence of various emission band maxima in CdS(b) and ZnO(a). According to [19.2, 4]

19.3 Biexcitons

The biexciton or excitonic molecule is, as already mentioned, a quasi-particle consisting of two electrons and two holes, in analogy to the H_2 or the positronium molecule.

Its dispersion relation is given, in the simplest approximation, by

$$E_{\text{biex}}(\boldsymbol{k}) = 2E_{\text{ex}}(n_{\text{B}} = 1, \boldsymbol{k} = 0) - E^{\text{b}}_{\text{biex}} + \frac{\hbar^2 \boldsymbol{k}^2}{4M}, \tag{19.4}$$

where one assumes that the envelope function in the ground state is symmetric under the exchange of equal particles, while the combination of the electrons and of the holes is antisymmetric including spin.

General calculations have been made of the biexciton binding energy normalized by the exciton binding energy as a function of the ratio of electron and hole mass $\sigma = m_{\text{e}}/m_{\text{h}}$ (Fig. 19.4). The experimental data points for E_{biex} are normalized to the experimentally determined values of E^{b}_{ex} or to the calculated excitonic Rydberg energy. We have seen already in Sect. 11.2 that these two quantities partly disagree. What is important from the theoretical point of view is that the biexciton exists as a bound state for all values of σ and that the curve $E^{\text{b}}_{\text{biex}}/\text{Ry}^*$ decreases monotonically in the range $0 \leqslant \sigma \leqslant 1$. The data points scatter around the theoretical predictions with an accuracy comparable to that of the various calculations.

In the following we concentrate on optical nonlinearities connected with the creation and the decay of biexcitons. For further reading see [19.2, 7–10].

The probability of creating two electron–hole pairs with one photon is very low, but the probability for two-photon excitation is rather high and is sometimes said to have a "giant" oscillator strength, since the virtually created intermediate state is almost resonant with the exciton if photons with $\hbar\omega_{\text{exc}} \approx E_{\text{biex}}/2$ are used [19.2–7].

In connection with Fig. 19.5 we discuss this process as a two-polariton transition. We give the real part of $\boldsymbol{k}$ and schematically the imaginary part. The latter is close to zero for low damping away from the exciton resonance (see Fig. 6.2.). If we shine (laser-) light of a frequency $\hbar\omega_{\text{exc}}$ onto the sample, we populate this state on the LPB. A second polariton with an energy

$$\hbar\omega_{\text{abs}} = E_{\text{biex}} - \hbar\omega_{\text{exc}} \tag{19.5}$$

can accomplish the transition from there to the biexciton. This means that we get an absorption dip or a peak in Im k at $\hbar\omega_{\text{abs}}$ with an oscillator-strength which increases with the population at $\hbar\omega_{\text{exc}}$ on the LPB, evidently an optically nonlinear effect. The peak in Im k is necessarily connected with a resonance-like structure in the real part of the dispersion relation. In the following we give examples of both phenomena, but first we should mention that there is another way to create biexcitons. This starts from exciton-like polaritons, which may even have an incoherent, e.g., thermal population. In the latter case an exciton is

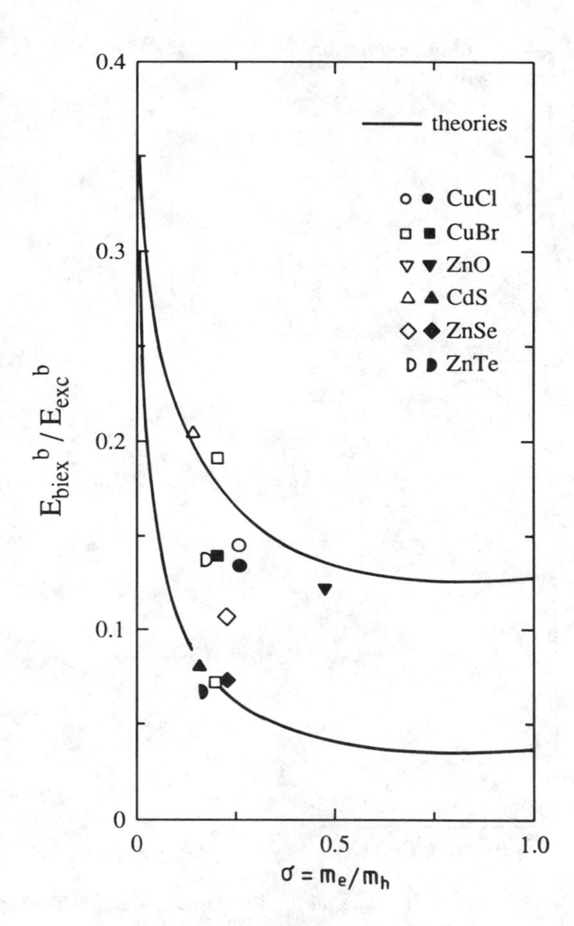

Fig. 19.4. Calculations of the binding energy of the biexciton in units of the binding energy of the exciton. Experimental data for various semiconductors, normalized to the experimentally observed binding energy (*open symbols*) and to the calculated Ry* energy (*full symbols*). According to [19.2]. Theories according to [19.9]

converted into a biexciton by absorption of a photon of energy $\hbar\omega'_{abs}$ fulfilling energy and momentum conservation. Considering the different curvatures, i.e., effective masses, of exciton and biexciton dispersion, this process yields an induced absorption at

$$\hbar\omega'_{abs} = E_{ex}(n_B = 1, \boldsymbol{k} = 0) - E^{b}_{biex} - \frac{\hbar^2 \boldsymbol{k}_i^2}{4M} , \tag{19.6}$$

where M is the translational exciton mass, $2M$ the biexciton mass and $\boldsymbol{k}_i$ the momentum of the exciton-like polariton in the initial state, which equals the momentum of the biexciton in the final state if we neglect the wave vector of the photon-like polariton. This process is also known as a two-step process to distinguish it from the two-polariton (or two-photon) transition discussed first.

Actually one difference lies in the fact that in the latter case one of the polaritons is either on the photon- or on the exciton-like branch, respectively. A second difference is that the laser at $\hbar\omega_{exc}$ generally produces a more-or-less coherent

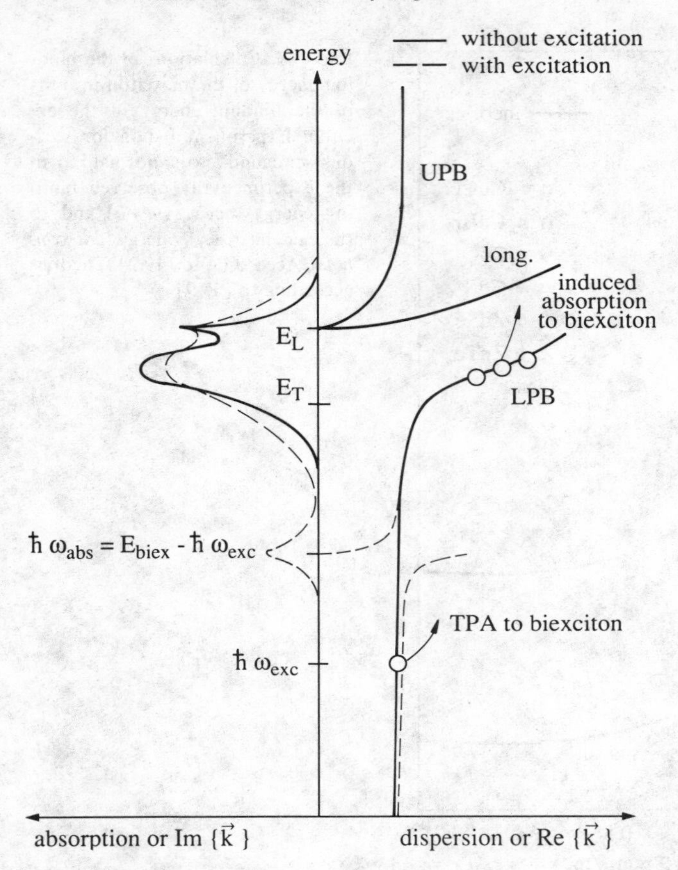

Fig. 19.5. The two-photon (or two-polariton) transition from the crystal ground state to the biexciton state and the two-step process starting from the exciton state

population on the LPB while the population on the exciton-like branch produced, e.g., by some resonant or non-resonant pump laser, tends to lose its coherence within a few ps (see Sect. 22.2).

In Fig. 19.6 we give an example of the various processes in a pump and probe beam experiment for CuCl. One can see the scattered, spectrally narrow pump laser light $\hbar\omega_{exc}$ and the absorption dips in the probe continuum, which correspond to a bound-exciton complex, the TPA process of (19.5) which shifts oppositely to the energy of the laser, and the two-step or induced absorption process of (19.6) which is fixed in energy and has a larger width due to the incoherent distribution of the excitons on their dispersion curve.

By varying the wave vectors of pump and probe beam relative to each other, it is even possible to measure the dispersion relation of the biexciton. An example is shown in Fig. 19.7 for CuBr. Due to the four-fold degeneracy of the Γ_8 valence band, the biexciton ground state splits into three states with symmetries Γ_1, Γ_5

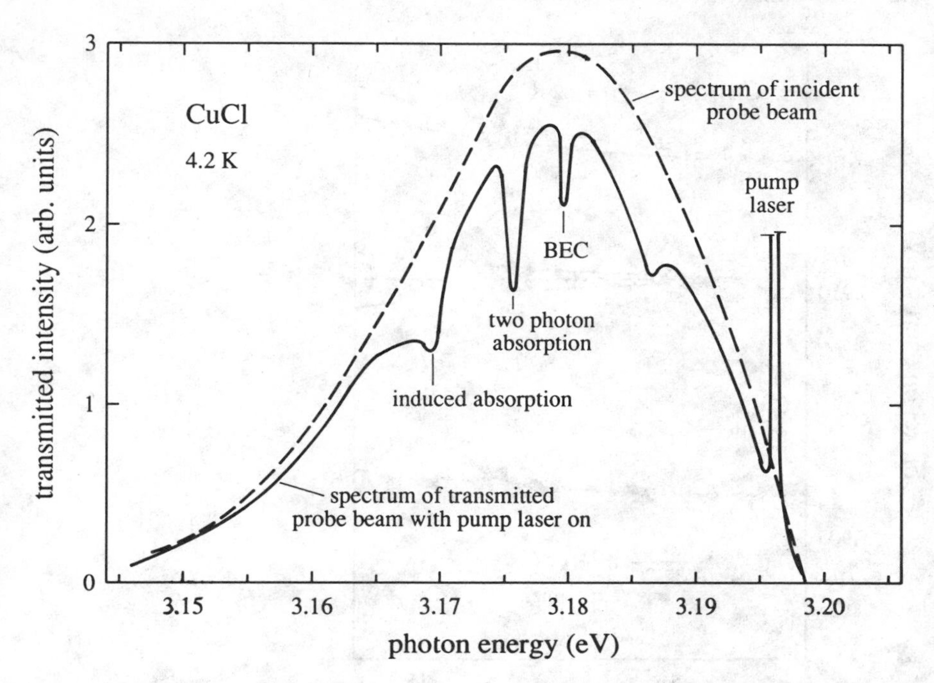

Fig. 19.6. Two-photon and two-step transitions to the biexciton observed in CuCl. According to [19.10]

and Γ_3. The Γ_1 state shows the simple dispersion of (19.4), while the higher states show some $\boldsymbol{k}$-dependent splitting.

The decay of biexcitons into an exciton-like and a photon-like polariton is just the reverse process of the two-step transition of (19.6).

Figure 19.8 shows biexciton luminescence spectra of CuCl under non-resonant and resonant high excitation. In the first case excitons are created, which form biexcitons with a thermal distribution. The temperature of the exciton and the biexciton gas can be higher than the lattice temperature T_L which describes the phonon system, and in the situation of Fig. 19.8 lies around 25 K.

In the decay process a longitudinal exciton or a transverse exciton-like polariton can appear in the final state together with the photon-like luminescence. Consequently in cubic materials one observes the so-called M_L and M_T bands. Under resonant two-photon excitation of the biexciton, one creates a biexciton gas with a narrow, non-thermal distribution, the decay of which gives rise to narrow emission structures N_T and N_L at the high energy sides of the M_L and M_T bands, respectively. This spectral position corresponds to the recombination of biexcitons with small wave vectors. Similar effects are also known for CuBr [19.7, 10].

The attempt to connect these narrow lines and some related phenomena with a Bose–Einstein condensation of biexcitons did not give a unique proof of this phenomenon [19.2, 10].

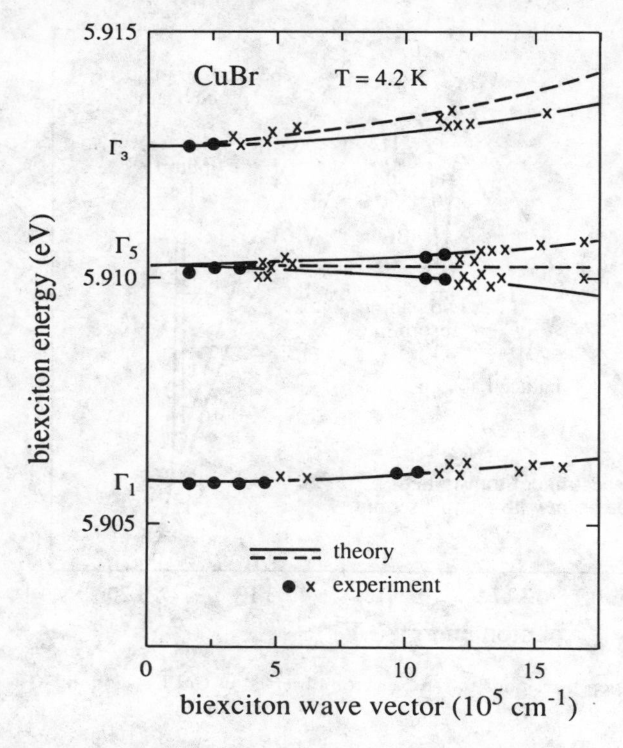

Fig. 19.7. The dispersion of the lowest biexciton states in CuBr. According to [19.10]

In the II–VI compounds, the emission bands of biexcitons are less structured due to the orientation-dependent mixed mode final states (see Chap. 14 and [19.2]). Furthermore, in many II–VI compounds the biexciton emission is spectrally almost degenerate with other recombination processes, especially ones involving bound-exciton complexes like inelastic BEC–free-carrier scattering or the acoustic phonon side-band of BEC. Therefore a clear-cut proof of the existence and the properties of biexcitons is difficult to obtain in these materials from luminescence alone. Details of this topic are given in [19.2].

Excitonic molecules have also been observed in the indirect materials Ge and Si in the intermediate density regime before the onset of plasma formation [19.11]. Furthermore, biexcitons have not only been found in I–VII, II–VI and group–IV materials, but also in other less common semiconductors like HgI_2 or AgBr, so that the formation of biexcitons can be considered a rather general feature in the intermediate density regime at low temperatures. At higher temperatures biexcitons are rapidly ionized and their resonances become so broad that they can generally no longer be detected.

In quantum wells like GaAs/AlGaAs biexcitons have been predicted theoretically and verified experimentally; see [19.12–16] and references therein. In

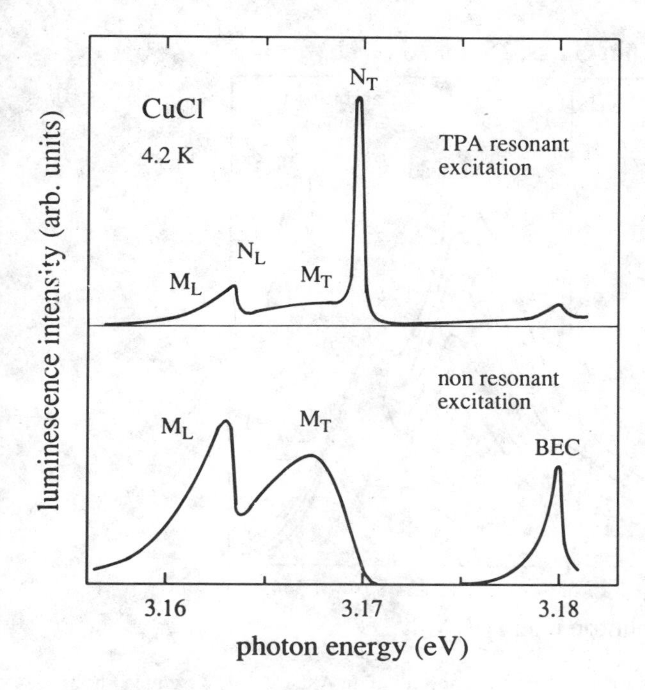

Fig. 19.8. Biexciton luminescence of CuCl under resonant and non-resonant excitation. According to [19.10]

Fig. 19.9 show emission spectra of an $l_z \approx 10\,\text{nm}$ GaAs/AlGaAs MQW sample, for various pump powers normalized at the emission of the $n_z = 1$ hh exciton. The peak on the low energy side, which grows more strongly than the exciton resonance, is due to the decay of biexcitons with a thermal distribution. The binding energy deduced from a line-shape fit is (1.75 ± 0.05) meV, in good agreement with data from coherent spectroscopy methods such as quantum beats (Chap. 22).

For quantum dots it has been found that two electron–hole pair states exist. These form a bound state with an energy lower than twice the energy of one electron–hole pair in one dot [19.18]. In addition, there are a large number of excited states of this two electron–hole pair system in the dot [19.19, 20]: In particular, transitions from the one-pair to the two-pair states have been observed by various authors as an excitation-induced absorption in differential transmission spectra; see [19.19–21] and references therein.

In Fig. 19.10a we show a set of emission spectra from CuBr dots in a glass matrix for increasing excitation intensity. The peak on the high-energy side is due to the recombination of single excitons in a quantum dot. The peak on the low energy side, which grows more rapidly with increasing I_{exc} than the other one, is attributed to the recombination of a biexciton, giving a photon and a single electron–hole pair in the dot. This interpretation is confirmed not only

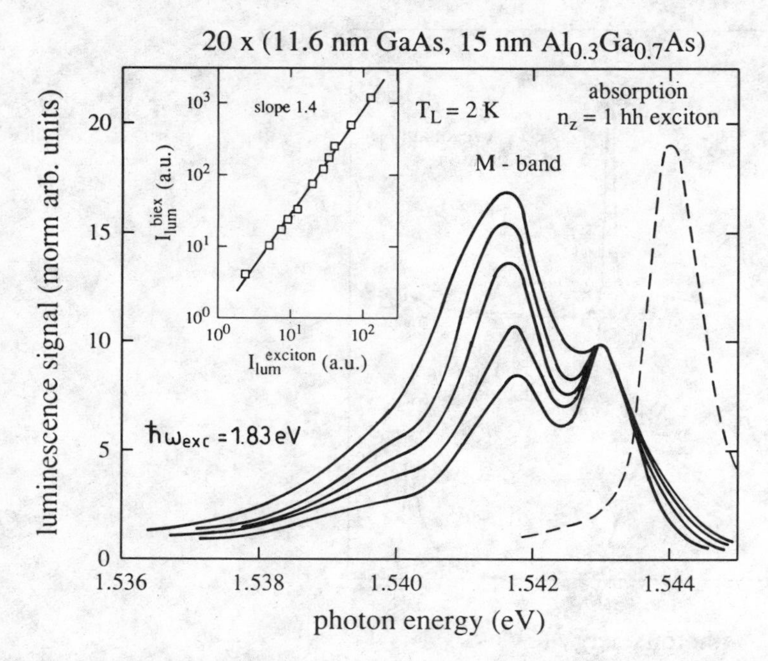

Fig. 19.9. The absorption and normalized emission spectra of an AlGaAs MQW sample, showing the biexciton luminescence band *M*, which grows superlinearly as shown in the insert. According to [19.15, 16]

by the superlinear increase of the emission but also by the dependence of the emission maxima on the average dot radius shown in Fig. 19.10b. The extrapolation to large dot radii converges to the values of the free exciton in the bulk and to thelowest biexciton emission band, respectively. The binding energy of the molecule as a function of *R* agrees well with theory (Fig. 19.10c).

In Fig. 19.11 we show the absorption spectra of two (photo-darkened) glass samples containing CdSe quantum dots and the change of the optical density (so-called differential transmission spectra DTS) observed for various pump-photon energies $\hbar\omega_{exc}$. In Fig. 19.11a we observe a bleaching or spectral hole burning in the inhomogeneously broadened absorption band at $\hbar\omega_{exc}$ and at a second energy for the dot size of Fig. 19.11a. These two bleaching features correspond to the saturation of the transitions from the two uppermost quantized hole states to the lowest conduction band state.

In addition there are two structures of induced absorption, also marked by arrows. They correspond to transitions in dots containing one electron–hole pair to the ground state and to excited states of the biexciton, respectively.

To conclude this section on biexcitons we describe in some detail for bulk samples a coherent process which is doubly resonant and which belongs into the group of $\chi^{(3)}$ phenomena. We refer to the two-photon or hyper-Raman

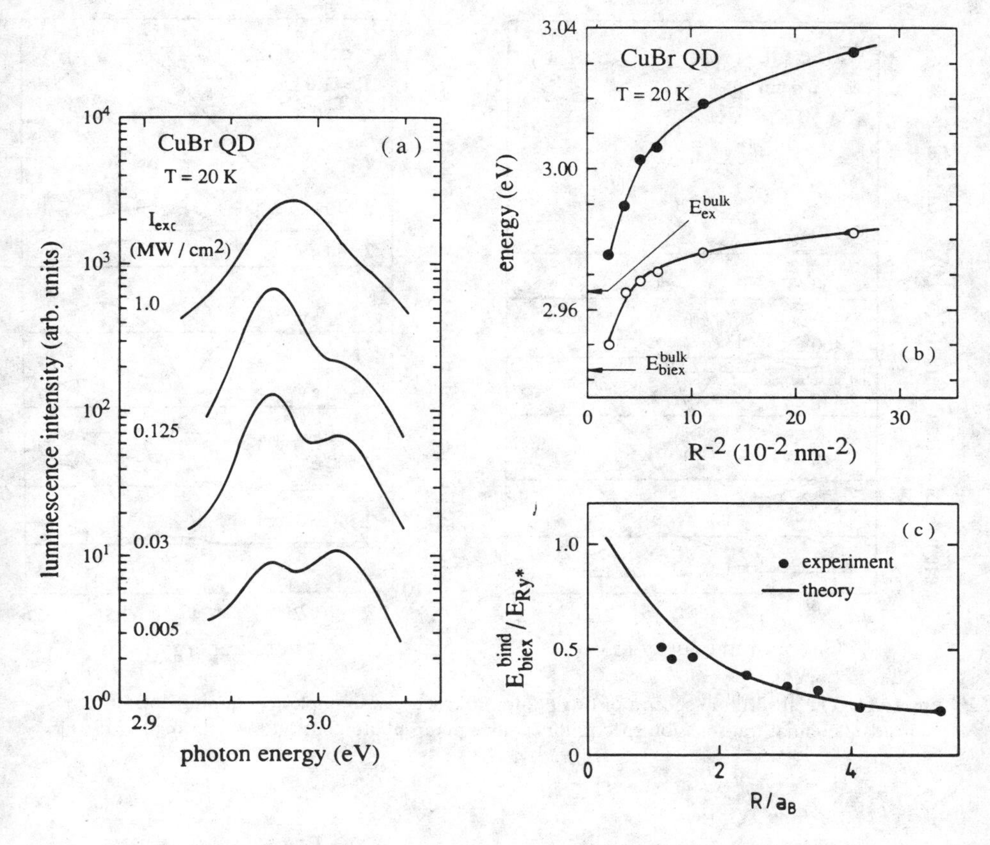

Fig. 19.10a–c. The emission of CuBr quantum dots for various excitation intensities (**a**); the spectral positions of the emission maxima as a function of the average dot radius $\bar{R}$(**b**); and the binding energy of the biexciton compared to theory (**c**). According to [19.17, 18]

scattering and the associated (non-) degenerate four-wave mixing. The idea is described in the following in both the weak and the strong coupling limits (see Fig. 19.12 and compare with Fig. 18.2).

An incident photon, which is almost resonant with the exciton state, is converted by a second photon into a virtually excited biexciton. This biexciton has to decay again after a time Δt determined by the energy–time uncertainty relation. If it decays into two photons – which are identical to the incident ones – this process would describe an intensity induced change of the phase velocity, i.e., a contribution to $\mathrm{Re}\{\tilde{n}_2\}$ in (18.3). However, there are many other decay processes which fulfill energy and momentum conservation.

In a backward scattering process, the virtually excited biexciton at wave vector $(\boldsymbol{k}_i^{(1)} + \boldsymbol{k}_i^{(2)})$ can decay into a transverse or a longitudinal exciton at $\boldsymbol{k}_f$

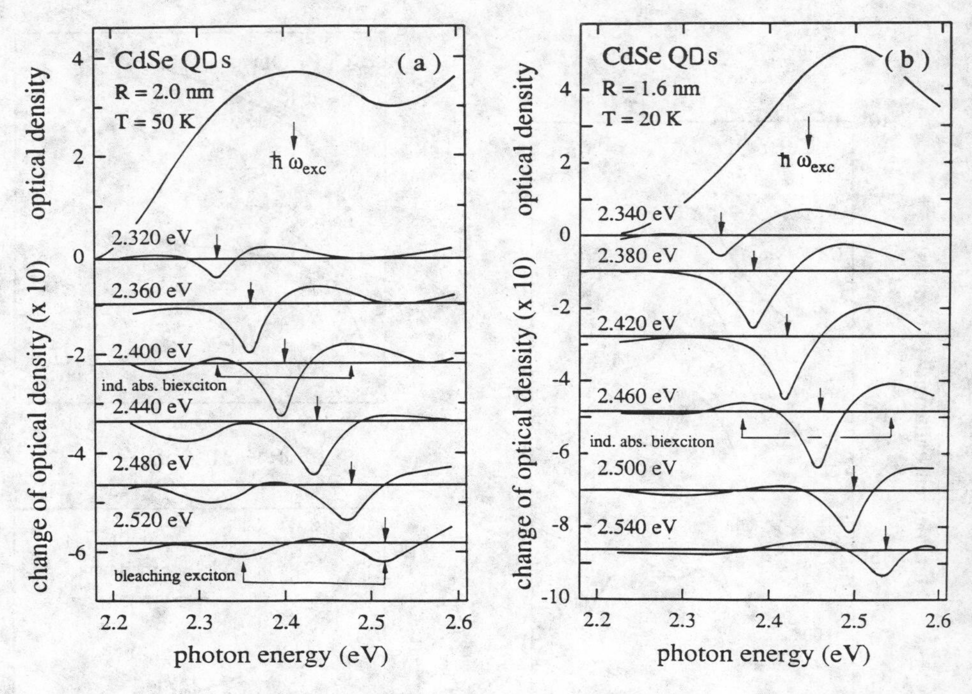

Fig. 19.11. The absorption spectra of two photo-darkened glass samples containing dots of CdSe and the differential transmission spectra for various pump-photon energies. According to [19.21]

and a Raman-like photon with $\hbar\omega_{\mathrm{R}_{t,l}}$ according to

$$\boldsymbol{k}_{\mathrm{R}} = \boldsymbol{k}_{\mathrm{i}}^{(1)} + \boldsymbol{k}_{\mathrm{f}}^{(2)} - \boldsymbol{k}_{\mathrm{f}} \,, \tag{19.7a}$$

$$\hbar\omega_{\mathrm{R}_{l,t}} = \hbar\omega_{\mathrm{i}}^{(1)} + \hbar\omega_{\mathrm{i}}^{(2)} - E_{1,\mathrm{t}} \,, \tag{19.7b}$$

Since the transverse and longitudinal eigenenergies do not vary much with k in this range, we get for $\hbar\omega_{\mathrm{i}}^{(1)} = \hbar\omega_{\mathrm{i}}^{(2)}$ simply

$$\hbar\omega_{\mathrm{R}} = +2\hbar\omega_{\mathrm{i}} - E_{1,\mathrm{t}} \,, \tag{19.7c}$$

where the factor 2 on the right explains the name two-photon Raman scattering.

In forward scattering and for all intermediate scattering geometries (19.7) would be the same in the weak coupling limit. In the polariton picture this statement is only true if one particle in the final state is a longitudinal exciton. Indeed, the two-polariton state created by the two incident beams and which is almost resonant with the biexciton can decay either into a longitudinal exciton and a transverse polariton or into two polaritons. In the latter situation both final state particles are on the transverse branches. One particle is exciton-like in the backward configuration, but moves down into the bottle neck if the

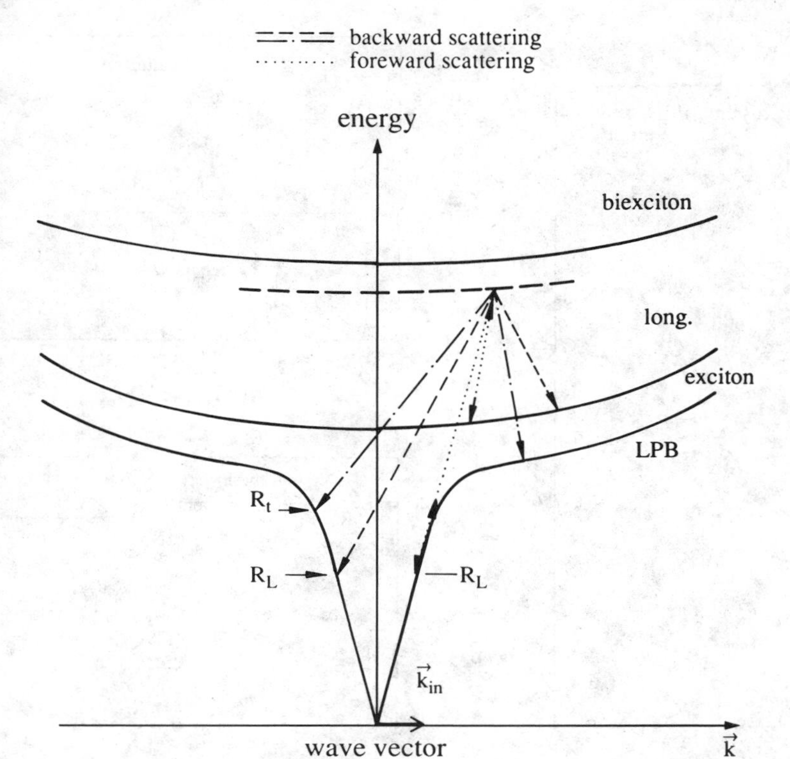

Fig. 19.12. The two-photon Raman scattering via intermediate exciton and biexciton states

scattering geometry is changed towards forward scattering. In this case both final state particles have a certain chance of leaving the sample as a photon. For $\omega_i^{(1)} = \omega_i^{(2)}$ they will be situated in scattering geometries close to forward scattering slightly above and below ω_i and they will be therefore called $\hbar\omega_R^+$ or $\hbar\omega_R^-$, respectively. The resulting relation between $\hbar\omega_R$ and $\hbar\omega_i$ deviates significantly from the slope-two relation of (19.7c).

Examples for forward and backward scattering are given in Fig. 19.13 for CuCl and various geometries. A self-consistent fit of the data with calculated dispersion curves for the exciton polaritons allows one to determine their dispersion curve with high accuracy, as was mentioned in Sect. 14.4 on the $\boldsymbol{k}$-space spectroscopy of excitonic polaritons. It should be stressed that the deviations from the slope-two relation in Fig. 19.13c can be only understood in the polariton or strong coupling picture.

The excitation-induced resonance of Fig. 19.5 shows up as an anomaly in the otherwise smooth relation between $\hbar\omega_i$ and $\hbar\omega_R$ if one of the incoming or outgoing energies coincides with it. An example is shown in Fig. 19.14.

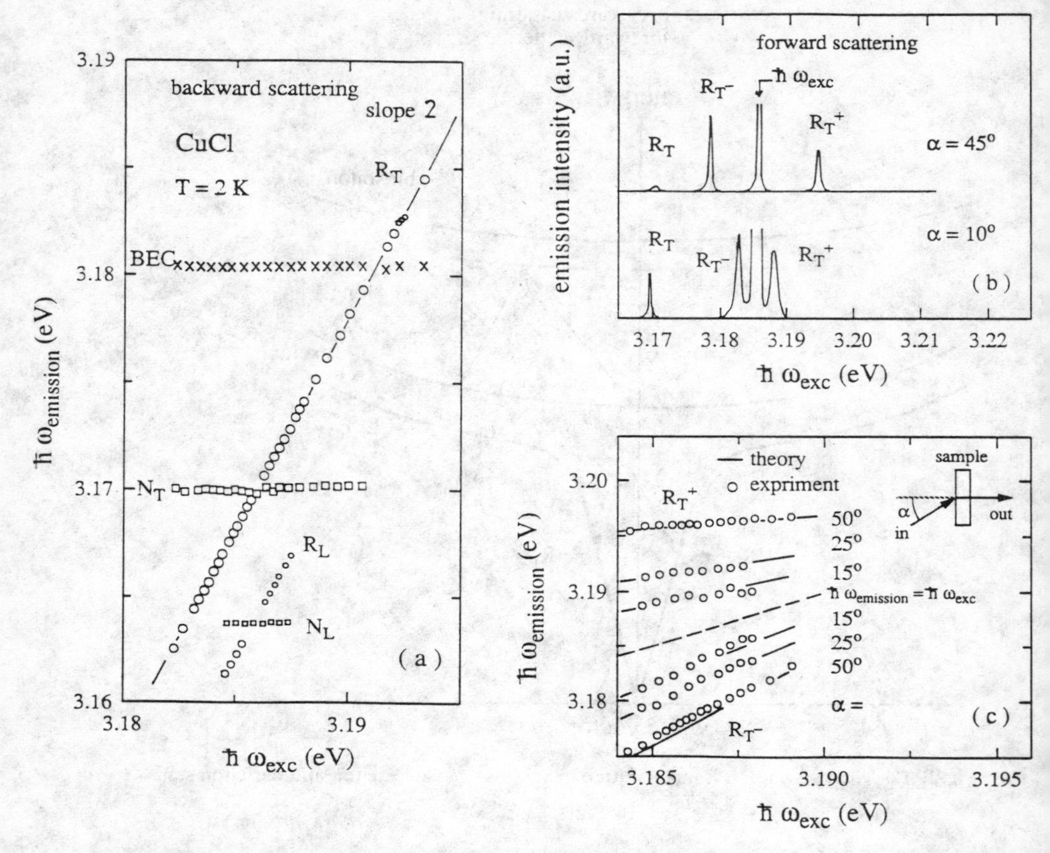

Fig. 19.13a–c. The dependence on $\hbar\omega_{exc}$ of the two-photon Raman emission via an intermediate biexciton state in CuCl for various scattering geometries (**a, c**) and a set of spectra for forward scattering (**b**). According to [19.10]

The fact that the splitting between $\hbar\omega_{exc}$ and $\hbar\omega_{R_t}$ does not converge to zero for higher intensities in exact forward scattering indicates that there are also self-renormalization effects of the dispersion at $\hbar\omega_{exc}$ [19.10].

Until now we have been considering the "spontaneous" decay of the virtually excited biexciton in the intermediate state. It is also possible to stimulate this decay by directing a third beam onto the sample which coincides in direction and energy with one of the particles in the final state. The emission of the other particle is then stimulated and we get a typical $\chi^{(3)}$ process in the sense of Fig. 18.2d. Alternatively we can describe this process as (non-) degenerate four-wave mixing (N)DFWM or as a diffraction of the third beam from a coherent laser-induced grating set up by the interference of the other two incident beams. Actually the close relation between two-photon Raman scattering and (N)DFWM has been verified by various groups, e.g., for CuCl. Details are given in [19.7, 10].

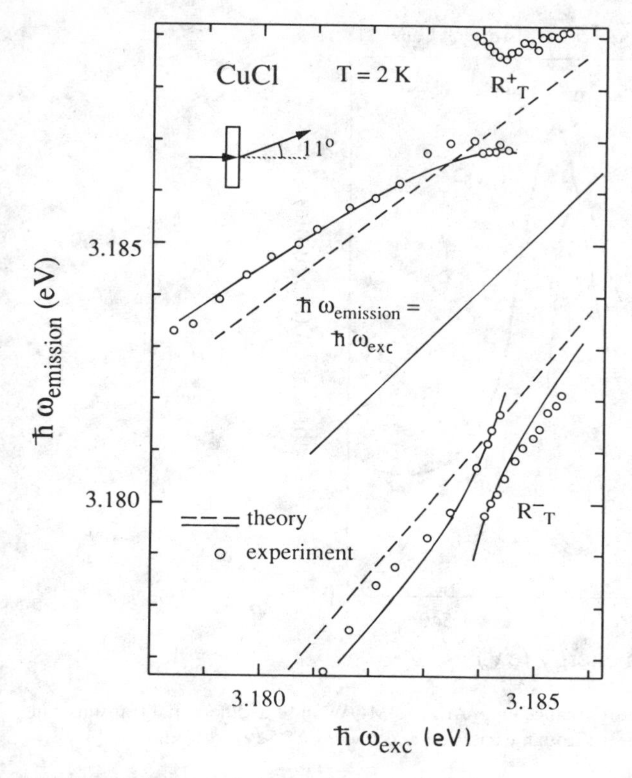

Fig. 19.14. The anomaly in the relation between $\hbar\omega_{exc}$ and $\hbar\omega_{emission}$ caused by the TPA resonance explained with Fig. 19.5. The dashed and solid lines give the theory without and with this resonance. According to [19.10]

19.4 Optical or ac Stark Effect

A further example of optical nonlinearities in the intermediate density regime is another coherent process, namely to optical, ac or dynamical Stark effect. The name implies that we are dealing with a shift of the exciton resonance caused by the electric field of a light beam in the sample. For recent reviews see [19.22, 23] and references therein.

This phenomenon is well known in atomic spectroscopy and is frequently described in the "dressed atom" picture, i.e., as an atom in the presence of photons. A simple explanation, outlined in the following, is in terms of level repulsion, a phenomenon which generally occurs in quantum mechanics for two energetically close lying levels which interact with each other.

We assume that the energy of the incident photons is chosen slightly below the exciton resonance and use the weak coupling picture. Then a state $|m, n = 0>$ containing m photons and $n = 0$ excitons is almost degenerate with

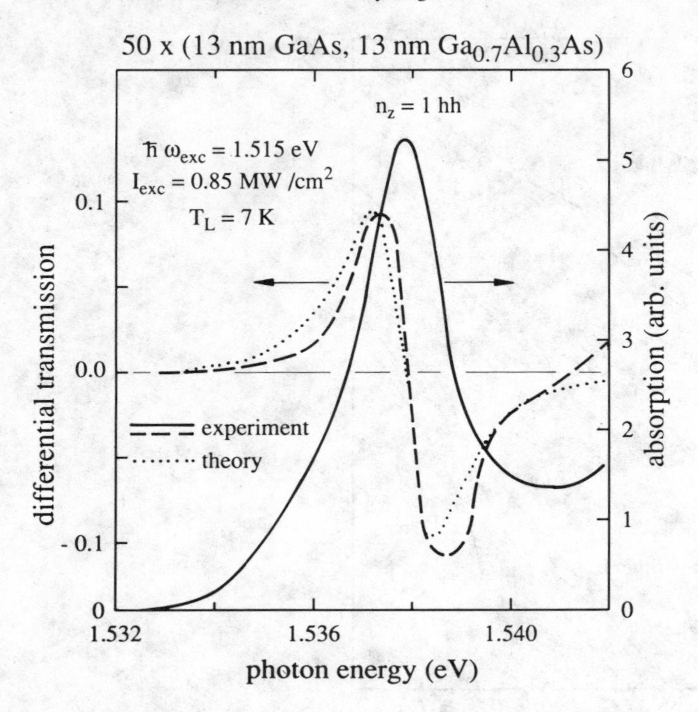

Fig. 19.15. The $n_z = 1$ hh excition resonance in an AlGaAs MQW and the differential transmission spectrum due to the ac Stark effect through excitation and $\hbar\omega_{\mathrm{exc}} = 15.5\,\mathrm{eV}$. According to [19.26]

a state $|m-1, n=1>$. Since the two states are coupled via dipole interaction, an energetic repulsion of the two levels occurs.

The frequency of the photons ω is held constant from the outside by the incident laser. Consequently the exciton resonance shifts to slightly higher energies for $\hbar\omega - E_{\mathrm{exc}} < 0$ and to lower energies for $\hbar\omega - E_{\mathrm{exc}} > 0$ by an amount δE. This quantity has been calculated by various authors [19.22–24]. We cite here the result from [19.24]:

$$\delta E = \frac{2|er_{\mathrm{cv}}E_{\mathrm{p}}|^2 |\phi_{1\mathrm{s}}(\boldsymbol{r}_{\mathrm{e}} - \boldsymbol{r}_{\mathrm{h}} = 0)|^2}{(E_{1\mathrm{s}} - \hbar\omega_{\mathrm{p}}) N_{\mathrm{S}}^{\mathrm{PSF}}}, \tag{19.8}$$

where er_{cv} is the band-to-band transition matrix element or polarizability, E_{P} the field amplitude of the incident laser at frequency ω_{P}, $|\phi_{1\mathrm{s}}(\boldsymbol{r}_{\mathrm{e}} - \boldsymbol{r}_{\mathrm{h}} = 0)|^2$ describes the (Sommerfeld) enhancement of the transition probability of the 1s exciton, $E_{1s} - \hbar\omega_{\mathrm{p}}$ the detuning, and $N_{\mathrm{S}}^{\mathrm{PSF}}$ the density of electron–hole pairs necessary to block the exciton resonance by phase-space filling (Chap. 20).

The term $er_{cv}E_P\hbar^{-1}$ gives the so-called Rabi frequency for the band-to-band transition. This is basically a field-dependent beat frequency which is the frequency with which the excitation oscillates coherently back and forth between the electronic excitation and the photon state.

A better description would again be in terms of polariton–polariton interaction [19.25]. This picture describes not only the shift of the exciton resonance by an amount δE but also the appearance of optical gain positioned symmetrically to the low energy side of the pump laser, i.e., at

$$\hbar\omega_g = \hbar\omega_P - (E_{exc} - \hbar\omega_P) \,. \tag{19.9}$$

Equation (19.9) follows from the simple picture that a state with two polaritons at $\hbar\omega_P$ decays into a Stark-shifted exciton-like polariton and a photon-like polariton at $\hbar\omega_g$.

Since the ac Stark effect is a coherent process, most of the successful attempts to observe it have used ps and sub ps laser pulses [19.22–24]. In this case the duration of the laser pulse τ_P is shorter than or comparable to the phase-relaxation time T_2 of the exciton resonance. The effect has been observed in bulk materials, quantum wells, and quantum dots. Since we deal with ultra short pulses only in Chap. 22, we choose here an example where the ac Stark effect has been measured under quasi-stationary conditions with ns pulses. In this case rather low pump intensities I_P and a large detuning are necessary to avoid real excitations, which can mask or destroy the ac Stark effect. Unfortunately, due to (19.8) this also implies only a small shift so that sensitive differential transmission spectroscopy (DTS) has to be used.

In Fig. 19.15 we show the $n_z = 1$ hh exciton resonance for an AlGaAs MQW structure, together with the DTS spectrum, i.e., the difference between the transmission spectrum without and with pump laser on. The sign and the symmetric shape of the DTS spectrum show that the exciton resonance shifts to the blue with virtually no change of its oscillator strength or damping. The quantity

$$\sigma = \frac{\delta E.(E_{1hh} - \hbar\omega_p)}{I_P} \,, \tag{19.10}$$

deduced from Fig. 19.15 is $\sigma_{exp} = 8.5 \cdot 10^{-8}\,(\mathrm{meV})^2\,\mathrm{cm}^2\,\mathrm{W}^{-1}$ and compares favorably with the theoretical value and ps experiments giving $\sigma_{theory} \approx 5.8 \cdot 10^{-8} (\mathrm{meV})^2\,\mathrm{cm}^2\,\mathrm{W}^{-1}$ [19.26].

If we pump the samples even harder we leave the intermediate density regime and enter the high density regime where the optical properties are determined by the electron–hole plasma. This concept was introduced in Sect. 18.2. However, before treating this aspect we discuss briefly the Bose–Einstein condensation of excitons and biexcitons and introduced another group of optical nonlinearities, the so-called photo-thermal nonlinearities which also generally belong to the intermediate density regime.

19.5 Bose–Einstein Condensation

Excitons and biexcitons are bosons at low densities. Ideal bosons exhibit a so-called Bose–Einstein condensation at sufficiently low temperature and high density. Bose–Einstein condensation (BEC) is a macroscopic population of one state in $\boldsymbol{k}$-space, generally $\boldsymbol{k} = 0$. Examples of BEC are superfluidity (e.g., of He) and superconductivity.

For ideal, non-interacting bosons one expects the onset of BEC if at constant temperature T the density n reaches a certain critical density n_c or if at constant n the temperature falls below a certain temperature T_c. The relation between n_c and T_c is given for effective-mass particles, which are ideal bosons, by the effective density of states (Sect. 10.7)

$$n_c = g\left(\frac{mk_B T_c}{2\pi\hbar^2}\right)^{3/2}. \tag{19.11}$$

This means that the density at which a gas of fermions becomes degenerate is just equal to the density at which a gas of ideal bosons starts to condense. The fraction of (quasi-) particles in the condensed state increases with increasing n and decreasing T. The particles in this state are separated by a small energy gap from the normal ones. For a detailed theoretical analysis of these topics as applied to excitons and biexcitons see [19.27].

According to the above statements, one has to look for narrow emission bands which are connected with the recombination of free excitons or biexcitons and which spectrally are situated close to the position at which quasiparticles at $\boldsymbol{k} = 0$ contribute to the luminescence.

We discuss first the situation for excitons. Excitons that couple strongly to the radiation field and form polaritons are not good candidates for a BEC, since for $\boldsymbol{k} = 0$ their dispersion goes to $E = 0$. Of the excitons that do not couple strongly to light, we can rule out longitudinal excitons because they tend to relax to lower lying triplet and transverse polariton states. The best candidates are excitons in semiconductors with a dipole-forbidden band-to-band transition since at least all excitons with S-like envelope and thus the ground-state excitons with $n_B = 1, l = 0$ are then dipole-forbidden, too. The most prominent representative of this group of semiconductors is Cu_2O. This material has the additional advantage that the binding energy of the biexciton is expected to be close to zero, so that a fusion of excitons into biexcitons is less probable. Indeed, some hints for BEC of the excitons at low temperature and high excitation in this material have been obtained from a detailed analysis of the exciton emission band, at least if biexciton effects are not included, but these findings are not unambiguous [19.28].

The model substance for biexcitons is CuCl. Therefore this material has been also investigated for BEC of biexcitons. Narrow emission bands, which occur under resonant two-photon excitation of the biexciton on the high energy side of the M-bands, i.e., at a spectral position to which biexcitons with small $\boldsymbol{k}$ contribute, have been found to be due to a cold gas of biexcitons [19.29] or to hyper-Raman scattering (Sect. 19.3) but not to BEC. Some indications of the Bose character of biexcitons have been drawn from the observation that thermal biexcitons are scattered preferentially into states which are strongly populated by an external laser pump source [19.30], but this is a situation which is far from thermodynamic quasi-equilibrium conditions and is not a spontaneous BEC with increasing n or decreasing T.

To summarize this section, we can state that there is some evidence for the Bose character of excitons and biexcitons at low temperatures, but no clear evidence for BEC. This situation can be basically understood by the following considerations. With increasing density, the temperature of a (bi-) exciton gas tends to increase above the lattice temperature, e.g., due to nonradiative or Auger-type recombination processes even under resonant excitation. Higher temperatures necessitate higher densities which result, on the one hand, in a further increase of T and, on the other, in an increasing deviation of the (bi-) excitons from their Bose character until a pure fermion phase, the electron–hole plasma, is reached. Some authors even claim that BEC of excitons cannot occur [19.31]. For more recent approaches to the problem of BEC in quantum wells see [19.32] and references therein.

19.6 Photo-thermal Optical Nonlinearities

The first interaction of visible light with matter is usually via the electronic system, i.e., an incoming photon is absorbed creating an electron–hole pair. This pair recombines after a while and, since most recombination processes in semiconductors are nonradiative, energy is transferred to the lattice, i.e., to the phonon system. In simpler words, the lattice is heated. An increase of the lattice temperature in turn results in changes of the absorption spectrum, e.g., via the Urbach rule of (14.12) and of the real part of the refractive index (Chap. 8).

We show an example in Fig. 19.16. The transmission spectrum (a) of a CdS sample is measured at RT with a weak cw probe beam from an incandescent lamp. We see Fabry–Perot modes of the platelet type sample in the transparent region and the onset of the absorption. When we illuminate the sample with the green (514.5 nm) line of a cw Ar^+ laser with a power of about $2 kW/cm^2$ we get the spectrum (b). The sample temperature increases in the laser spot by roughly 50 K and the consequences are seen in the spectrum. The absorption edge becomes less steep and shifts to the red. The Fabry–Perot modes also show a slight red shift revealing an increase of the refractive index.

At the position of the Ar^+-laser line, which is indicated by an arrow, we find a strong excitation-induced increase of absorption.

The photo-thermal optical nonlinearities are thus much less complex than some of the electronic effects of preceding sections and they usually have rather long relaxation time constants, often in the ms range. But, as we shall see, it is precisely because of these properties that they can be used as model

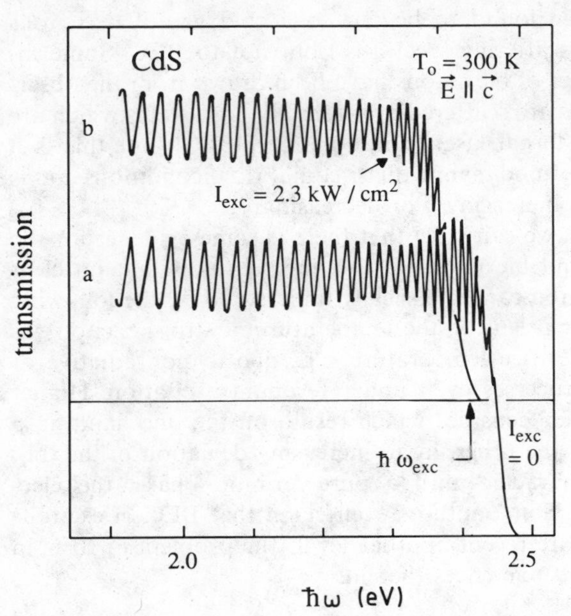

Fig. 19.16. Transmission spectra of a CdS platelet at a surrounding temperature of 300 K without (**a**) and with (**b**) excitation by the green line of an Ar^+ laser. According to [19.33]

systems for the applicability of (electro-) optical nonlinearities in optical bistability or in (electro-) optic data handling. We come back to these aspects in Chap. 23.

19.7 Problems

1. If a crystal has inversion symmetry, then the parity of the eigenstates is well defined. Which states can you reach from the crystal ground state (even parity) in one-, two- and three-photon absorption processes? Compare your results later with the information given in Chap. 25.

2. Assume that the scattering cross section of excitons is determined simply by their Bohr radius. Try to estimate the density at which an exciton has a probability of 0.1 of hitting another exciton during its lifetime. Which other parameters of the exciton do you need to make this estimate? Compare with values of the Mott density in Chap. 20.

3. Calculate the spectra of the luminescence of the biexciton and of the induced absorption. From which feature can you deduce the binding energy of the biexciton?

4. What are the best conditions for observing the optical Stark effect? Check if they have been met in some of the experiments described in the references.

5. Calculate the increase of the lattice temperature of a semiconductor under pulsed excitation: $I_{exc} = 10^6$ W/cm^2 ; $\hbar\omega_{exc} = 2$ eV; pulse duration = 10 ns repetion rate = 10 s^{-1}; $\alpha(\hbar\omega_{exc}) = 10^5$ cm^{-1}; diffusion length of excitons $l_d = 1$ μm. Consider bath temperatures of 5 K and of 300 K and use Debye's approximation for the specific heat with $T_D = 300$ K. Can you make a modified guess if you know that the velocity of sound in solids is of the order of 310^3 m/s?

6. Calculate the critical density for the onset of BEC in a typical semiconductor (reduced mass $\mu = 0.2\,m_0$, dielectric constant $\varepsilon = 10$) for $T = 5, 20$ and 100 K. Campare with the densities for the transition to an electron–hole plasma given in Chap. 20.

20 The Electron–Hole Plasma

Having introduced the basic idea of the electron–hole plasma (EHP) in Sect. 18.2, we now give details of some of its properties, e.g., the density at which the transition from an exciton gas to an EHP occurs, the renormalization of the band gap in the EHP and its thermodynamic properties. Finally we present results for indirect and direct-gap semiconductors showing some characteristic differences in these properties. Recent reviews and further reading which treat various aspects of the EHP are [20.1–10].

20.1 The Mott Density

Equation (18.20) gave a very crude approximation of the density of electron–hole pairs n_c at which the transition from an excitonic system to an EHP can be expected. In this section we give some more refined considerations.

The transition to the EHP can be tackled in the following way. We consider one exciton in a sea of free carriers (electrons and holes) of density n_P. The free carriers screen the Coulomb potential of the exciton transforming it to a Yukawa-type potential. For the derivation of this and the following formula see [20.3, 6, 7] and references therein.

$$\frac{1}{4\pi\varepsilon_0\varepsilon}\frac{e^2}{|\boldsymbol{r}_e-\boldsymbol{r}_h|}\Rightarrow\frac{1}{4\pi\varepsilon_0\varepsilon(n_P)}\exp\left\{-\frac{|\boldsymbol{r}_e-\boldsymbol{r}_h|}{l}\right\}\cdot\frac{1}{|\mathbf{r}_e-\mathbf{r}_h|} \tag{20.1}$$

where some thought is necessary concerning the value of ε entering in this equation. In the simplest approximation one could use the static dielectric constant (static screening). A better approximation is obtained if both the dependence of ε on frequency and on n_P are taken into account (dynamic screening) [20.2, 4]. If the screening length l falls below a certain value l_c, the Yukawa potential no longer has a bound state, at least in three-dimensional systems. The excitonic Bohr radius and l_c are connected with each other by

$$a_B\, l_c^{-1} = 1.19\ . \tag{20.2}$$

By considering an electron–hole gas described by classical Boltzmann statistics, one can find the so-called Debye–Hückel screening length l_{DH} [20.11] and the density n_M at which the EHP starts to exist is given by

$$l_{DH} = \left(\frac{\varepsilon_0 \varepsilon k_B T}{4\pi e^2 n_P}\right)^{1/2},$$

$$n_M = (1.19)^2 \frac{\varepsilon \varepsilon_0 k_B T}{e^2 a_B^2} = \frac{k_B T}{2 a_B^3 \mathrm{Ry}^*}, \quad (20.3)$$

since we have

$$a_B \mathrm{Ry}^* = \frac{\hbar^2 \varepsilon \varepsilon_0}{\mu e^2} \cdot \frac{e^4 \mu}{2(\varepsilon \varepsilon_0)^2 \hbar^2} = \frac{e^2}{2\varepsilon \varepsilon_0},$$

where μ is the reduced mass of electron and hole.

Though this is a reasonable approach at higher temperatures, it gives the physically unreasonable result that in the limit $T \Rightarrow 0$ l_c is reached already for vanishing values of n_P, which can be orders of magnitude smaller than what is given by (18.20).

For the case of a degenerate electron–hole plasma, which is more likely at low temperatures, (see (10.33)), one gets the Thomas–Fermi screening length l_{TF} [20.12] and correspondingly a different value for n_M

$$l_{TF}^{-2} = \frac{e^2}{\varepsilon \varepsilon_0} \sum_{i=e,h} \frac{\partial n_i}{\partial E_F^i}, \quad (20.4a)$$

where E_F^i stands for the quasi Fermi levels of electrons and holes, respectively. In the effective mass approximation (20.4a) simplifies for isotropic, non-degenerate bands to:

$$l_{TF}^{-1} = \left[\frac{3e^2}{\varepsilon \varepsilon_0 \hbar^2}(m_e + m_h)\left(\frac{1}{3\pi^2}\right)^{2/3} n_P^{1/3}\right]^{1/2}$$

and

$$n_M = (1.19)^6 a_B^{-6} \left(\frac{\varepsilon \varepsilon_0 \hbar^2}{3e^2} \frac{1}{m_e + m_h}\right)^3 (3\pi^2)^2 . \quad (20.4b)$$

The density n_M given by (20.3, 4) is also known as the Mott density, because it describes the transition from an insulating gas of excitons at lower densities to the metallike state of an EHP at higher densities [20.13].

A description of the screening on a more sophisticated theoretical level involves the so-called random-phase approximation (RPA) or the simplified version of the single plasmon pole (SSP) approximation. In the latter case the contribution of the free carriers in the EHP to $\varepsilon(\omega)$ is described by a pole at the plasma frequency ω_{PL} (Chap. 12) or more precisely by the plasmon–phonon mixed state (Sects. 12.1 and 13.6). These topics however are beyond the scope of this book and we refer the reader to [20.1–7] and references therein.

20.2 Band Gap Renormalization and Phase Diagram

In addition to the screening of the Coulomb interaction in the exciton, there are further important renormalization effects of the electronic eigenstates in an EHP. They will be treated in this section and we shall show how they result in the formation of a liquid-like state of the plasma below some critical temperature T_c, at least under thermodynamic quasi-equilibrium conditions.

The width of the forbidden gap is a monotonically decreasing function of the electron–hole pair density n_P in the plasma, due to exchange and correlation effects as shown in Fig. 20.1, where we give various energies as a function of n_P. This statement can be explained qualitatively in the following way. In the plasma we have Coulomb energies which attract carriers of opposite charge and repel those of like charge. If electrons and holes were completely randomly distributed in the sample, then the Coulomb attraction and repulsion energies would cancel exactly and the width of the forbidden gap would be independent of n_P.

In reality, the carriers are not randomly distributed. The Pauliprinciple which is a consequence of the exchange interaction of identical fermions

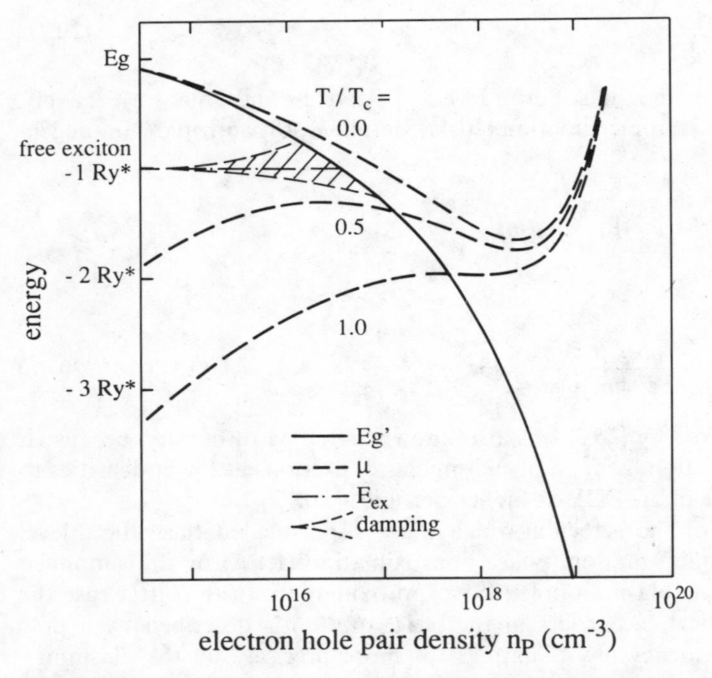

Fig. 20.1. Schematic dependence of the width of the forbidden gap E'_g, of the exciton energies, and of the chemical potential of the electron–hole pair system μ on the electron–hole pair density n_p. Numerical data for CdS. According to [20.6]

forbids two electrons with parallel spin from sitting in the same unit cell. Since this situation would occur for a random distribution, but does not for fermions, we can conclude that the exchange energy increases the average distance between electrons with parallel spin and consequently reduces their total repulsive Coulomb energy. The reduction of a repulsive energy term means a lowering of the total energy of the electron system. The same arguments also hold for the holes.

The correlation energy is spin independent and descibes the fact that the electron–hole pair system can lower its energy further, if the distribution of electrons and holes relative to each other is not random, but if in the vicinity of a hole an electron is found with higher probability than another hole and vice versa. This correlation really occurs. It is a "reminder" of the Coulomb interaction between electron and hole which is responsible for the formation of excitons at low densities.

There are some universal formula that describe the reduction of the band gap ΔE_g normalized by the excitonic Rydberg energy as a function of the normalized plasma density n_p including the exchange (or Hartree-Fock) and the correlation energy [20.6, 14]. Usually one chooses a dimensionless quantity called r_s in which the volume occupied by a carrier pair in the plasma is compared with the volume of an exciton

$$r_s = \left(\frac{4\pi a_B^3}{3} n_P \right)^{-1/3} . \tag{20.5}$$

One should note that r_s decreases with increasing n_P due to the negative exponent. In an EHP in semiconductors values of r_s generally lie between 1 and 4.

In Fig. 20.2 we give approaches to the universal behavior of

$$\Delta E_g / \mathrm{Ry}^* = f(r_s) \tag{20.6}$$

from the two authors of [20.14]. It can be stated that the band-gap renormalization is almost temperature independent, and that the universal curves describe the experimental situation very well in the more covalently bound semiconductors like Si, Ge, GaAs, GaP etc. For clarity these are not shown in Fig. 20.2 since the experimental data for these substances coincide with the theoretical curves. In contrast, further corrections have to be considered for substances with increasing degree of ionic binding like the II–VI materials CdS or ZnSe. We come back to this aspect later.

In addition to E'_g (n_P) Fig. 20.1 shows some exciton energies. It has been found that the excitons roughly maintain their absolute energy with increasing n_p until the binding energy vanishes at the Mott density. Sometimes a small blue-shift is observed with increasing n_P, especially in narrow QW. This finding is due to the compensation of two effects. One is the decrease of E'_g with increasing n_P which should shift the exciton energy to the red. The other is the decrease of the exciton binding energy with increasing n_P due to the screening

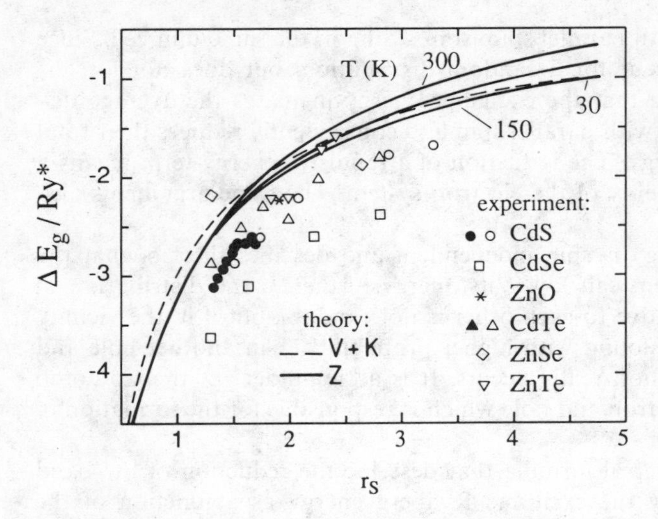

Fig. 20.2. The universal relation between the band-gap renormalization ΔE_g measured in units of the excitonic Rydberg energy Ry* and the normalized carrier density $r_s(n_p)$ together with data for various II–VI compounds. According to [20.14, 16]

of the Coulomb energy mentioned in Sect. 20.1 which shifts the exciton closer to the gap, i.e., towards the blue.

This screening of the binding energy leads to an increase of the excitonic radius and thus to a decrease of the overlap of electron and hole wavefunction and a decrease of oscillator-strength with increasing n_P. Furthermore, the damping of the exciton resonances increases with n_P due to an increasing scattering rate which reduces the phase relaxation time T_2. This effect is indicated schematically by the hatched areas in Fig. 20.1. The Mott density introduced in Sect. 20.1 is, in the representation of Fig. 20.1, the density at which the curves $E_{ex}(n_P)$ and $E_g'(n_P)$ cross. This density decreases with increasing quantum number n_B. Usually one concentrates on the lowest exciton, i.e., on $n_B = 1$.

Until now we have discussed mainly the screening of the Coulomb interaction between electrons and holes as a reason for the disappearance of the exciton as a bound electron-hole pair state. Actually there is another process which contributes, namely phase-space filling [20.15]. As has been explained in connection with (10.17), (11.3) and (11.5), one needs electron and hole wavefunctions from a certain regime in $\boldsymbol{k}$-space to build up the exciton wavefunction. If these states are occupied by electrons and holes, they are blocked and can no longer be used for the construction of the exciton. This phase-space filling is again a consequence of the fermion character of the constituents of the exciton. It is already present in three-dimensional semiconductors but plays a more important role in QW where the screening of the Coulomb interaction is reduced, since it is not so easy to screen the electric field lines between electron and hole which propagate through the barriers.

The next quantity given in Fig. 20.1 is the chemical potential μ of the electron–hole pair system. In a pumped system away from thermodynamic equilibrium the distribution of electrons and holes in their bands can no longer be described by a single Fermi energy (or chemical potential) E_F. Instead individual quasi Fermi energies E_F^e and E_F^h have to be introduced for electrons and holes, respectively. The chemical potential of the electron–hole pair system is just the energetic distance between E_F^e and E_F^h. If we measure them from the extrema of the renormalized bands, we get

$$\mu(n_P, T_P) = E_F^e(n_P, T_P) + E_F^h(n_P, T_P) + E_g'(n_P) , \tag{20.7}$$

where we assume that the carriers have a thermal distribution in their respective bands and that the temperatures of electrons T_e and of holes T_h are equal. We call this temperature then the plasma temperature T_P:

$$T_e = T_h = T_P . \tag{20.8}$$

Depending on the material and the excitation conditions, T_P can be higher than the lattice temperature T_L. This means that a thermal distribution of electrons and holes is established by mutual scattering (this generally happens in an EHP on a ps time scale) but that this distribution is not in equilibrium with the phonon system described by T_L.

Without pumping, μ is zero. It increases with increasing density. At some density, μ will exceed $E_g'(n_P)$. In this case the quasi Fermi level of the electrons and possibly also of the holes is situated in the band. This is the onset of population inversion between conduction and valence band. We speak in this situation of a degenerate EHP and the use of Fermi statistics is mandatory. When neither E_F^e nor E_F^h are situated in the respective bands, Boltzman statistics are a good approximation.

In the degenerate case, the mean kinetic energy of the particles increases rapidly with density. At zero temperature a simple relation holds in the effective mass approximation

$$\bar{E}_{kin}^{e,h} = \frac{3}{5} E_F^{e,h} . \tag{20.9}$$

The steep increase of μ with increasing n_P is due to this effect and at high densities (20.9) overcomes the decrease of $E_g'(n_P)$.
The free energy $F(n, T_P)$ of the EHP is connected with $\mu(n, T)$ by

$$\mu(n, T_P) = \left(\frac{\partial F}{\partial n}\right)_{T_{0,v}} . \tag{20.10}$$

On the other hand, it is given by

$$F(n_P, T) = U - TS \tag{20.11}$$

with S entropy and U internal energy of the EHP.

Below a certain critical temperature T_c the increase of μ is not monotonic but may go through a maximum and a minimum as shown in Fig. 20.1. In this situation quasi-equilibrium thermodynamics predict a first-order phase transition to an electron–hole liquid (EHL) [20.1–7]. Thus, similar to the case of a real or van der Waals gas, below T_c we expect a phase separation into a liquid-like EHL surrounded by a gas phase of excitons and free carriers. Liquid-like means essentially that the density of the plasma is constant for constant T_P. An increase of the pump-power, i.e., of the average density then merely increases the ratio of the volumes filled by the EHL and the gas. The phase diagram of this transition follows from a Maxwell-type construction and is shown in Fig. 20.3 for the simplest case, where we have a coexistence region of gas and liquid below T_c while a distinction between gas and liquid is not physically meaningful above T_c. Modifications of this simple phase diagram may be introduced if one considers excitons and biexcitons, possible regions where these quasi-particles might undergo a Bose condensation, the influence of the finite lifetime, etc. These complications are beyond the scope of this book; for references see [20.3, 6, 17, 18].

The EHL has indeed been observed in some semiconductors in the form of small liquid-like EHP droplets (EHD) when the lifetime of the carriers is long enough that the spatial separation into a liquid and a gas phase can develop. This is often the case for indirect gap materials like Si, Ge, GaP or $Al_{1-y}Ga_yAs$ ($y > 0.45$) [20.1, 3, 5, 10].

In the direct gap materials with dipole allowed band-to-band transition the lifetime of the carriers is generally so short ($\tau \approx 0.1$ ns) that no liquid phase can develop, though an EHP is created under sufficiently strong pumping. We

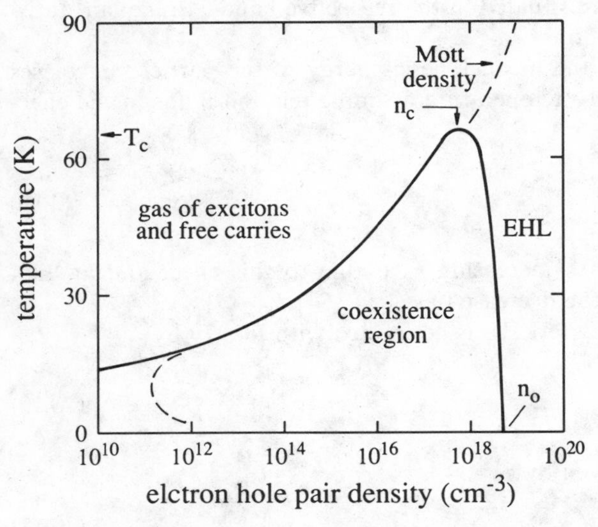

Fig. 20.3. The phase diagram for an EHP under thermodynamic quasi-equilibrium conditions. From [20.6]

shall see examples of both cases (the formation of a liquid-like phase or not) in the next sections.

Before that we comment briefly on the description of the EHP used so far. We considered an exciton in a sea of electrons and holes. This description is adequate at higher temperatures, where a significant fraction of the excitons are thermally ionized, or at densities around n_M, or under band-to-band excitation where primarily free carriers are created which form excitons only after relaxation from the continuum to the bound states. If we excite excitons resonantly, we should consider the screening and phase-space filling of excitons by excitons and not by free carriers. Furthermore, the curve of $\mu(n_p)$ should start for $T_p = 0$K and $n_p \to 0$ not from E_g but from E_{ex}. The corresponding calculations and experiments have been made and indicate that the above–outlined scenario of the EHP formation remains qualitatively valid except for a possible shift of n_M to higher values [20.19, 20].

20.3 Electron–Hole Plasmas in Indirect-Gap Semiconductors

The first observation of an EHP was in Si. The "history" of its investigation is outlined briefly in reviews [20.1, 5]. In Fig. 20.4 we show a luminescence spectrum of Ge under high excitation. Due to the indirect bandstructure, the emission bands shown involve a momentum-conserving LA phonon. We see the recombination of free excitons and a broad band which is the emission of the EHL. Due to the long carrier lifetime in pure samples, which is in the μs to ms regime, the phase separation can develop and the carriers can cool down to equilibrium with the lattice so that very low values of T_p close to T_L can be reached. As expected for the liquid phase the shape of this emission does not change with increasing pump intensity below T_c but the ratio of free exciton to

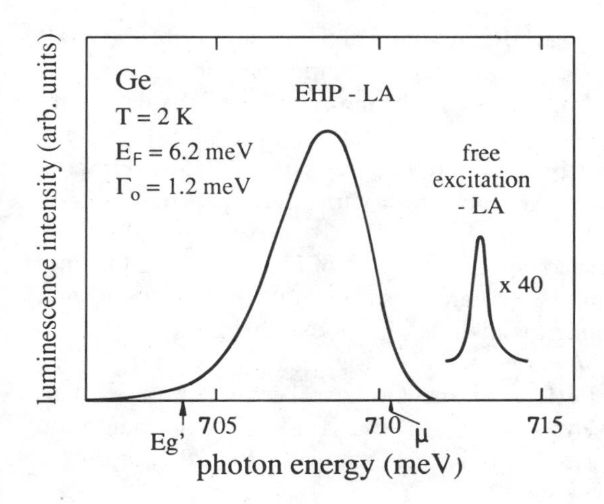

Fig. 20.4. The luminescence of Ge in the presence of an EHP (LA phonon replica). According to [20.21]

EHP emission changes according to the changing volume fractions that the two phases occupy. The density of electron–hole pairs can be estimated from the generation rate and the lifetime. The onset of plasma luminescence for a fixed T_L then gives the low density side of the coexistence region in the phase diagram. The high density side can be deduced more easily from an analysis of the lineshape of the plasma luminescence.

Since a momentum-conserving phonon is involved, the emission at a certain energy $\hbar\omega$ is given simply by an integral over all transitions separated by $\hbar\omega$, independent of the wave vectors of electron and hole, weighted by the density of states and the occupation probabilities of electrons and of holes in their respective bands.

For spontaneous emission we get in the simplest approximation

$$I_{\text{lum}}(\hbar\omega) = c_1 \int_0^\infty \int_0^\infty (E_{\text{kin}}^{\text{e}})^{1/2} (E_{\text{kin}}^{\text{h}})^{1/2} f_{\text{e}} f_{\text{h}} \delta[\hbar\omega - (E_{\text{g}}' + E_{\text{kin}}^{\text{e}} + E_{\text{kin}}^{\text{h}} - \hbar\Omega_{\text{LA}})] \, dE_{\text{kin}}^{\text{e}} \, dE_{\text{kin}}^{\text{h}}, \tag{20.12}$$

where we have assumed parabolic bands, i.e., $E_{\text{kin}}^{\text{e,h}} = \hbar^2 k_{\text{e,h}}^2 / 2m_{\text{e,h}}$ resulting in the square-root dependence of the density of states and the energy conservation contained in the δ-function. The quasi Fermi functions are given by

$$f_{\text{e,h}} = \left[1 + \exp\left(\pm \frac{E_{\text{kin}}^{\text{e,h}} - E_{\text{F}}^{\text{e,h}}}{k_{\text{B}} T} \right) \right]^{-1}. \tag{20.13}$$

Furthermore the transition probability is assumed in (20.12) to be energy independent and to be contained in c_1 together with other constant terms.

The term for stimulated emission or reabsorption looks similar, except that the product $f_{\text{e}} f_{\text{h}}$ in (20.12) is replaced by $(1 - f_{\text{e}} - f_{\text{h}})$. Stimulated emission however is of no importance for most indirect gap materials since the transition probability is very small. This prevents the standard semiconductor Si from being used as a material for laser diodes.

Recently it has been shown, however, that stimulated emission from the EHP can be obtained in indirect $Al_{1-x}Ga_xAs$ and similar compounds for compositions x close to the direct–indirect transition due to a strong coupling of the direct and the indirect conduction band minima due to alloy scattering [20.10, 22].

Some modifications can be introduced in (20.12), e.g., by final state damping Γ_0. We come back to these effects in connection with direct gap materials in Sect. 20.4.

A detailed analysis of luminescence spectra like that of Fig. 20.4 along the lines outlined above gives the values of T_{p}, $E_{\text{F}}^{\text{e,h}}$ and consequently the densities n_{p} and $E_{\text{g}}'(n_{\text{p}})$. This allows one to determine the high density side of the phase diagram of the EHL. Examples of such results are shown in Fig. 20.5 for Ge and Si. The critical temperatures T_{c} for the EHL in Si and Ge have been determined to be around 23 and 6.5 K, respectively. In addition, the band-gap renormalization $E_{\text{g}}'(n_{\text{p}})$ is found. It agrees nicely with the universal formula of Fig. 20.2.

The multi-valley structure of the conduction band reduces the kinetic energy and helps to stabilize the plasma. By applying homogeneous uniaxial stress to the samples it is possible to lift the degeneracy of the multi-valley conduction band structure. The results elucidated the influence of the band peculiarity on the kinetic energy and on $\mu(n_p, T_p)$ [20.23].

The size of, and the number of carriers in the EHD has been determined by various techniques including disintegrating them in the internal field of a pn-junction and measuring the current pulses connected with this event. Usually one finds about 10^3 carrier pairs per droplet [20.1, 5].

The EHD can be pinned by impurities such as those which produce the multi-exciton complexes and which may act thus as nucleation centers [20.25]. In pure materials the EHP may move through the sample in a diffusive way or be driven by a heat flow (phonon wind) if there is a spatial gradient in the phonon populations [20. 26]. In Fig. 20.6a and b we show an example. The phonons are created in the excitation spot by the nonradiative recombination of carriers. Since the phonons propagate (ballistically) into some preferential directions determined by their dispersion, they push the droplets also into these directions. The expansion of the droplet cloud is therefore inhomogeneous as seen in Fig. 20.6a where the luminescence from the EHD is recorded directly and even better in b where contours of equal luminescence brightness are given.

We want to conclude the discussion of indirect bulk semiconductors with a beautiful experiment concerning the so-called γ-drops in Fig. 20.6 c, d. The

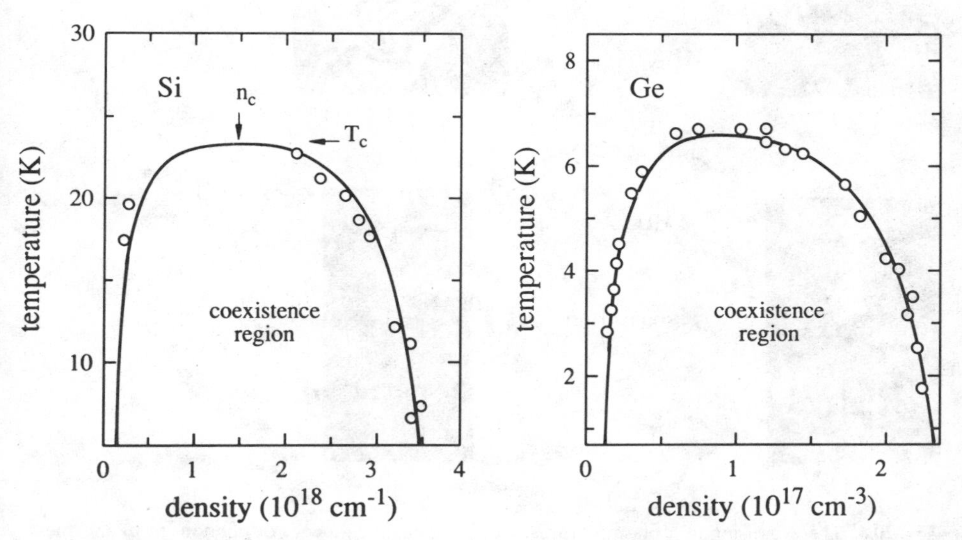

Fig. 20.5. The phase diagrams of the electron–hole liquid in Si and Ge. According to [20.23] and [20.24], respectively. Note the linear and logarithmic scales on the x-axes of Figs. 20.3 and 20.5

basic idea is the following. The width of the forbidden gap depends on strain as already mentioned above. If inhomogeneous stress is applied to a sample in an appropriate orientation, a situation can be created in which a minimum of E_g appears in the volume of a sample and not at its surface. If the sample is strongly pumped at the surface (e.g., by a blue laser) then the excitons and the droplets move under the influence of the gradient of E_g and accumulate in the potential minimum, forming there a macroscopic drop called a γ-drop. The diameter of such drops can be as large as one millimeter. In Fig. 20.6c, d this situation is seen recorded again by imaging the luminescence of the electron–hole pair recombination.

Very recently, the phase diagram of an EHL has been determined in a semiconductor structure of reduced dimensionality, namely in a quantum-well-wire superlattice, which is formed from wire-like structures of GaAs and AlAs. This system is of type-II and therefore "indirect" both in real and in momentum space. The resulting long carrier lifetime favors the occurrence of phase separation. For details of this exciting new topic see [20.10].

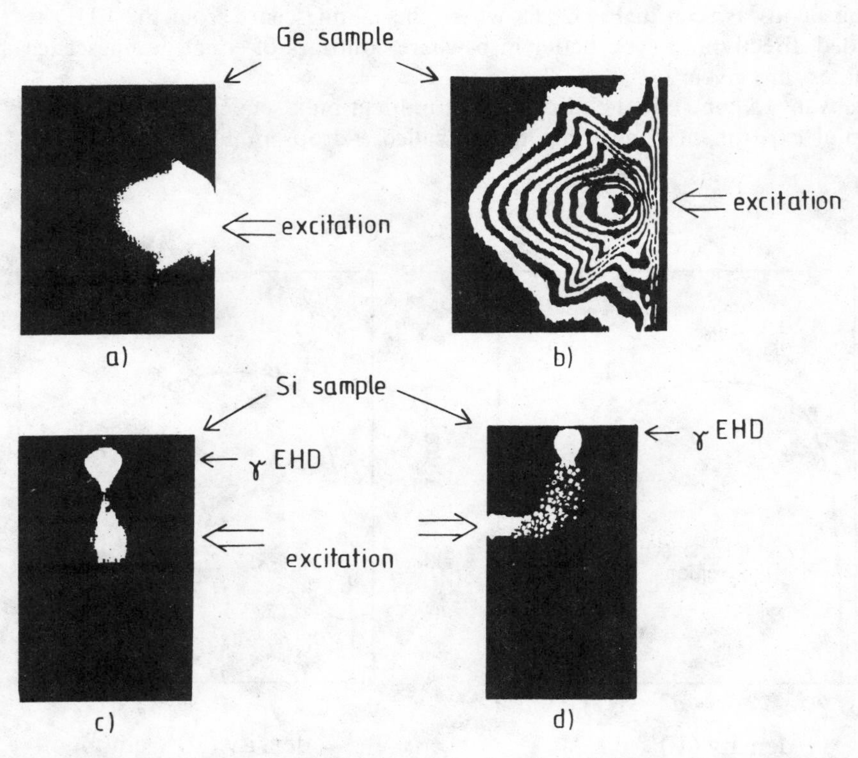

Fig. 20.6. The expansion of a plasma droplet cloud under the influence of phonons (a, b) and the flow of excitons and plasma droplets into a γ-drop. Front and side view of the sample c, d. According to [20.26, 27]

20.4 Electron – Hole Plasmas in Direct-Gap Semiconductors

Many properties of the EHP in direct-gap materials are comparable to those in indirect-gap semiconductors, for example, the renormalization of the gap and the transition from a degenerate to a non-degenerate EHP. Calculations in the limit of quasi-equilibrium thermodynamics predict for direct gap materials, too, a first-order phase separation into an EHL. For CdS these calculations yield a value of T_c of 64 K [20.6, 2]

However, there are also characteristic differences between the EHP in direct and indirect gap materials. One is the appearance of strong stimulated emission in a degenerate EHP in semiconductors with a direct, dipole-allowed band-to-band transition. Gain values of up to $10^4\,\mathrm{cm}^{-1}$, i.e., $1\,\mu\mathrm{m}^{-1}$ cm can be expected. Another difference is the short carrier lifetime, which is in the (sub)-ns regime for a nondegenerate EHP and may be as short as 100 ps in the degenerate case.

We now discuss the consequences of these two differences and other aspects of the EHP in direct gap semiconductors. The EHP has been observed in many direct gap III–V, II–VI and I–VII compounds [20.2, 9]. We concentrate here on CdS and GaAs for bulk materials and on AlGaAs and InGaAs MQW structures as representatives of quasi-two-dimensional systems.

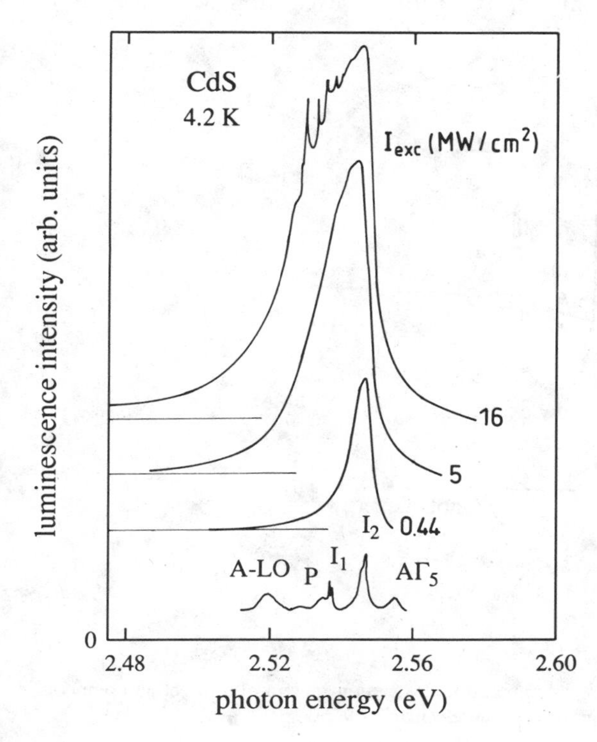

Fig. 20.7. The plasma emission of CdS for increasing excitation using an extremely small excited volume. According to [20.28]

The gain in the degenerate EHP leads to stimulated emission which distorts the luminescence spectra so that their evaluation becomes very difficult. Only if extremely small volumes are excited with a few μm length in all three dimensions is it possible to observe sometimes the spontaneous emission of the EHP. In Fig. 20.7 we give an example where an excitation spot of a few μm^2 has been used and even there some laser modes start to show up. We stress that the spectral width of the emission increases with increasing pump intensity, in contrast to the situation in Si or Ge below T_c.

In order to overcome the problem of stimulated emission, the gain spectra have been investigated directly. The experimental technique is pump-and-probe beam spectroscopy. One measures the transmission or reflection spectra with a weak, spectrally broad probe beam, once without and once with a spectrally narrow, intense pump beam on. The difference between the two spectra gives information about optical nonlinearities and renormalization effects. With pulses of several ns duration one observes a quasi-stationary situation, while measurements with ps excitation allow the investigation of the decay dynamics of the EHP. A tuning of the pump energy $\hbar\omega_{exc}$ can be used for excitation spectroscopy of the plasma gain, and even spatially resolved pump and probe beam experiments are possible [20.29].

We discuss now in connection with Fig. 20.8 the changes of the absorption spectra which we expect in a direct gap semiconductor when we go from a low-density exciton gas to an EHP. Without excitation we observe in

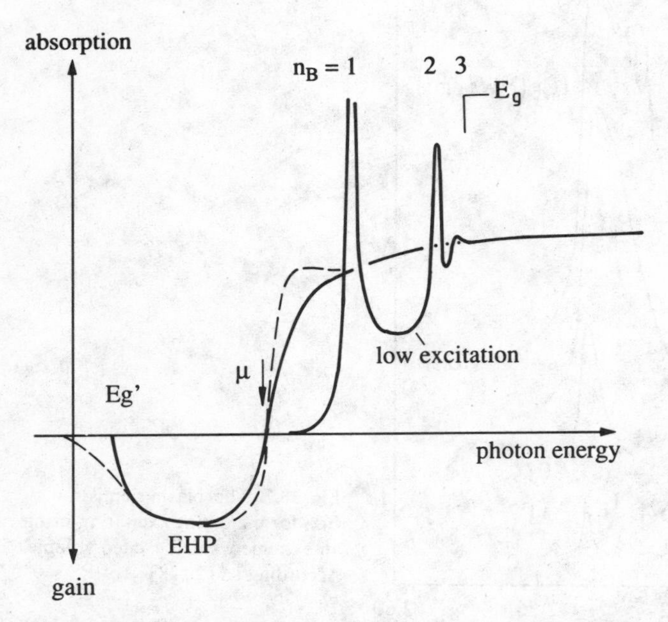

Fig. 20.8. Schematic drawing of the low temperature absorption spectra of a direct-gap semiconductor at low excitation and in the presence of an electron–hole plasma

absorption (or reflection) the series of exciton resonances and the continuum transitions above E_g. In the presence of a plasma the exciton resonances are gone, the forbidden gap $E_g'(n_p)$ is renormalized and the states are filled in the degenerate casc up to the chemical potential $\mu(n_p, T_p)$. As will be explained in more detail below, we expect optical gain between E_g' and μ and absorption due to band-to-band transitions above. For a detailed analysis of the gain spectra we consider the recombination rate $r(\hbar\omega)$ at a certain photon energy $\hbar\omega$. Since in direct gap semiconductors no momentum-conserving phonon is involved, we have to consider here recombination under energy and momentum conservation in contrast to the indirect semiconductors. For the recombination rate, which contributes to the emission at a certain photon energy $\hbar\omega$ we get

$$r(\hbar\omega) \propto f_e f_h (1+N) - (f_h - 1)(f_e - 1)N \,. \tag{20.14}$$

The first term describes spontaneous and stimulated emission, where N is the density of photons in the (laser) mode with frequency ω under consideration. The second term describes the reabsorption.

Equation (20.14) can be rewritten as a term independent of N giving the spontaneous emission and a term linear in N giving the net rate of stimulated emission and of reabsorption

$$r \propto f_e f_h + N(1 - f_e - f_h) \tag{20.15a}$$

$$\propto r_{\text{spont}} + r_{\text{stim}} \,. \tag{20.15b}$$

Net gain evidently results for $f_e + f_h > 1$ and this simply means

$$\mu(n_p, T_p) > E_g'(n_p) \tag{20.16}$$

as mentioned above.

Using now the square-root dependence of the combined density of states (Sects. 10.7 and 14.8) and also momentum conservation, we get for the spectra of spontaneous emission $I_{\text{lum}}(\hbar\omega)$ and of gain $g(\hbar\omega)$

$$I_{\text{lum}}(\hbar\omega) = c_2 [\hbar\omega - E_g'(n_p)]^{1/2} f_e f_h \,, \tag{20.17a}$$

$$g(\hbar\omega) = c_2 [\hbar\omega - E_g'(n_p)]^{1/2} (f_e + f_h - 1) \,, \tag{20.17b}$$

where $\hbar\omega$ is given by

$$\hbar\omega = E_g'(n_p) + \frac{\hbar^2 \boldsymbol{k}_e^2}{2m_e} + \frac{\hbar^2 \boldsymbol{k}_h^2}{2m_h} \tag{20.18}$$

with $\boldsymbol{k}_e = -\boldsymbol{k}_h$, and c_2 is a proportionality constant including once more the transition matrix element, which is assumed to be $\boldsymbol{k}$-independent. The expression (20.17) leads to a square-root increase of the gain spectra at the low energy side and a transition to absorption at $\hbar\omega = \mu$; see Fig. 20.8.

Two modifications of the simple shape of the spectra given in (20.17) are necessary. One is the so-called Landsberg or final-state damping explained on the right of Fig. 20.9. If an electron recombines with a hole there is an empty state in the Fermi sea of electrons in the conduction band and an occupied state in the sea of holes in the valence band. Both states relax very rapidly as indicated by the dashed arrows. This short lifetime enters as an energy-dependent broadening which is also known as Landsberg damping. Consequently the gain spectrum has to be convoluted with a Lorentzian. The parameter Γ in this Lorentzian depends on energy [20.2]. It has a minimum at the quasi Fermi energies and increases below and above roughly linearly with energy; see Fig. 20.10a. This final-state damping is the reason why the gain and luminescence spectra do not actually start at E'_g with a square-root dependence but with a smooth tail extending below E'_g as indicated in Fig. 20.8. Further contributions to the broadening include the scattering with other quasiparticles such as carriers or plamsons. This phenomenon is also known as "Fermi-sea shake up". It has been stressed recently in [20.30] and references therein.

The other correction is the so-called excitonic enhancement [20.2, 6] which, for quasi two-dimensional systems, is also known as Fermi-edge singularity, see [20.31] and references therein. It describes an enhancement of the oscillator strength around μ as a multiplicative term $\rho(\hbar\omega)$ to the gain spectrum

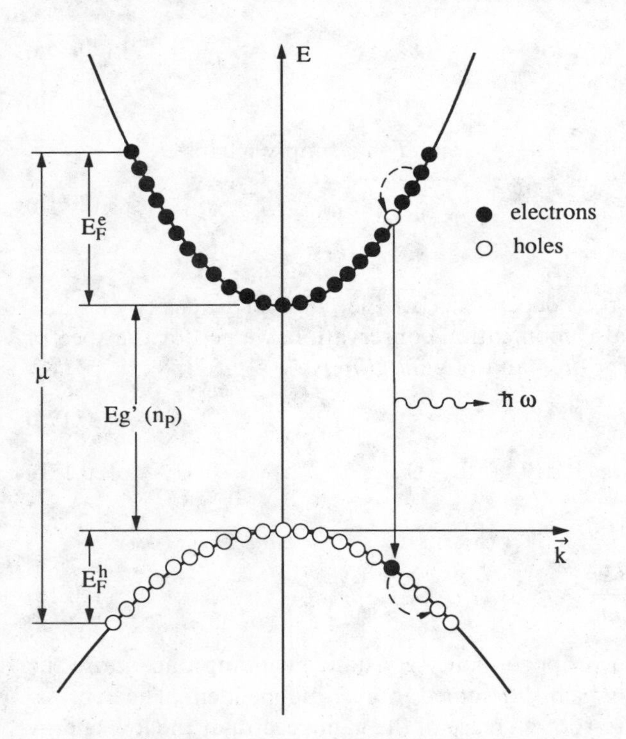

Fig. 20.9. Schematic explanation of the recombination and of the final state damping in a plasma

Fig. 20.10b. The physical origin is the following. Around the quasi Fermi levels there are close-lying occupied and empty states. Electrons and holes in the EHP can therefore perform around μ under their mutual Coulomb attraction at least pieces of a Bohr-orbit-like motion. This correlation effect is reminiscent of the exciton and enchances the oscillator strength. In Fig. 20.10 we give as examples the k-dependence of the damping for the conduction and valence bands in GaAs and the excitonic enhancement $\rho(\hbar\omega)$.

After these general considerations we want to see some examples. In Fig. 20.11 we show transmission spectra of a CdS single crystal platelet with a thickness of a few μm without and with additional excitation in the band-to-band transition region.

The plane parallel surfaces of the sample give rise to Fabry Perot modes in the transparent spectral regions. At low temperatures we see without excitation the absorption dip of the $A\Gamma_5$ exciton (cf. Fig. 14.9), the transparent window above, and the onset of the $B\Gamma_5$ exciton resonance. For higher lattice temperatures the absorption becomes less steep and shifts to the red according to the Urbach rule of (14.12) or Fig. 14.10. With pump, we see that the exciton resonance disappears at low temperatures. This fact is also confirmed in the reflection spectra not shown here [20.29]. In addition we clearly see optical amplification. In Fig.. 20.11a the chemical potential μ coinciding with the crossover from gain to absorption is situated around 2.54 eV and the reduced gap is at 2.49 eV compared to 2.58 eV in the unexcited sample. For higher temperatures, population inversion is no longer reached (see the curve for $T_L = 300$ K) because of the temperature dependence of the effective density of states (10.33) which gives the onset of degeneracy. Nevertheless it is still possible to bleach the tail of the absorption.

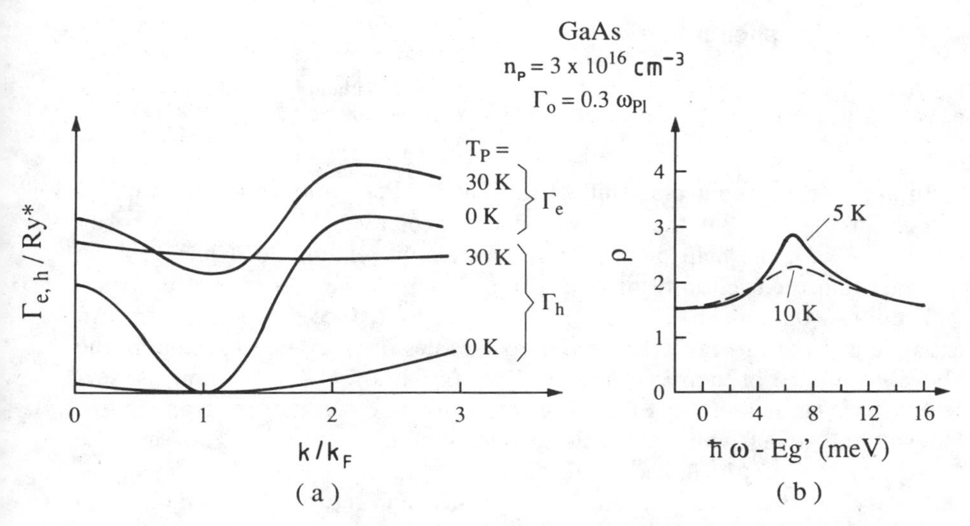

Fig. 20.10a, b. The damping in the conduction and valence bands of GaAs in the presence of a plasma (**a**) and the excitonic enhancement (**b**). From [20.2]

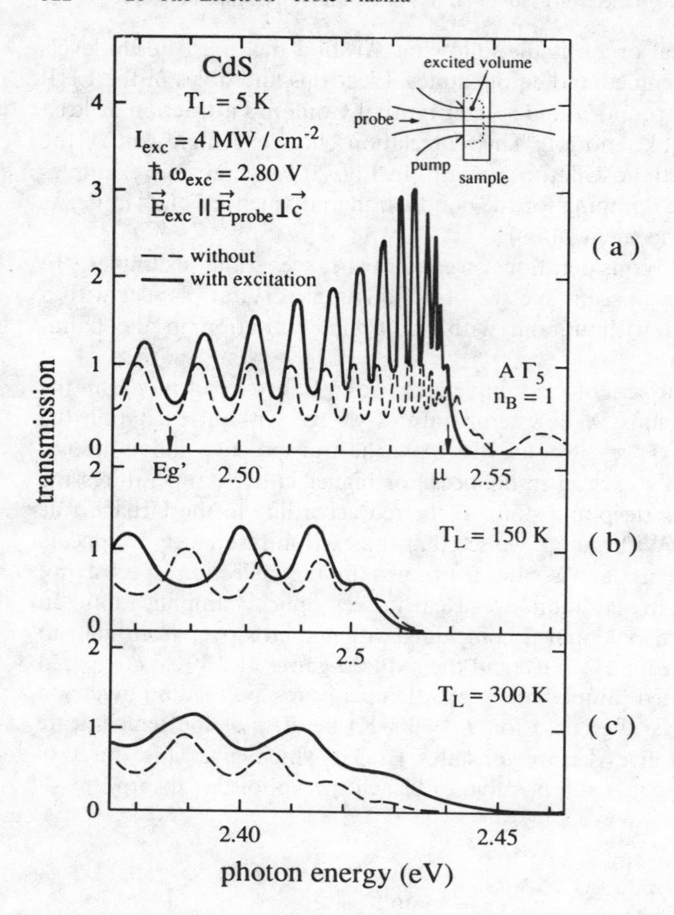

Fig. 20.11. Transmission spectra of a CdS platelet-type sample without (---) and with (–) additional excitation and for various lattice temperatures. According to [20.32]

In addition we see a blue-shift of the Fabry – Perot modes at all temperatures. This means that the refractive index decreases. This finding is easily intelligible with the help of the Kramers – Kronig relations of Chap. 8, if we consider that the exciton resonance disappears from the absorption spectrum.

From raw data as in Fig. 20.11 it is possible to deduce the gain spectra. An example is given in Fig. 20.12 for different values of I_{exc}. The fit including the above mentioned phenomena gives n_p, T_p, $E'_g(n_p)$ and $\mu(n_p T_p)$. We now list several typical features of the EHP observed in many direct gap semiconductors, but without giving figures or detailed references in every case. They may be found in [20.2–4, 6, 7, 9]. The width of the gain spectra increases under quasi-stationary nanosecond excitation with the generation rate as long as there is no pinning of the quasi Fermi levels by stimulated emission, and decreases in time after picosecond excitation. This means that the density is not constant,

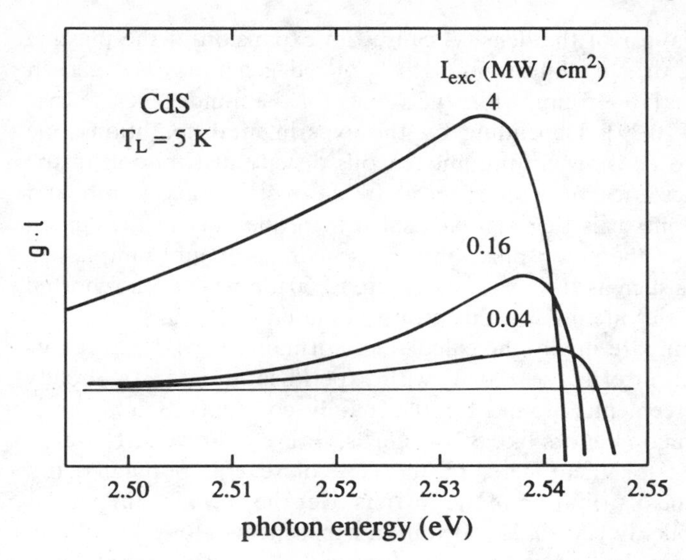

Fig. 20.12. Gain spectra of CdS at low temperatures for various pump intensities. According to [20.29]

although T_p is clearly below T_c. In Fig. 20.12 the values for the highest excitation are $n_p = 3.10^{18}\,\mathrm{cm}^2$; $T_p = 20\,\mathrm{K}$ ($T_L = 7\,\mathrm{K}$, $T_c = 64\,\mathrm{K}$). Evidently no liquid phase is formed. The reason is the short lifetime of the carriers of about 100 to 200 ps in a degenerate EHP [20.29]. This time is not long enough for the formation of a spatial phase separation into liquid like droplets and a gas phase. There will be strong spatio-temporal density fluctuations and in some regions (e.g., in the vicinity of fast recombination centers in the volume or at the surface) the density will stay lower than in others, possibly even below the Mott density. No clear evidence of a phase separation has been observed in a binary direct gap semiconductor up to now. A pinning of n_p with increasing excitation and spatial density inhomogeneities are often results of stimulated emission and not of a phase separation.

The plasma temperature T_p usually lies above the lattice temperature T_L. The difference is only a few ten K in the more ionic II–VI compounds with good coupling of the carriers to the lattice. However, after pulsed excitation the difference can reach several tens of K up to around a hundred K in a first transient cooling period. In the more covalent III–V compounds the differences between T_p and T_L are generally higher. This temperature difference also increases with the excess energy of the excitation, i.e., with $\hbar\omega_{exc} - E_{ex}$. A rather cool electron–hole plasma can consequently be created under resonant excitation in the exciton region or even slightly below.

Since the absorption coefficient is rather high in direct gap semiconductors (10^4–$10^6\,\mathrm{cm}^{-1}$) the penetration depth of the pump light is $\lesssim 1\,\mu\mathrm{m}$. Consequently a strongly inhomogeneous spatial density distribution is created. The gradient of the chemical potential (or of the density) causes an expansion of the plasma. Typical values of the drift or diffusion length l_D of a degenerate EHP, e.g., in CdS or CdSe, are $l_D \approx (10 \pm 5)\,\mu m$, while the values for the nondegenerate case are closer to $1\,\mu m$ [20.29]. Depending on the experimental conditions one therefore often has to consider an inhomogeneous density distribution in the evaluation of luminescence or gain spectra. In quasi-stationary pump and probe beam experiments it is therefore advisable to probe only in the spatial and temporal center of the pump pulse and to use thin ($\lesssim 10\,\mu m$) samples. In narrow gap materials such as InSb values of l_D up to $60\,\mu m$ have been reported due to the longer carrier lifetimes in this group of materials [20.33].

In Fig. 20.2 we compare finally the calculated, normalized band-gap renormalization as a function of r_s [see (20.5)] with experimental data. As already mentioned, good agreement is found for the mainly covalenty bound direct and indirect gap semiconductors like Si, Ge, GaAs, GaP etc. For clarity we do not give data points for them here. In the case of several populated (in) equivalent minima, the distribution of the carriers over the various minima has to be considered explicitly [20.10, 22]. Beautiful experiments along these lines exist for $Al_{1-y}Ga_yAs$ in the transition region from a direct to an indirect band structure around $y = 0.43$ [20.22].

The data for various II–VI semiconductors are shown explictly in Fig. 20.2. They follow the trend of the theory, but with an average shift of about 0.5 Ry* to lower energies. This discrepancy has two main causes. The ionic part of the binding in the II–VI stabilizes the plasma, e.g., via plasma–phonon mixed states. This contribution is not contained in the general formula. Additionally the values of Ry* and of the experimentally determined exciton binding energies do not exactly coincide, as was mentioned in Sects. 11.2 and 19.3 in connection with the biexciton. Similar arguments also hold for a_B which appears in r_s.

In mixed crystals such as $CdS_{1-x}Se_x$ and $Al_{1-y}Ga_yAs$ it has been found that the many-particle effects are strongly reduced as long as the electron–hole pairs occupy only the localized states. One finds essentially a bleaching of the absorption tail of the localized states with increasing density, but not much of a band-gap renormalization [20.34]. If the extended states are populated, however, the properties of the band renormalization and band filling are very similar to those of the binary compounds CdS, CdSe and GaAs [20.10, 34–36]. The population of the extended states occurs e.g. under sufficiently strong pumping so that the total density of electron–hole pairs exceeds the density of localized states. N_{loc}. This situation can be achieved quite easily in $Al_{1-y}Ga_yAs$ but not in $CdS_{1-x}Se_x$ where N_{loc} is between 10^{18} and $10^{19}\,cm^{-3}$. In the latter case, however, the extended states are also populated to a considerable degree for $T_p \gtrsim T_L > \varepsilon_{loc}/k_B$ where ε_{loc} is the tailing parameter introduced in (15.7). In Fig. 20.13 we show the shift of the absorption edge under high excitation for CdS (see also Fig. 20.11), CdSe, and $CdS_{1-x}Se_x$. In $CdS_{1-x}Se_x$ one finds at low temperatures a blue shift, due to state filling without renormalization, but a higher temperatures a behavior similar to that for the EHP in the binary compounds.

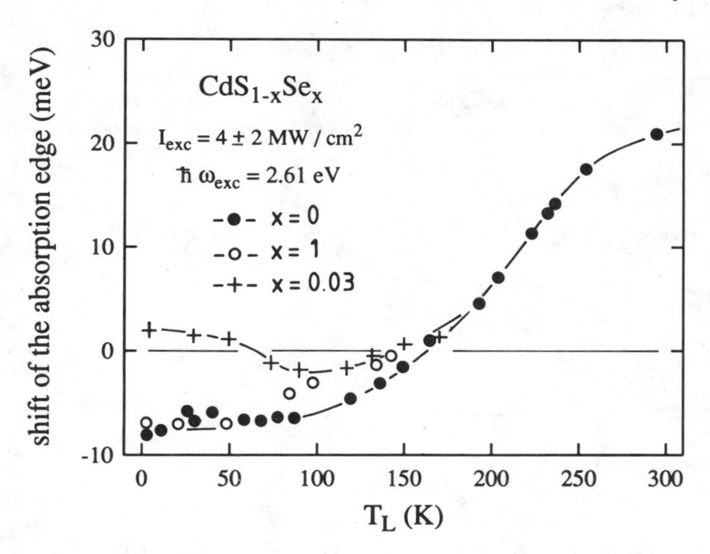

Fig. 20.13. The shift of the absorption edge in CdS, CdSe, and $CdS_{1-x}Se_x$ under high excitation as a function of temperature. According to [20.35]

In the remainder of the section, we look at the electron–hole plasma in systems of reduced dimensionality. For recent reviews of this topic see [20.7–10] and references therein.

In Fig. 20.14a we show pump and probe spectra of an $Al_{1-y}Ga_yAs/GaAs$ MQW sample, the GaAs substrate of which has been removed by a selective etching technique. Without excitation we see the now familiar features of the $n_z = 1$ and 2 hh and lh exciton resonances. With increasing pump power, again using quasi-stationary ns excitation, a scenario develops similar to that for the three-dimensional system: The exciton resonances disappear, and there is population inversion between the reduced gap and the chemical potential. The weak modulation in the resulting gain spectra comes from residual Fabry-Perot modes of the structure. The slope of the transition from gain to absorption gives the plasma temperature if the excitonic enhancement is also properly taken into account. The approximately constant gain value reflects the constant density of states for effective mass particles in a two-dimensional system [see (2.77) and (10.29)]. In agreement with the statements above, T_p is higher than in the more ionic II–VI compounds. In Fig. 20.14b we show the density dependence of E_g' and of μ and a calculation of E_g' for a strictly two-dimensional system using some effective parameters of the excitonic Rydberg energy to account for the finite well thickness l_z.

Looking closer, we find some differences that are characteristic of the two-dimensional system. The $n_z = 2$ hh resonance is still seen quite clearly at densities where the $n_z = 1$ hh feature has already vanished. The screening of the Coulomb

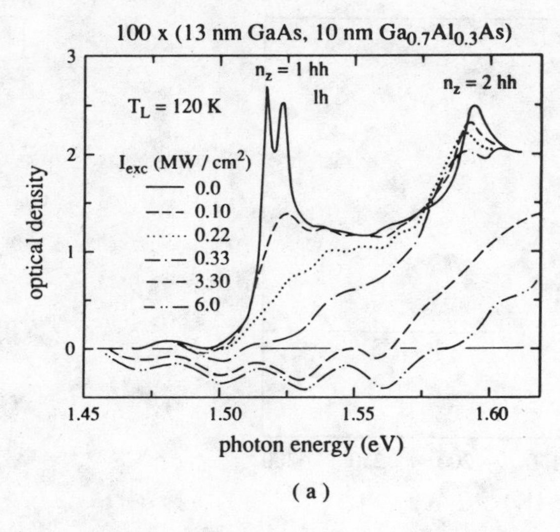

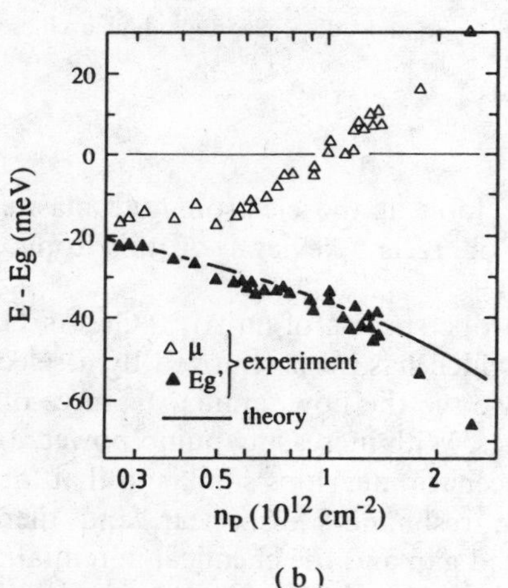

Fig. 20.14a, b. Transmission spectra of an $Al_{1-y}Ga_yAs$/GaAs MQW sample for various pump intensities (**a**) and the resulting dependence of the reduced gap E_g' (**b**) and of the chemical potential μ on the electron–hole pair density per unit area n_p. According to [20.37]

interaction by the free carriers would act on occupied and empty states in the same way. Consequently we may conclude that phase-space filling, which depends on the occupation of the states, is more important in quasi-2d systems than in 3d ones. This interpretation is confirmed by the fact that the $n_z = 2$ hh resonance disappears only when the chemical potential comes close to it, i.e., when the occupation of the higher subbands starts.

Qualitatively this finding can be understood as follows: The Coulomb interaction between electron and hole can be screened in a 3d system in all directions of space, whereas the screening in a QW affects preferenentially the field lines in the well. It is more difficult to screen the field lines which go through

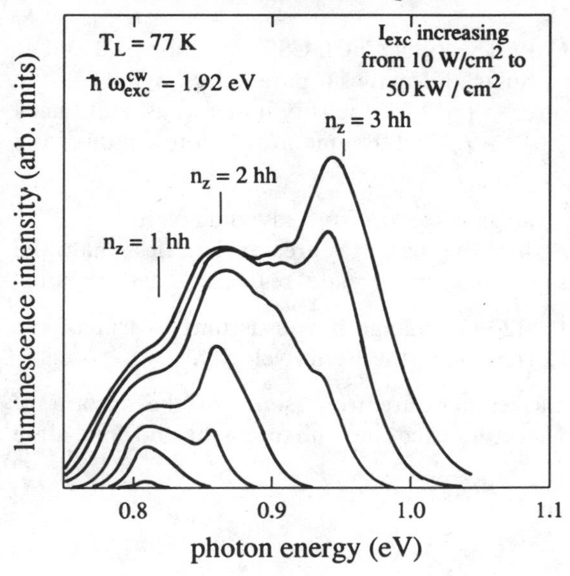

Fig. 20.15. Luminescence spectra of small mesas of InGaAs MQW samples. According to [20.38]

the barrier since here the probability of finding carriers is reduced to the exponential tails shown in Fig. 10.14b.

The other point concerns the fact that the renormalization of the higher subbands ($n_z \geqslant 2$) in a QW system is smaller than that of the fundamental gap, as long as the higher subbands are not populated. One reason for this finding [20.15, 37, 38] is that electrons (and holes) in different subbands are no longer subject to the exchange interaction, which forms an important contribution to the renormalization of the gap, as outlined in Sect. 20.2.

In Fig. 20.15 we give as a further example for the above phenomena luminescence spectra of an $In_{1-y}Ga_yAs$ MQW sample. In this case the plasma has been confined laterally by the formation of mesa-structures of about 10 μm × 10μm, resulting in a very homogeneous excitation. The band renormalization and the successive filling of states and of higher subbands with increasing pump power are clearly visible even under cw excitation.

Recent results on the EHP in quasi one-dimensional quantum wires and in quantum well wire superlattices are compiled in [20.8–10].

In this section we have mentioned several times the appearance of optical gain or of stimulated emission in an EHP. Actually this phenomenon does not only appear in an EHP but also in connection with other high-excitation processes. We give more information about these cases in the next chapter.

20.5 Problems

1. Compare the values of the Mott density of CuCl, CdS, and InSb at 0 and at 300 K using the appropriate formulas and material parameters. What are the excitation intensities needed to reach the Mott density under quasi-stationary conditions if you excite with $\hbar\omega_{exc} \gtrsim E_g$ and assume a diffusion length of the carrier of 1, 10 or 50 μm?

2. Develop formulas to deduce the gain spectra from raw data as in Fig. 20.11 assuming that the diffusion length of the plasma is greater than (less than) the thickness of the sample. If you get stuck, try to make reasonable assumptions.

3. How would the gain spectra (20.17) change if you assume a drift of the carriers with a velocity $v = \frac{1}{2}v_F$? Here, v_F is the Fermi velocity.

4. Deduce the effective plasma temperature from some of the spectra of Figs. 20.12, 20.14, and 20.15, neglecting excitonic enhancement and final-state damping.

21 Stimulated Emission

Stimulated emission is usually identified in experiments by the occurrence of one or several of the following criteria: a strongly superlinear increase of the optical luminescence output I_{lum} as a function of the pump power I_{exc} above a certain threshold $I_{\mathrm{exc}}^{\mathrm{th}}$ with slopes α in $I_{\mathrm{lum}} \propto I_{\mathrm{exc}}^{\alpha}$ of three and more; a simultaneous spectral narrowing of the emission, often accompanied by the appearance of laser modes imposed by some cavity length; and /or a spatially directed emission above $I_{\mathrm{exc}}^{\mathrm{th}}$.

21.1 Intrinsic Processes

Many of the high excitation effects outlined in Chaps. 18–20 can lead to stimulated emission under suitable conditions; see [21.1, 2] and references therein. In this chapter we treat those aspects of the high excitation phenomena of Chaps. 18–20 which are specific for stimulated emission.

Pumping can be accomplished by excitation with intense (laser-) light [21.1, 2], ($I_{\mathrm{exc}} \gtrsim 10^4\,\mathrm{W/cm^2}$ depending on the material and the pump conditions), with electron beams [21.3–5] ($j \gtrsim 10\,\mathrm{A/cm^2}$, $U \gtrsim 30\,\mathrm{keV}$), by a flashlamp [21.6] or by a forward biased *pn* junction [21.7, 8], provided that it is possible to grow the material in the desired highly *p*- and *n*-doped versions. Optical pumping is usually the proper choice for scientific investigations of the gain processes, since it allows resonant excitation of some species, e.g., certain exciton or biexciton levels. Pumping by electron beams, on the other hand, is unselective and an energy of about $3E_{\mathrm{g}}$ is deposited for the creation of one electron–hole pair. Electron-beam pumping was widely used before high power lasers were available [21.3, 4] and more recently for the realization of color projection TV tubes with high brightness [21.5].

Pumping by a *pn* (hetero-) junction biased in the forward direction is obviously the best choice for most technical applications in laser diodes, which are found, for example, in every CD player.

Stimulated emission has, until now been almost exclusively limited to direct gap semiconductors. One exception, namely $\mathrm{Al}_{1-y}\mathrm{Ga}_y\mathrm{As}$ close to the direct-indirect crossover, was mentioned in Sect. 20.3 and we shall come back to this later. Our considerations are restricted therefore to direct gap materials where not stated otherwise.

The processes which lead to stimulated emission can either be of intrinsic nature, i.e., they involve only excitations of the perfect lattice, or of extrinsic nature, involving some impurity or defect states. We shall see in the following examples of both groups.

We use the weak coupling limit and start with intrinsic processes in the intermediate density regime, namely the recombination of an exciton with emission of one or more LO phonons, or with inelastic scattering by another exciton or a free carrier, or finally the decay of a biexciton into a photon and an exciton. All these processes were already introduced in Chap. 19. In all four cases some initial state decays under emission of a photon and leaves behind in the crystal another excitation, for example an exciton in its ground state in the case of biexciton decay, an LO phonon in the case of the phonon replica, or an exciton in an excited state $n_B \gtrsim 2$ or a free carrier at higher kinetic energy in the cases of inelastic exciton–exciton scattering or-free carrier scattering, respectively. All these processes have the low threshold for population inversion typical for four-level systems as long as this excited final state in the sample is not (thermally) populated. We consider now the rate equations for these processes, assuming that for a given photon energy $\hbar\omega$ there is just one (laser-) mode containing photons with a density N_{ph}. The general rate equation which holds is [21.1]:

$$\frac{dN_{Ph}}{dt} = -2\kappa N_{Ph} + \sum \frac{2\pi}{\hbar} \delta(\Delta E) |W|^2 Q , \tag{21.1}$$

where $-2\kappa N_{Ph}$ contains all losses of the resonator, e.g., due to finite reflection or diffraction. The second term on the right of (21.1) contains a sum over all (scattering) processes contributing to the emission at $\hbar\omega$, $\delta(\Delta E)$ stands for the energy conservation, Q is a population factor into which we can integrate the $\boldsymbol{k}$-conservation, and $|W|^2$ finally is the transition matrix element. As a first approach to the strong coupling limit one can also incorporate into $|W|^2$ the probabilities that the exciton polariton is exciton-like in the initial state and photon-like in the final state.

We consider here as an example the P_2 line, i.e., the inelastic scattering between two excitons in the $n_B = 1$ state, from which one is scattered under energy and momentum conservation into a state with $n_B = 2$, while the other one appears as a photon.

In this case we have

$$\Delta E = E_{ex}(n_B = 1, \boldsymbol{k}_1) + E_{ex}(n_B = 1, \boldsymbol{k}_2) - \hbar\omega - E_{ex}(n_B = 2, \boldsymbol{k}_1 + \boldsymbol{k}_2) , \tag{21.2a}$$

and the population factor Q reads

$$\begin{aligned} Q = {} & N_{ex}(\boldsymbol{k}_1, n_B = 1) N_{ex}(\boldsymbol{k}_2, n_B = 1) [1 + N_{ex}(\boldsymbol{k}_1 + \boldsymbol{k}_2, n_B = 2)] (1 + N_{Ph}) \\ & - N_{Ph} N_{ex}(\boldsymbol{k}_1 + \boldsymbol{k}_2, n_B = 2) [1 + N_{ex}(\boldsymbol{k}_1, n_B = 1)] [1 + N_{ex}(\boldsymbol{k}_2, n_B = 1)] , \end{aligned} \tag{21.2b}$$

where we consider photons and excitons as bosons.

We can now decompose Q into two terms. One which is independent of N_{Ph} and describes the spontaneous emission and another one which is linear in N_{Ph} and gives the net gain or absorption. For the latter, Q_{stim}, we find

$$Q_{\mathrm{stim}} = N_{\mathrm{Ph}}\{N_{\mathrm{ex}}(\boldsymbol{k}_1, n_{\mathrm{B}}=1)\,N_{\mathrm{ex}}(\boldsymbol{k}_2, n_{\mathrm{B}}=1) - N_{\mathrm{ex}}(\boldsymbol{k}_1+\boldsymbol{k}_2, n_{\mathrm{B}}=2)$$
$$\times\,[1 + N_{\mathrm{ex}}(\boldsymbol{k}_1, n_{\mathrm{B}}=1) + N_{\mathrm{ex}}(\boldsymbol{k}_2, n_{\mathrm{B}}=1)]\}\,. \tag{21.2c}$$

If we assume thermal equilibrium between the excitons in the various states, we see that inversion occurs at low temperatures, even if we have only two excitons which collide.

The gain increases roughly quadratically with the exciton density and thus superlinearly with the pump intensity at low temperature until it overcomes the losses and stimulated emission sets in. At higher temperatures the $n_{\mathrm{B}}=2$ states will be also populated, reabsorption occurs and may even overcome the gain in some spectral regions, resulting in excitation-induced absorption.

The biexciton decay and the exciton-free carrier scattering also have gain values which increase superlinearly with density, while the gain of the ex-n LO process grows essentially linearly with the generation rate until laser emission sets in.

If Fig. 21.1 we show the calculated temperature dependences of the laser threshold density for three of the above-mentioned processes for a given constant value of κ. A variation of κ will shift the various quadratic processes with respect to the linear one. The increase of the thresholds with increasing temperature comes from the thermal population of the final states as indicated above. The high threshold at low temperatures of the process involving free carriers originates from the assumption of thermal equilibrium. In this case no free carriers (i.e., excitons in the continuum state) are present at low temperatures. The exciton-free carrier-scattering process can be influenced by doping. The fact that the calculated carrier densities are relatively high and may exceed the Mott density (Chap. 20) is due to the use of rather high losses in the calculations. Lower loss values reduce the calculated threshold densities.

In Fig. 21.1 we give also experimental data for the excitation intensity at threshold for CdS which show the trends predicted by theory. The low value of the threshold around 80 K comes from a cooperative effect, since various processes are spectrally degenerate in CdS at this temperature; see Fig. 19.3.

Currently, similar calculations and experiments are being performed for systems of reduced dimensionality, especially for II–VI quantum wells [21.1, 8]. Very recently stimulated emission from quantum dots has been reported [21.10, 11].

In Fig. 21.2 we show an example of the optical input-output curves, showing clearly the laser threshold. As an example, the exciton LO process in ZnO has been chosen [21.1, 9].

Another group of laser processes, still in the intermediate density regime, involves only interaction processes of virtually and coherently created

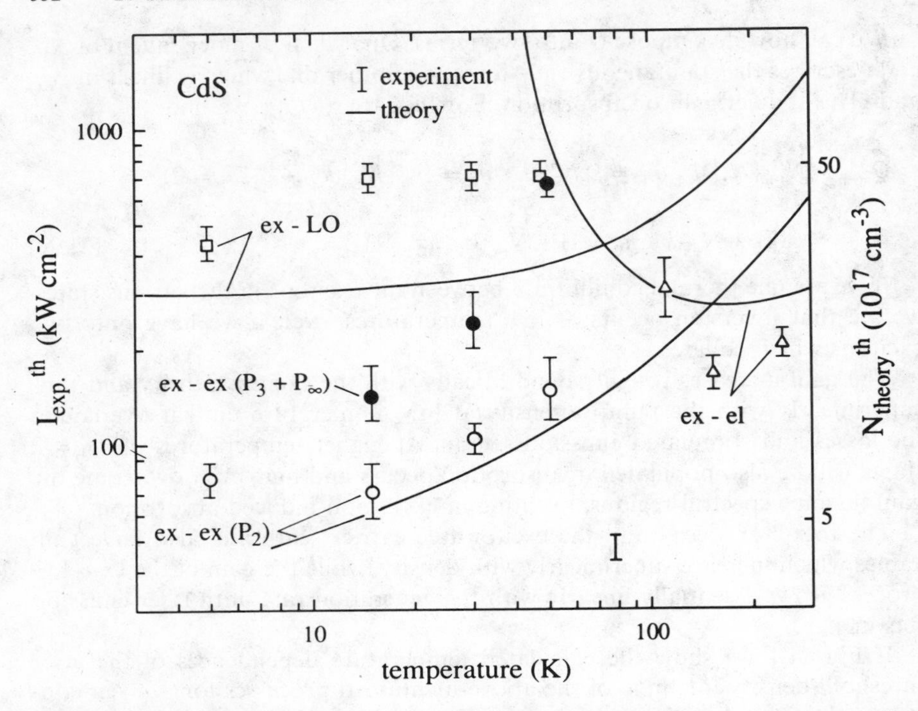

Fig. 21.1. The calculated threshold density $N^{\mathrm{th}}_{\mathrm{theory}}$ for various gain processes in the intermediate density regime of CdS as a function of temperature, and the experimentally observed excitation intensity at threshold $I^{\mathrm{th}}_{\mathrm{exp}}$ as a function of the lattice temperature. According to [21.1]

particles. As an example we take the two-photon or hyper-Raman scattering already introduced in Sect. 19.3. There we presented a process where a biexciton is created virtually by two quanta and decays under energy and momentum conservation.

If we now send an additional third quantum $\hbar\omega_s$ into the sample with momentum and energy coinciding with those of a possible decay channel, this quantum can stimulate the decay of the virtually excited biexciton into another photon $\hbar\omega_s$, $\boldsymbol{k}_s$. A second photon $\hbar\omega_f$, $\boldsymbol{k}_f$ must then necessarily be simultaneously emitted according to (19.7) [21.12].

If the third quantum lies energetically below the two pump quanta, $\hbar\omega_f$ is necessarily located above, and the whole process is an example of electronic CARS (coherent anti-Stokes Raman scattering). This latter process also occurs with optical phonons. The sample in this case is again illuminated with some pump photon and with quanta corresponding to the Stokes emission, resulting in stimulated emission of the anti-Stokes line.

Alternatively these processes can be named non-degenerate four-wave mixing (NDFWM) and can also be considered as diffraction from a moving

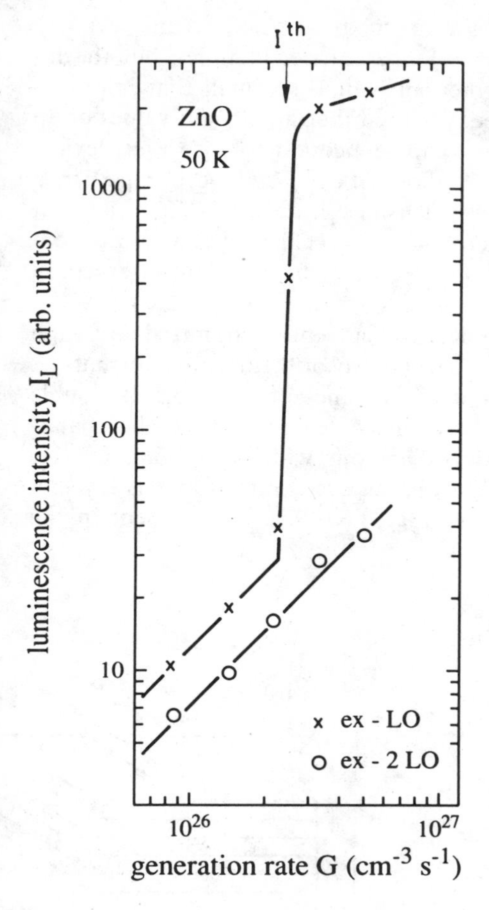

Fig. 21.2. The luminescence intensity of the first two LO phonon replicas of ZnO as a function of the generation rate G, showing the laser threshold. According to [21.9]

laser-induced grating written by the interference of one pump and one stimulating quantum, $\hbar\omega_p$ and $\hbar\omega_s$, respectively, and read by one of the two incident beams resulting in a Doppler-shifted diffracted signal.

The laser process which has presently the highest importance with respect to technical applications in laser diodes is the stimulated emission of a degenerate electron–hole plasma as outlined in Sect. 20.4. Population inversion occurs when the chemical potential μ of the electron–hole pair system is located above the reduced gap E_g' [Fig. 20.1 or (20.16)]. It should be mentioned that some of the inelastic scattering processes or recombination under emission of a phonon or of a plasmon–phonon mixed state quasiparticle mentioned above may also occur in a plasma, contributing to the long-wavelength part of the gain spectra, or at densities below those fulfilling (20.16).

Population inversion can be achieved much more easily in indirect gap semiconductors like Si or Ge due to the longer carrier lifetimes, but the indirect nature of the transition makes the gain values so small that impracticably large volumes have to be pumped. On the other hand, gain values of up to 10^4 cm^{-1} can be reached with direct gap semiconductors, so that devices with a length of the active material of about 100 μm can be pumped in a forward-biased *pn* junction to give power densities in the 10^5 W/cm^2 range at the surface of the laser. Differential internal quantum efficiencies deduced from the slope of the light power output versus electrical current input in excess of 50% have been reported.

The aims towards which laser diodes are currently optimized are high output power, good mode stability, and particularly, low threshold currents. A significant step in this direction was made by the introduction of double heterostructures [21.7]. In this case an intrinsic region with smaller bandgap is grown between the *p*- and *n*-doped regions with larger gap. By this technique both the carriers and the light quanta are confined in this layer in one direction as shown schematically in Fig. 21.3a. The confinement in the

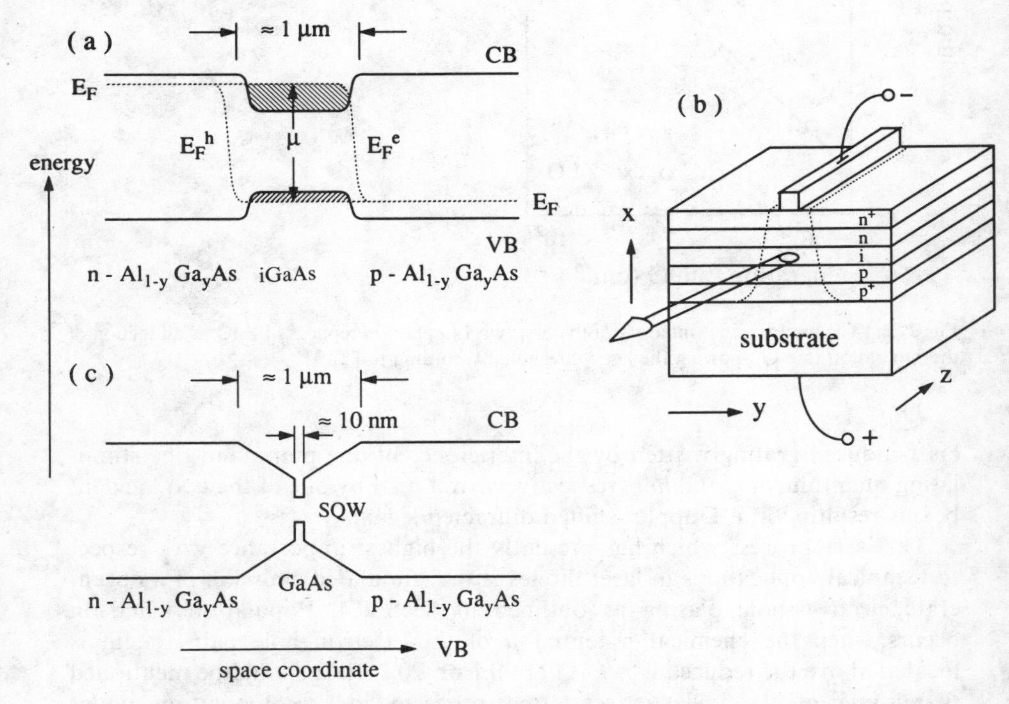

Fig. 21.3a–c. Schematic drawing of the bandstructure of a double heterostructure pin diode biased in the forward direction (**a**), a layout for a device with a strip-like contact for lateral confinement of the light by "gain-guiding" (**b**), and the bandstructure of a graded index separate confinement heterostructure (GRINSCH) laser diode (**c**)

y-direction is achieved (Fig. 21.3b) by a strip-like contact in the z-direction and/or by a deep mesa-etch of the structure followed sometimes by a regrowth with material of higher band gap (dashed lines).

Since the threshold current in the forward-biased pin laser diode depends, among other things, on the volume in which inversion has to be reached, a more advanced design has been realized in which the electrons and holes are confined in one (or a few) quantum wells with a typical width of about 10 nm, while the light quanta are guided in a structure with a width comparable to their wavelength, i.e., about 1 μm [21.7, 8]. These graded index separate confinement heterostructure (GRINSCH) structures are shown in Fig. 21.3c. The funnel-shaped bandstructure of the optical waveguide helps to collect the injected carriers in the quantum well. Minimum threshold currents reached with these and similar structures are in the 1 mA range. A certain drawback of these structures is the limited spatial overlap of light-field and electron wavefunction.

Another currently very lively field of research concerns the development of surface-emitting laser diodes which can be arranged in one- and two-dimensional arrays and addressed individually. These arrays are very important ingredients in parallel electro-optic data handling (Chap. 23).

The laser diodes are at present based on the material system $GaAs/Al_{1-x}Ga_xAs$ emitting around 0.8 μm and on $InP/In_{1-y}Al_yAs/In_{1-x}Ga_xAs$ for the regions around 1.3 μm and 1.5 μm, which are very attractive in connection with optical data-transfer through glass fibers.

For visible displays, or to increase the packing density in optical memories like CDs, diode lasers with shorter emission wavelength in the visible or even in the UV part of the spectrum are desirable. First progress in this direction has been made very recently with II–VI materials based on ZnSe after the problem of p and n doping has been solved to some extent. Laser emission has been reported in the blue-green part of the spectrum with $Zn_{1-y}Cd_ySe$ quantum wells [21.8]. The emission process in this case was not an EHP recombination but rather an excitonic process [21.8] such as the emission from localized tail states caused by fluctuations of the well width and of the composition [21.1, 8]. Reasons for this difference are the higher effective masses in the wide-gap II–VI compounds compared to the III–V materials, which increase the densities necessary to reach a degenerate electron–hole plasma, and the generally lower perfection of the II–VI quantum wells, which favors the appearance of localized tail states.

21.2 Localized and Extrinsic Processes

We now give two examples of stimulated emission in disordered systems. The tail of localized states in $CdS_{1-x}Se_x$ typically has a maximum tailing parameter $E_0(x)$ of about 5 meV and contains roughly 10^{18} states per cm^3. So the lower portion of this tail can be easily filled by optical pumping at low temperatures where thermal excitation into the extended states with a much higher density of states is prevented. If the gain value, i.e., the inversion, of the filled states is sufficiently large, laser emission sets in.

We show schematically in Fig. 21.4 (left-hand side) the density of states with the mobility edge ME and the chemical potential, μ, which indicates the energy up to which the states are filled at the highest excitation intensity. The right-hand side gives observed emission spectra showing the spikes of laser modes slightly above threshold.

Another aspect of disordered systems has been exploited recently to observe stimulated emission in indirect gap $Al_{1-y}Ga_yAs$ and similar materials as mentioned briefly in Chap. 20 [21.14].

GaAs is a direct-gap material with $E_g \approx 1.4\,\text{eV}$ at 4K and AlAs an indirect one with $E_g \approx 2.2\,\text{eV}$. There exist alloys of all compositions y. For $y < y_c = 0.57$, the minimum in the conduction band at the X point in the first Brillouin zone becomes lower than that at Γ. Under high (optical) pumping most of the electrons therefore sit in the X minimum. However, the alloy disorder couples the states and Γ and at X so that the electron–hole pairs can recombine without participation of phonons. This fact enhances the transition rate so strongly that stimulated emission has been reached for $y < 0.57$ at a wavelength of 620nm, i.e., already in the orange part of the spectrum. An example is given in Fig. 21.5. Part (a) shows emission spectra for various excitation strip lengths with constant excitation intensity I_{exc}, while (b) gives the resulting gain spectra. For details of this technique see Chap. 24. The slight

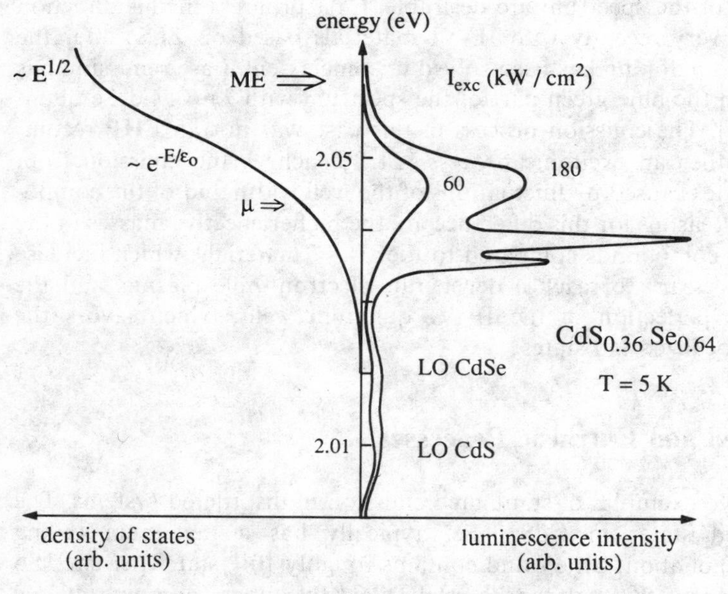

Fig. 21.4. *Left*: The density of localized and extended states in $CdS_{1-x}Se_x$ (schematic). *Right*: Observed emission spectra below and above threshold. According to [21.13]

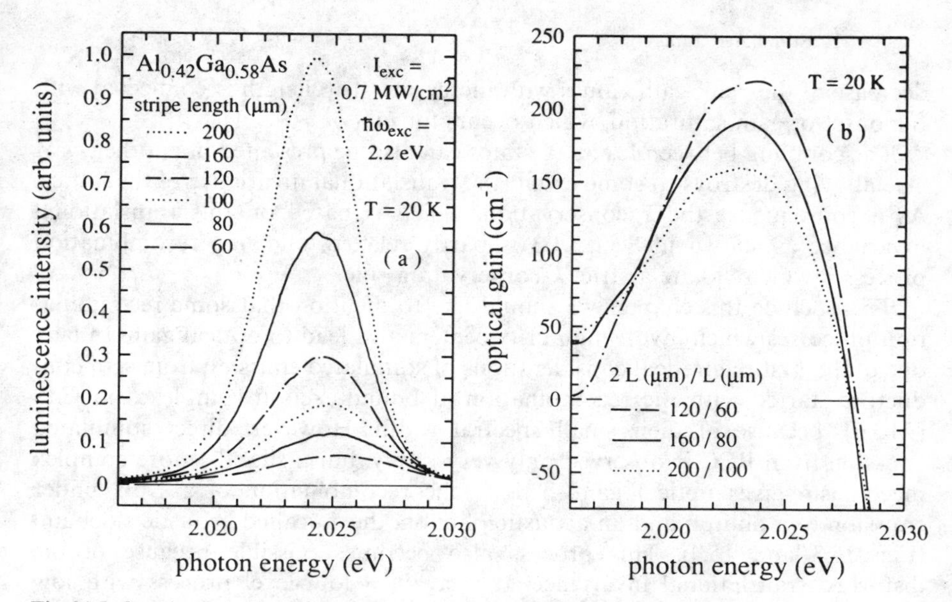

Fig. 21.5a,b. Stimulated emission in indirect $Al_{1-y}Ga_yAs$. Emission spectra for various excitation strip lengths (**a**) and the resulting gain spectra (**b**). According to [21.15]

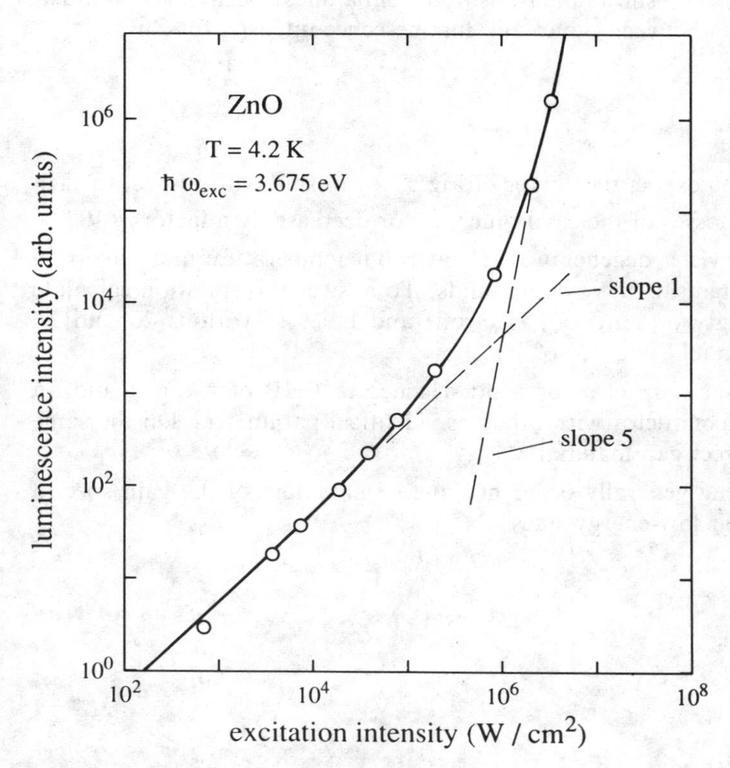

Fig. 21.6. The relation between the excitation and the luminescence intensity of the two-electron transition in ZnO. According to [21.17]

decrease of the peak gain values with increasing strip length is connected with the onset of gain saturation, which occurs for $g(\hbar\omega)l > 1$.

The coupling between Γ and X states can also be presented in another way. The alloying destroys to some extent the translational invariance of the lattice. As a consequence the $\boldsymbol{k}$ conservation, which is based on this translational invariance as shown in Sect. 3.1.3 is partly relaxed, allowing recombination processes which violate a strict $\boldsymbol{k}$-conservation rule.

To conclude this chapter we should like to mention that some recombination processes which involve impurity centers also lead to optical gain. In fact, one of the first theoretical considerations of stimulated emission from semiconductors started with the recombination of bound-exciton complexes in CdS [21.16], because of their small spectral width. However, direct stimulated emission from BEC is observed only very rarely, but a slightly more complex mechanism gives optical gain. This is the recombination of a BEC under emission of a photon and an acoustic phonon, the so-called acoustic sideband (Figs. 15.3 and 15.4). This process also becomes possible because of the disturbed translational invariance. It is again a four-level process with low threshold as long as the acoustic phonon states are not (thermally) populated.

The two-electron transition explained in connection with Fig. 15.5 also yields gain. Figure 21.6 shows the transition from a linear behavior to stimulation in the relation between pump and luminescence intensity for ZnO.

21.3 Problems

1. How would you expect the curves of Fig. 21.1a to shift with respect to each other if the total losses of the cavity increase or decrease by a factor $\sqrt{10}$?

2. Why is lasing via a degenerate EHP at room temperature more likely in standard III–V than in II–VI compounds? To answer this question calculate the effective density of states for electrons and holes in various 2d and 3d materials. Why should you do so?

3. Calculate the gain spectra for a 3d degenerate EHP of a direct and an indirect-gap semiconductor with otherwise identical parameters. Do the same for a quasi-2d direct gap material.

4. Why does lasing generally occur not at the maximum of the gain spectra but rather on their low-energy sides.

22 Ultrafast Spectroscopy

Until now we have considered the linear and nonlinear optical properties of semiconductors under quasi-stationary conditions, i.e., the excitation or measuring pulse lengths were assumed to be longer than the lifetime of the excited species.

The development of ps and sub ps lasers over the last two decades has allowed the dynamics of the optical excitations in semiconductors to be measured directly in the time domain. Since the characteristic time constants can be as short as a few tens of fs, this field of research is also known as "ultrafast spectroscopy". A lot of work has been devoted to this topic by various groups. The progress in this field can be followed, e.g., from the Proceedings of the International Conference Series on "The Physics of Semiconductors" [22.1], on "Ultrafast Phenomena" [22.2] or other conferences like [22.3].

The material accumulated in this rapidly developing field is too extensive by far to be treated here exhaustively. What we shall do instead is to first introduce on a very basic level the essential relaxation and recombination processes, and then give some representative examples including some experimental techniques. More details can be found in the above-mentioned conference proceedings [22.1–3].

22.1 The Basic Time Constants

We concentrate again on the properties of electronic excitations in semiconductors and present the basic mechanisms with the help of Fig. 22.1, again using the weak-coupling limit.

With a short laser pulse an electronic excitation is created, here excitons in the continuum states.

The polarization, $\boldsymbol{P}$, produced by the incident electromagnetic field of this pulse is initially in phase with this field, i.e. the two waves are coherent. The first scattering processes which occur destroy this coherence. The characteristic time in which the fraction of the polarization that is still in phase with the exciting pulse decays to e^{-1} is called the phase-relaxation time T_2 (see also Chap. 4 and Sect. 3.1.5). For the part of the polarization wave which is still coherent with the exciting pulse we can write, to a first approximation

$$\boldsymbol{P}_{coh} = \boldsymbol{P}_0 \exp(-t/T_2)\,. \tag{22.1}$$

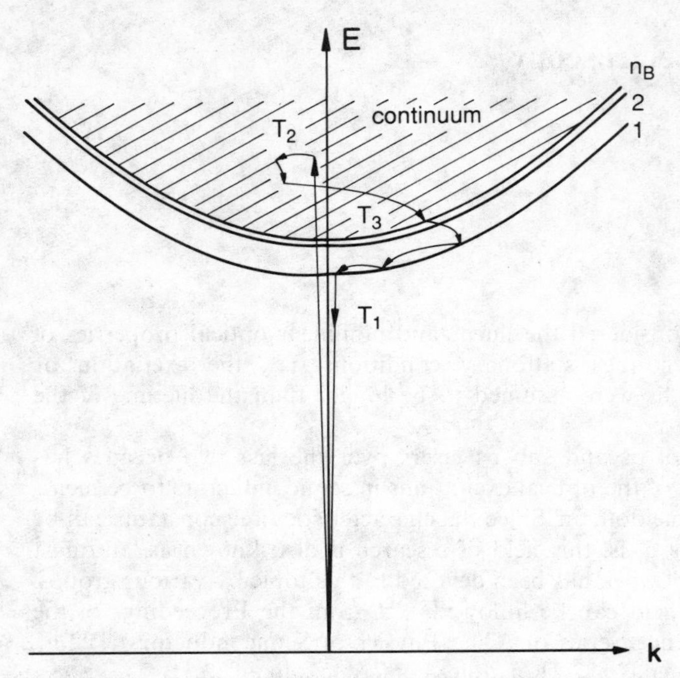

Fig. 22.1. A schematic drawing of the dispersion relation of excitons and of the main time constants

The scattering processes which destroy the phase include:

- scattering at impurities or other lattice defects. These processes also include the interface roughness scattering in quantum wells and scattering with alloy fluctuations in mixed crystals,
- scattering with phonons. The interaction mechanisms are the Fröhlich coupling with the electric field of (preferentially longitudinal) optical phonons, the deformation potential and the piezo-coupling to acoustic and optical phonons. For the interaction processes with phonons see sect. 10.5.
- scattering with other electronic excitations, e.g., elastic and inelastic exciton-exciton or exciton-electron scattering,
- recombination. If there are no other phase-destroying processes, the phase will ultimately be lost in the recombination process, which can be described by a lifetime T_1. In this case we have as an upper limit

$$N = N_0 e^{-t/T_1} \propto \boldsymbol{P}^2 \propto [\boldsymbol{P}_0 \exp(-t/T_2)]^2 \tag{22.2}$$

since the number-density N is proportional to the amplitude squared. From (22.2) follows the inequality

$$2T_2 \leqslant T_1 . \tag{22.3}$$

The "$<$" sign is valid if there are phase disturbing processes other than recombination, the equality is valid if recombination is the only phase-destroying process. We shall see in the future examples of both cases.

The first and the third of the above mentioned processes are especially important for free excitons in ordered structures and for excitons in extended states, while these processes can be considerably reduced for the spatially localized excitons occurring in bound exciton complexes or in the localized tail states in disordered system (Chap. 15). The second process increases significantly with increasing lattice temperature T_L, i.e., with increasing population of the phonon modes, and the third one evidently with increasing excitation intensity.

The first scattering process which limits T_2 may or may not change the energy of the state or the direction or the magnitude of its $\boldsymbol{k}$ vector. It is already sufficient that the phase correlation to the exciting pulse is destroyed.

A next step during the lifetime of an electron–hole pair is the intraband relaxation. This process is sometimes desbribed by a time constant T_3. It is of special importance when the excess energy with which the pair has been created is considerably larger than $k_B T$. If we create the exciton directly optically at the bottom of the band, there is not much reason to speak about intraband relaxation unless we also regard the increase of the kinetic energy from the energy connected with the very small momentum $\boldsymbol{k}_{exc}$ of the exciting photon to the larger kinetic energy of $3/2 k_B T$ as a thermalization or intraband-relaxation. In this latter case "thermalization" means an increase of energy.

The usual intraband relaxation for excess energies above $\hbar\omega_{LO}$ takes place by emission of optical phonons. This process is rather fast with typical time constants in the (sub–) ps regime. The rest of the excess energy has to be dissipated by emission of acoustic phonons. This process is usually much slower and takes progressively more time with decreasing energy and increasing momentum transfer.

After a few scattering processes among themselves and with the lattice, free electron–hole pairs (excitons) reach a thermal distribution which can be described by a temperature T_e and (usually) by Boltzmann statistics. (This statement is sometimes not true for excitons in localized tail states [20.4]).

If the lifetime of the exciton is sufficiently long and the coupling to phonons sufficiently strong, the excitons thermalize with the lattice. This means that the temperatures T_e and T_L describing the distribution of excitons and of phonons in their respective bands become equal.

If the excitons are created with some excess energy $E_{excess} \gg k_B T_L$, thermalization means a decrease of the average kinetic energy towards a value of

$3/2 k_B T_L$. If, in contrast, excitons are created resonantly of the bottom of the lowest band, thermalization means an increase of energy and a spreading in $\boldsymbol{k}$ space, as mentioned above, since the kinetic energy connected with the photon momentum $\boldsymbol{k}_{ph}$ is usually much smaller than $k_B T_L$ even for $T_L \approx 4.2$K, i.e., at liquid He temperature. The quantity k_B "times" 4.2 K is roughly 0.4 meV while $\hbar^2 \boldsymbol{k}_{ph}^2 / 2M$ is in the 10 μeV range depending on the material parameters.

Finally the excitons recombine radiatively or non-radiatively with a time constant T_1. T_1 is generally in the ns regime for direct gap semiconductors and reaches values in the μs or even ms range for indirect materials.

In most semiconductors, the recombination process is predominantly non-radiative. Only for some high quality quantum well samples has the luminescence yield η (i.e., the ratio of the number of emitted photons to excited electron–hole pairs) been claimed to reach unity. Otherwise typical values of η are $10^{-3} \lesssim \eta \lesssim 10^{-1}$ for good direct gap bulk materials. For materials containing a lot of non-radiative centers (so-called luminescence killers like Fe or Cu ions in some compounds) or in indirect gap materials, η can be even considerably smaller.

Apart from the "mono-molecular" recombination processes which result in an exponential decay described by T_1, there are other recombination processes like inelastic scattering described in Sect. 19.2 or the nonradiative Auger recombination. In the latter process one electron–hole pair (exciton) recombines and transfers all of its energy to a third particle, e.g., a free electron, as additional kinetic energy. This process can limit the quantum efficiency in high density electron–hole pair systems, e.g., in laser diodes [22.5].

After this general introduction, we present some selected and we trust, representative examples together with some information about the relevant experimental techniques.

22.2 Phase Relaxation

We start with the measurement of the phase-relaxation time T_2 since this is the first relaxation process for an optical excitation. T_2 can in principle be determined in the frequency and in the time domain. T_2 in the simplest approximation is directly connected with the damping γ in the resonance denominator of $\varepsilon(\omega)$, e.g., in (4.19), by [22.6, 7]

$$T_2 = \frac{1}{2\gamma} \,. \tag{22.4}$$

Therefore it is possible in principle to determine γ from the analysis of the transmission and/or reflection spectra, which depend in turn on $\tilde{n}(\omega)$ or on $\varepsilon(\omega)$. This however, is a rather indirect way and in the case of exciton resonances involves in addition the problem of additional boundary conditions as presented in Sects. 14.2, 3. To give a typical value, one finds from the analysis of the lowest free exciton resonance in the reflection spectra of high quality samples at low temperature ($T \lesssim 4.2$K) a damping value of less than or about 1 meV, corresponding to $T_2 \gtrsim 10$ ps.

The invention in recent years of tunable short pulse lasers in the ps and fs regime allows the investigation of T_2 directly in the time domain. The most widely used technique in this field is the so-called photon-echo technique, which will be considered first. It is usually applied in laser-induced grating arrangements.

22.2.1 The Photon-Echo Technique

The standard mathematical description of what follows is in terms of optical Bloch equations and the density matrix formulation. These aspects have already been presented many times, and we refer the reader to [22.8–10] and references therein, giving here an intuitive description as befits the philosophy of this book.

In Fig. 22.2a–c we show the principle setup and in (b) the temporal evolution of various quantities for an homogeneously broadened situation. In this case all oscillators in the medium have exactly the same eigenfrequency ω_0 and the finite width of a resonance comes exclusively from its finite T_2 time. In Fig. 22.2c we give the more realistic case of inhomogeneous broadening. In this case the eigenfrequencies are additionally distributed over a certain frequency range $\Delta\omega_{\text{inh}}$ around ω_0.

We start with the homogeneous case. A first incident pulse with wave vector $\boldsymbol{k}_1$ and frequency ω_1 produces in the sample a polarization $\boldsymbol{P}_1$, the coherent part of which decays with T_2. During its decay, this polarization radiates according to the well-known law

$$I_{\text{rad}} \propto |\ddot{\boldsymbol{P}}|^2 . \tag{22.5}$$

This radiation is known as "free induction decay" in analogy to corresponding techniques in magnetic (spin–) systems. A more appropriate name is "free polarization decay" as proposed by R. Ulbrich. Usually it has only a minor influence on the damping, i.e., T_2 is usually limited by scattering processes rather than by the radiative lifetime.

After a delay τ a second pulse $\boldsymbol{k}_2$ arrives which is coherent with the first one and (usually) has the same frequency but a different direction. To achieve this coherence the two pulses are generally produced from a single pulse using a beam splitter and a variable optical delay for one of them.

The second pulse creates a polarization $\boldsymbol{P}_2$, which interferes with the coherent part left over from the first one. This interference produces a laser-induced grating (LIG) of the polarization which radiates its first orders in the directions given by (22.6), assuming a thin grating, i.e., the Raman–Nath regime:

$$\boldsymbol{k}_{\text{diff}}^{(1)} = 2\boldsymbol{k}_2 - \boldsymbol{k}_1;\ \boldsymbol{k}_{\text{diff}}^{(-1)} = \boldsymbol{k}_1 . \tag{22.6}$$

The Raman–Nath regime is characterized to a first approximation by the following inequality between the thickness of the sample d, the spacing of the

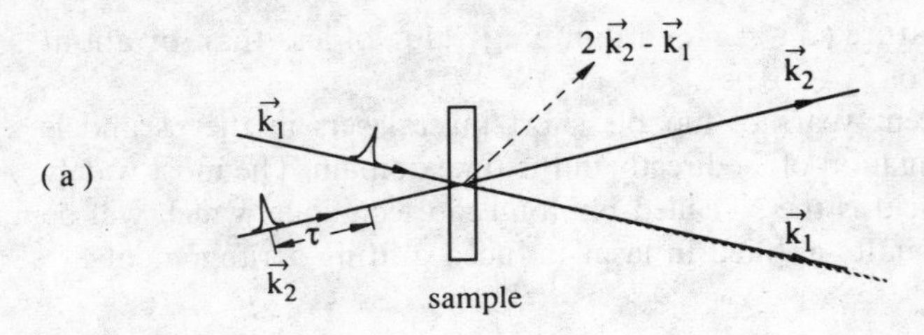

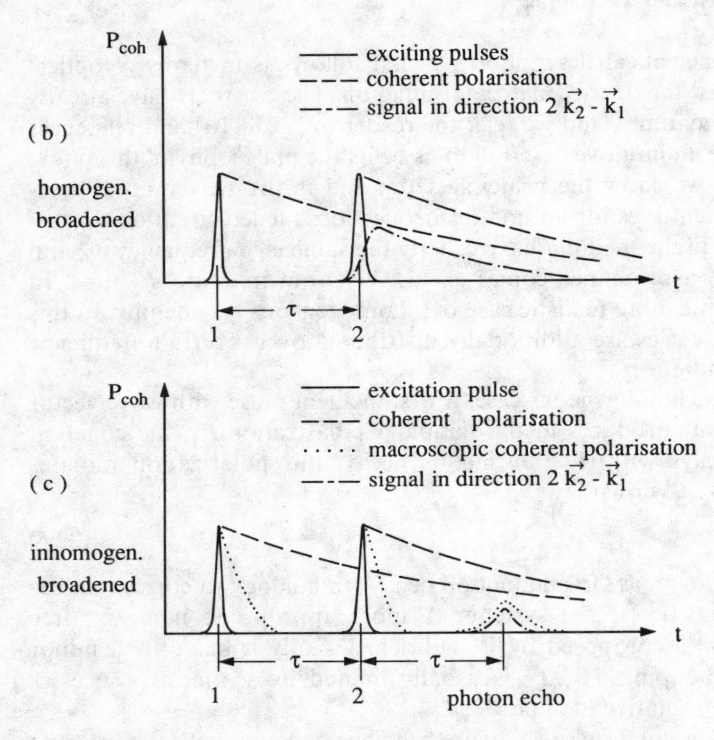

Fig. 22.2a–c. The setup for a photon-echo experiment (**a**), and the time dependence of the polarization for a homogeneously (**b**) and an inhomogeneously (**c**) broadened two-level system

grating Λ and the wavelength of the incident beams λ:

$$d^2 < \lambda\Lambda . \tag{22.7}$$

The negative first-order signal is difficult to detect since it coincides with the direction of the transmitted part of the pulse 1. So we concentrate on the other, background-free direction.

Since, in the homogeneously broadened case, all oscillators have the same eigenfrequency, the coherent parts of the polarizations all have the same phase and the scattered signal I_s starts to develop immediately when the second pulse arrives. It reaches its maximum I_{sp} when the second pulse is over, assuming that the duration τ_p of both pulses is much shorter than T_2, i.e.,

$$\tau_p \ll T_2 . \tag{22.8}$$

The diffracted signal in the case of homogeneous broadening depends on the delay time τ in the following way

$$I_{+1}^{hom} \propto |\boldsymbol{P}(t=\tau)|^2 \propto [\exp(-\tau/T_2)]^2 \propto \exp(2\tau/T_2) , \tag{22.9}$$

and the same relation holds for the time-integrated signal as shown below in Fig. 22.3a.

For the inhomogeneously broadened case of Fig. 22.2b, the part of the polarization amplitude of every oscillator which is still coherent with pulse 1 again decays with T_2. But since all oscillators have (slightly) different eigenfrequencies, they lose phase with one other in a time inversely proportional to the inhomogeneous broadening $(\Delta\omega_{inh})^{-1}$, resulting essentially in destructive interference of the radiation of all oscillators. The emission of the free polarization decay is then also limited by this quantity as shown in Fig. 22.2b. When the second pulse arrives after τ, a rephasing starts in the sense of phase conjugation or cum grano salis of a time reversal. After another time τ all oscillators are in phase again and radiate the so-called photon-echo (Figs. 22.2c and 22.3b). Its temporal width is limited by the free polarization decay and thus by τ_p or by $(\Delta\omega_{inh})^{-1}$, and the (time-integrated) signal intensity decays with τ according to

$$I_{s(0)}^{inhom} \propto |\boldsymbol{P}(t=2\tau)|^2 \propto [\exp(-2\tau/T_2)]^2 \propto \exp(4\tau/T_2) . \tag{22.10}$$

The appearance of an echo as shown in Figs. 22.2c and 22.3b is thus a clear indication of an inhomogeneous broadening.

A peculiarity of the description in the weak coupling limit is the fact that no signal appears in the direction $2\boldsymbol{k}_2$-$\boldsymbol{k}_1$ for negative delay, i.e., if pulse 2 arrives before pulse 1. If the polariton picture or the semi conductor bloch equations are used, a signal is also expected for negative delays. First calculations show [22.11] that the temporal buildup of the signal for negative delays has twice the slope of the decay for positive τ.

We have mentioned already that most transitions in semiconductors are inhomogeneously broadened, at least at low temperatures and densities. For free excitons or free polaritons a different eigenenergy belongs to every $\boldsymbol{k}$ vector due to their dispersion. Bound exciton complexes are inhomogeneously broadened since the surroundings of every impurity are slightly different due, e.g., to other impurities in the neighborhood. For the localized tail states in disordered systems the inhomogeneous broadening is obvious. At higher densities or temperatures the T_2 times get shorter and may become so short that they dominate the broadening, i.e., we get a transition to homogeneous broadening.

We now present a first experimental result, namely photon-echo measurements in the $(n_B=1)$ $A\Gamma_5$ free exciton resonance of CdSe at low temperature. Since even samples of a few μm thickness are opaque in this regime, the

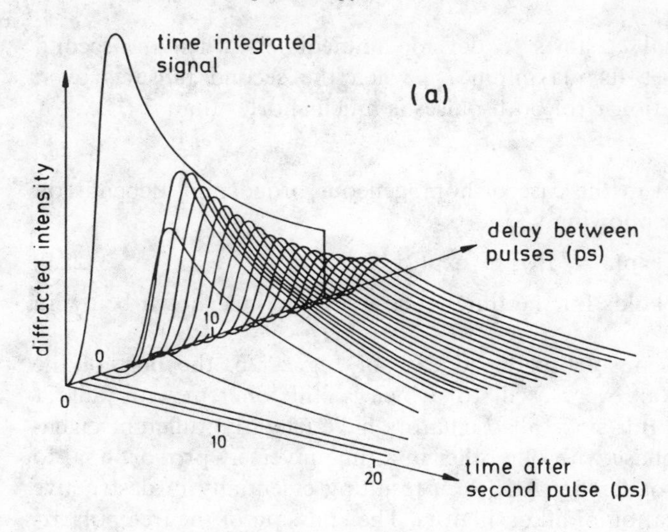

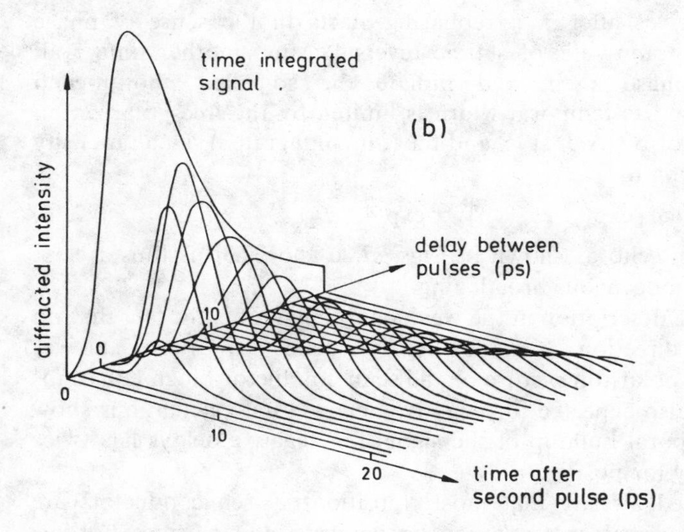

Fig. 22.3a, b. The temporal evolution of the diffracted order as a function of the delay between the two pump pulses and the time after the second pump pulse for a homogeneously (**a**) and an inhomogeneously broadened system (**b**). The time-integrated diffracted signal is also given. According [22.12]

diffracted signal in reflection was used (Fig. 22.4a). With increasing pump power, the signal increases, too, but the decay gets faster. This is due to inelastic exciton–exciton scattering. Using these data and others obtained in the exciton to biexciton transition (Fig. 22.4b), the curve of Fig. 22.4c is obtained. The extrapolation to zero density (and temperature) gives T_2 values around 50 ps. The slope of the curve in Fig. 22.4c allows one to determine additionally the exciton–exciton scattering cross-section via

$$T_2^{-1} = T_{02}^{-1} + \delta N_{ex} \tag{22.11a}$$

and one finds with [22.12]

$$\sigma_{ex-ex} = \delta v_{th} \tag{22.11b}$$

$\sigma_{ex-ex} = 6.8\,\pi a_B^2$, i.e., a value slightly larger than the geometrical one.

The onset of the signal for negative delays is not covered by the simple theory, as mentioned above, but by the improvements given in [22.11].

Photon-echo experiments at low temperature in GaAs revealed T_2 values around 10 ps [22.13, 14]. Diffusion measurements with laser-induced population gratings in CdS gave a diffusion constant $D(5\,\mathrm{K}) \approx 20\,\mathrm{cm}^2.\mathrm{s}$ and with the Nernst–Townsend–Einstein relations,

$$D = \mu \frac{k_B T}{e}, \quad \mu = e \frac{T'_2}{M}, \quad D = \frac{k_B T}{M} T'_2, \tag{22.12}$$

for classical effective mass particles, a time between scattering events $T'_2 \approx 30$ ps [22.15]. Here, μ is the mobility of excitons and T'_2 the time between two scattering processes. Evidently this quantity is closely related to the phase relaxation time. We shall describe a more detailed experiment concerning this point later in connection with MQW. Here, we can state that T_2 times in the range 10–50 ps are obviously typical for the lowest free exciton states in high quality bulk (and MQW, see below) samples at low temperature and density in agreement with data from reflection spectroscopy. The T_1 values under these conditions are in the 0.3–3 ns regime, i.e., we have $T_2 \ll T_1$. T_2 decreases with increasing density and with increasing temperature. In the latter case T_2 values in the range of 100 fs are reached at room temperature, so that the homogeneous dominates the inhomogeneous broadening.

Now we consider, still in 3d semiconductors, excitons which cannot move freely through the sample. We start with an example of a bound-exciton complex (Sects. 11.4, 15.1), more precisely with an exciton bound to a neutral acceptor (A^0X) in CdSe.

In Fig. 22.5a we show directly the time-resolved photon-echo for the laser tuned into the A^0X resonance of CdSe at low temperatures. One sees clearly the scattered intensities of the two pump pulses $P_{1,2}$ and the photon-echo P_s. A first inspection already shows that the T_2 times of this BEC are much longer than for free excitons. Extrapolation of the data to $T = 0$ K (Fig. 22.5b) gives $T_2 \approx 600$ ps. Since the lifetime T_1 of this BEC in this same sample has been determined to be around 400 ps, we come here close to the limit given by (22.3). The comparison of this value, which corresponds roughly to a homogeneous width of 20 μeV, with the spectral width of the BEC luminescence, which in good samples is between 0.1 and 1 meV, together with the observation of a photon-echo, clearly shows the inhomogeneous broadening of the BEC resonances.

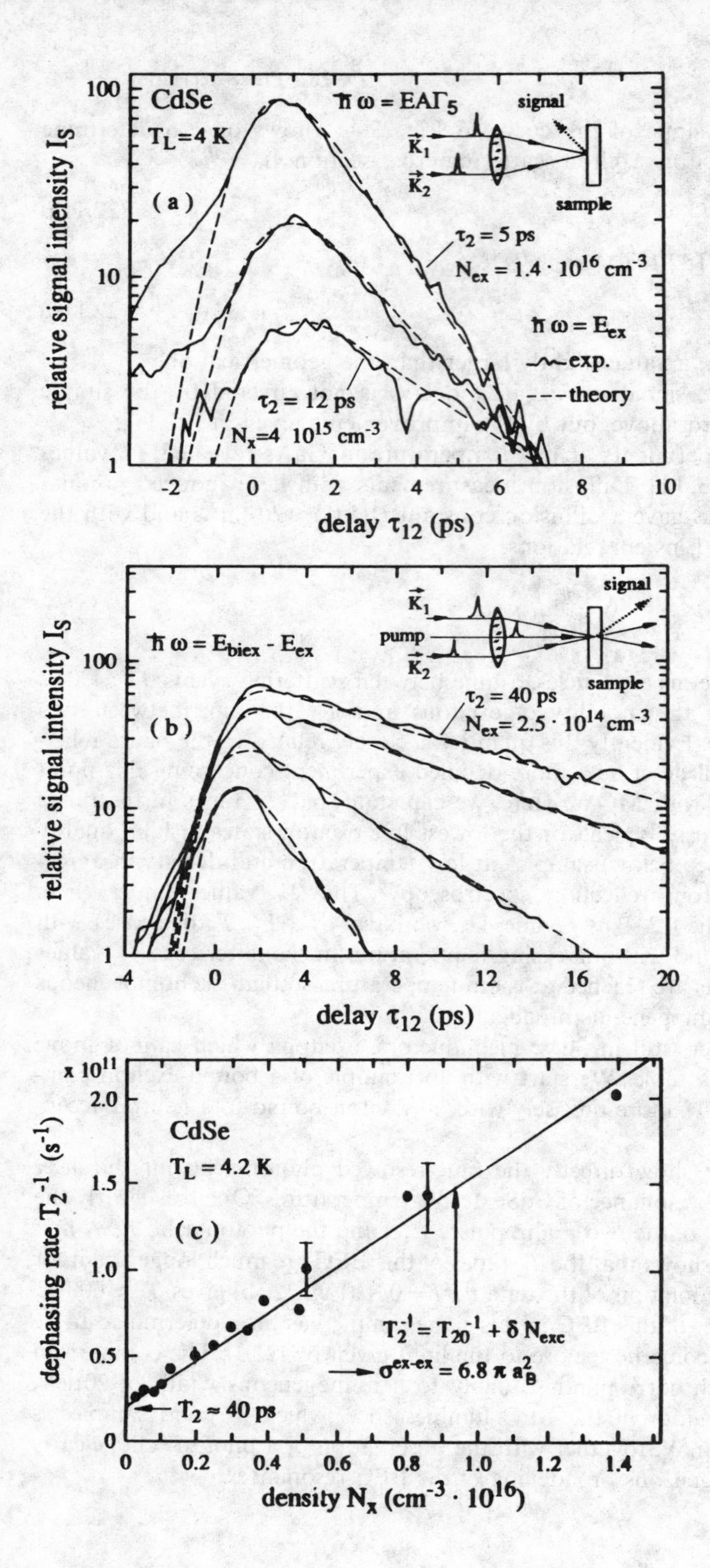

CdSe
T_L = 4 K
ħ ω = EAΓ5
signal
K1
K2
sample
(a)
τ2 = 5 ps
N_ex = 1.4 · 10^16 cm^-3
ħ ω = E_ex
exp.
theory
τ2 = 12 ps
N_x=4 10^15 cm^-3
relative signal intensity I_S
delay τ12 (ps)
ħ ω = E_biex - E_ex
pump
signal
K1
K2
sample
τ2 = 40 ps
N_ex = 2.5 · 10^14 cm^-3
(b)
relative signal intensity I_S
delay τ12 (ps)
x 10^11
CdSe
T_L = 4.2 K
(c)
T2^-1 = T20^-1 + δ N_exc
σ^ex-ex = 6.8 π a_B^2
T2 ≈ 40 ps
dephasing rate T2^-1 (s^-1)
density N_x (cm^-3 · 10^16)

The temperature dependence of the T_2 time is given in Fig. 22.5b. It can be fitted by an activation law

$$T_2^{-1} = T_{02}^{-1} + \nu_0 \exp(-E_a/k_B T) \tag{22.13}$$

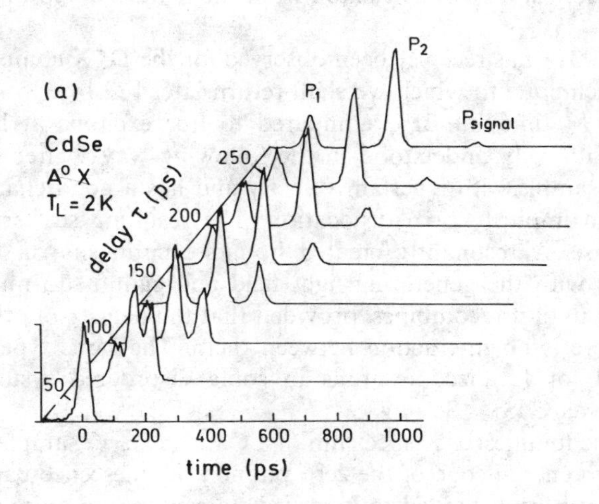

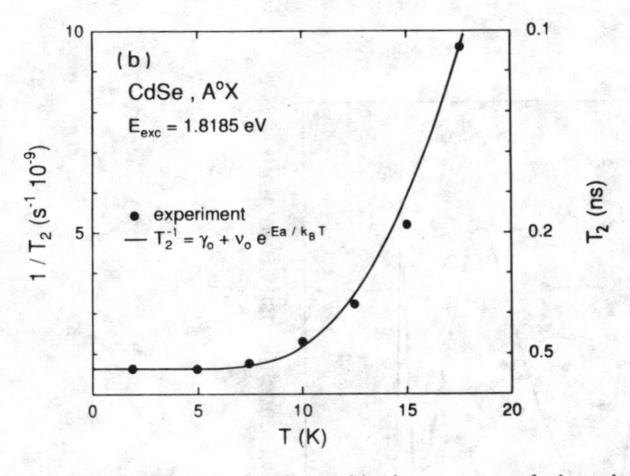

Fig. 22.5a, b. Photon-echo observed in the resonance of a bound exciton complex in CdSe (**a**) and the dephasing rate as a function of temperature (**b**). According to [22.16]

Fig. 22.4a–c. The evolution of the time-integrated photon echo as a function of τ in the $n_B = 1$ $A\Gamma_5$ exciton resonance of CdSe for different pump powers (**a**) in the exciton → biexciton transition, where a well-defined exciton density is created by a prepulse (**b**) and the dependence of T_2^{-1} on the exciton density (**c**). According to [22.12] with $T_{02} \approx 600\,\mathrm{ps}$, $\nu_0 = 3 \times 10^{11}\,\mathrm{s}^{-1}$ and an activation energy of $E_a \approx 5.6\,\mathrm{meV}$. This value corresponds approximately to the binding energy of the exciton to the complex. Since T_{02} is slightly above the observed luminescence decay time and thus of T_1 [22.17], we can conclude that the main phase destroying processes for this complex at low temperature are the recombination and at higher temperatures the thermal ionization of the exciton from the neutral acceptor by absorption of a phonon.

In CdS a T_2 value of 400 ps has recently been observed for the D^+X complex with the quantum-beat technique to which we shall return later [22.18].

The difference in the T_2 times of BEC compared to free excitons at low temperatures can be qualitatively understood in the following way. A free exciton moves through the sample with a certain velocity and has a good chance of hitting a defect like an impurity or a dislocation. The resulting scattering process destroys the phase. A resonantly created bound exciton sits on the defect, oscillates in phase with the generating light field and cannot do much more at low temperatures until it recombines, provided that the density of other centers is so low that there is no interaction between them. The same type of argument seems to hold for localized excitons in some disordered systems (Sects. 11.5 and 15.4) such as $CdS_{1-x}Se_x$.

In Fig. 22.6 we show the luminescence spectrum of a $CdS_{0.65}Se_{0.35}$ sample at $T = 2\,\mathrm{K}$ excited at the high energy edge of the zero phonon luminescence band, i.e., in the transition region from extended to localized states. One observes the

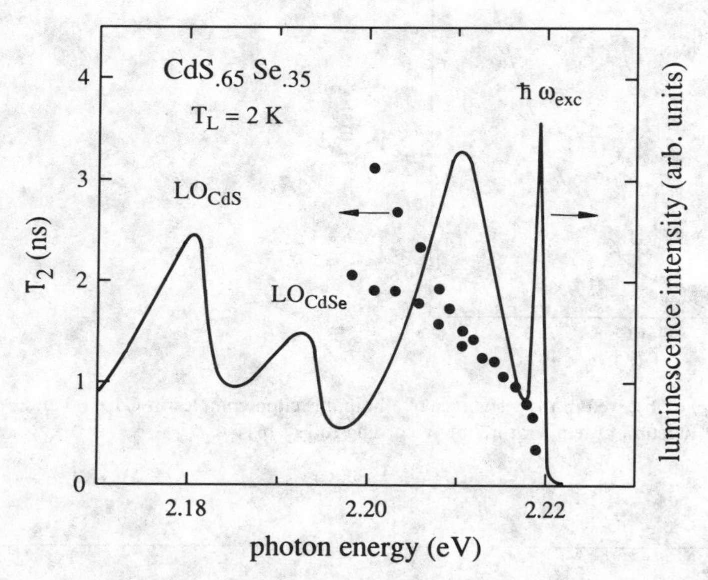

Fig. 22.6. The low temperature luminescence spectrum of a $CdS_{1-x}Se_x$ sample and the energy dependence of the T_2 times deduced from photon-echo experiments. According to [22.19]

zero phonon emission from the localized states around 2.21 eV and the CdSe and CdS LO phonon replica around 2.19 and 2.18 eV, respectively. The full dots give T_2 values at the respective energies measured with the photon-echo technique. The values range from $T_2 \lesssim 400$ ps close to the "mobility edge" up to $T_2 \approx 3$ ns for deeper localized states. Again these values are comparable with the intraband relaxation times T_3 and the interband recombination times T_1. Data for these latter quantities have been determined from time- and spectrally resolved luminescence and range from some hundreds of ps to several ns; see [22.19] and references therein. An increase of the lattice temperature leads once more, via an increasing phonon scattering, to a decrease of T_2.

An interesting aspect of ultrafast time-resolved spectroscopy can be treated in connection with Fig. 22.7, where we give the temporal decay curves of the time-integrated photon echo for $CdS_{1-x}Se_x$ using laser pulse durations from 0.12 to 10 ps at low temperatures ($4.2\,K \leqslant T \leqslant 10\,K$). The density of excited carrier pairs is in all cases roughly equal and in two cases even the same sample has been used.

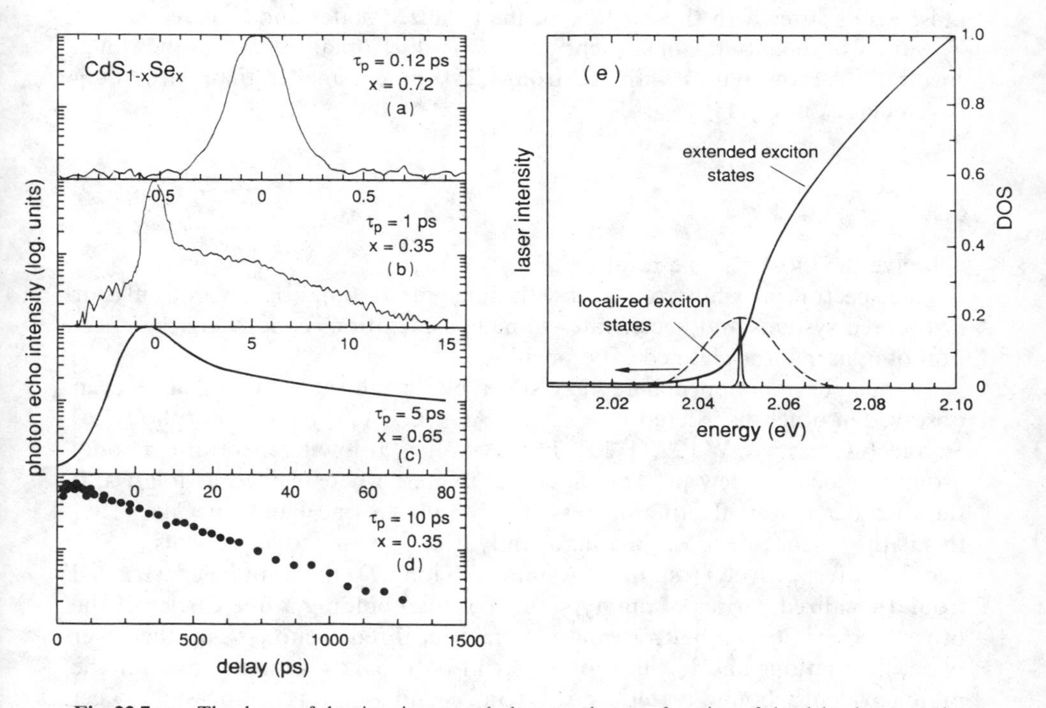

Fig. 22.7a–e. The decay of the time-integrated photon-echo as a function of the delay between the two pump pulses for various durations of the pump laser (**a–d**) and the density of localized states and the minimum spectral width of the laser pulses for $\tau_{laser} = 10$ ps and $\tau_{laser} = 0.12$ ps, respectively (**e**). According to [22.19, 21]

Two effects are striking. The exponential tail from which we deduce T_2 is missing for $\tau_p = 0.12$ ps. However, it becomes increasingly stronger and longer for increasing τ_p. Furthermore the ratio of the first initial spike, which corresponds roughly to the autocorrelation function of the laser pulse and which is also known as the "coherent artefact", to the slower decaying part decreases with increasing τ_p. The interpretation follows Fig. 22.7e where we show the density of states for extended and for localized states, and the spectral shape of a short ($\tau_p = 0.12$ ps) and a long ($\tau_p = 10$ ps) laser pulse imposed by the relation

$$\Delta E \cdot \tau_p \geqslant \hbar \, . \tag{22.14}$$

A spectrally narrow, rather long pump pulse tuned to the spectral range of localized states allows one to measure the true T_2 of these states. If the pulse length is reduced, the high energy tail of the laser extends into the region of extended states. In alloy semiconductors even at low temperatures these states have a very short T_2 time of the order of 0.1 ps due to alloy disorder scattering [22.19]. This value explains the short autocorrelation spike. Furthermore, the excitons in the extended states which are created by a short, spectrally broad pulse also scatter with the excitons in the localized states and thus reduce the T_2 values of the latter. Consequently the T_2 values obtained with rather long, spectrally narrow pulses with ΔE from (22.4) much smaller than the tailing parameter ε_0 in (10.48)

$$\Delta E \ll \varepsilon_0 \tag{22.15}$$

will give the most reliable results.

The spectral proximity of states with different T_2 times occurs not only for disordered systems and necessitates some considerations concerning the selection of a laser for a planned experiment.

It must be mentioned here that short phase relaxation times have been observed in other disordered systems like $Al_{1-y}Ga_yAs$ crystals, amorphous (α-) Si, and InGaAs MQW [22.11, 20]. The T_2 values at low temperature are only around or below a few ps. The difference is presently explained as follows. If the disorder causes fluctuations essentially only in one band (in $CdS_{1-x}Se_x$ this is the valence band originating mainly from the 3p and 4p orbitals of S^{2-} and Se^{2-}, respectively) a stationary state with long T_2 time can be constructed from a localized carrier of one type (here a hole) binding a free carrier of the other type by Coulomb attraction to form a localized exciton. If, on the other hand, fluctuations and localization occur in both bands, as is the case for the more covalently bound materials $Al_{1-y}Ga_yAs$ and α-Si, it is not possible to get a stationary state with two localized carriers plus Coulomb interaction. Here alloy disorder scattering always appears as a phase-destroying process. More details about these topics are found in [22.20] and references therein.

22.2.2 Further Techniques to Determine the Phase Relaxation Times

We proceed now to other techniques to determine T_2 or momentum relaxation times and to the comparison of these techniques with photon-echo experiments. We start with an example for a quasi two-dimensional MQW system.

In Fig. 22.8 we compare the results for phase-destroying processes measured with two different techniques. The crosses are data for the scattering rate, i.e., for T_2^{-1}, obtained from photon-echo experiments. They start with values around 30 ps at 2 K and T_2 decreases with increasing temperature essentially due to scattering with phonons. The open circles are obtained from diffusion measurements. The idea of this experiment is the following: two coherent pulses falling simultaneously on the sample interfere and produce a spatially modulated population grating (laser-induced grating, LIG). If this population produces any optical nonlinearity, light will be diffracted from this grating. The grating efficiency decays for times longer than T_2 due to recombination and due to diffusion. The first process does not depend on the grating period Λ of the LIG, but the second one does. By variation of Λ the two contributions can be separated. If two short pulses ($\tau_p \ll T_1$) are used, the diffracted signal intensity, which is measured by a third delayed pulse, decays with a characteristic time τ_s given by

$$\frac{1}{\tau_s} = \frac{2}{T_1} + \frac{8\pi^2 D}{\Lambda^2}\,, \tag{22.16}$$

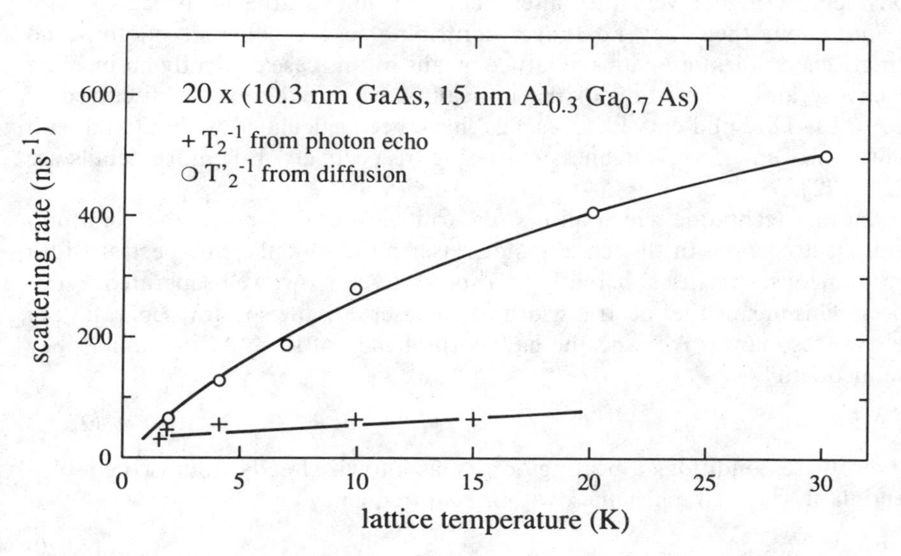

Fig. 22.8. The inverse T_2 times deduced by photon echo experiments and the inverse momentum relaxation times T'_2 or scattering rates deduced from diffusion experiments for $Al_{1-y}Ga_yAs$ MQW samples. From [22.22]

which allows the determination of D and via (22.12) of T'_2. The values given by open circles in Fig. 22.8 have been obtained in this way.

As a short aside, we mention here the following alternate method: For quasistationary conditions (i.e., $\tau_p > T_1$), one can state as a rule of thumb that the grating efficiency (i.e., the intensity of the diffracted order) decays with varying Λ to one half for $\Lambda_{1/2}$ (see e.g. [22.15])

$$\Lambda_{1/2} = l_D 2\pi \, , \tag{22.17}$$

where l_D is the diffusion length of the relevant particles, i.e., excitons. This quantity is connected with the diffusion constant D via

$$l_D \approx (D T_1)^{1/2} \tag{22.18}$$

and (22.12).

We return now to the ps experiments.

As can be seen in Fig. 22.8, the values of T_2 and T'_2 coincide approximately for T around 5 K. For higher temperatures T'_2 becomes significantly shorter than T_2.

The reason is that at least one additional scattering process contributes to T'_2, which is relevant for transport properties and which is absent or much smaller in photon-echo experiments. One of these processes is the interface roughness scattering in the quantum wells. The scattering rate increases with the (thermal) velocity v_{th} of the excitons, i.e., with $T^{1/2}$. In photon-echo experiments the excitons are created optically and have only extremely small in-plane velocities. The excitons acquire thermal equilibrium with the lattice and considerably higher velocities after one (or a few) scattering processes with phonons, but then they no longer contribute to the coherent photon-echo signal. Thermalization with the lattice means in this case evidently an increase of energy and momentum of the optically created excitons, as discussed in Sect. 22.1. The solid curves in Fig. 22.8 have been calculated without and with inclusion of interface roughness scattering, respectively. For more details see [22.14, 22].

Another technique which allows one to measure T_2 times is the quantum-beat spectroscopy. In this case a short laser pulse of sufficient spectral width simultaneously excites coherently two spectrally narrow, well-separated resonances. This means the spectral width of the laser ΔE, the spectral separation of the two resonances ΔE_r and the halfwidth of each of them ΔE_{inh} should obey the inequality

$$\Delta E \gtrsim \Delta E_r \gg \Delta E_{inh} \, . \tag{22.19}$$

Under these conditions a beating occurs, as known already from two coupled pendula in classical mechanics, with a beat frequency ω_b

$$\hbar\omega_b = \Delta E_r \, . \tag{22.20}$$

In our case a modulation of the intensity with ω_b can be observed, e.g., in a LIG experiment or in the free polarization decay signal. In Fig. 22.9 we give examples of the beating between the $n_z = 1$ heavy- and light-hole exciton resonances (*a, b*) and the beating between the exciton resonance and the induced transition exciton → biexciton (*c, d*). Since the beating is a coherent process, the decay of the envelope of the beat signal contains information on the T_2 of the resonance with the lower T_2 value.

The beat frequency just fulfils (22.20). These quantum beats have recently been observed in many systems, such as excitons in the indirect gap semiconductor AgBr [22.18], bound exciton complexes in CdS and CdSe [22.16, 18, 23, 25] and even between the lower and upper polariton branches of the lowest dipole-forbidden $n_B = 1$ exciton resonance in Cu_2O, which has a longitudinal transverse splitting of a few μeV only [22.24]. Other experiments concern the beating between various polariton branches in CdSe [22.25] or in InGaAs MQW [22.11] or between the exciton and the biexciton state in AlGaAs MQW; see [22.14, 25] and references therein.

Recently it has been possible to distinguish between a beating of two elec-electronic transitions and a mere interference of the electromagnetic waves

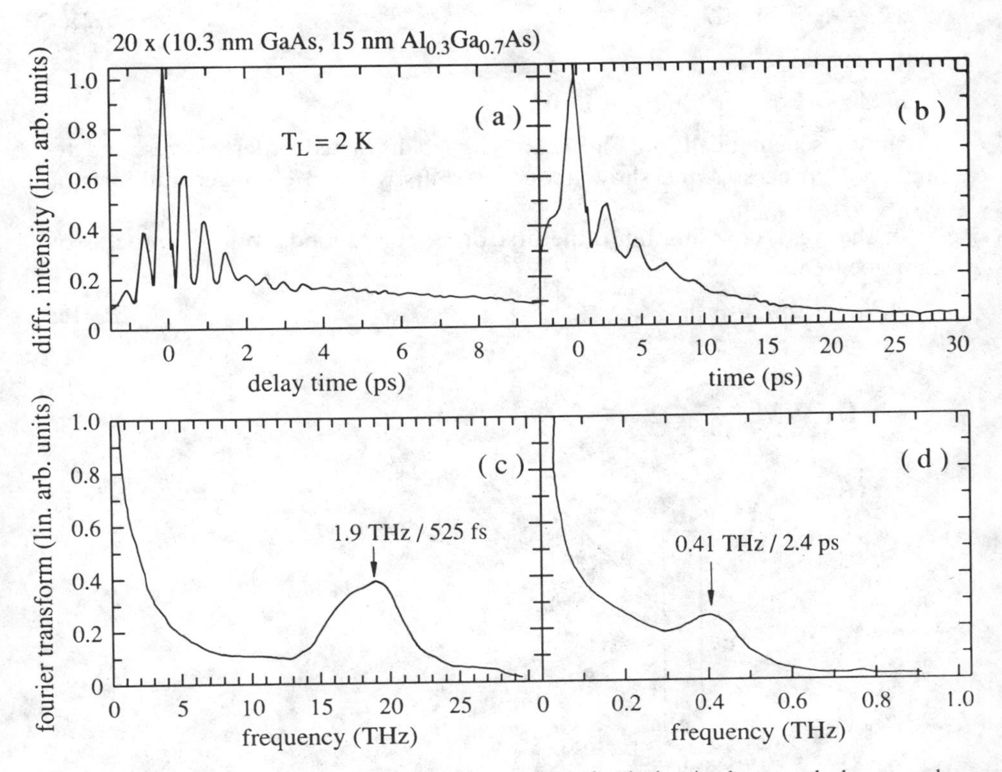

Fig. 22.9a–d. Quantum beats and their Fourier transforms for the beating between the heavy- and light-hole exciton in an $Al_{1-y}Ga_yAs/GaAs$ MQW (**a, c**) sample and for the exciton → biexciton transition (**b, d**). According to [22.14]

radiated from two independent, slightly detuned oscillators [22.16, 23, 26, 27]. In the picture of strong exciton–photon coupling a generalized description of the quantum beats can be expected to include both the above-mentioned aspects as limiting cases.

We move on now to a third experimental technique to determine T_2 which is again connected with laser-induced gratings. This is nondegenerate four-wave mixing (NDFWM), and the system to which we apply it are quantum dots. It is a method to determine T_2 from spectroscopy, i.e., in the frequency rather than in the time domain.

In this case one writes a grating with two laser beams of different frequencies ω_1 and ω_2. The coherence time of each laser must be long enough to provide a sufficiently coherent overlap between them. One generally uses ns dye lasers with an additional intracavity etalon to reduce their linewidths. In the polariton picture one would describe the diffraction process in the following way: Two quanta of beam 1 (or 2) create an intermediate two-polariton state, the decay of which is stimulated by a polariton from beam 2 (or 2) resulting in the emission of a quantum in the direction

$$2\boldsymbol{k}_1 - \boldsymbol{k}_2 \quad (\text{or} \quad 2\boldsymbol{k}_2 - \boldsymbol{k}_1) \tag{22.21a}$$

with energies

$$\hbar(2\omega_1 - \omega_2) \quad (\text{or} \quad \hbar(2\omega_2 - \omega_1)), \tag{22.21b}$$

as shown schematically in Fig. 22.10. Higher diffracted orders correspond to higher $\chi^{(n)}$ processes and show frequency shifts which are integer multiples of $(\omega_1 - \omega_2)$.

In the weak coupling limit, the first order corresponds to a $\chi^{(3)}$ process of the type (see Sect. 18.1)

$$\chi^{(3)}(\omega\colon \omega_1, \omega_1, -\omega_2). \tag{22.21c}$$

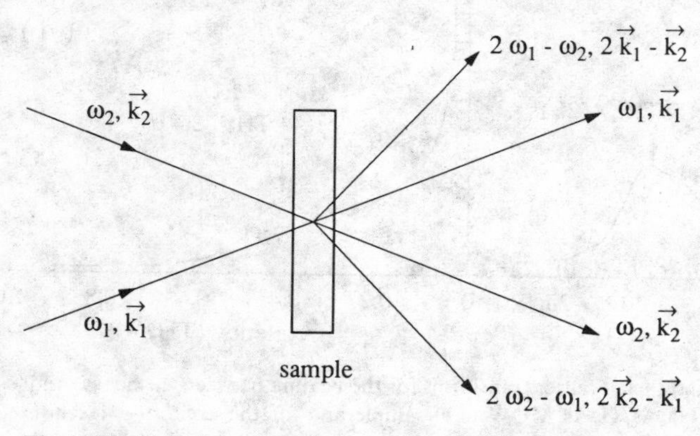

Fig. 22.10. The schematics of a non-degenerate four-wave mixing (NDFWM) experiment

In a classical wave picture we have the following description. The two incident waves interfere to form a moving interference pattern. It moves laterally with a speed v_G given by

$$v_G = \frac{\omega_1 - \omega_2}{|\boldsymbol{k}_1 - \boldsymbol{k}_2|} . \tag{22.22}$$

The orders diffracted from this moving grating are Doppler shifted and a quantitative analysis of these ideas leads to exactly the same results as given in (22.21).

If there is an optical nonlinearity due to the real or virtual excitation of some species with a lifetime T_1 or T_2, respectively, then the grating will be washed out if it moves over one period Λ in a time shorter than $T_{1,2}$. Consequently the efficiency drops, when

$$\Lambda v_G^{-1} < T_1 \quad \text{or} \quad T_2 . \tag{22.23}$$

A quantitative analysis leads to the following relation for the efficiency $\eta_{\pm 1}$ of the $\pm$ first diffracted order [22.28].

$$\eta_{\pm 1} \propto |\chi^{(3)}|^2 \propto [(1 + \Omega^2 T_1^2)(1 + \Omega^2 T_2^2)]^{-1} \tag{22.24}$$

with $\Omega = \omega_1 - \omega_2$. This means that there is a narrow central spike, whose decay with increasing detuning Ω is determined by T_1 and wider wings from which one can deduce T_2 in the case $T_1 \gg T_2$.

In Fig. 22.11 we give as an example results for a semiconductor-doped glass with CdS crystallites with an average radius of 7.5 nm. The narrow central spike is limited rather by the spectral width of the lasers than by T_1. The T_1 time of a few hundred ps typical for such a system was measured independently by time-resolved luminescence and would give a width below 0.1 meV. The wider wing can be fitted with (22.24) to give a T_2 time of the resonantly excited electron–hole pairs in quantum dots of about 100 fs, decreasing with increasing pump power as shown in the inset. These values increase only slightly when going to lower temperatures. The reason why this value is so short in comparison with bound exciton complexes is not completely clear at present. The stronger coupling to the optical phonons in QD compared to the bulk material may be of importance or the scattering at, or the relaxation of the carriers into, some interface states between the semiconductor and the surrounding matrix.

A final but no less powerful technique to determine T_2 times is spectral hole burning in an inhomogeneously broadened absorption band. Under favorable conditions, the spectral width of the hole in the absorption band is connected with T_2. This is the method of choice in atomic and molecular spectroscopy but has only more recently been applied in semiconductors. We shall not go into details here, but refer the reader to [22.6, 7, 29–31] and references therein.

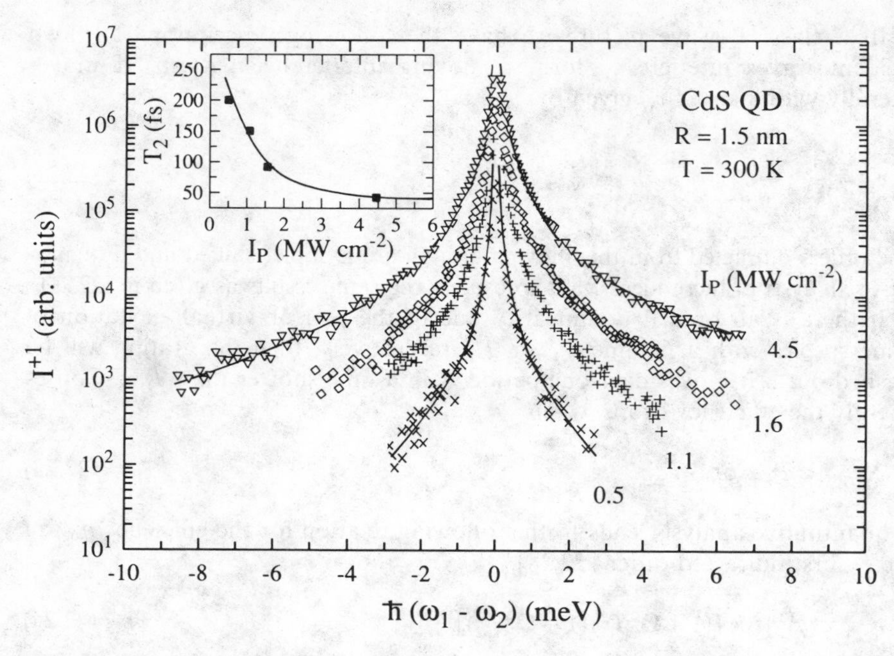

Fig. 22.11. The NDFWM signal as a function of the detuning in a silicate glass doped with CdS nanocrystals for various pump powers I_p. According to [22.29]

22.3 Intraband Relaxation

Now we proceed to two examples of intraband relaxation. We start with rather old measurements on the exciton polariton in CdS [20.32]. Figure 22.12 shows the polariton dispersion of the $n_B = 1$ $A\Gamma_5$ exciton, the luminescence spectra for two polarizations, and the lifetime measured from the luminescence decay at various energies after band-to-band excitation. The excitons relax in about 1 ps by LO-phonon emission onto the LPB. There they can relax further, e.g., by emission of acoustic phonons. This process becomes slower and slower with decreasing slope of the dispersion, since less and less energy transfer is allowed per unit of momentum transfer. For this reason the excitons accumulate in the indicated area, which is known as the "bottle neck" (Chap. 14). As a loss mechanism, polaritons can be transmitted from the UPB and the LPB through the surface, but only if the parallel component of their wave vector $\boldsymbol{k}$ satisfies

$$k_{\parallel} \lesssim \frac{2\pi}{\lambda_{\text{vac}}} \tag{22.25}$$

with a probability depending on the reflection coefficient and on the squared

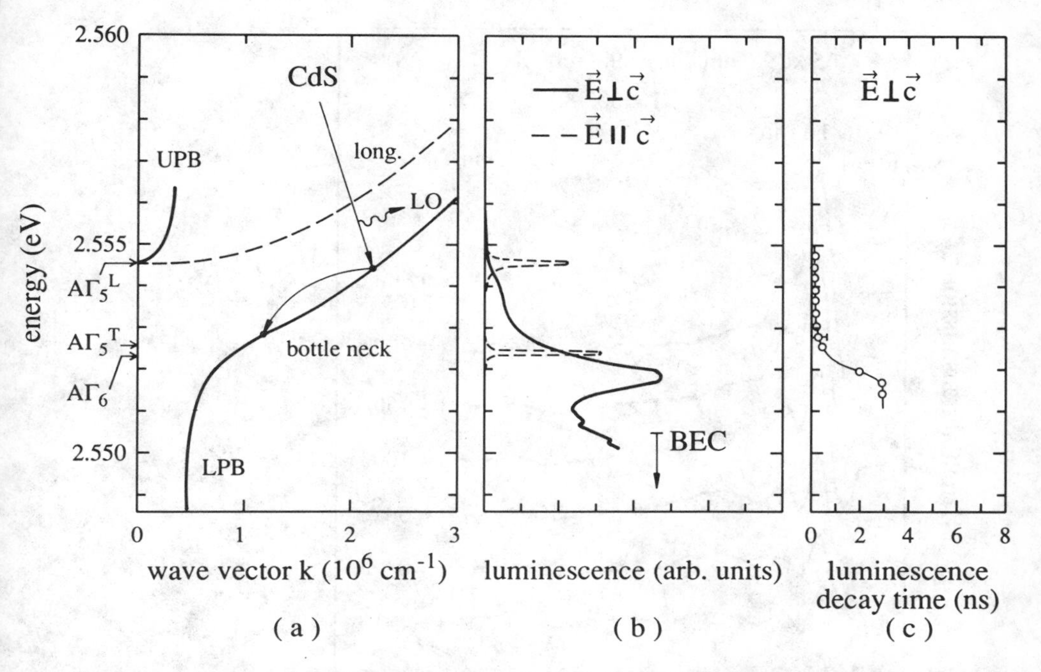

Fig. 22.12a–c. The dispersion of the $n_B = 1$ AΓ_5 exciton polariton in CdS in the bottle-neck region (**a**); the luminescence spectra for the dipole allowed orientation $\boldsymbol{E} \perp \boldsymbol{c}$ and the forbidden one $\boldsymbol{E} \parallel \boldsymbol{c}$ (**b**); and the decay time of the luminescence as a function of photon energy (**c**). According to [20.32] compare also with Figs. 14.4, 15, 22

amplitude of the photon-like part of the wavefunction. Additionally the excitons can be trapped at defects or scattered onto the photon-like branch of the dispersion curve under emission of one or more LO phonons (Sect. 14.3). A further relaxation to the regime below the bottle-neck by emission of acoustic phonons becomes rather improbable due to the rapidly decreasing density of final states; see Sect. 3.2 and Fermi's golden rule.

The accumulation of the exciton-like polaritons in the bottle-neck during the relaxation in the band directly explains the increase of their radiative decay times in Fig. 22.12c at these energies.

In Fig. 22.13 we give another example of intraband relaxation, this time of a AlGaAs MQW sample. The differential transmission between unexcited and excited sample is measured with 50 fs increments when the sample is excited around $t = 0$ with a pump pulse of 80 fs duration. There is some bleaching due to partial blocking of the band-to-band transitions by the bunch of electron–hole pairs excited in the continuum (dotted areas). The temporal evolution of the intraband relaxation can be seen nicely. There is a decrease of excitonic absorption in the region of the $n_z = 2$ hh transition around 1.57 eV. This change is rather small and constant on the time scales shown here ($\ll T_1$)

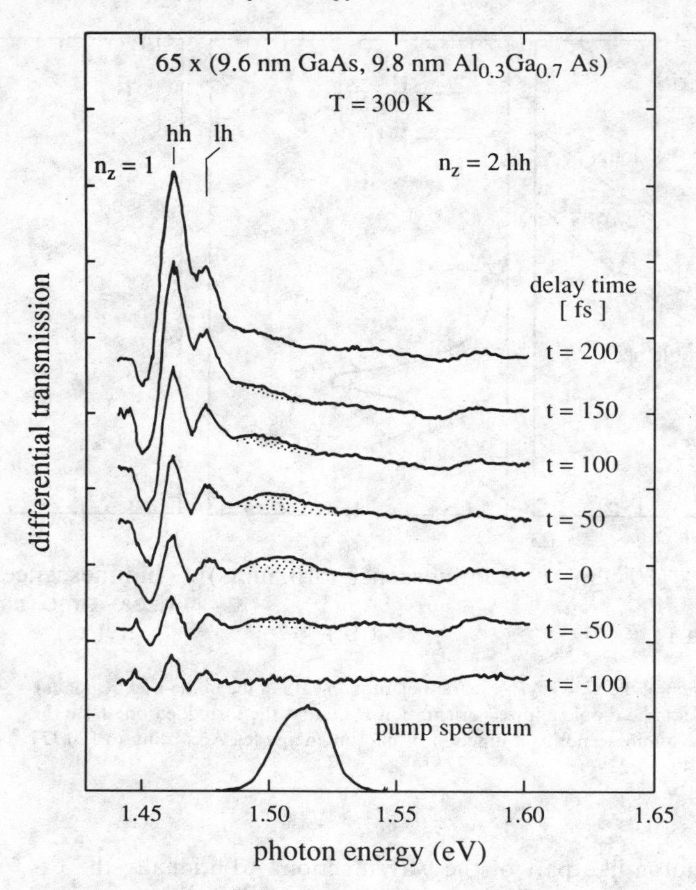

Fig. 22.13. Intraband relaxation of electron–hole pairs in the $Al_{1-y}Ga_yAs$/GaAs MQW measured by time-resolved pump and probe experiments. According to [22.33]

and is due to direct screening of the Coulomb interaction in the exciton by the carriers. This screening is largely independent of the energy and distribution of the free carriers in the bands. A similar behavior is found for the $n_z = 1$ hh and 1h resonances around 1.46 eV in the three lowest traces. When the excited electrons relax down into the exciton states, the transmission increases drastically. This finding is attributed to phase-space filling and exchange interactions which become effective for these exciton resonances only after the excited species start to populate the respective states. Figure 22.13 thus demonstrates qualitatively that direct screening of the Coulomb interaction is less efficient in quasi two- (and one-)dimensional systems than exchange interaction and phase-space filling. This difference to three-dimensional materials has already been pointed out Sect. 20.4 in connection with electron–hole plasmas.

22.4 Interband Recombination

To conclude this chapter we discuss some measurements of the interband recombination time T_1.

To measure this quantity it is best in principle to observe effects to which all excited species contribute by roughly the same amount. One could follow, e.g., the further temporal evolution of the bleaching of the exciton resonance in Fig. 22.13 and one would find that the differential transmission signal disappears with a time constant of about 0.5 ns.

Very often one relies, however, simply on the temporal evolution of the luminescence, e.g., of excitons or biexcitons, as a monitor of the population of the respective species, although it is well known that the luminescence yield η for most semiconductors is considerably smaller than 1. For high quality direct gap samples one finds, as already mentioned,

$$10^{-1} \gtrsim \eta \gtrsim 10^{-3} ; \tag{22.26}$$

only for some laser diodes and quantum well samples have values of η close to one been reported. For indirect gap materials or direct ones containing "luminescence killers" like Fe or Cu ions η can even be much smaller than in (22.26).

On the other hand, it is experimentally quite easy to measure the temporal decay of a luminescence signal, and therefore many data on T_1 of electronic excitations of semiconductors are obtained by this technique. If one remains aware of the fact that only a small fraction of the excited species are monitored in this case, it is a useful technique.

In the following example (Fig. 22.14) we consider a MQW sample of high quality with narrow absorption and emission lines and virtually no Stokes shift between emission and absorption. In emission under cw excitation this sample shows the $n_z = 1$ hh exciton and weakly, on its high energy side, the $n_z = 1$ lh exciton. A shoulder on the low energy side is presumably due to a bound exciton complex or to an excitonic molecule (Sect. 19.3) and a band around 1.53 eV to a free-to-bound transition (Sect. 15.2). The latter structure saturates rapidly with increasing pump power. In Fig. 22.14a, b we see the decay of the $n_z = 1$hh luminescence after resonant and non-resonant ps excitation for various pump powers. The most striking feature is that a simple exponential decay is almost never observed.

For low, resonant excitation (curve 1) one observes a rapid decay of the luminescence, which can be attributed to a rapid capture of excitons into some deep centers. For increasing excitation this process saturates and for $t > 250$ ps one gets in curve 2 a simple exponential decay with $T_1 \approx 200$ ps. This order of magnitude seems to be representative for the radiative and nonradiative decay of excitons in good MQW samples at low temperatures ($T \lesssim 10$ K). For even higher pump power (curve 3) this value is reached only at later times and a

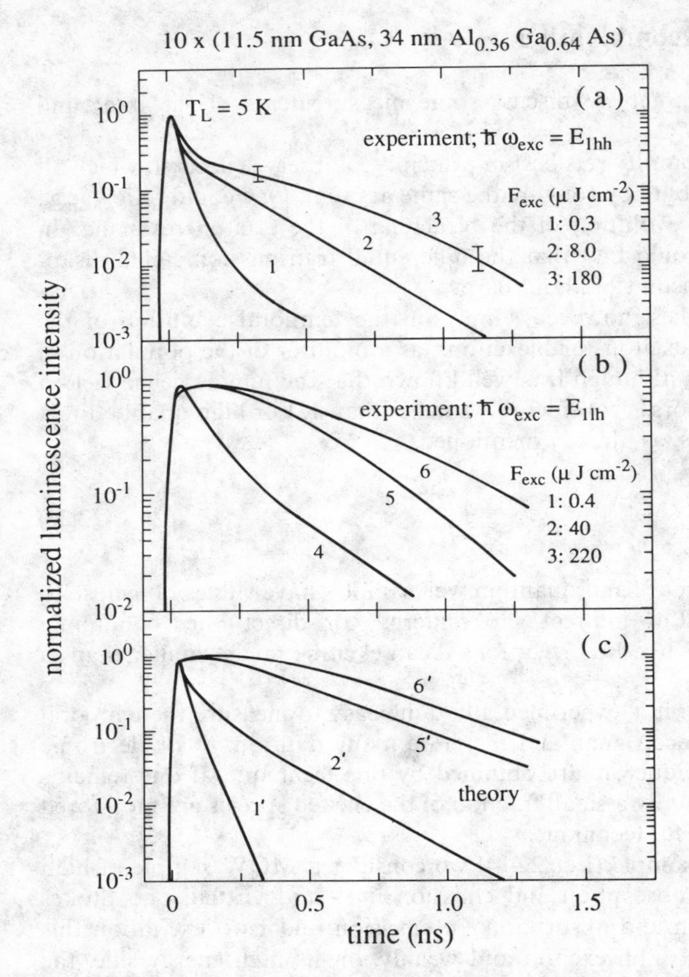

Fig. 22.14a–c. The decay curves of the 1hh exciton luminescence for various excitation fluences for resonant (**a**) and non-resonant (**b**) excitation of an AlGaAs MQW sample, and some calculated decay curves (**c**). According to [22.14]

hump develops in the decay curves, which is even more pronounced in the case of non-resonant excitation with some excess energy (curves 4 to 6).

This feature can be interpreted as a thermalization (or T_3) process. Increasing I_{exc} or $\hbar\omega_{exc}$ leads to the creation of a hot gas of excitons with a certain average kinetic energy and with finite $\boldsymbol{k}_\parallel$ vectors. In contrast to the three-dimensional case (22.25), $\boldsymbol{k}_\parallel$ is the total momentum $\boldsymbol{k}$ of the excitons in quasi two-dimensional systems, whereas $\boldsymbol{k}$ can be decomposed into $\boldsymbol{k} = \boldsymbol{k}_\parallel + \boldsymbol{k}_\perp$ in the three-dimensional case. Due to the conservation of $\boldsymbol{k}_\parallel$ in the radiative recombination process only excitons with $\boldsymbol{k}_\parallel$ around zero (or excitons in the small tail of localized states caused by well-width fluctuations) can participate in radiative recombination. The humps or plateaux in curves 7 and 8 reflect the cooling process of the exciton gas from states with larger towards states with smaller $\boldsymbol{k} = \boldsymbol{k}_{\parallel}$.

A modeling of the decay along the lines above gives the curves 1′, 2′, 5′, and 6′, in Fig. 22.14c, which correspond to the unprimed experimental ones. The good agreement gives some support to these ideas.

A related experimental finding concerns the fact that in quantum wells one often observes a linear increase of the T_1 time with lattice temperature for $T_L \gtrsim 10\,\mathrm{K}$ up to around 100 K. Examples are given in Fig. 22.15 or in [22.24 a, b].

$$T_1 = \frac{T'_1 k_B T_L}{E_m} (10\,\mathrm{K} \leqslant T_L \lesssim 100\,\mathrm{K}) \,. \tag{22.27}$$

where T_1'/E_m is for the moment only a characteristic parameter of the material. We will present a simple model which accounts for this effect. For details and partly similar, partly complementary approaches see [22.14, 34].

The relation to the effect discussed above comes from the consideration that the exciton temperature T_{ex} has the lattice temperature T_L as a lower limit and is close to this limit for low excitation intensities and small excess energies. In the following we thus assume $T_{ex} \approx T_L$. It is also reasonable to assume that only excitons with small (kinetic) energy $0 \leqslant E \leqslant E_m$ participate in the recombination process. This condition is fulfilled for radiative decay for free excitons

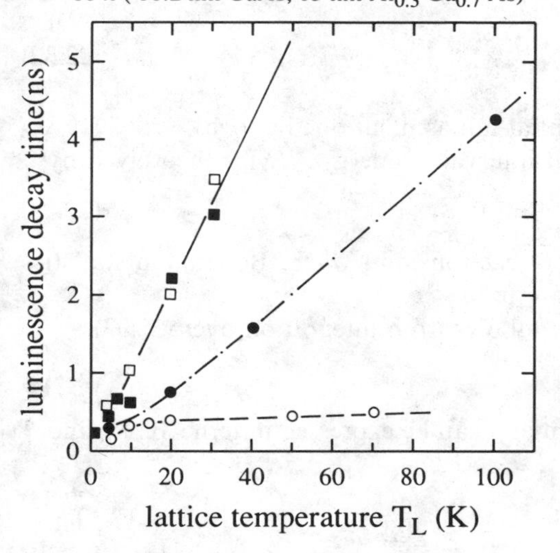

Fig. 22.15. The decay time of the luminescence in various $Al_{1-y}Ga_yAs/GaAs$ MQW samples as a function of the lattice temperature. According to [22.14]

with [see also (22.25)]

$$|\boldsymbol{k}_{\parallel}| \lesssim \frac{2\pi}{\lambda_v} \approx 8 \cdot 10^4 \,\mathrm{cm}^{-1} \tag{22.28}$$

where $\boldsymbol{k}_{\parallel}$ is the in-plane wave vector and λ_v the luminescence wavelength in vacuum, due to the conservation of the parallel component of $\boldsymbol{k}$.

The numerical values refer to $Al_{1-y}Ga_yAs/GaAs$(M)QW and $\boldsymbol{k}_{\parallel} = 0$ and $\boldsymbol{k}_{\parallel} = 2\pi/\lambda_v$ correspond to normal and grazing emission, respectively.

The contribution to E_m resulting from (22.28) is

$$E_m^0 = \frac{\hbar^2 \boldsymbol{k}_{\parallel}^2}{2M} = \frac{\hbar^2 2\pi^2}{M\lambda_v^2} \approx 10 \,\mu\mathrm{eV} \tag{22.29}$$

The value of E_m^0 can be increased by damping and by inhomogeneous broadening. This means that we can include a tail of localized states, provided that its tailing parameter E_0 fulfills the condition $E_0 \lesssim k_B T_L$, i.e., the localized excitons are in equilibrium with the free ones. From the absorption and luminescence linewidth of good samples (Fig. 11.14a) we can conclude that

$$E_0 < E_m \approx 0.5\text{–}1 \,\mathrm{meV} \tag{22.30}$$

so that the above conditions are fulfilled for $T_L \gtrsim 10\,\mathrm{K}$.

For $k_B T_L < E_m$ the excitons decay with an average lifetime around 200 ps (Fig. 22.14b curve 2). For $T > 10\,\mathrm{K}$ we give the following treatment.

The two-dimensional density of states in effective mass approximation $D(E)$ is given according to Sect. 2.6 by

$$D(E) = \begin{cases} D_0 & \text{for } E \geqslant 0 \\ 0 & \text{for } E < 0 \end{cases} \tag{22.31}$$

with

$$D_0 = g \cdot \frac{1}{2\pi} \frac{2M}{\hbar^2} \approx 10^{12} (\mathrm{cm}^2 \,\mathrm{meV})^{-1} \tag{22.32}$$

possibly including a small exponential tail as mentioned above.

The number of excitons per unit interval of energy $N(E)$ is then given by

$$N(E) = D_0 \exp[-(E-\mu)/k_B T] \quad E \geqslant 0 \,, \tag{22.33}$$

where μ is the chemical potential of excitons and where Boltzmann statistics and the weak coupling approach are used.

The total density N_0 at a time t follows from integration over (22.33):

$$N_0(t) = k_B T D_0 \exp[\mu(t)/k_B T] \tag{22.34}$$

This means that the chemical potential μ can be expressed in terms of N_0 and T

$$\mu = k_B T \ln \frac{N_0}{D k_B T} \,, \tag{22.35}$$

or

$$N(E) = D_0 \exp\left(-\frac{E}{k_B T} + \ln\frac{N_0}{D_0 k_B T}\right). \tag{22.36}$$

For the Boltzmann approximation to be valid we must have $\mu < 0$ or

$$N_0 < D_0 k_B T \approx 10^{12}\,\text{cm}^2 \quad \text{for} \quad T = 10\,\text{K}. \tag{22.37}$$

This condition is very well fulfilled for low excitation.

For a "monatomic" recombination process one generally uses the ansatz

$$-\frac{dN_0}{dt} = \frac{1}{T_1'} N_0. \tag{22.38}$$

This ansatz has to be modified because of momentum conservation since only a fraction γ of the excitons can participate in the recombination, namely those which are sitting in the interval E_m of (22.30) around the bottom of the band:

$$\gamma = \frac{N(E=0)E_m}{N_0}. \tag{22.39}$$

Equation (22.38) has then to be rewritten as

$$-\frac{dN_0}{dt} = \frac{1}{T_1'} \gamma N_0 = \frac{1}{T_1'} \frac{N(E=0)E_m}{N_0} N_0, \tag{22.40}$$

which means we have an effective lifetime T_1

$$T_1 = T'_1 N(E=0)\, E_m / N_0. \tag{22.41}$$

With (22.30) and (22.34–36) we finally get

$$T_1 = T'_1 \frac{k_B T}{E_m} \approx 200\,\text{ps}\,\frac{k_B T}{1\text{meV}} \quad \text{for} \quad T \gtrsim 10\,\text{K}, \tag{22.42}$$

in agreement with the experimental finding.

The linear increase of T_1 with temperature is sometimes used to argue that the recombination of excitons in quantum wells is essentially radiative [22.34]. This assumption may hold in some selected samples in which the spectral width of the free exciton emission is governed by E_m^0 in (22.29). For other samples, the above approach seems to be more realistic, especially if one recalls that MQW samples with comparable parameters can show different slopes in the relation $T_1 = f(T_L)$ (Fig. 22.15) and that some II–VI superlattices with extremely low luminescence yields also show a linear relation [22.35].

We refer the reader for more details and examples again to e.g. [22.2, 3, 16, 18–20] and references therein.

22.5 Bloch Oscillations

The Bloch oscillations or the Wannier–Stark ladder, which are observed in a superlattice (Sect. 10.8) if a dc electric field is applied in growth direction, are currently a very intensively investigated topic in semiconductor optics both from the experimental and the theoretical point of view. Therefore this section has been included in this book almost during proof-reading. Though the topic belongs basically to those presented in Chap. 16, we discuss it here, since it also involves aspects of nonlinear optics and of ultrafast spectroscopy.

The idea rather old and goes back to the 1930s [22.36–38]. We shall first introduce the concept of Bloch oscillations, explain why they have been observed only recently, and then present various approaches to understand the phenomenon and give experimental data. The presentation here is essentially based on various recent review articles [22.39–43]. Finally we outline some analogies and differences between the concept of Bloch oscillations in the presence of an electric field and the cyclotron resonance in the presence of a magnetic field. For the latter topic see also Sect. 16.1.

The basic concept of Bloch oscillations is based on the motion of a wave packet centered around $\boldsymbol{k}$ in a periodic lattice under the influence of an external force $\boldsymbol{F}$, e.g., an electric field $\boldsymbol{E}$, neglecting for the moment all scattering processes; see [22.36] or Sect. 10.4

$$\hbar \dot{\boldsymbol{k}} = \boldsymbol{F} = e\boldsymbol{E} \ . \tag{22.43}$$

Under the above conditions the electron will perform a periodic orbit in $\boldsymbol{k}$-space in the direction of $\boldsymbol{E}$ as shown schematically in Fig. 22.16.

The carrier starts, e.g., at point A at rest, and is accelerated by the external field moving thus towards B. At B the sign of the effective mass changes and the carrier is decelerated by the field but still increasing its momentum and energy according to (22.43) and reaching the right-hand edge of the first Brillouin zone at point C with zero group velocity. Instead of considering wave vectors beyond the first Brillouin zone, we can assume that the electron is transferred by a reciprocal lattice vector $\boldsymbol{G}$ from π/a to $-\pi/a$, i.e., to point D; see Sects. 9.2 and 10.1. From D, the wave packet is accelerated until it reaches point Ê and decelerated from E to A reaching point A again with zero group velocity. The duration of one orbit in $\boldsymbol{k}$-space T_B and the frequency ω_B are given by [22.39].

$$\omega = 2\pi / T_B = ae\,|\boldsymbol{E}|\,\hbar \ . \tag{22.44}$$

This periodic motion in $\boldsymbol{k}$-space is necessarily connected with a periodic motion in real space in the direction of $\boldsymbol{E}$ with an amplitude l_B [22.39]:

$$l_B = B/2e\,|\boldsymbol{E}| \ , \tag{22.45}$$

where B is the width of the band. The periodic motions in real and in momentum space are known as Bloch oscillations or as "Zener Pendeln" [22.36–38].

This phenomenon has been used to explain (partly erroneously) the I–V characterisitcs of Zener diodes in the blocking direction. It was assumed that the electrons in the valence band under the influence of the blocking voltage over the pn junction execute such a periodic orbit, and tunnel with a certain probability to the conduction band whenever they hit the top of the valence band.

Actually, the carriers in bulk semiconductors have no chance to complete such an orbit in a partly occupied band (in a completely filled band, there is no net motion at all due to Pauli's principle), since the T_2 times are too short. Even at low temperatures the T_2 values at the bottom of the band are only a few tens of ps and they tend towards ten fs deeper in the band; see Sect. 22.2. In order to fulfill the condition

$$\omega_B \cdot T_2 \geqslant 1 \tag{22.46}$$

which would be necessary to be able to observe Bloch oscillation one would need with (22.44) dc electric fields in excess of 10^7 V/m which would result in rapid destruction of the sample.

The way out of this dilemma follows from an inspection of (22.44). For a given T_2 an increase of the spatial period of the system results in a decrease of the electric field strength $\boldsymbol{E}$ necessary to fulfill (22.46). Since the lattice constant a of all simple semiconductors is comparable to the atomic diameters and thus of the order of 0.5 nm only the use of man-made artificial superlattices can help. With periods d around 10 nm we can expect a reduction of the field strength by more than an order of magnitude. In connection with Figs. 22.17 and 22.18, we recall and extend what we have learned about superlattices in Sect. 10.8.

The solid line in Fig. 22.17 gives the lower edge of the conduction band of the two materials forming the superlattice. The width of the barriers is so narrow that the electron wavefunctions of adjacent wells overlap. This overlap integral results in a finite band width (Sect. 10.1). The width of these minibands is directly proportional to the overlap integral and can thus be varied in this Kronig–Penney-like system by the thickness of the barriers. Typical widths of

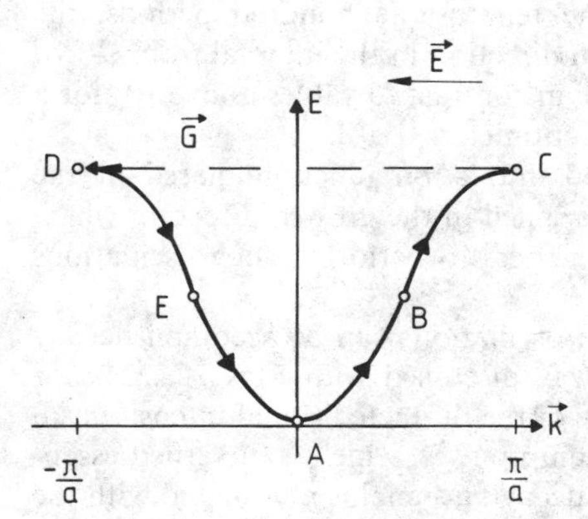

Fig. 22.16. Schematic presentation of the Bloch oscillation or "Zener Pendeln" of a charge carrier in an electric field in the absence of scattering

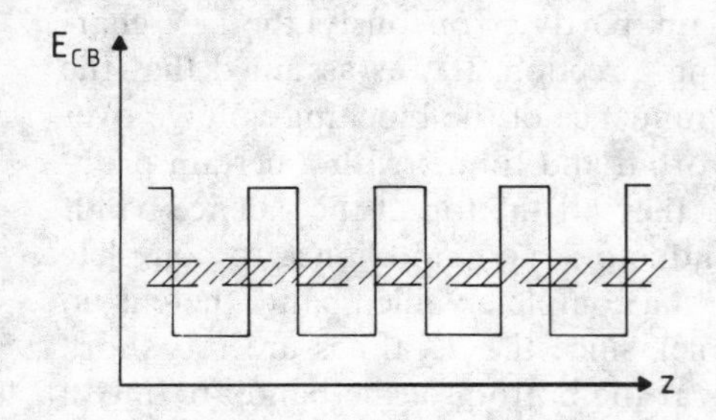

Fig. 22.17. The conduction-band edge of a superlattice and the resulting miniband for $k_x = k_y = 0$ indicated schematically by the *hatched area*

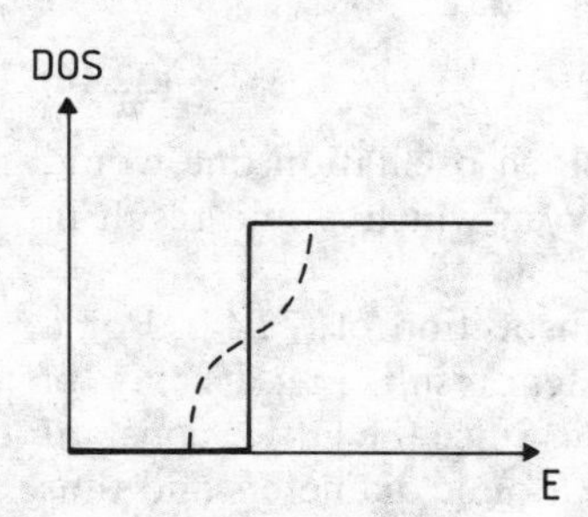

Fig. 22.18. The density of states (DOS) for a single or multiple quantum-well system (*solid line*) and for a superlattice (*dashed line*)

the minibands are of the order of a few meV, in contrast to about 10 eV for the width of the bands in bulk material. Note that the dispersion $E(\boldsymbol{k})$ and the motion of the carriers in the plane normal to the growth direction is still almost bulk-like. The formation of minibands in the valence band is generally much less pronounced, due to the greater effective masses of the holes, which reduces the tunneling into the barriers and consequently the overlap integral. The appearance of the minibands in the conduction bands also modifies the density of states in the effective-mass approximation from the step-like function of a two-dimensional system to the one given by the dashed line in Fig. 22.18.

Since superlattices usually have some ten to one hundred periods, the number of different $\boldsymbol{k}$-states in the growth direction in the interval from $-\pi/d$ to $+\pi/d$ is also some ten to one hundred, in contrast to values above 10^7 for a bulk crystal with a length of about one centimeter.

If such a superlattice is incorporated into a pn junction biased in the blocking direction, electric fields can be applied in the growth direction which are sufficient to cause optically excited carriers to perform Bloch oscillations with frequencies ω_B which fulfil (22.46).

The experimental observation of these oscillations can be accomplished by time-resolved four-wave mixing in the way discussed with Figs. 22.2, 24.8 or 24.10. There is an overall temporal decay of the diffracted signal intensity as a function of the delay between the two pump pulses, which is governed essentially by the dephasing time T_2, but this decaying signal is modulated with the

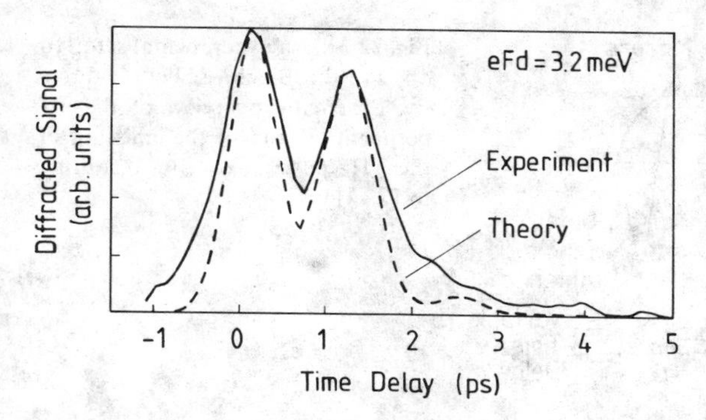

Fig. 22.19. The temporal evolution of the FWM signal in a GaAs/$Al_{1-y}Ga_yAs$ superlattice with a period of 11 nm under the influence of an electric field of 0.3×10^7 V/m. According to [22.39]

frequency ω_B. A rather early example for such an experiment is given in Fig. 22.19.

The frequency of the modulation varies in this type of experiment linearly with the applied field according to (22.44) giving support to the underlying concepts.

In a more elaborate model, detailed in [22.39, 43], the occurrence of the oscillations can be understood along the following lines. A pump pulse of duration of about 100 fs has a (Fourier-limited) spectral width comparable to, or larger than, the width of the minibands. It thus excites coherently all the several ten to one hundred band-to-band transitions in the miniband. Since each of these transitions has a different frequency, the polarizations connected with the transitions run out of phase with respect to each other. It can be shown, however, that the influence of an external field is such that the polarization of all these transitions rephase again after integer multiples of T_B. Consequently the signal peaks for delay times of the two pump beams which are integer multiples of T_B.

This picture has another consequence: If there is a macroscopic polarization in the medium, which oscillates with frequency ω_B, this frequency must be radiated. Indeed it was recently possible to observe the THz emission at frequency ω_B direclty [22.41, 42] by the use of a micro-dipole antenna which is gated for a sampling technique by a part of the 100 fs pump pulse. The experimental setup is shown in Fig. 22.20. In Fig. 22.21 the signal of the antenna is shown as a function of the delay time of the gate pulse.

The temperature of the GaAs/$Al_{1-y}Ga_yAs$ superlattice with a period of 11.4 nm was 10 K resulting in a reasonably long T_2 time. Beautiful oscillations with frequency ω_B can be seen. With increasing temperature, the decay of the envelope gets faster due to a decrease of T_2. More details of this experiment are given in [22.41, 42].

Before we proceed to have a look at Bloch oscillations from another point of view we make the following statement. Unitl now we presented the Bloch oscillations in the picture of free carriers. This is the original approach and it is easily intelligible from the didactic point of view. Actually we should

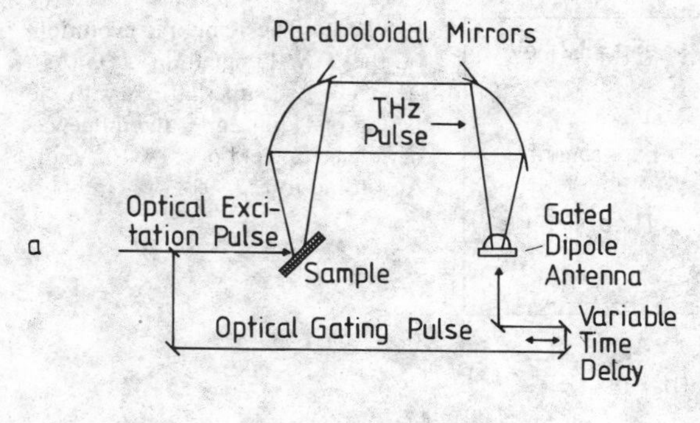

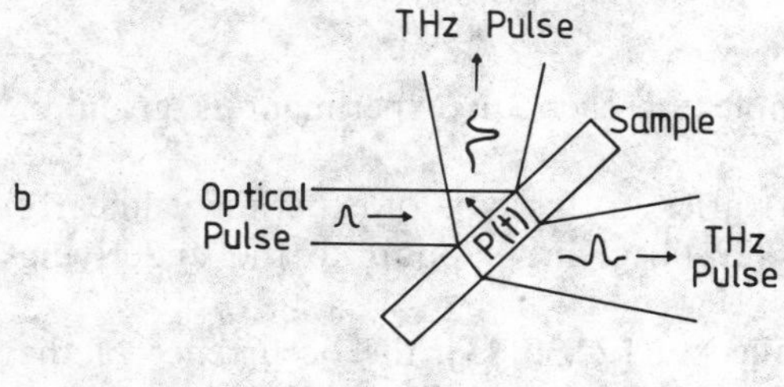

Fig. 22.20. An experimental setup to observe the Bloch oscillation directly. Schematic overview of the experiment (a) and of the emission of the THz pulse at ω_B (b). According to [22.41]

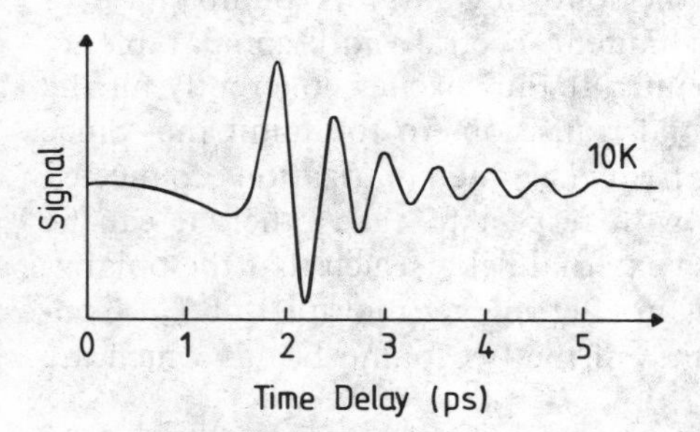

Fig. 22.21. Electric field at the antenna as a function of the delay of the gate pulse. According to [22.41]

remember, however, that by optical pumping, electron–hole pairs or excitons are created. At a first glance it seems difficult to understand the concept of Bloch oscillations for the neutral quasi-particle exciton. However, it can be shown theoretically [22.44] that the frequency ω_B survives when the exciton wave function is constructed from a superposition of electron and hole wave packets. This statement is a good example for the motto given at the beginning of this book.

Now we come back to the formation of the minibands. They appear, as already stated, from the overlap of the wavefunctions in the various quantum wells which would have all the same energy without this overlap in the some way as the bands appear from atomic orbitals in the bulk (Fig. 10.1).

If we apply a constant electric field along the structure of Fig. 22.17 the whole structure is tilted (Fig. 22.22), and the eigenstates in the various wells no longer have the some eigenenergy.

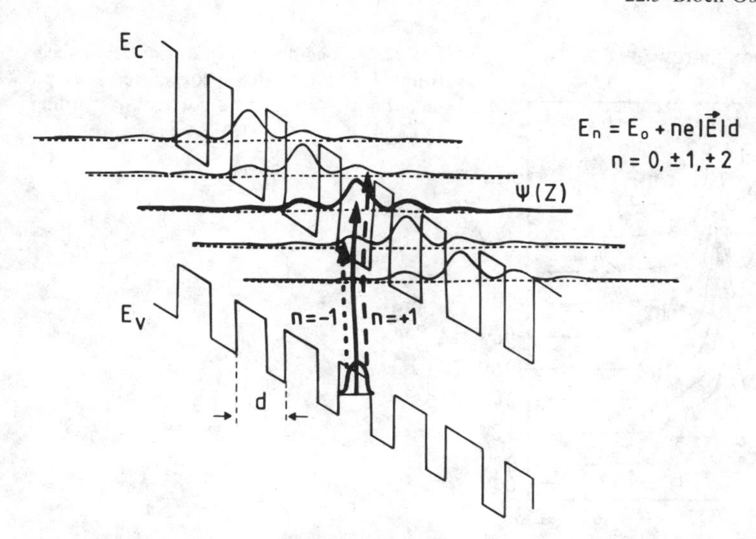

Fig. 22.22. The conduction bandstructure (similar to Fig. 22.16) and the valence bandstructure of a type-I superlattice, however now under the influence of a constant electric field in the growth direction. A wave function in the valence band is shown and several in the conduction band and three different transitions with energies E_0 and $E_0 \pm e|\boldsymbol{E}|\, d$. According to [22.41, 43]

As a consequence, a miniband is no longer formed and the wavefunctions, which are extended and of the Bloch-type without electric field, become localized in the growth direction extending only over a few periods of the superlattice. They have discrete energy levels.

$$E_n = E_0 \pm ne|\boldsymbol{E}|d, \quad \text{with} \quad n = 0,\ \pm 1, \pm 2, \ldots \tag{22.47}$$

This picture is especially appropriate for electric fields, which produce a potential drop over one superlattice period $e|\boldsymbol{E}|d$ which is larger than the width of the miniband without field.

The formation of the localized states is also known as Stark localization and the equally spaced ladder of eigenenergies of (22.47) is known as the Wannier–Stark ladder. In this picture, the temporal modulation of the FWM signal in Fig. 22.19 can be understood as quantum beats between adjacent band-to-band transitions with $\Delta n = \pm 1$. The (coherent) emission of quanta $\hbar\omega_B$ in Fig. 22.21 is then just a result of transitions between adjacent rungs of the ladder.

In addition, the transitions indicated in Fig. 22.22 suggest that the structures should also be visible in simple linear absorption spectroscopy and indeed they are. An example is given in Fig. 22.23. Photocurrent excitation spectra are shown instead of absorption because the substrate of the $GaAs/Al_{1-y}Ga_yAs$ superlattice which is incorporated in a pn junction is opaque; see Sect. 24.1.

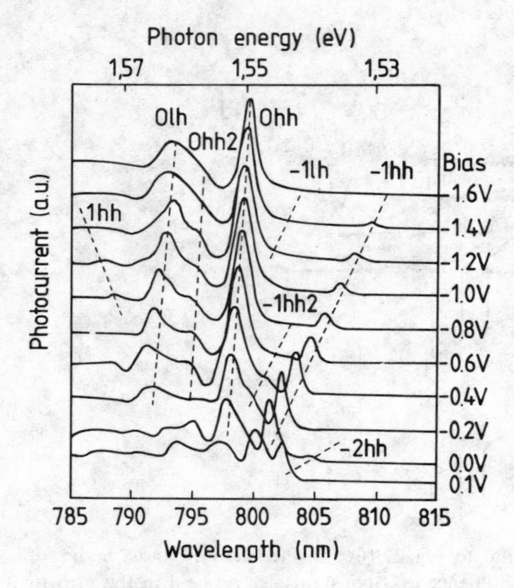

Fig. 22.23. The photocurrent excitation spectrum of a superlattice incorporated in a pn junction for various bias voltages. According to [22.43]

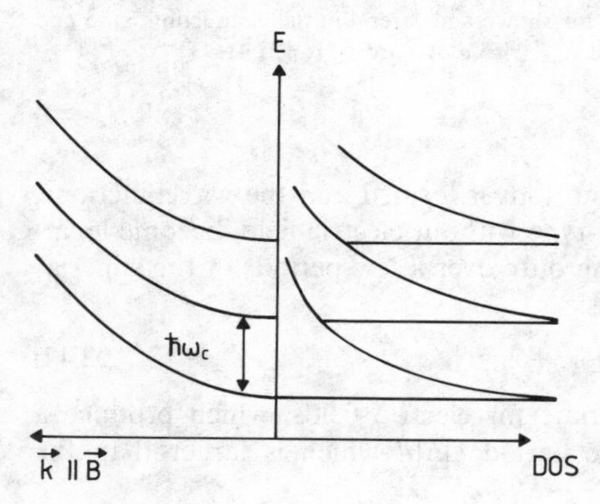

Fig. 22.24. The dispersion of the subbands resulting from the Landau levels and the corresponding density of states

The parameter in Fig. 22.23 is the bias voltage over the diode which is closely related to the electric field strength over the depletion region. A rather complex fan of optical transitions evolves with increasing blocking voltage. The following transitions can be identified: the quantized heavy hole to electron transition for $n = 0, \pm 1, -2$ denoted by 1hh, 0hh, -1hh and -2hh, two transitions involving the second quantized hh level denoted by 0hh2 and -1hh2 and one transition starting from the first quantized light hole level. The fan of the various transitions at least for higher fields, again follows the linear relations of (22.44) (with a being replaced by the superlattice period d) and (22.47).

The Bloch oscillation and the Wannier–Stark ladder which we have discussed now from various points of view evidently allow one to perform beautiful experiments and they may also offer prospects for application, especially concerning the creation of THz radiation. More information on these topics is found in [22.39–43] and references therein.

Now we address a rather well-known magnetic-field-induced effect, which has a lot of similarities but also some differences to the above-discussed electric-field-induced effects, namely the cyclotron resonance. Under the influence of a magnetic field $\boldsymbol{B}$ a particle with a velocity $\boldsymbol{v}$ performs a periodic orbit in the plane perpendicular to $\boldsymbol{B}$. For particles with an isotropic effective mass this orbit is simply a circle (again assuming negligible scattering) and the frequency around this circle is independent of $\boldsymbol{v}$ (we assume $|\boldsymbol{v}| \ll c$) and given by the cylotron resonance ω_c

$$\omega_c = 2\pi/T_c = e|\boldsymbol{B}|/m_{\text{eff}} \,. \tag{22.48}$$

The circular motion in real space is necessarily connected with a periodic orbit in $\boldsymbol{k}$ space. In contrast to the Bloch oscillations, however, these orbits occur at constant energy since the force $e\boldsymbol{v} \times \boldsymbol{B}$ is always normal to the orbit. If scattering processes are included, the condition for the observability of the cyclotron orbit is, in analogy with (22.46),

$$\omega_c T_2 \geqslant 1 \,. \tag{22.49}$$

In analogy to the THz emission of the Bloch oscillation, the cyclotron resonance can be observed in absorption or emission at frequency ω_c depending on the relative phase of the electromagnetic radiation and the particle in the cyclotron orbit.

In the regime of the excitonic continuum or of the band-to-band transition, the combination of electron and hole cyclotron energies can be seen, as shown in Fig. 16.6, which corresponds in this comparison to Fig. 22.23.

The analog of the Wannier–Stark ladder are here the Landau levels. The motion of the carrier is quantized in the plane normal to $\boldsymbol{B}$ resulting in a harmonic-oscillator-like ladder

$$E = E_0 + (n + 1/2)\hbar\omega_c + \hbar^2 \boldsymbol{k}_{\parallel B}/2m_{\text{eff}} \quad n = 0, 1, 2, \ldots \,. \tag{22.50}$$

The quasi one-dimensional free motion parallel to $\boldsymbol{B}$ results, in the effective mass approximation, in a DOS for every level of the ladder which varies with the inverse square-root of the kinetic energy, as found in Sect. 2.6 and shown in Fig. 22.24.

Quantum beats induced by a magnetic field are described in [22.45], though the explanation is in most cases not quite as simple as outlined here in connection with the Landau levels.

Further aspects of intersubband transition in a magnetic field and related topics are treated in [22.46] and references therein.

Exciting and beautiful new results can be expected in the field of ultrafast spectroscopy of semiconductors in the near future. They will comprise e.g. the first elementary interaction processes of the electromagnetic radiation field with the electronic system, where memory effects or non-Markovian behaviour is expected, or the coherently driven polarization field in the sense of Rabi flopping or the consistent description of the phenomena in the strong coupling limit, i.e. in the polariton picture. An important step into this direction is the transition from the optical Bloch equations to the semiconductor Bloch equations, though this approach involves an extremely high computational effort. First results into these directions will be found e.g. in [22.3] in the proceedings of NOEKS IV.

22.6 Problems

1. If you prefer the mathematical description of physics to the intuitive picture given here, then go through the concept of the optical Bloch equations, e.g., with the help of [22.10], and find out what the Rabi frequency, a $\pi/2$- and a π-pulse are, and what the relation to a spin system is (spin echo). By the way, the photon echo also works if pulses other than $\pi/2$ and π are used, but the description is more difficult.

2. Find some data for the electrical conductivity of standard semiconductors. Use the equations given in this chapter to deduce the momentum relaxation of free carriers in electrical transport at various temperatures and compare with the data for excitons. Are the above-mentioned formulas also adequate to analyse the electrical conductivity of a metal?

23 Optical Bistability

The widespread interest in semiconductors arises not only from the beautiful physics that can be done with them, but also from the fact that there are massive applications in present and future key technologies such as data processing. At present, the applications are based essentially on electronic transport devices like diodes, transistors, and thyristors which rely on the unique possibility to influence the conductivity of semiconductors by n and p doping. However, we are nowadays witnessing electro-optics and opto-electronics, i.e., the linking of electronic devices and light as a clear up-stream technology. A laser diode, as discussed in Sect. 21.1, is found in every CD player. In the future, optics may gain increasing importance in data handling and processing, starting with the transmission of optical signals through glass fibers and eventually extending to "optical computing". This term implies the use of pulses of light instead of voltage or current to represent, store, and handle data bits. Essential ingredients of this development are optically bistable memories, switches, and modulators which exploit the (electro-)optical nonlinearities that we treated in Chaps. 18–21.

As an example of the application of these nonlinearities we present here the optical bistability and some related phenomena.

We start with the basic concepts, proceed to some examples, and then discuss some aspects of the use of optical bistability both in applied and fundamental research. For more information on this topic or on the development of this field during its first decades we refer the reader to [23.1–7] and references therein.

23.1 Basic Concepts and Mechanisms

An optically bistable element shows a static hysteresis loop in the relation between incident and transmitted (or reflected) light intensities. In Fig. 23.1 we show two possible input–output characteristics. In the case Fig. 23.1a the device switches with increasing incident power at $I\uparrow$ from a state of low, into a state of high transmission and stays there for further increasing incident intensity I_{in}, but also for decreasing intensity. Only at $I_{\text{in}} = I\downarrow < I\uparrow$ does the system switch back to the state of low transmission.

In Fig. 23.1b the device switches with increasing incident intensity at $I\downarrow$ from a high transmission a lower one and back at $I\uparrow < I\downarrow$. This means that the

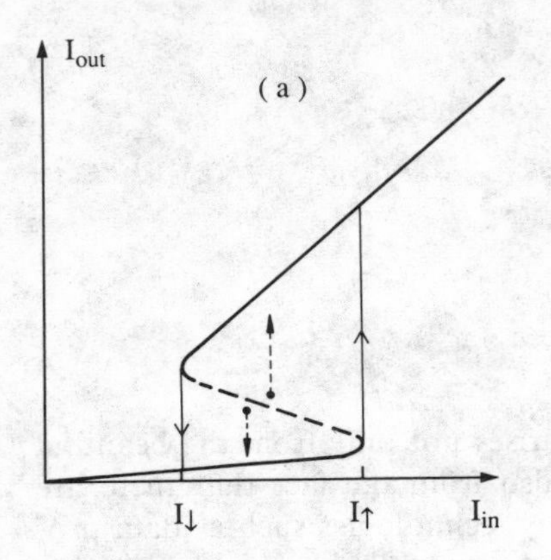

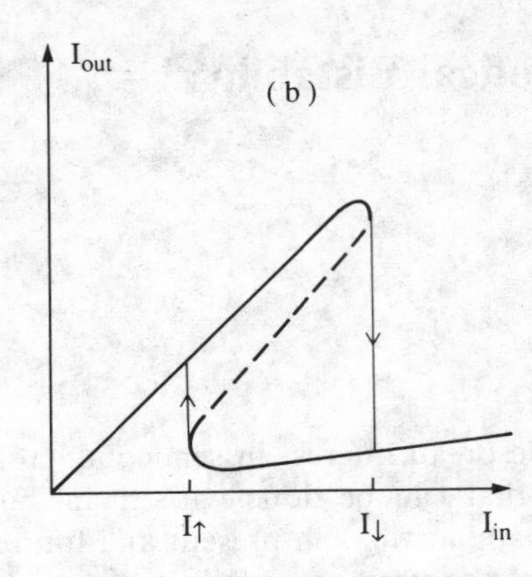

Fig. 23.1a, b. Input–output characteristics of optically bistable elements (**a**) anticlockwise; (**b**) clockwise

hysteresis loops in Fig. 23.1a and b are rotated anticlockwise and clockwise, respectively.

In both cases one has two stable output states between $I\uparrow$ and $I\downarrow$ and which one of them is realized depends on the history, i.e., on whether one comes from higher or lower intensities. This is the essence of optical bistability.

The dashed lines in Fig. 23.1 show a third, unstable branch. If the system is prepared by some means on this branch it can in principle stay there forever. However, the slightest deviation from this unstable branch (e.g., due to fluctuations) causes the system to move into a state on the upper or lower stable branch, as indicated by the dotted arrows. The dotted line thus separates the "basins of attraction" of upper and lower branches and is therefore also called "separatrix".

Only two ingredients are necessary to create optical bistability, namely a sufficiently strong (electro-)optical nonlinearity, e.g., of the types we have seen in Chaps. 18–20 or Sect. 16.2, and a suitable feedback.

It is not necessary that the optically nonlinear material is a semiconductor, it can equally be a gas like sodium vapor, but in line with the title of this book we concentrate here exclusively on semiconductors.

A hysteresis loop which is rotated anticlockwise as in Fig. 23.1a can be obtained in transmission with a dispersive nonlinearity and the feedback from a Fabry–Perot resonator. In reflection one then observes a hysteresis which is rotated clockwise as in Fig.23.1b. The hysteresis of Fig. 23.1a in transmission can be also obtained with bleaching of absorption and again a Fabry–Perot, while a hysteresis like Fig. 23.1b is obtained in transmission with excitation-induced absorption. In this case the feedback is "built-in", as we shall see later.

To further clarify the terminology we can say that dispersive nonlinearity means a change of the real part of the refractive index with increasing light intensity $I[\Delta n(I) \gtrless 0]$, bleaching means an excitation-induced decrease of the absorption coefficient or of the imaginary part of $\tilde{n}$, i.e., $\Delta\alpha(I) < 0$ or $\Delta\kappa(I) < 0$, and induced absorption means finally an increase of α or κ with increasing excitation, i.e., $\Delta\alpha(I) > 0$ or $\Delta\kappa(I) > 0$.

There are also some other mechanisms which can lead to optical bistability, but those mentioned above are the most important ones and we shall concentrate on these in the following examples.

23.2 Dispersive Optical Bistability

For dispersive optical bistability one starts with an incident wavelength corresponding to a transmission minimum of the Fabry–Perot (FP) resonator (Sect. 3.1.6). In this situation one has destructive interference of the partial waves reflected from the partly transmitting mirrors at the front and back of the FP. Consequently most of the incident intensity I_{in} is reflected and the intracavity intensity I_{intra} as well as the transmitted intensity I_{out} are much smaller then I_{in}. Now we increase I_{in}. Then I_{intra} increases, too. If I_{intra} becomes so large that it changes the refractive index n of the medium in the cavity, the positive feedback sets in: a change on n causes a change of the phase shift per round trip δ. As a consequence, the interference of the partial waves becomes more constructive whatever the sign of Δn, since we started with maximum destructive interference. Thus, I_{intra} increases, which changes δ even more, and so on. If this positive feedback is sufficiently strong, at a certain incident intensity $I\uparrow$, it can cause an abrupt transition to a state of high transmission close to a transmission maximum of the FP. Then we have a high I_{intra} which can even exceed I_{in}. We can then lower I_{in} below $I\uparrow$ and still keep the system in the highly transmitting state. Only at some value $I\downarrow$ of I_{in} with $I\downarrow < I\uparrow$ does the system return to the low transmission state.

To complement these intuitive arguments we now describe the process mathematically. We have two equations for the transmission of the Fabry–Perot resonator which have to be solved simultaneously. One is the Airy function (3.42, 43), repeated here for convenience:

$$T_{FP} = (1-R)^2\{[\exp(\alpha L/2) - R\exp(-\alpha L/2)]^2 + 4R\sin^2\delta\}^{-1} = \frac{I_{out}}{I_{in}}, \quad (23.1a)$$

$$T_{FP} \approx \frac{(1-R)^2}{(1-R)^2 + 4R\sin^2\delta} \quad \text{for} \quad \alpha L \ll 1, \quad (23.1b)$$

where R is the reflectivity of one mirror.

Equation (23.1b) holds for weak absorption, and we assume for the moment that α is small and does not change much with I_{intra} so that (23.1b) can be used.

The refractive index and thus the phase shift δ is assumed to depend on I_{intra} in the simplest case in a linear way; see (18.3):

$$\delta = \delta_0 + \Delta\delta(I_{\text{intra}}) = n\omega L_c^{-1}(1 + n_2 I_{\text{intra}}) \tag{23.2a}$$

$$\text{or} \quad \Delta\delta \propto I_{\text{intra}} \,. \tag{23.2b}$$

The second equation for the transmission reads

$$T_{\text{FP}} = I_{\text{intra}}\left(\frac{1-R}{1+R}\right)\frac{1}{I_{\text{in}}} \,. \tag{23.3}$$

Equations (23.1) and (23.2) give the transmission as a function of I_{intra}, and (23.3) gives a straight line with a slope inversely proportional to I_{in}.

In Fig. 23.2a we show both curves. The solutions are the intersections and are shown in Fig. 23.2b. If we start with small I_{in} we have an almost vertical line from (23.3) and only one solution. With increasing I_{in} two new solutions appear, but the system still remains in the low transmission state. At $I_{\text{in}} = I\uparrow$ this lower solution disappears (point A) and the FP jumps to the new solution (point B). The spike is a transient feature which can be observed if the transition from A to B is so slow that the system can be followed through the transmission maximum. For further increasing I_{in} we remain on the upper branch. For decreasing I_{in} this remains true down to $I\downarrow$ where this solution disappears (point C) and the system has to return to the lower state at point D.

The intermediate solution appearing in Fig. 23.2a, e.g., at point E and indicated in Figs. 23.1a and 23.2b by dashed lines, can be shown to be unstable in the sense mentioned above.

Since for low absorption the intensity reflected from a FP is just the complement of the transmitted intensity, i.e.,

$$\frac{I_r}{I_0} = 1 - \frac{I_t}{I_{\text{in}}} \,, \tag{23.4}$$

one can observe hysteresis which is rotated clockwise in reflection.

In Fig. 23.3a, b we show this situation for an idealized, lossless FP and Fig. 23.3c, give an example. The nonlinear medium is a semiconductor-doped glass, coated on both plane parallel surfaces by dielectric mirrors to form the FP. The optical nonlinearity is a photo-thermal one (Sect. 19.6), namely the increase of the refractive index with increasing temperature caused by the small absorbed fraction of the intracavity intensity.

Figure 23.4 shows another example of dispersive optical bistability. In this case the dispersive nonlinearity of CdS observed under high excitation is exploited (Fig. 20.11a). The surfaces of the platelet type sample are coated with dielectric mirrors with $R \approx 0.6$. In this case pulses of a duration of only 15 ns are used. The switching processes, which occur on a (sub) ns time scale, are clearly seen in Fig. 23.4a, b. The resulting hysteresis loop is given in Fig. 23.4c. Many other examples are given in [23.1–7].

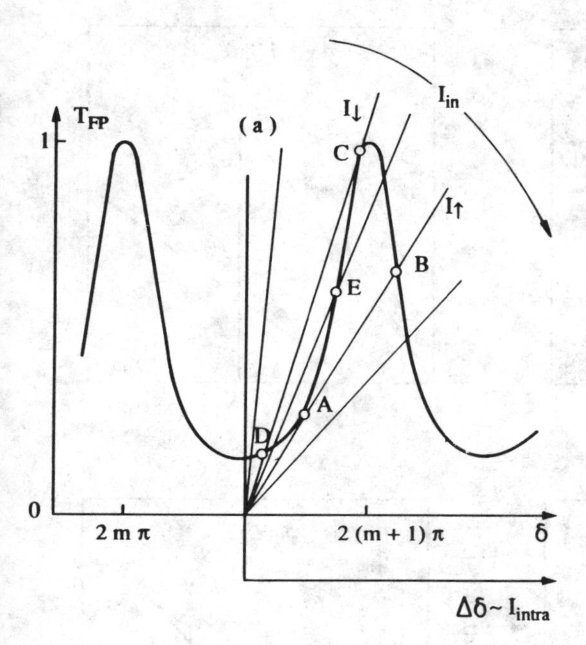

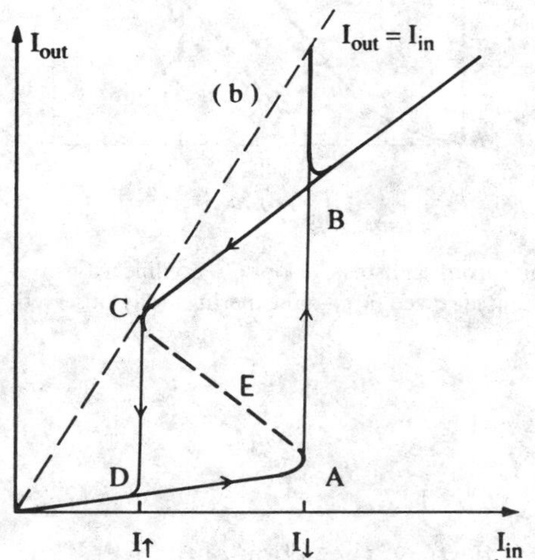

Fig. 23.2a, b. The graphical solution of (23.1b, 2, 3) (**a**) and the resulting input–output characteristic (**b**)

23.3 Optical Bistability Due to Bleaching

A hysteresis loop similar to those of Figs. 23.1a, 23.2b, and 23.4c can be created by bleaching of absorption.

We start with a FP resonator with an absorbing medium and use a wavelength for which a transmission maximum would occur if the medium in the resonator were transparent.

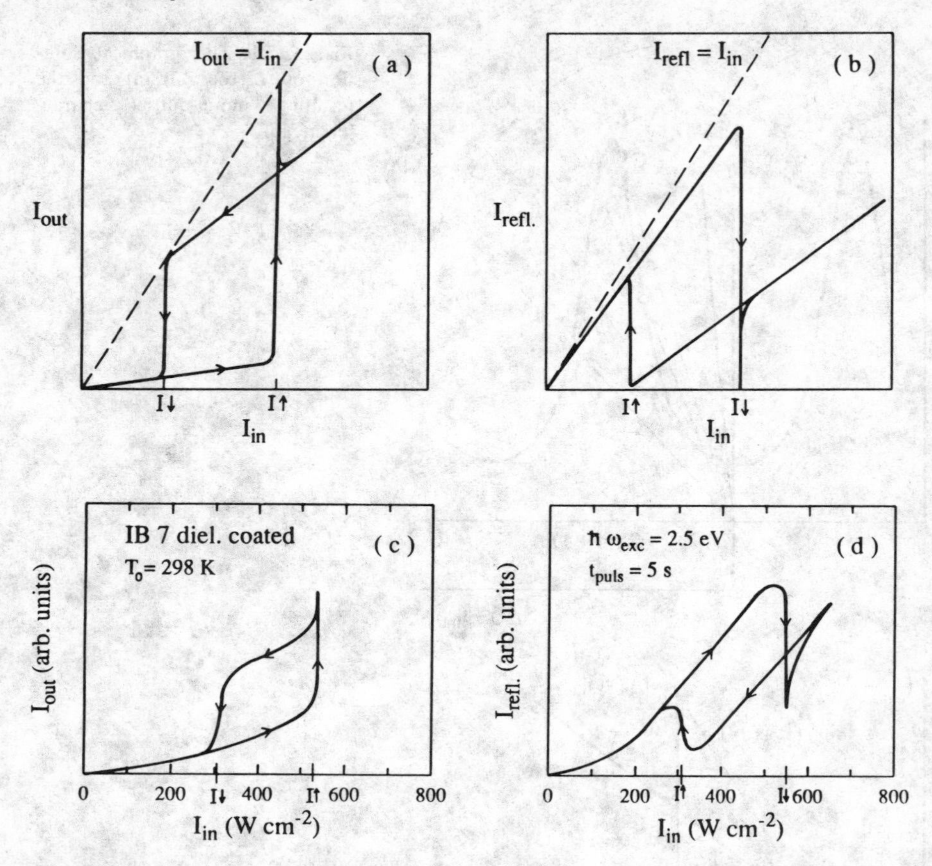

Fig. 23.3a–d. The bistable hysteresis loops resulting from a dispersive optical nonlinearity in a Fabry–Perot resonator in transmission and reflection; idealized (**a, b**) and experimentally observed (**c, d**). According to [23.8]

This means

$$\sin^2\delta = 0 \quad \text{but} \quad T_{\mathrm{FP}} \approx (1-R)^2 \exp(-\alpha L) \ll 1 \,. \tag{23.5}$$

The absorption of the medium with $\alpha L \gg 1$ destroys the action of the FP. If the medium starts to bleach with increasing I_{in}, however, the constructive interference in the FP comes into play and increases I_{intra} even further, causing an even stronger bleaching, and so on.

I_{intra} can again become larger than I_0 and T_{FP} reaches unity.

The construction of the bistable loop is similar to that described in Sect. 23.2 above with the main difference that the nonlinearity now occurs in the $\alpha(I_{\mathrm{intra}})$ instead of the $\delta(I_{\mathrm{intra}})$. In Fig. 23.5 we give an experimental example. The nonlinearity is the bleaching of the absorption edge in CdS at room temperature due to the formation of an electron–hole plasma

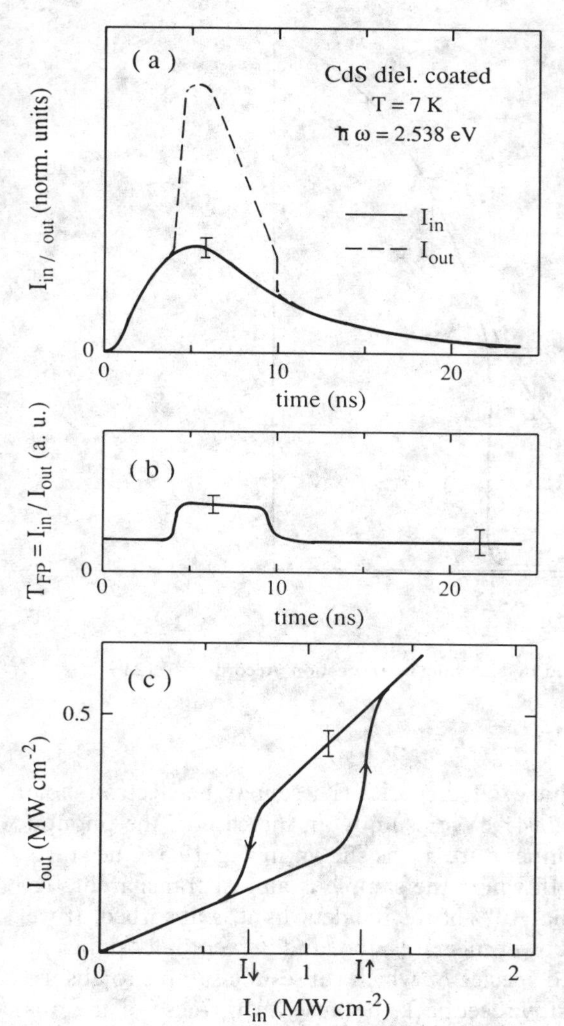

Fig. 23.4a–c. The dispersive optical bistability of a CdS platelet with dielectric coatings observed in the temporal evolution of the incident and transmitted pulses (**a**) and in the transmision (**b**). Part (**c**) shows the resulting hysteresis loop. According to [23.9]

(Fig. 20.11c). Dielectric mirrors have been evaporated directly onto the plane parallel surfaces of the platelet type crystal.

Since bleaching is often connected with dispersive changes due to Kramers–Kronig relations, hysteresis loops like that in Fig. 23.5 are often the result of a cooperative effect of bleaching and dispersive nonlinearities.

23.4 Induced Absorptive Bistability

We first explain the process of induced absorptive optical bistability in words. The idea is the following: A sample is used whose absorption increases as a

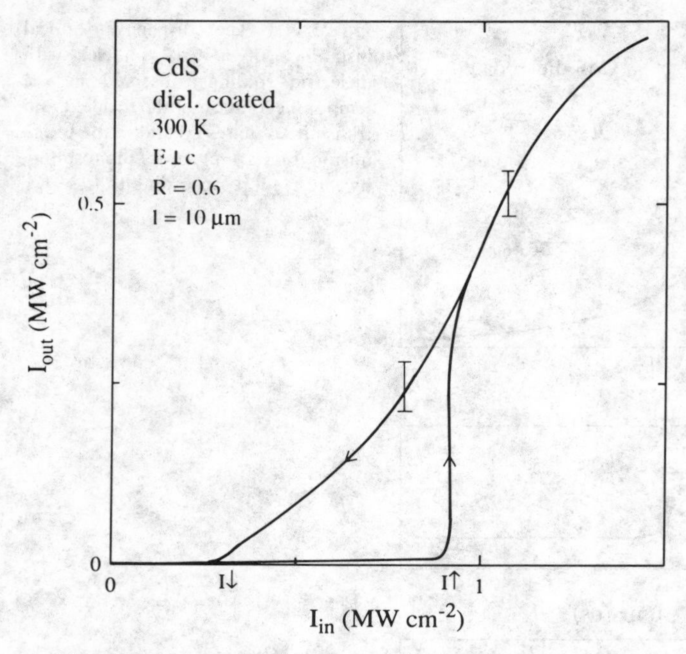

Fig. 23.5. Bistable hysteresis loop caused by bleaching of absorption. According to [23.9]

function of the density N of some excited species. These may be electron–hole pairs as in Fig. 20.11a around 2.541 eV or simply an increase of the phonon population, i.e., of the lattice temperature T_L as shown in Fig. 19.5. One starts at low excitation at a wavelength where the sample is almost transparent. To give a number, let us assume that 10% of the incident light is absorbed. If we increase I_{in}, then the absorbed power increases, too, and is eventually sufficient to raise the density of the excited species N, which causes the sample to absorb more. This gives a built-in positive feedback, because an increase of the absorption in turn increases N. If the effect is sufficiently strong the sample can switch at $I\downarrow$ in Fig. 23.1b to a state of low transmission and high absorption.

We assume now that 80% of the incident light is absorbed. Because of this high value, we can lower I_{in} to values below $I\downarrow$ and still keep the sample on the low transmission branch down to $I\uparrow(<I\downarrow)$ where the sample switches back again. In our example the two intensities would obviously be connected by

$$I\downarrow \cdot 0.1 = I\uparrow \cdot 0.8 \,. \tag{23.6}$$

Note the interchange of $I\downarrow$ and $I\uparrow$ between Figs. 23.1a and b and between Figs. 23.2b and 23.6b.

The most important difference between the induced absorptive bistability and those relying on bleaching or on dispersive nonlinearities is the fact that

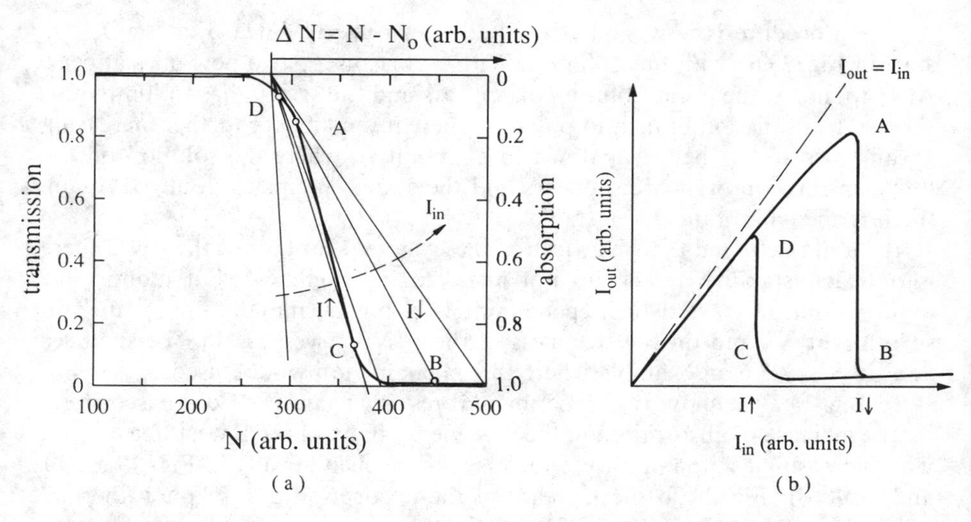

Fig. 23.6a, b. The graphical solution of (23.7–9) (**a**) and the resulting input–output characteristic (**b**). In this case N is the lattice temperature and the data refer approximately to a CdS platelet with $d < 10\,\mu\mathrm{m}$, $\boldsymbol{E} \perp \boldsymbol{c}$ and $\hbar\omega \approx 2.4\,\mathrm{eV}$. According to [23.10]

no external FP resonator is necessary to produce the positive feedback since it is "built-in".

To calculate the hysteresis we again have to solve two simultaneous equations, as shown in Fig. 23.6a. One is the transmission as a function of N:

$$T = \exp\{-\alpha(\hbar\omega, N)L\} = \frac{I_t}{I_0} \tag{23.7}$$

represented in Fig. 23.6 by the heavy line.

The other equation states that the increase of N, i.e.,

$$\Delta N = N - N_0 \tag{23.8}$$

is a function of the absorbed fraction of the incident light. In the simplest case a linear relation can again be assumed

$$\Delta N = A I_{\mathrm{in}}\,, \tag{23.9a}$$

with

$$A = 1 - T = 1 - \exp[-\alpha(\hbar\omega, N)L] \tag{23.9b}$$

or

$$T = 1 - \frac{\Delta N}{I_{\mathrm{in}}}\,. \tag{23.9c}$$

So we once more have a straight line with a negative slope inversely proportional to I_{in}.

The procedure is now similar to that shown in Fig. 23.2a. For low I_{in} we start in Fig. 23.6a with one solution. With increasing I_{in} two new ones appear. At $I\downarrow$ (point A) the initial solution disappears and the system has to jump onto the low transmission branch to point B. There it remains for further increasing I_{in} and also for I_{in} below $I\downarrow$ down to $I\uparrow$ (point C) where the solution on the low transmission branch disappears and the system jumps to point D. Again, the intermediate branch is unstable.

It should be noted that the width of the hysteresis, or indeed the fact if there is optical bistability at all and not just strongly nonlinear, but monostable input–output characteristic, depends critically on the initial value of the absorption at N_0 and on the steepness of the $T(N)$ curve [23.11]. We shall see later in Sect. 23.6 an example where this phenomenon will be used.

In Fig. 23.7 we show two experimental results, again for CdS, based in (a) on the increase of absorption on the low energy flank of the lowest free exciton with increasing exciton density, which is shown schematically in Figs 19.5, 20.1 and 22.4c. In Fig. 23.7b the increase of the absorption due to photothermal nonlinearities is exploited. (see Fig 19.16)

In Fig. 19.16 we have also seen that photothermal nonlinearities can lead to an increase of both n and of κ. In Fig. 23.8 this fact is used to demonstrate a "butterfly" hysteresis of a CdS sample with dielectric reflecting coatings.

The sample is first in a transmission minimum of the FP. With increasing I_0 it switches to the highly transmitting state at $I_1\uparrow$. A further increase of I_0 results at $I_1\downarrow$ in an induced-absorptive switch down, which simultaneously switches off the FP. If we now decrease I_0 we have at $I_2\uparrow$ the switch back of the induced-absorptive bistability and the system is again on the upper branch of the dispersive bistability from which it switches down at $I_2\downarrow$. For more details of this phenomenon see [23.13–15].

There are several further topics connected with (photothermal induced-absorptive) optical bistability which we mention here only briefly, referring the reader for more details to the literature.

Lateral structure formation occurs , e.g., when a spatially wide holding beam with an intensity in the bistable regime is switched in a spot and the resulting "switching waves" has been treated in [23.15]. The use of photothermal optical bistability as a temperature sensor is investigated in [23.15] and the relation between the dynamic width of the hysteresis loop and the duration of the incident pulse is analyzed in [23.16] and references therein.

23.5 Electro-optic Bistability and Devices

As a further example of optical bistability, we present an electro-optic device that relies on the quantum-confined Stark effect (QCSE) introduced in the Sect. 16.2. The mechanism is the excitation-induced increase of absorption [23.17].

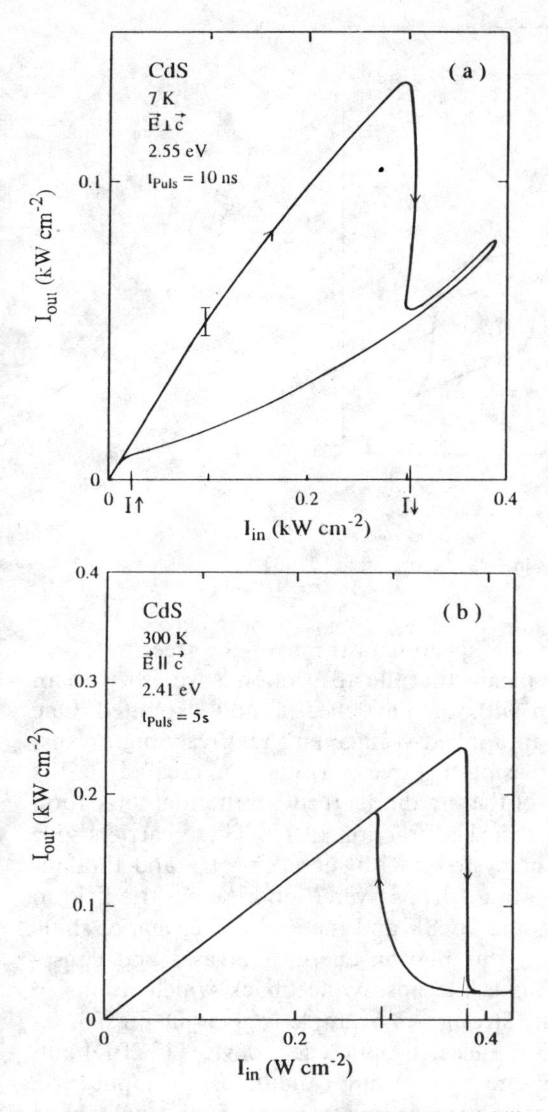

Fig. 23.7a, b. The induced absorptive optical bistability demonstrated with a photo-electronic (**a**) and a photothermal nonlinearity (**b**). According to [23.12] and [23.13], respectively

The basic concept is outlined in Fig. 23.9a. A constant voltage u_0 is applied in the blocking direction to a MQW pin structure through a resistor *R*. If the semiconductor is not illuminated, or only weakly illuminated, it has an internal resistivity $R_i \gg R$ so that most of the voltage drops over the diode. Consequently, the exciton resonances are red-shifted due to the QCSE as shown in Fig. 16.10 or in Fig. 23.9b. The photon energy $\hbar\omega$ of the incident light is

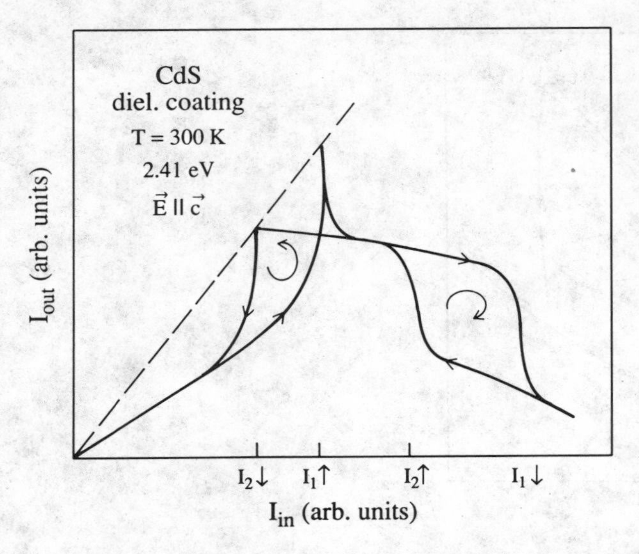

Fig. 23.8. "Butterfly" type hysteresis loops in CdS. According to [23.13]

chosen so that it corresponds to the spectral position of the free hh $n_z = 1$ exciton without electric field. This means that the absorption is weaker with an electric field over the MQW than without. This effect is now exploited. One starts at low light input I_{in} with an applied voltage and relatively high transmission (Fig. 23.9b, c). With increasing I_{in} free carriers are created in the MQW, i.e., in the intrinsic region of the pin diode, through thermal ionization of the excitons produced by the absorbed fraction of I_{in}. These carriers give rise to a photocurrent through the system. With increasing I_{in}, and thus increasing photocurrent, a greater voltage drops over R and the electric field in the MQW decreases. This reduces the QCSE and the exciton resonance shifts towards $\hbar\omega$, i.e., the absorption at this photon energy increases and consequently the photocurrent, too. This is the positive feedback which results in optical bistability if it is sufficiently strong. An example is given in Fig. 23.9c.

These devices have been called self-electro-optic effect devices (SEED) and are presently considered to be the most promising candidates for application in optical data handling (see below). The devices used in this field usually have a more complex structure. We show an example in Fig. 23.9d. The device contains two pin diodes and a field effect transistor (FET). The first diode is essentially used as a photodetector. The photocurrent is amplified in the FET and the output voltage modulates the transmission of the second pin diode via the QCSE. These relatively complex devices have better characteristics for quantities such as the constrast ratio, the width of the hysteresis loop, etc., as compared to the simple device in Fig. 23.9a, c. They can be integrated on a chip in two-dimensional arrays. A recent review of various types of these SEED devices of the "second generation" is found in [23.6, 17].

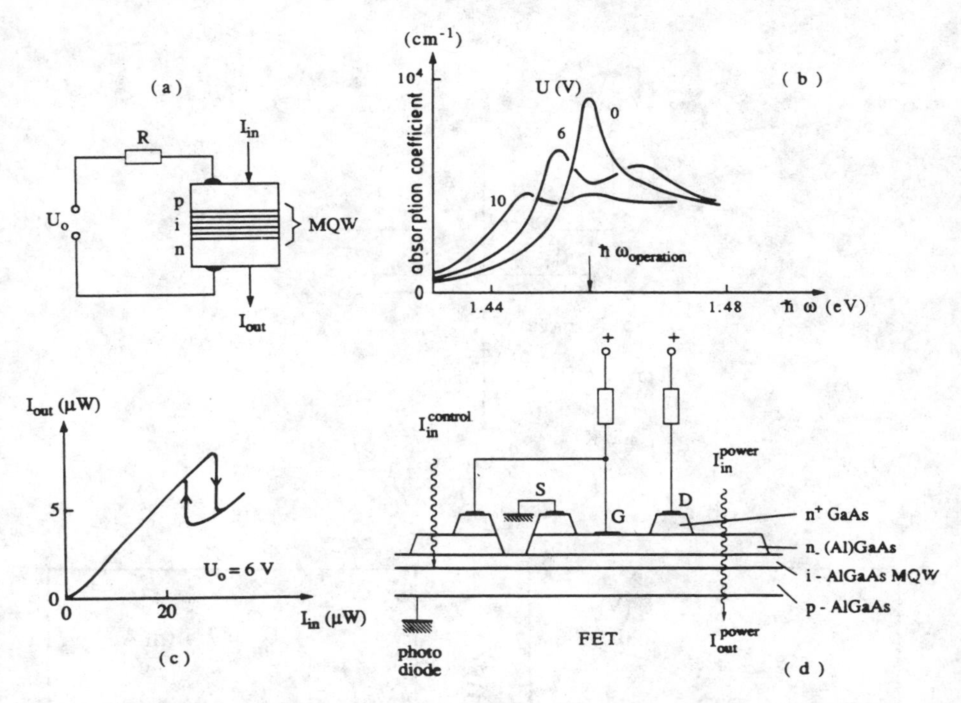

Fig. 23.9a–d. The basic concept of a self-electro-optic effect device (SEED) (**a**), the QCSE in an AlGaAs MQW pin diode (**b**), its input–output characteristics (**c**), and a schematic drawing of an advanced device (**d**). According to [23.17]

Finally it should be mentioned that photo-thermal SEEDs have been developed on the basis of Si, CdS, and other semiconductors [23.18–21]. One again uses the induced absorptive optical bistability shown in Fig. 23.6b and evaporates some transparent (ohmic) contacts onto the sample. An applied voltage helps, via the photocurrent, to increase the sample temperature. This means that the battery delivers a part of the energy necessary to switch the sample. The incident light intensity can be correspondingly reduced. Furthermore, the device can be driven through a bistable loop by varying the incident light intensity at constant applied voltage and vice versa.

The switching process shows up both in the transmitted optical power and in the photocurrent. An example of the bistability is shown in Fig. 23.10.

To conclude this section, we mention two cases of strongly nonlinear optical systems in which internal electric fields play a dominant role.

The first are nipi structures (Sect. 14.7 and Fig. 10.14i). The opposite space charges from ionized donors and acceptors in the nipi structures produce internal electric fields, and a bandstructure which is indirect in real space. If carriers are brought into the spatial minima of the bandstructure, they will start to compensate the space charges of the ionized donors and acceptors

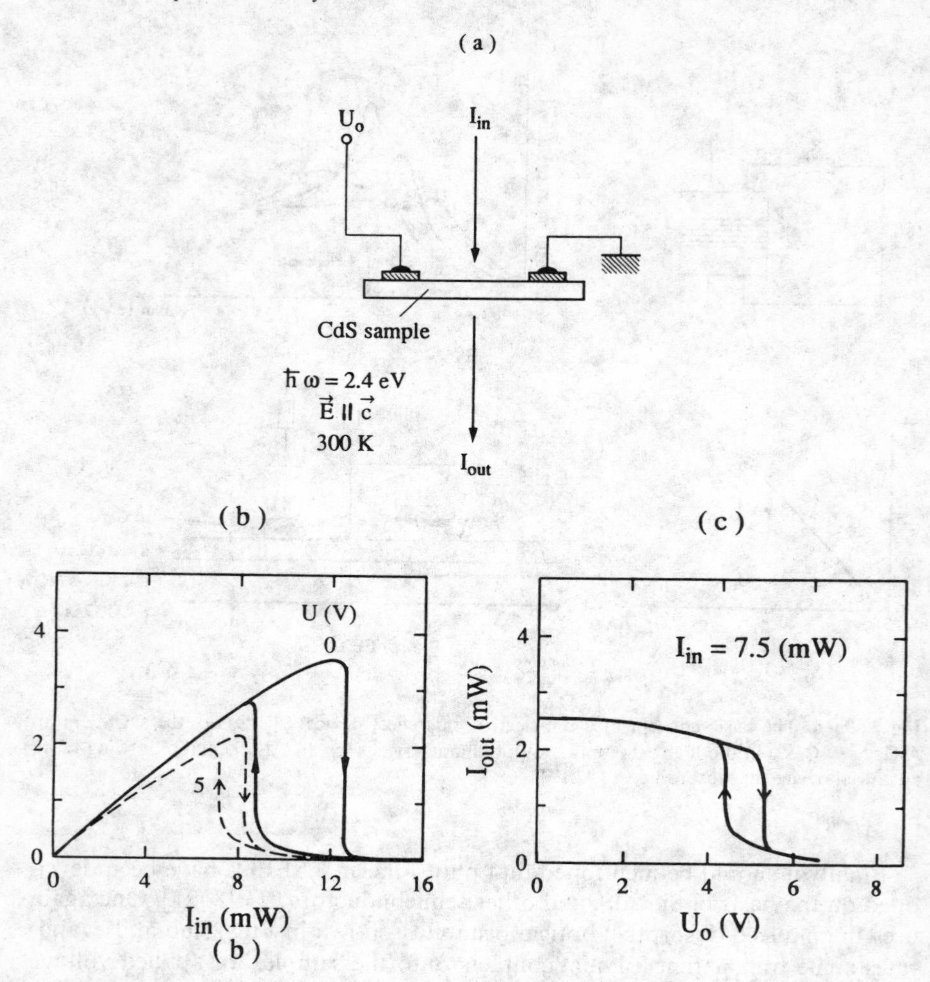

Fig. 23.10a–c. Some of the electro-optic input–output characteristics of photothermal SEED devices. According to [23.19]. Schematics of the setup (**a**), the optical input–output characteristics for two different applied voltages u (**b**), and the optical output as a function of the applied voltage for constant optical input (**c**)

and the width of the spatially indirect, forbidden gap will increase since the band-structure develops towards the usual flat-band situation and the absorption edge shifts towards higher photon energies. The above-mentioned carriers can be injected electrically by selective contacts to the n and p regions and a forward bias voltage, or by optical excitation. The combination of these two ways to change the absorption and the possibility to integrate quantum wells into the nipi structures gives the imagination of the physicists a wide field to play, to plan, to propose, and to realize various types of electro-optic switches, memories, or modulators. For recent results see [23.22] and references therein.

The other example concerns piezo-electric devices. Semiconductors with at least partly ionic binding and without inversion symmetry usually manifest the piezo-electric effect, i.e., the appearance of a macroscopic polarization and of an electric field as a consequence of a lattice deformation (and vice versa).

In strained-layer superlattices, layers of different lattice constants are grown on top of each other. As long as the thickness of these layers is below a critical thickness for the formation of misfit dislocations, strong piezo-electric fields appear in the strained layers; see [23.23, 24] and references therein. An especially instructive example is provided by CdS/CdSe layers grown on GaAs(111) with the hexagonal ***c***-axis in the growth direction, since the piezo-electric effect is quite strong in this material. Due to the common cation rule, the conduction band of the CdS/CdSe is relatively flat with a small type-II offset of about (0.1 ± 0.1) eV [23.24]. The rest of the band offset of 0.6 eV is located in the valence band.

Due to the alternating piezo-electric field, the bandstructure at low carrier density appears as in Fig. 23.11a and the transition between the highest valence band state and the lowest conduction band state is significantly red-shifted with respect to both bulk CdSe and the situation of Fig. 23.11b. This red-shift can be understood in analogy to the quantum-confined Stark effect of Sect. 16.2 and used in the SEED devices presented above, with the main difference that the field in the latter case is applied from outside.

If carriers are now created in this structure, the piezo-electric field is screened and the bandstructure develops from the situation of Fig. 23.11a towards that of Fig. 23.11b. As a consequence, the spatially indirect gap shifts to higher photon energies similar to the case of nipi structures or SEED devices. This shift manifests itself in a blue-shift of the emission. The effect starts even at the very low light intensities produced by a 1 mW HeNe laser, a prerequiste for technical applications. At higher, pulsed excitation even stimulated emission begins. A total blue shift of the emission of 0.4 eV with increasing excitation has been reported in [23.24]. Since the CdS/CdSe system can be integrated on GaAS and matches the wavelength of $Al_{1-y}Ga_yAS$/GaAs MQW, exciting prospects for application exist in the future.

23.6 Applications

As mentioned in the introduction, semiconductors have a very wide field of applications especially in electronics and opto-electronics. It is not the aim of this book to cover this field, but we refer the reader to [23.25] and references therein.

What we want to do here is to outline two aspects of applications of (electro-) optical bistability in (electro-) optic data processing, also known as "optical computing" [23.3, 6] and in the realisation of nonlinear-dynamical or "synergetic" systems [23.7].

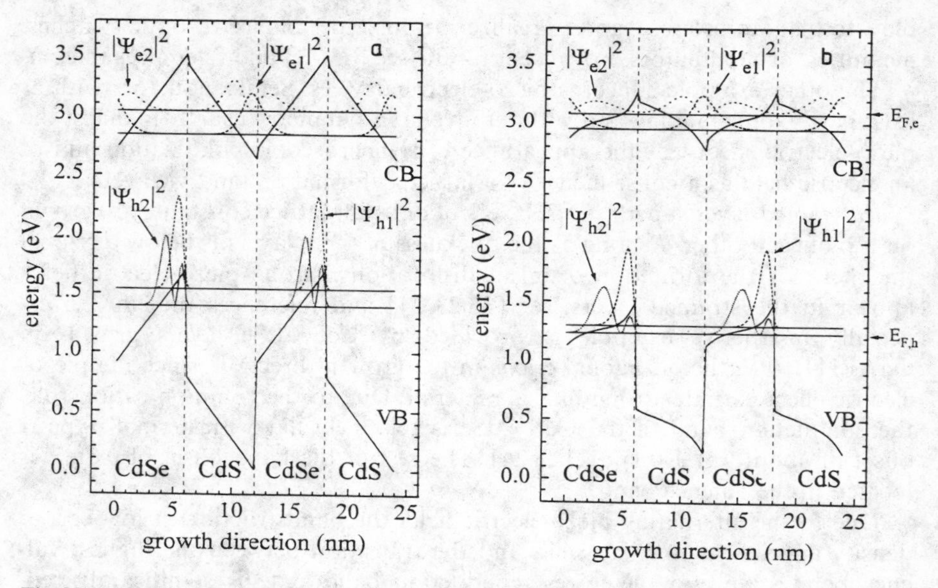

Fig. 23.11a, b. The band structures and envelope wave functions of a 6 nm CdS/6 nm CdSe superlattice under the influence of the piezo-electric fields (**a**) and the screening of these fields for high carrier concentration around $10^{13}\,cm^{-2}$(**b**). According to [23.24]

23.6.1 Optical Computing

An optically bistable device with an input–output characteristic such as those shown in Fig. 23.1 can be considered as a binary memory if, in the bistable region, we identify the states of low and high transmission with the logic states "zero" and "one", respectively.

By a suitable choice of the nonlinearity and the excitation conditions, the width of the bistable loop can be made rather narrow. In this case the devices can be used as "AND", "OR" and "NOT" gates (or inverters) as shown schematically in Fig. 23.12. These three functions, together with a memory, are the basic ingredients needed to construct a computer. When this fact became clear in the early nineteen-seventies, there was a very enthusiastic period, where scientists hoped that it would be relatively easy to build an all-optical computer based on the above devices [23.2, 3].

Indeed, such an optical computer would in principle have one big advantage over electronic computers and this is massive parallel data handling.

What one wants to do basically in a computer is the following (Fig. 23.13): One has a certain input signal which is processed in a logic unit. The processing depends on the information stored in a large number of memories and results in an output signal.

In electronic computers one faces the following problem: If there are N memory cells, the processing unit has to be connected with $2N$ wires if one

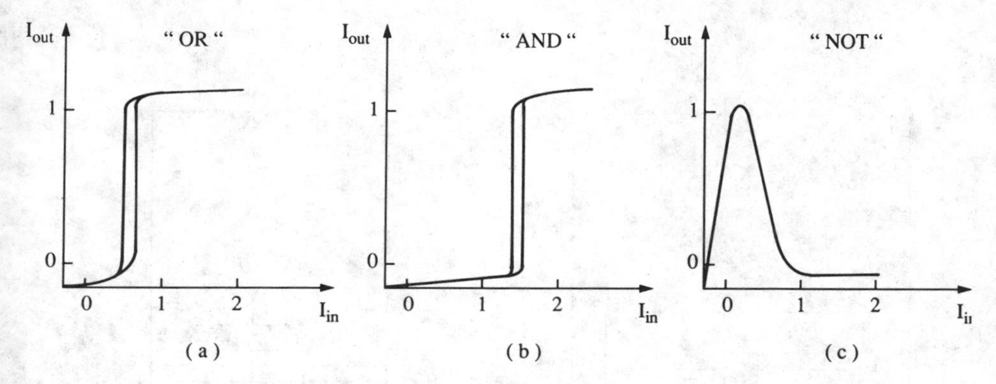

Fig. 23.12. The use of nonlinear optical devices as logic gates

wants to simultaneously know the contents in all memories for the logic decision, since in one wire only one signal can propagate at one (cycle-)time (Fig. 23.13a). If N becomes large ($N \geqslant 10^6$) this concept breaks down because it is impossible to have 2×10^6 pins on one chip. The ingenious way out of this problem was serial data processing. Here one has only two connections between the logic unit and the memory and one uses an address and requests the information in the memories serially, i.e., one after the other (Fig. 23.13b). Unfortunately this takes time and, together with the propagation time in the connections, it is the limiting factor for presently available "calculation speed" in electronic computers. This problem is also known as the "von Neumann" bottleneck.

In optics, on the other hand, one simple lens can be used to handle many data-streams in parallel, since photons in vacuum usually do not interact. So one could imagine a two-dimensional array with $N = 10^3 \times 10^3$ data points (Fig. 23.14). All these data points can be focussed simultaneously by one lens or by a holographic grating (not shown here) onto a logic plane which simultaneously reads data from a memory and performs some logic.

One (or several) further planes can be used to interchange the pixels, e.g., with the help of permanent or reconfigurable holographic gratings, to obtain a "perfect shuffle" function. The output obtained in this way can be processed further or fed back as input to go several times through a loop.

Another approach to optical data handling involves data transport through (glass) fiber optics. In this case, data have to be brought into an optical form anyhow and one can try to do progressively more operations in the optical regime. Here the demand for parallelism is reduced, but the switching speed has to be high.

At the end of the nineteen-seventies, expectations, especially in all-optical computing, were strongly damped for two reasons. Amplification of optical signals with high frequency is very difficult. But this is a crucial requirement

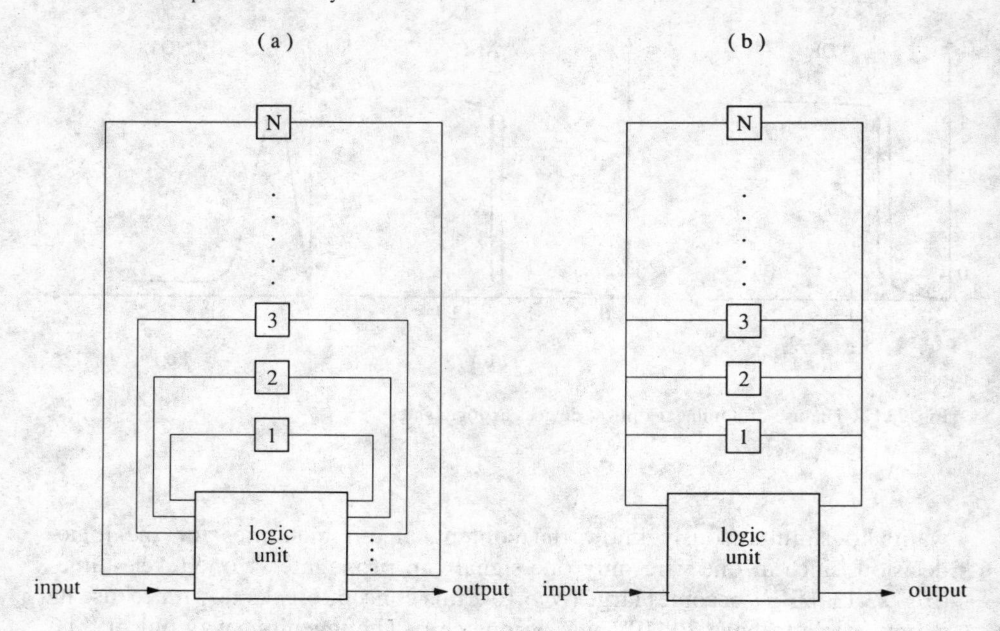

Fig. 23.13a, b. Parallel (**a**) and serial (**b**) data handling

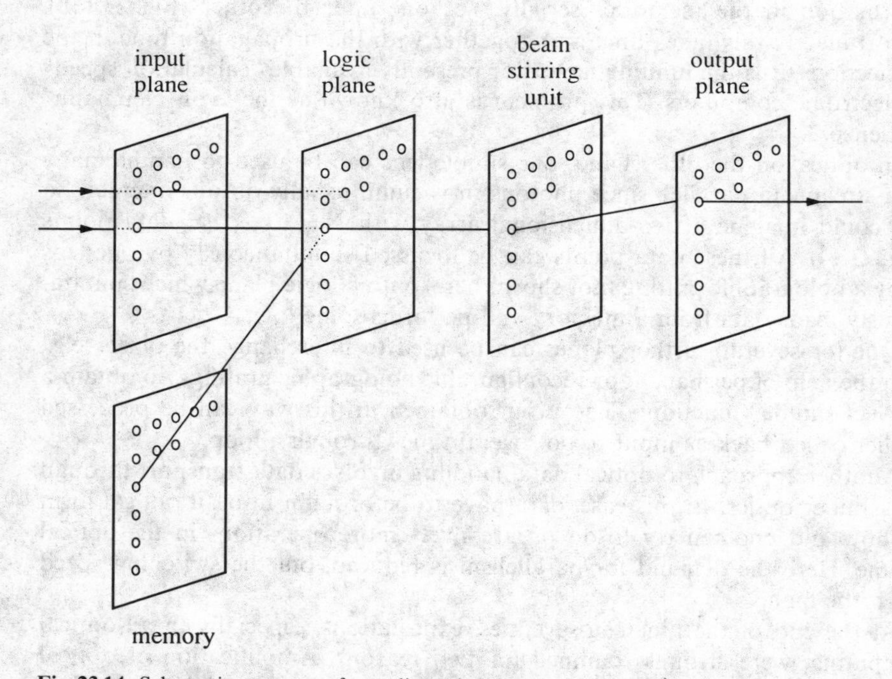

Fig. 23.14. Schematic sequence of two-dimensional optical logic arrays

for fan-out, i.e., the output of one device must be able to drive several others in the next logic plane. The first way found to solve this problem uses "threshold devices" which, with a constant optical holding beam, are kept just below the switching point and can then be switched with a small input signal. However, it was found that the stability of such a "threshold logic" is not sufficient for practical applications. The other disappointment was that optical switches were found to be inferior to electronic switches concerning speed and power consumption. Optical switches which need low input powers are usually slow (μs to ms) like the photothermal ones. The fast ones, exploiting for example the formation of an electron–hole plasma can have switching times in the ps to ns regime but need powers that are too high for dc operation and sometimes also low temperatures.

As a consequence of these findings, it is now the general consensus that electro-optic hybrid systems are the most promising ones to introduce the advantages of optics into data handling. Many scientists in research laboratories all over the world are now trying to develop this concept. Current fields of interest include the development of one- and two-dimensional arrays of surface-emitting laser diodes (Chap. 21), which have low threshold current and can be addressed individually. Other scientists are developing logic planes, e.g., by the integration of SEEDs (Sect. 23.5) and related elements. Others work on two-dimensional lenslet arrays, permanent or reconfigurable holographic gratings for beam stirring, or input and output devices.

Another important issue is the architecture for electro-optic data handling. It clearly does not make sense to copy the architecture of serial electronic computers and merely use optical pulses instead of electric ones. What one needs is a genuinely new architecture that makes use of the massive parallelism offered by optics. In some approaches, optics is used only for the distribution (or communication) of the data between the various chips on a board or from board to board, while the logic operations are still done electronically in the chips.

Interesting developments can be expected in this field during the next one or two decades.

23.6.2 Nonlinear Dynamics

A field that in recent years has developed partly in parallel to, and partly independent of optics is that of nonlinear dynamics and synergetics. In this field one considers, e.g., phase transitions in driven systems, the formation of temporal and/or spatial patterns in dissipative systems under a constant inflow of energy, and other cooperative effects.

Such investigations are of rather general nature and, in addition to physical systems, may also involve biological and sociological systems. References to this fascinating field are [23.7, 26–28].

Examples in which the concepts of nonlinear dynamics are applied to solids have been collected in a recent school [23.7]. We concentrate here on some selected examples in which the nonlinear optical properties of semiconductors and especially optical bistability play an essential role.

First, we repeat schematically in Fig. 23.15a the phase diagram of a real or van der Waals gas in the p–V plane with the coexistence region below T_c and Maxwell's construction.

A first example of a pumped system was already discussed in Chap. 20. The transition from an exciton gas to an electron–hole plasma below T_c is a first-order phase transition in a driven system with a coexistence region (Fig. 23.15b) similar to that of a van der Waals gas (Fig. 23.15a).

Another example is the optical bistability. The hysteresis is again a typical indication of a first-order phase transition in a driven system; it is shown in Fig. 23.15c and d. The bistable or "coexistence region" depends on some control parameter such as the initial temperature of the sample in the case of Figs. 23.2b or 23.6b. We compare it in Fig. 23.15 with the usual p–V phase diagram of a "real" or van der Waals gas. The usual isothermic way through the coexistence region follows the Maxwell construction and is shown by a thin horizontal line. In principle however, one could go, on the curve given by the van der Waals equation along the solid lines to the points A and B,

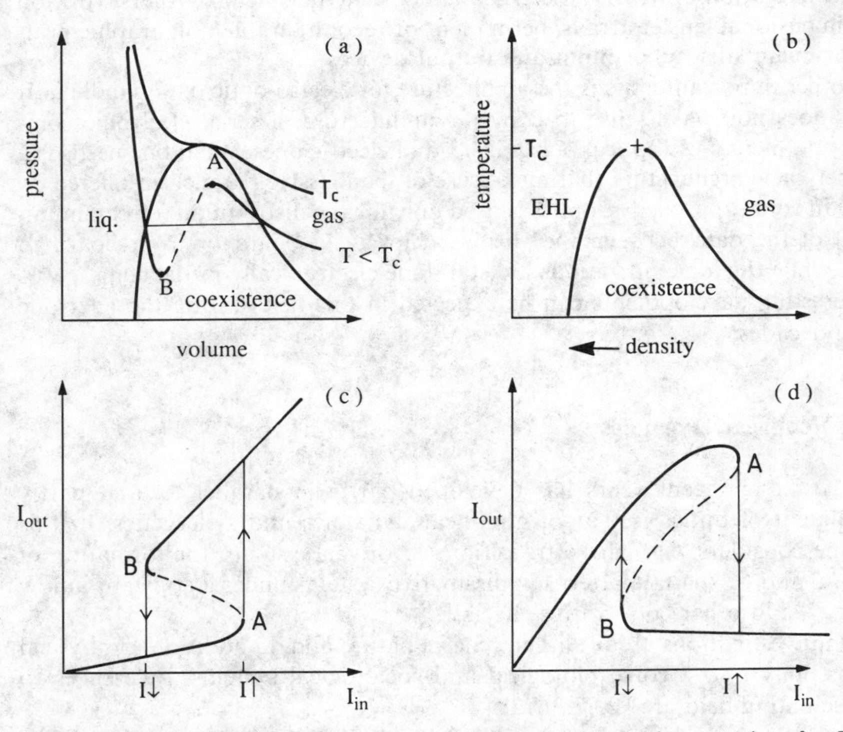

Fig. 23.15a–d. The phase diagram of a van der Waals gas (**a**) the co-existence region of an EHP (**b**) and the input–output characteristics of optically bistable devices (**c, d**)

resulting in a hysteresis loop similar to Figs. 23.15c and d. These states correspond to a supersaturated vapor or superheated liquid. They can occur if care has been taken to remove all nucleation centers. The dotted line, however, is really unstable; if the system could be prepared on this branch by some trick, the slightest fluctuation would cause it to develop into a stable state.

In the case of optical bistability, which is an open, dissipative, driven system one often has transitions from the extrema and not the "mean-field" solution of the van der Waals gas. However, the mean-field solution pertains if a sufficient amount of "noise" is added to the optical input beam. For more details of these aspects, see [23.27].

A characteristic of transitions from the extrema (and also of second-order phase transitions) is the so-called critical slowing down. This term denotes the following: If we prepare the system in a state, e.g., on the upper branch in Fig. 23.13d, and increase the incident intensity at $t=0$ to a value above $I\downarrow$ by an amount δ, it takes some time τ until the system switches to the lower branch. For small values of δ ($\ll I\downarrow$) one obtains a logarithmic singularity of τ with δ, i.e.,

$$\tau \propto -\ln \delta/I\downarrow \quad \text{for} \quad \delta/I\downarrow \ll 1\,. \tag{23.10}$$

In Fig. 23.16 we show as an example for the critical slowing the case of induced absorptive photothermal optical bistability of Fig. 23.6b. The experimental data points coincide nicely with the logarithmic singularity. The singularity can be understood in this case at least qualitatively in the following way. As long as the system is on the upper branch, all power which is deposited (i.e., absorbed) from the input beam can be dissipated while the system stays on the upper branch. If we now proceed stepwise to an input power $I\downarrow + \delta$, only the small fraction δ will be used to drive the system to the new stable state on the lower branch. Since a finite amount of energy is needed to go from one state to the other, i.e., to go to a higher temperature, the transition time τ increases with decreasing δ. The quantitative analysis then gives the logarithmic singularity of (23.10).

Other features of this driven or non-equilibrium phase transition include the dynamical blowing-up of the hysteresis loop, the measurement of the unstable branch, and the temporal expansion of the switching front in the case of inhomogeneous excitation. We do not want to go into details here but refer the reader to [23.13, 14–19] and references therein.

Instead we now introduce a setup used to study another group of effects that are characteristic of nonlinear dynamics, namely self-oscillations. The setup is a ring resonator containing an induced absorptive element, which can be bistable or monostable. This system is thus complementary to the Ikeda resonator [23.29], which contains a dispersive nonlinear element in a ring resonator.

The basic idea [23.13, 30, 31] is shown in Fig. 23.17a. We have a ring resonator whose roundtrip time τ_R is long compared to the switching time of the nonlinear device. The incident intensity I_0 is chosen such that the part

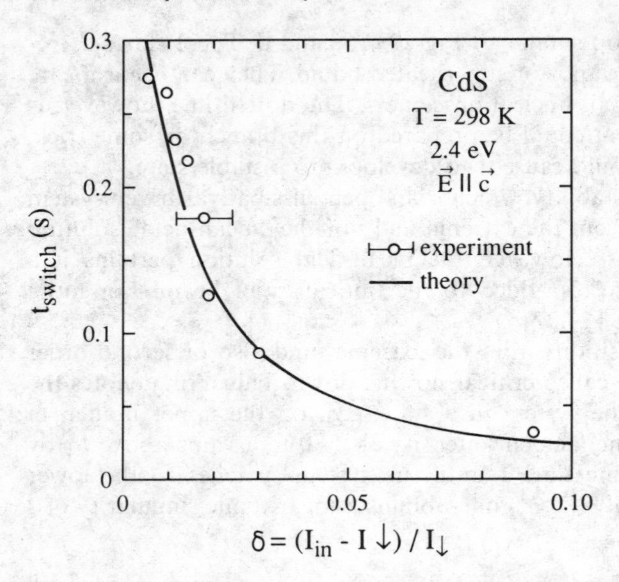

Fig. 23.16. The critical slowing down for the induced absorptive optical bistability of Fig. 23.6b. According to [23.13]

transmitted through the first beam splitter TI_0 is below the switch back intensity $I\uparrow$ of the bistable device and the sample is assumed to be in the highly transmitting state at the beginning. This means that the intensity falling on the sample $I(t)$ is almost completely transmitted. (We neglect the reflectivity of the sample itself for the moment for reasons of simplicity). A small fraction of the transmitted intensity I_t is coupled through the next mirror and detected to see what is going on in the resonator. After one round trip, the incident and the transmitted intensity fall on the sample, and we assume, again for simplicity, that we have to add intensities, i.e., that τ_R is much larger than the coherence length of the laser. The above sequence repeats itself several times, each with a step-like increase of the intensities falling on the sample and being transmitted through it, I and I_t, respectively. After several round trips I eventually exceeds $I\downarrow$ and the crystal switches to the absorbing state. This means, that almost no more light is transmitted. Consequently only TI_0 falls on the sample after another round trip time and the crystal switches again into the transmitting state and the above described sequence begins anew, i.e., for constant incident intensity we observe self-oscillation (a temporal structure) with a period which is an integer multiple of τ_R.

For photothermal optical nonlinearities the condition $\tau_R \gg \tau$ would necessitate ring resonators with a delay time of at least 100 ms, equivalent to approximately one trip around the globe. Such devices tend to get unwieldy and therefore the electro-optic hybrid system of Fig. 23.17d has been set up. The incident light passes through an electro-optic modulator (EOM), operated in

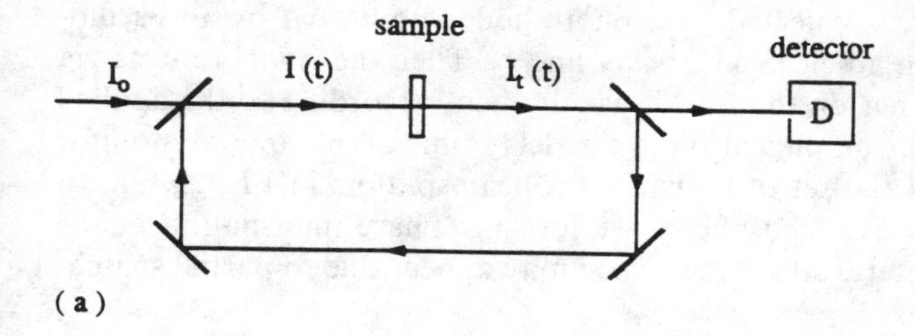

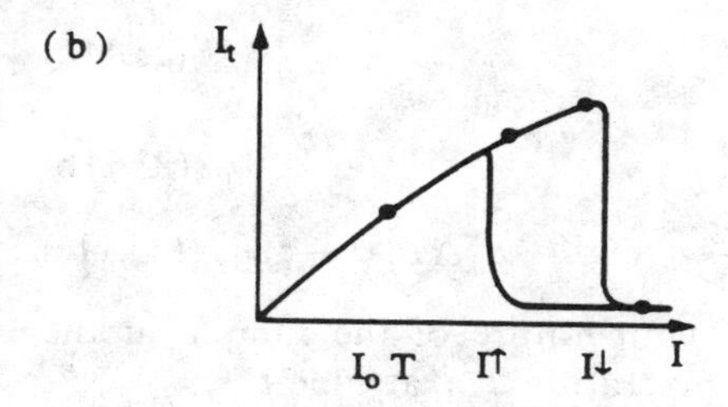

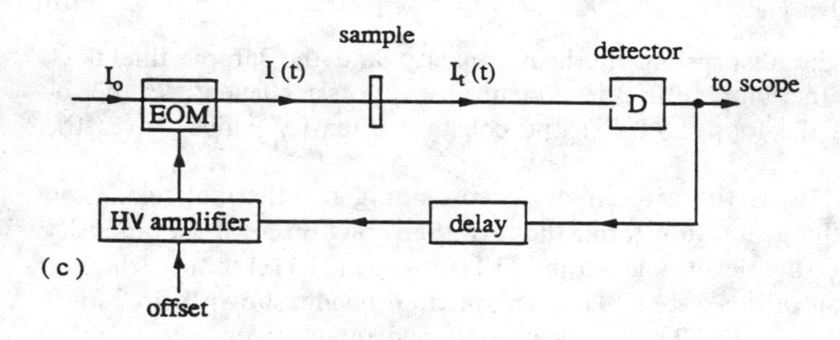

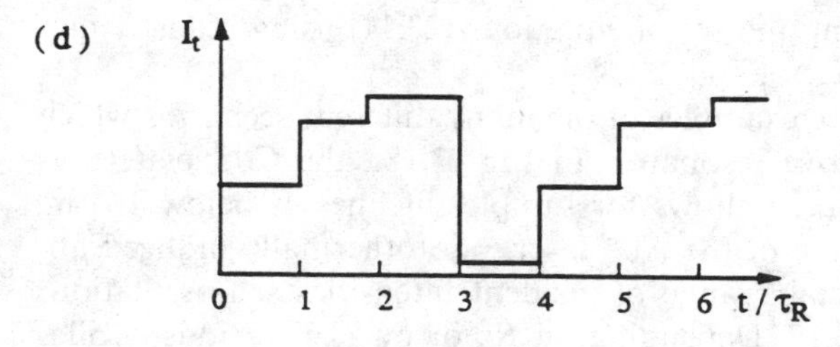

Fig. 23.17a–d. The ring resonator with a bistable element (**a**); the input–output characteristics of this element (**b**); the realisation of (**a**) by a electro-optic hybrid resonator (**c**); and the resulting self-oscillations (**d**)

the linear regime. A constant offset U_0 on the high-voltage amplifier (HVA), which drives the EOM models the constant incident intensity I_0T of Fig. 23.17c. The optical nonlinearity is really the photothermal nonlinearity of a CdS platelet type sample using the green line (514.5 nm) of an Ar^+ laser. The transmitted intensity I_t is detected by a photodiode, monitored by an oscilloscope and sent through an electronic delay line τ_R. Then the signal returns via the high voltage amplifier on the EOM, which opens according to the applied signal. The choice of the amplification in the delay time allows one to monitor the total reflectivity of the set of four mirrors/beamsplitters in Fig. 23.17a. In this ring resonator, as assumed above, we lose the phase information of the light, and we neglect lateral structures which may appear due to partial switching of the sample.

The system is described by the following set of equations:

$$\frac{d}{dt}\Delta T_L = -\frac{\Delta T_L}{\tau_0} + A(\Delta T_L) I(t)\frac{1}{c'L}, \tag{23.11a}$$

$$A(\Delta T_L) = 1 - \exp[-\alpha(\Delta T_L) L], \tag{23.11b}$$

$$I(t) = I_0 T + R^2 I_t(t-\tau_R) = I_0 T + R^2 I(t-\tau_R)\exp\{-\alpha[\Delta T_L(t-\tau_R)]\}, \tag{23.11c}$$

where ΔT_L is the increase of the lattice temperature of the sample in the illuminated spot above the surrounding temperature T_{0L} at time t

$$\Delta T_L(t) = T_L(t) - T_{0L} \tag{23.11d}$$

A, c', and L are the absorption, the heat capacity, and the sample thickness, respectively; α is the temperature-dependent absorption coefficient; R^2 the total reflectivity of the loop, and $I_0 T$ the constant intensity which drives the whole system.

Equation (23.11a) is the rate equation containing the thermal relaxation time τ_0 and, in the generation term, the absorbed fraction A of the intensity $I(t)$ falling on the sample. A is given in (23.11b) while (23.11c) is the delay (or iteration) equation of the system. The experimental results shown below for a system such as that of Fig. 23.17c can be obtained qualitatively and in most cases quantitatively by solving the set of equations (23.11) using realistic material parameters for $\alpha(T_L)$, τ_0, c', and L.

In Fig. 23.18 we show three examples of the many different scenarios which can be realized with the above resonator. In Fig. 23.18a the CdS platelet is used in the polarization $\boldsymbol{E} \parallel \boldsymbol{c}$. It shows for sample thicknesses below 10 μm, $T_{0L} \lesssim 300$ K and the green line of the Ar^+ laser, photothermally induced absorption bistability. In a certain range of incident intensities self-oscillations occur as described in Fig. 23.17. Depending on R^2 or on $I_0 T$ various oscillation modes can be observed, which lock into multiples of τ_R for $\tau_R \gg \tau_0$. A mode with a total period $n\tau_R$ and m different maxima per period is designated an n/m mode. With this definition we see in the row (a) of Fig. 23.18 for different incident intensities a 3/1 and a 2/1 mode. For intermediate incident intensity we get a 5/2 mode. This new mode can be predicted from those of the preceding generation by independently adding the numerators and denominators of the mode. This is exactly the way to construct a Farey tree and indeed it can be shown using (23.11) that the modes follow this Farey tree pattern and

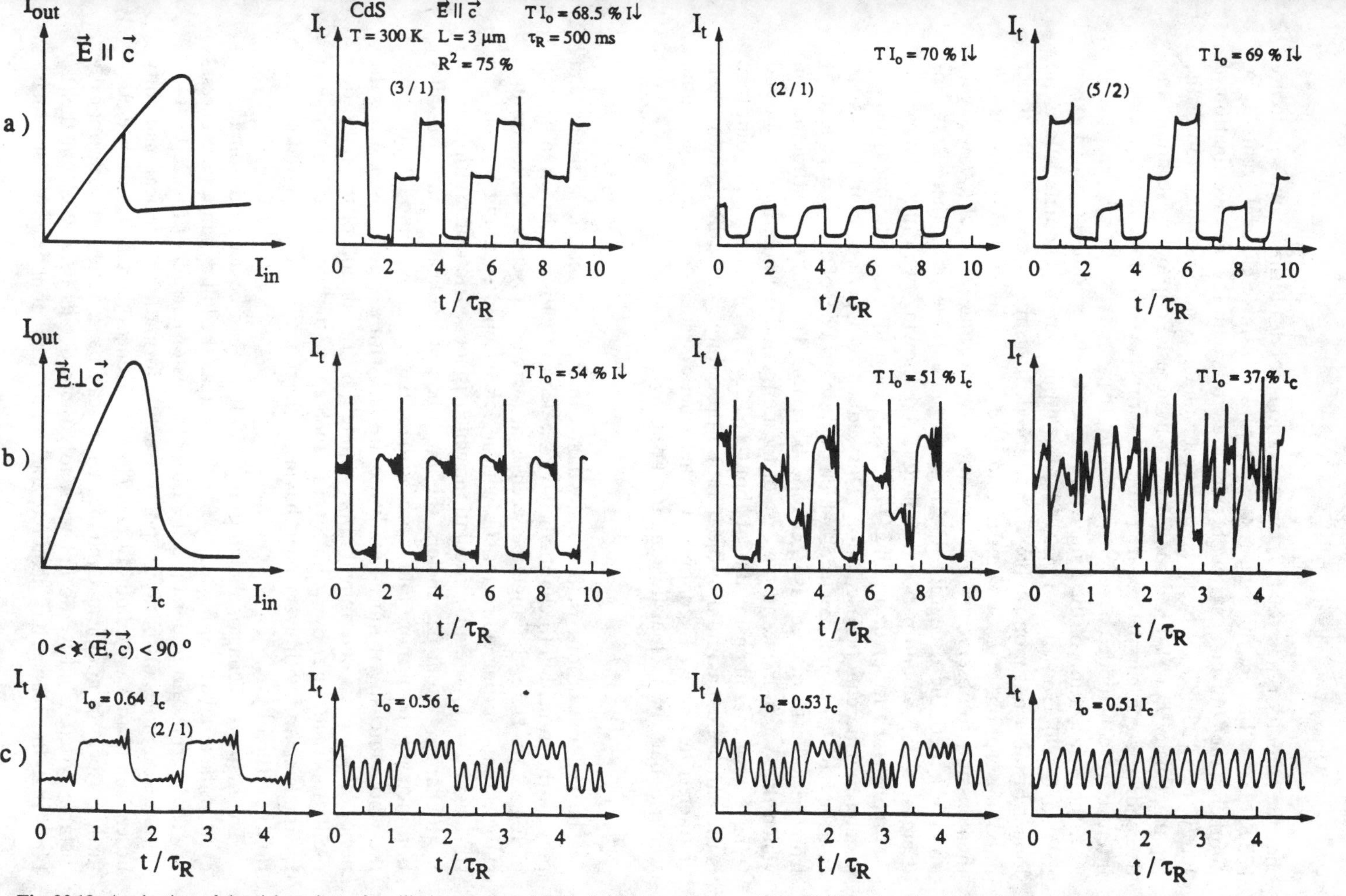

Fig. 23.18. A selection of the rich variety of oscillation modes of the hybrid ring resonator of Fig. 23.17c using bistable and monostable input–output characteristics of the optically nonlinear element. According to [23.13, 15]

that the stability ranges of $I_0 T$ for the various modes form a devil's staircase [23.8, 13, 31]. Indeed, this ring resonator with a bistable element seems to be another rather universal system besides the circle map for the investigation of Farey trees [23.28]. Farey trees also occur under periodic modulation of the input intensity $I_0 T$. With other forms of hysteresis loop both ascending and descending steps can be realized and a short perturbation of the incident intensity reveals mode coexistence [23.8, 13, 15].

If the polarization of the light with respect to the crystallographic $\boldsymbol{c}$ axis is rotated by 90° [row(b) of Fig. 23.18], the input–output characteristic is no longer bistable but monostable, due to the dichroism of CdS and the higher initial absorption for $\boldsymbol{E} \perp \boldsymbol{c}$ compared to $\boldsymbol{E} \parallel \boldsymbol{c}$. Under favorable conditions it is then possible to observe another scenario. For a certain value of $I_0 T$ we again observe a periodic oscillation locking into a multiple of τ_R (here $2\tau_R$). For decreasing TI_0 we find period doubling to $4\tau_R$ showing two different maxima.

Finally, for further decreasing TI_0, we reach a non periodic behavior, i.e., a deterministic chaotic oscillation. This sequence of period doublings is the so-called Feigenbaum scenario, which represents one of the routes leading from periodic to deterministic chaotic behavior. The "strange attractor" on which the system moves in the case of the nonperiodic oscillations was found to have a fractal dimensionality of about 2.6 [23.13]. The important point here is the parabolic maximum of the iteration procedure, i.e., of the input–output characteristic, which is common to the generic system for the Feigenbaum scenario, namely the logistic equation given by

$$x_{n+1} = a x_n (1 - x_n) . \tag{23.12}$$

The fast relaxation oscillations which can be seen on the plateaux can prevent the transition to chaotic behavior. Such a situation is shown in part (c) of Fig. 23.18. Here, the angle between $\boldsymbol{E}$ and $\boldsymbol{c}$ was chosen to yield a monostable input–output characteristic with a very steep descending branch. We start again with a regular oscillation, which is locked into a multiple of τ_R. With changing input parameters, the amplitude of the fast relaxation oscillation increases until more and more of the spikes result in a modulation over the whole intensity regime. Finally we are left with a regular oscillation, the period of which is determined approximately by the period of the relaxation oscillation τ_0 and not by τ_R, however in such a way that an integer multiple of τ_0 equals τ_R and with every peak having a shape similar to the initial oscillation governed by τ_R.

The first two scenarios shown in Figs. 23.18a, b can be deduced from (23.11) even in the adiabatic limit, i.e., $\tau_0/\tau_R \Rightarrow 0$, whereas the full dynamics are necessary to describe and understand the observations of part (c) of Fig. 23.18.

Some examples of spatio-temporal structure formation in optical bistability in semiconductors are found in [23.4, 7] and references therein. They represent a further aspect of nonlinear dynamics which can be investigated with optically bistable systems, but we do not want to go into details here. We, nonetheless, hope that the reader has acquired the feeling that nonlinear semiconductor optics can contribute significantly to the fascinating field of nonlinear dynamics and synergetics.

23.7 Problems

1. Find a simple relation for the "sufficiently strong" excitation induced increase of the absorption (Sect. 23.4) needed to produce optical bistability.

2. Is it possible for longitudinal structures to appear in the switching process of induced absorptive optical bistability in thick samples, under the assumption that the excited species responsible for the increase of absorption have negligible diffusion length?

3. Can induced absorptive optical bistability occur if the increase of the absorption depends not on the density of some excited species, but directly and instantaneously on the amplitude or intensity of the light field? Is your answer also true for dispersive and bleaching optical bistability in a Fabry–Perot resonator?

4. Give the conditions on the finesse of the Fabry–Perot and the bleaching required to get optical bistability from the combination of both.

5. Iterate the logistic equation (23.12) to get the Feigenbaum scenario.

6. Find some additional information about the "circle map" and the Farey tree, e.g. in [23.28].

24 Experimental Techniques

In this chapter, we outline some aspects of the experimental techniques frequently used in the spectroscopy of semiconductors. Some examples have already been encountered in the preceding chapters, especially Chaps. 19–23. The aim here is not to give an exhaustive description but to mention some of the advantages and pitfalls associated with various techniques. We start with methods which are frequently, but not exclusively, used in linear optics—luminescence, reflection and transmission spectroscopy—proceed then to techniques for time-resolved spectroscopy, and finally describe some methods which are utilized in nonlinear optics. Additional information may be found in [24.1–7] and references therein.

24.1 Linear Spectroscopy

The most widely used techniques of linear spectroscopy are transmission (or absorption), reflection, luminescence, and luminescence-excitation spectroscopy. In the following we treat these topics in this sequence.

24.1.1 Transmission Spectroscopy

For the evaluation of transmission and reflection spectra one should, in principle, use the equations given in Sect. 3.1.6, depending on the quality and parallelism of the surfaces of the sample and the coherence length of the light. Generally, the full use of these equations is rather difficult. If one has a reasonably well established idea of the reflectivity of the material and if one does not observe Fabry–Perot modes, then (3.40) can be used. A very simple method is the following: One measures the spectrum of the incident light without sample (Fig. 24.1a) and the spectrum of the transmitted light (Fig. 24.1b). If the sample shows significant luminescence excited by the absorbed fraction of the incident light, this contribution has to be recorded separately and subtracted (Fig. 24.1c). The same is true for the dark current of the detector. Then one can calculate the ratio $I_0/(I_t - I_{lum})$. This ratio still contains the reflection and scattering losses of the surface. If one knows for sure that the sample is transparent $[\alpha(\omega)d \ll 1]$ in a certain spectral range, and that the above-mentioned

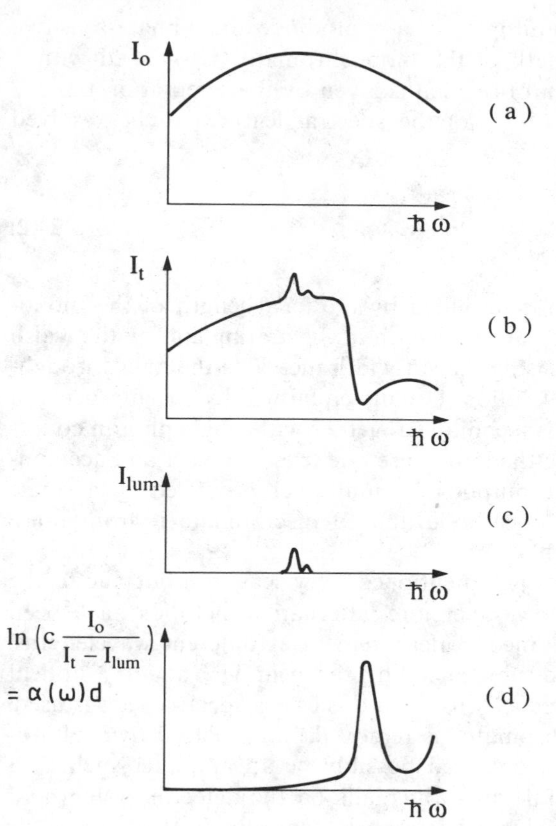

Fig. 24.1 a–d. Schematic spectra for transmission measurements. The incident lamp spectrum (**a**), the transmitted spectrum (**b**), a luminescence band of the sample which has to be subtracted (**c**), and the resulting spectrum for the optical density (**d**)

losses are roughly frequency independent, one can normalize the above ratio in this spectral region to 1 by multiplication with a constant c_n. The quantity

$$l_n\left(c_n \frac{I_0(\omega)}{I_t(\omega) - I_{lum}(\omega)}\right) = \alpha(\omega)d \tag{24.1}$$

then gives the optical density αd (Fig. 13.1d).

The various spectra entering in (24.1) can be recorded by different techniques. The classical method is to image the light source after passage of the beam through the sample onto the entrance slit of a monochromator with an aperture slightly smaller than that the aperture or f number of this monochromator, to chop it by a mechanical chopper at a frequency around 1 kHz which is not an integer multiple of the line frequencies of 50 or 60 Hz or of another significant frequency appearing in the setup to avoid spurious signals, and to detect the output of the monochromator after the exit slit with a photodetector such as a photomultiplier or a photodiode and a phase-sensitive lock-in amplifier. The wavelength of the monochromator is scanned with a speed $v_{sc} = d\lambda/dt$, the upper limit of which is given by the integration time t_{in} of the lock-in and the width $\Delta\omega_{res}$ of the spectral features to be resolved according to

$$v_{sc} \lesssim \frac{2\pi c}{5\omega^2} \frac{\Delta\omega_{res}}{T_{in}} . \tag{24.2}$$

The spectral resolution is also determined by the focal length of the monochromator, the number of lines per unit length of the grating and by the width of entrance and exit slits. Increasing the slit width increases the light throughput but reduces the spectral resolution. Usually equal widths for entrance and exit slit are the best choice. This technique, together with single photon counting is the most sensitive one. In the latter case one tries to detect and accumulate every current pulse at the output of a multiplier produced by a single photon impinging on the photocathode and to discriminate it from noise current pulses of different height.

The above techniques are, as just mentioned, very sensitive, but due to the wavelength scanning rather slow. Therefore, alternative methods have been developed which enable simultaneous measurement at different wavelengths. In the visible part of the spectrum (including the near UV and IR) optical multichannel techniques are generally used. In this case a spectrometer is used, i.e., the exit slit of the monochromator is removed and replaced by a photodiode array or a vidicon tube, preceded possibly be an amplifier such as a channel plate. A certain spectral range thus falls on the detector system and the spatially resolved readout signal is stored in typically $500-10^3$ channels corresponding to different values of $\hbar\omega$. This data-acquisition system is usually linked to a computer which allows the spectra to be stored and analyzed.

The spectral resolution is again determined by the width of the entrance slit, the grating and the focal length of the spectrometer, and by the width and the cross-talk of neighboring pixels or channels of the detector. The latter quantity cannot usually be brought below three channels with a typical width of pixels or channels around 25 μm. Due to the given geometrical width of the detector, the spectral resolution and the spectral range covered are inversely proportional to each other. The use of photographic plates and photodensitometry after exposure and processing, instead of an opto-electronic detector array, is nowadays rather outdated because of the nonlinear response characteristic of the former.

Another approach to wavelength multiplexing involves Fourier transforms [24.6]. The basic principle is the following: A cw beam whose spectrum $I(\omega)$ is to be measured is sent as a parallel beam into a Michelson interferometer (Fig. 24.2). The mirror of one arm is translated parallel to itself in the z-direction at constant speed and the signal intensity $I(z)$ of the interference structure is measured as a function of this displacement. $I(z)$ contains the

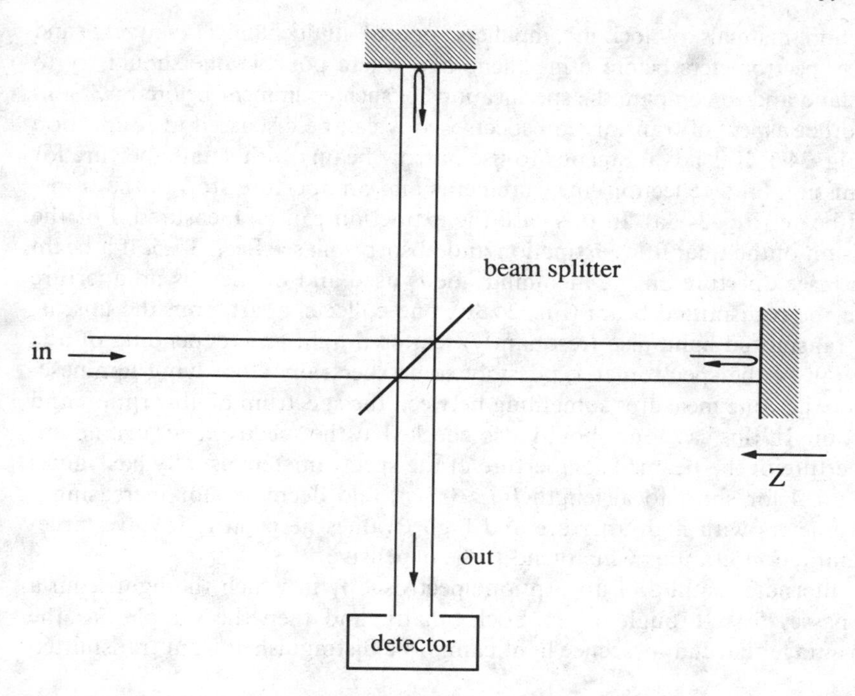

Fig. 24.2. The Michelson interferometer used in Fourier spectrometers

information about the spectrum $I(\omega)$. It can be obtained essentially via a Fourier transform. This is why these spectrometers are often called FTIR (Fourier transform infrared) spectrometers. A monochromatic incident beam, for example, results in a sinusoidal dependence $I(z)$ which in the Fourier transform, i.e., in frequency space, indeed gives a single contribution and vice versa. More complicated input spectra result in more complicated $I(z)$ dependences. The spectral resolution increases in this case with the distance z over which the mirror is moved. The Fourier transform is usually performed with standardized computer programs.

The FTIR setup imposes stringent requirements on the mechanical stability. Maximum acceptable deviations of the beam in all directions when moving the mirror are below one tenth of the shortest wavelength to be analysed. This is the reason why the device is used mainly in the longer wavelength regime, i.e., in the IR. Optical multichannel analyzers and FTIR spectrometers are thus complimentary with regard to their spectral ranges. In addition, a multichannel analyser can also detect very short pulses in a time-integrating manner. Sometimes it is useful in such a situation to gate the amplifier in front of the detector array during the times when a signal is actually arriving in order to improve the signal-to-noise ratio. It is advisable to carefully study the instruction manuals of lock-in amplifiers, optical multichannel analyzers, and Fourier spectrometers before using them, and where possible one should try to understand and to compare the specifications of such equipment before buying it.

Another aspect of transmission spectroscopy can be discussed in connection with Fig. 24.3. It is advantageous to use narrow beams with small aperture for transmission (and reflection) measurements and an aperture stop in the transmitted beam (Fig. 24.3a). In this case the extinction can be measured. For the discussion of the quantities extinction and absorption, see Sect. 3.1.5. If a beam with a large aperture on the incoming side is used and/or there is no aperture stop in the transmitted beam (Fig. 24.3b), one collects, apart from the unscattered transmitted light, also (resonantly) scattered light and, depending on the resolution of the spectrometer, possibly some spectrally close lying luminescence too, i.e., one measures something between the spectrum of absorption and extinction. In this case one should also check that the spectrometer can accept the aperture of the beam. The aperture of the spectrometers usually has values around 1:4 for short focal length f ($f \approx 0.3$ m) and decreases for increasing f. Spectrometers with high aperture and high f values necessarily involve large-area diffraction gratings which tend to be expensive.

An alternate method of absorption spectroscopy in which the light from a lamp passes first through the monochromator and then the sample has the disadvantage that luminescence light cannot be distinguished from transmitted light.

Some further aspects which deserve consideration are the dependence of the efficiency of the gratings on wavelength and on polarization (!) and the stray

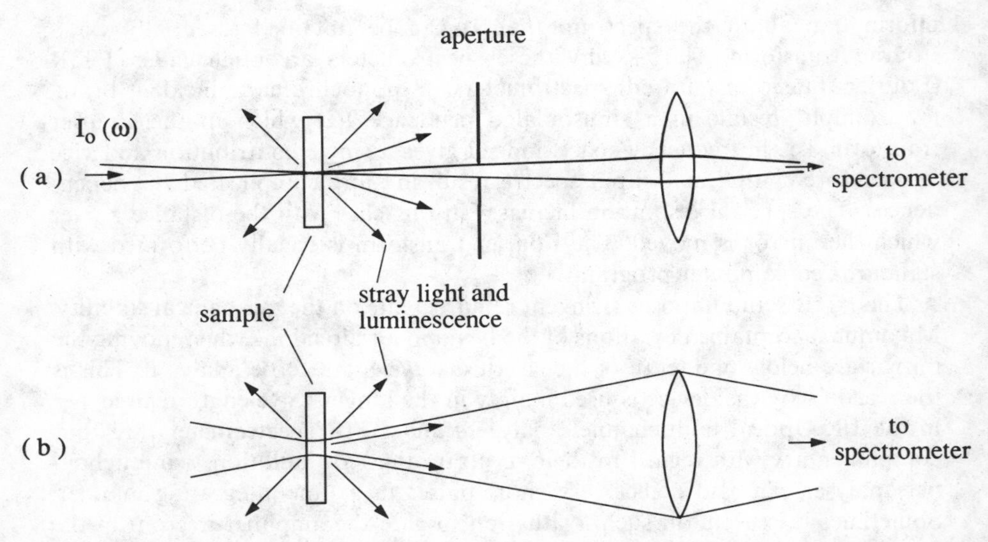

Fig. 24.3a, b. Experimental arrangements for transmission spectroscopy to reduce stray light and luminescence, i.e., to measure the extinction of light (**a**) or to measure the absorption (**b**)

light. Holographic gratings tend to have lower stray light but a stronger polarization dependence of their efficiency as compared to mechanically ruled gratings.

As a rule of thumb one can say that a grating which is blazed at a wavelength λ_B can be used in first order in a wavelength interval

$$\frac{2}{3}\lambda_B \leqslant \lambda \leqslant \frac{3}{2}\lambda_B \,. \tag{24.3}$$

A grating blazed at λ_B for the first diffracted order can also be used at $\lambda_B/2$ in second order. If the wavelength range which falls on the monochromator or spectrometer extends over more than a factor of two, care has to be taken to distinguish between a wavelength λ in first order and a wavelength $\lambda/2$ in second order. Band-edge filters allow this to be clarified easily.

A special group of techniques to measure the genuine absorption spectrum of a sample are photothermal and photo-acoustic methods. In these cases, a signal is deduced from the increase of the sample temperature upon absorption of temporally modulated light and its transformation into heat. This heat raises the temperature of the sample and may slightly deform its surface, or heat a gas over the sample resulting in a gradient of the refractive index of the gas or in a deformation of an elastic window above the gas. These deformations can be measured very sensitively by the deflection of a second (laser) beam falling under grazing incidence on the sample or passing through the gas directly over the sample. These methods are very sensitive and also allow one to measure αd values much smaller than unity.

In another system, which works at very low temperatures only, the heat dissipated in the sample is fed into a superconducting bolometer, which is very sensitive around T_c.

More details of these thermal techniques are given in [24.7] and references therein.

24.1.2 Reflection Spectroscopy

We continue now with some discussion of reflection spectroscopy. Most of the statements made about transmission spectroscopy also apply here. Normal incidence is the simplest geometry for reflection spectroscopy. This geometry can be implemented using either a small deviation from normal incidence to separate incident and reflected beams (Fig. 24.4a) or a beam splitter (Fig. 24.4b). The quantitative evaluation should again follow equations like those given in Sect. 3.1.6. However here too simplifications are possible, see Fig. 24.5. One can record the spectrum $I_0(\omega)$ for $R = 1$ by replacing the sample by an (almost) perfect 100% mirror.

This could be an Ag surface mirror for the visible (but not for the near UV!) or an Al mirror (preferentially with a protective coating) for the visible and the near UV. Then the spectrum reflected off the sample, $I_R(\omega)$, is recorded. The

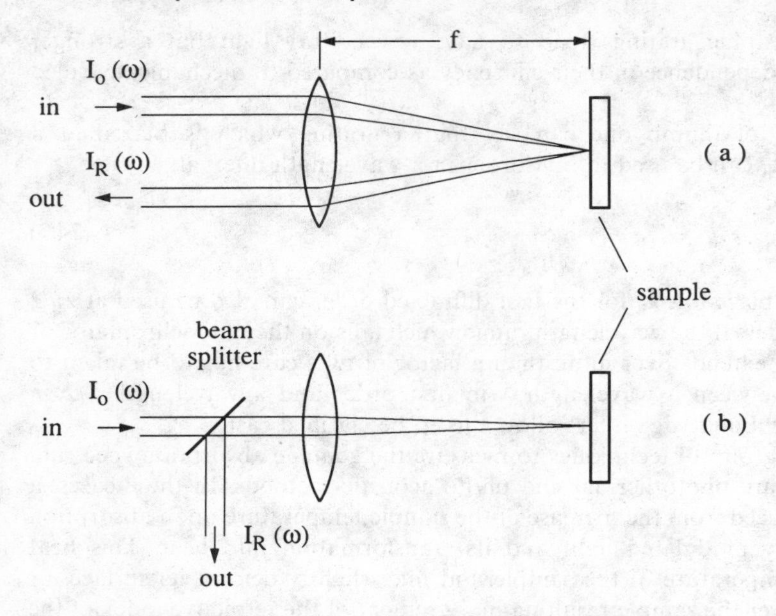

Fig. 24.4. Geometries for (near) normal incidence in reflection spectroscopy

ratio of these two corrected again for luminescence and/or dark current if necessary gives the true reflection spectrum (Fig. 24.5d). Within the resonance-like structure (i.e., in the restrahlbande) the absorption is so high that in most cases only the reflected beam from the front surface contributes to the signal. For photoenergies away from the resonance in spectral regions with $\alpha d \leqslant 1$, reflection from both the front and rear surfaces contributes, resulting either in Fabry–Perot modes (dashed lines in Figs. 24.5b, d) or simply in an enhancement as shown schematically by the solid line in Fig. 24.5d below $\hbar\omega_{\mathrm{abs}}$. Care has to be taken that such structures are not misinterpreted as resonances.

The interpretation of reflection spectra at oblique incidence (possibly close to the Brewster angle [24.8]) are more complicated in their evaluation. Therefore it is strongly recommended, especially for anisotropic materials (e.g., uniaxial ones), to begin with simple geometries like $\boldsymbol{k} \perp \boldsymbol{c}$ or $\boldsymbol{k} \parallel \boldsymbol{c}$, $\boldsymbol{E} \parallel \boldsymbol{c}$ or $\boldsymbol{E} \perp \boldsymbol{c}$ in both transmission and reflection spectroscopy. Oblique incidence or polarization should only be chosen deliberately after some consideration concerning ordinary and extraordinary beams or mixed mode polaritons (Sects. 3.1.4 and 3.1.7). It should be pointed out that for $\boldsymbol{k} \neq 0$ even materials with cubic symmetry have just cubic and not spherical symmetry and may show some anisotropies. Crystals with the cubic symmetry T_d, e.g., show a slight birefringence for $\boldsymbol{k} \parallel [110]$ and polarizations $\boldsymbol{E} \parallel [\bar{1}10]$ and [001], essentially due to the corresponding warping of the bandstructure, i.e., a directional dependence of the effective masses (Sect. 10.6).

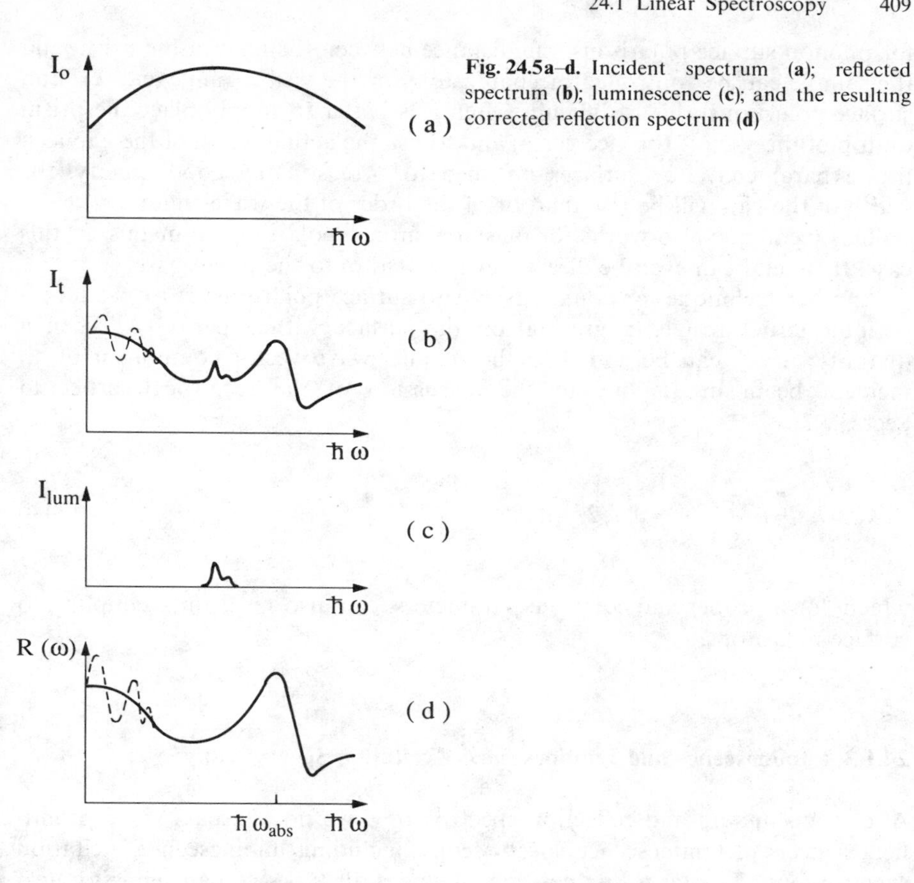

Fig. 24.5a–d. Incident spectrum (**a**); reflected spectrum (**b**); luminescence (**c**); and the resulting corrected reflection spectrum (**d**)

Finally a few words shall be said about attenuated total reflection (ATR), which is often used to study surface polaritons (examples were given in Fig. 14.21 and in sections 13.4 and 14.5).

A beam is sent into a prism such that it is totally reflected at its base (Fig. 14.21a). In this case an evanescent wave propagates along the base on the vacuum side. The frequency ω of this wave is given simply by the frequency of the incident light, the $\boldsymbol{k}$-vector corresponds to the component of the $\boldsymbol{k}$-vector of the beam in the prism parallel to the interface. This quantity can be varied for a fixed ω over a certain range by changing the angle of incidence α, as long as α is above the limiting value for the onset of total internal reflection. The evanescent wave propagates parallel to the base of the prism but its amplitude decays normal to it over a length scale given by the vacuum wavelength. If the sample is brought up to the base of the prism to within a comparable distance, the evanescent wave can couple to the surface polariton if both ω and $\boldsymbol{k}_\parallel$ coincide. In this case, the total reflection is attenuated ($\Rightarrow$ ATR) since some energy is coupled into the surface polariton mode. For resonances in the IR, as for phonon surface polaritons, the distance between the base of the prism and the sample can be controlled mechanically. For the visible range (e.g., exciton surface polaritons) this is difficult. Often it is better to simply place the prism on top of the sample (or vice versa) and to use the actual width of the gap as a fitting parameter. For surfaces polished to a reasonably good quality, the width of the gap will be less than or of the order of the wavelength.

This technique also works for plasmon surface polaritons in metals. In this case, the metal can even be directly evaporated onto the prism base.

Another technique for coupling light to surface polaritons is to produce a periodic structure (i.e., a grating) on the surface with a period Λ. Then a quantity $n2\pi/\Lambda$ can be added to the parallel wave vector component of an incident beam in analogy to the discussion of the reciprocal lattice in Sect. 3.1.3

$$\boldsymbol{k}'_{\parallel} = \boldsymbol{k}_{\parallel} \pm n\frac{2\pi}{\Lambda}; \quad n = 0, 1, 2 \ldots \tag{24.4}$$

which, for a proper choice of the parameters, can also result in a coupling to surface polaritons.

24.1.3 Luminescence and Luminescence-Excitation Spectroscopy

After transmission and reflection spectroscopy, we now discuss some prominent aspects of luminescence spectroscopy, including luminescence excitation spectroscopy. The term luminescence includes all types of light emission (especially fluorescence and phosphorescence) appearing as a consequence of some input of energy into the sample, except incandescent emission.

There are many ways to excite luminescence; For example, with electron beams or X-rays, or by application of high electric fields via impact ionization. These techniques are rather unselective concerning the excitation process. In the case of electron bombardment one can say, as a rule of thumb, that an energy of roughly $3E_g$ is needed to create one electron–hole pair.

The situation is much more favorable for the case of optical excitation resulting in photoluminescence. Therefore this technique is the one most widely used for the investigation of semiconductors.

The excitation can be via one- or two-photon absorption as already shown in Sects. 19.1 or 19.3. One inherent problem of luminescence spectroscopy is that the luminescence yield in most semiconductors is considerably smaller than one. This means that the luminescence gives direct information only about the decay of a minority of the electron–hole pairs created and some care has to be used in drawing conclusions from their fate about that of the majority. For free particles like excitons, only those that hit the surface of the sample with a parallel component of their wave vector

$$\boldsymbol{k}_{\parallel} \lesssim \frac{2\pi}{\lambda_v} \tag{24.5}$$

can give rise to luminescence. Of these a certain fraction will be reflected at the surface back into the sample.

For recombination processes from spatially localized states like bound-exciton complexes, or from quantum dots for which the $\boldsymbol{k}$-conservation is relaxed, this condition may be less stringent. Similar arguments hold for recombination processes under emission or excitation of another quasiparticle, like a LO phonon or another exciton (see Chap. 19).

A further crucial point is reabsorption, especially when the luminescence occurs in spectral regions where the sample is strongly absorbing. To avoid a distortion of the luminescence spectrum, the inverse of the absorption coefficient at the emission wavelength $\alpha^{-1}(\omega_{\text{lum}})$, i.e., the "escape depth" of the luminescence should be larger than the sum of the inverse absorption coefficient at the excitation wavelength and the diffusion length of the excited species l_{diff}:

$$\alpha^{-1}(\omega_{\text{lum}}) > \alpha^{-1}(\omega_{\text{exc}}) + l_{\text{diff}} \,. \tag{24.6}$$

To come close to this condition it is better to collect the luminescence in a backward geometry (configuration 1 in Fig. 24.6) than in transmission (configuration 2). Additionally the reflected part of the excitation should be blocked so that it does not enter the spectrometer, where it can produce unwanted stray light. Even then, the condition (24.6) cannot always be fulfilled. To give an example, we present numbers for the free exciton luminescence which are valid for many direct-gap semiconductors.

We excite in the continuum states of the excitons where frequently $\alpha(\omega_{\text{exc}}) \gtrsim 10^4\,\text{cm}^{-1}$. The excited carriers relax at low lattice temperatures in a time T_3 of a few ps towards the $n_B = 1$ exciton state and during their lifetime T_1 diffuse over a distance of about $l_D = 1\,\mu\text{m}$. The absorption coefficient at the free exciton resonance $\alpha(\omega_{\text{lum}})$ can be as high as 10^5–$10^6\,\text{cm}^{-1}$ resulting in an escape depth of 0.1–0.01 μm only. This is one of the reasons why free exciton emission is often very weak in bulk samples. See Ref. [24.9] where the luminescence of CdS has been corrected for reabsorption. This problem does not arise if very thin samples are used with a thickness smaller than the "escape depth" of the luminescence, as is the case in single quantum wells. (But see also Sect. 22.4).

Under high excitation, stimulated emission is a frequent occurrence in direct gap semiconductors (Chap. 21). The emission leaves the sample predominantly at its edges (configuration 3 in Fig. 24.6) especially in thin platelet type samples which act as a waveguide. If this emission is to be studied, it is advantageous to use the geometry 4 to shift the excitation spot close to the edge and to give it an elongated shape which directs superradiance towards this edge as shown schematically in Fig. 24.6b. Sometimes laser modes can be

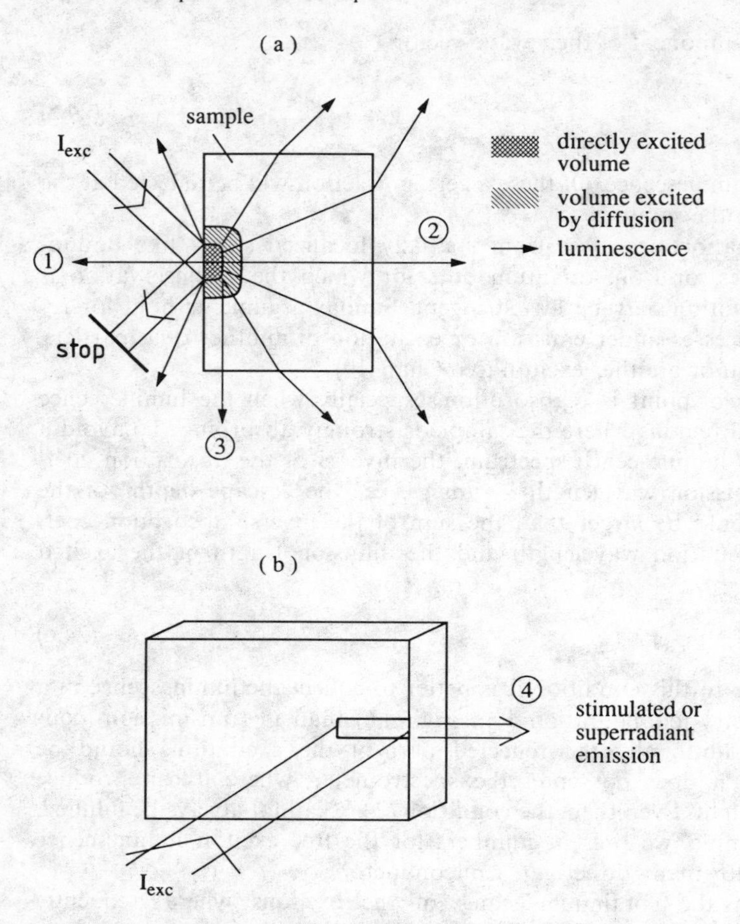

Fig. 24.6. Various geometries for luminescence spectroscopy

seen in the emission spectra if some suitable surfaces of the sample form a resonator.

It should be mentioned that (stimulated) emission also leaves the sample preferentially at surface steps and other imperfections. Such phenomena might be misleading if the diffusion of excited species is deduced from the spatial distribution of the emission around the excitation spot and care must be taken.

A versatile technique that is also sometimes used instead of absorption spectroscopy is the so-called photoluminescence excitation (PLE) spectroscopy.

The principle is the following: one records the intensity I_{lum} of the emission of a certain species, e.g., of a bound exciton complex, and varies the photon energy of the excitation $\hbar\omega_{exc}$ at a constant incident intensity I_{exc} or, even better, at a constant incident photon flux. The curves $I_{lum} = f(\hbar\omega_{exc})$ are then the PLE spectra. The PLE spectra usually exhibit some peaks, which indicate that the emission can be particularly strongly excited at certain photon energies. This is the case, when the absorption coefficient is higher at a certain energy than at others, or if an excited state of the emitting complex is resonantly excited and the relaxation from this excited state to the emitting level is very efficient.

It is clear from the above statement that the PLE spectrum is "somehow" related to the absorption spectrum, but there is no one-to-one correspondence. The fulfillment of (24.6) can vary with $\hbar\omega_{exc}$; fast nonradiative recombination centers which often exist at the surface may become important if $\hbar\omega_{exc}$ lies in a region of strong absorption and the relaxation process from the excited to the emitting state enters. All these aspects necessitate special care in the interpretation of PLE spectra. If possible, one should record not only at a fixed $\hbar\omega_{lum}$, e.g., behind a monochromator: Instead it is advisable to record a larger part of the luminescence spectrum with a spectrometer so that one can recognize whether, for example, a Raman line or a reabsorption dip shift over $\hbar\omega_{lum}$ with varying $\hbar\omega_{exc}$, since such phenomena produce spurious signals in the PLE spectra which can easily lead to misinterpretation.

A field where PLE spectroscopy has proved to be very powerful is the investigation of excited states of bound-exciton complexes. An example was shown in Fig. 15.6.

A case where PLE was necessary was the early stage of spectroscopy of $Al_{1-y}Ga_yAs/GaAs$ quantum wells grown on GaAs substrates. Since the substrate is opaque at the energies of the quasi two-dimensional quantized exciton states, direct absorption spectroscopy was impossible. This difficulty has now been overcome because one has learned to remove the substrates by selective etching [24.10] so that simple transmission spectroscopy also became possible for $Al_{1-y}Ga_yAs/GaAs$ (multiple) quantum wells.

Sometimes the detection of the luminescence in PLE spectra is replaced by a measurement of the photocurrent. Most of the statements made above for the PLE spectra are also valid for the interpretation of photocurrent excitation spectra; see [24.9, 11, 12] for examples. However, especially at low temperatures, one must also consider the dissociation of optically excited excitons to give free carriers that contribute to the electrical conductivity.

24.2 Time-Resolved Spectroscopy

In time-resolved spectroscopy one monitors the temporal evolution of an optical signal, e.g., the luminescence spectrum $I_{lum}(\hbar\omega, t)$ or the change of absorption $\Delta\alpha(\hbar\omega, t)$. We have already seen several examples in Chap. 22. Here we want to outline some of the basic concepts and requirements for this group of spectroscopic methods.

The first thing necessary for time-resolved spectroscopy in the time domain is an (optical) excitation source which is modulated in time with a period or time constant T_M which is at least comparable to the decay time τ which is to be measured, i.e.,

$$T_M \lesssim \tau . \tag{24.7}$$

Single or repeating pulses can be used or a sinusoidal or rectangular modulation. The light sources that fulfill the above condition in the various time domains are very different. We shall not go into details because this would fill a separate book, but give only the keywords to allow the reader to find more detailed information.

For times $T_M \gtrsim 50$ ns one can use cw light sources modulated by electro-optic or acoustic-optic modulators. Pulsed flashlamps have been developed to emit pulses down to a few ns, and $\boldsymbol{Q}$-switched ruby or Nd lasers have typical pulse durations in the 10–50 ns range. Some lasers, such as N_2 or excimer lasers, typically emit pulses of about 10 ns duration (with a lower limit of about 0.5 ns), and tunable dye lasers pumped by them have comparable or slightly shorter pulse lengths in the whole range from the near IR to the near UV or, after frequency doubling or tripling down even to the onset of vacuum UV (VUV).

Pulses of around 100 ps are obtained either by mode-locking of cw lasers (preferably Ar^+ lasers) or of Nd lasers, or by quenched cavity dye lasers.

In this regime the spectral width of the gain profile starts to be important. In the Fourier limit, the spectral width of the optical gain of the laser material $\Delta\omega_g$ and the shortest pulse length τ_p obtainable from this laser are coupled via

$$\Delta\omega_g \, \tau_p \approx 1 . \tag{24.8}$$

To reach the lower ps and the fs regime demands lasers with a broader gain spectrum than Ar^+ or Nd^{3+} can offer.

Lasers which are synchronously pumped by a mode-locked Ar^+ or Nd^{3+} laser and use laser dyes, special color centers, or more recently Ti-doped sapphire as active medium, reach typical pulse durations in the lower ps or sub ps regime.

The same is true for some specially designed distributed feedback dye lasers, which can be pumped with N_2 or excimerlasers.

More sophisticated setups like colliding pulse mode-locked (CPM) dye lasers or the self-mode-locking of Ti-doped sapphire lasers allow the sub-ps or even the 50–100 fs regime to be reached. A subsequent pulse compression can lead to even shorter values of τ_p. The record stands presently at around 6 fs. This means only two or three periods of light and consequently, with (24.8), a spectral width which extends over a substantial fraction of the visible spectrum. For some considerations about the ultimate limit of the duration of laser pulses see [24.13].

Broad spectra can be also obtained for longer, i.e., ps, pulses by focusing them into glass, water, or some organic liquids through the phenomenon of "self-phase-modulation". In this case one gets a spectrum which decays on both sides of the pump laser frequency approximately exponentially and which is known as "white light continuum". The crucial parameters for the appearance of this effect are the peak intensity of the incident pulse but – even more important – its temporal change. As a rule of thumb one can state that values of $\mathrm{d}I/\mathrm{d}t$

$$\frac{\mathrm{d}I}{\mathrm{d}t} \approx 10^{20}\,\mathrm{W/cm^2 s^{-1}\ d.h.s^{-1}} \tag{24.9}$$

are necessary to create a white light continuum.

Special care has to be used in handling fs pulses because every passage through a lens, a cryostat window, or even several meters of air, tends to increase the pulse length by dispersion. Another aspect of short-pulse lasers is chirp, i.e., frequency shift during the pulse.

Once we have a light source with the desired temporal properties, its light beam is sent onto the sample and some response is measured, such as the decay of a luminescence signal or a change in transmission or reflection.

Sometimes it is necessary to measure these signals with time resolution, sometimes a slow detector is sufficient, but in these cases at least the duration of the incoming pulse must be known. Therefore we mention in the following some of the corresponding techniques. The first type rely on the fast response of a detector, the second type uses some kind of (auto) correlation technique.

Photomultipliers typically have a lower limit for the temporal resolution of one or a few ns (with a transit time of the signal of some 10 ns) and channel plates can reach into the sub-ns regime. Fast vacuum and semiconductor diodes have typical response times in the 100 ps regime. Sometimes the rise-time of the latter is faster than the decay time. To record the signals of these devices one generally uses oscilloscopes. Fast oscilloscopes have upper frequency limits around 300 MHz to 1 GHz corresponding to rise and decay times in the lower ns regime. They can also be used for single-shot experiments. Sampling oscilloscopes need repeating events but can resolve temporal evolution down to about 0.1 ns provided there is no jitter. The amplifier stages of optical multichannel analyzers can be partly gated in the lower ns regime. The same is true for the time windows which can be set in fast box-car integrators.

A fundamentally different approach can be used with single photon counting. One measures the distribution $N(t_\mathrm{d})$ of the delays t_d between the excitation pulse and the first emitted photon. For a simple exponential decay the curve $N(t_\mathrm{d})$ is proportional to $I_\mathrm{lum}(t)$.

Detection in the time regime from several tens of ns down to about 5 ps is very conveniently covered by streak cameras. The basic principle of a streak camera is shown in Fig. 24.7. An incident photon produces a photo-electron on the photo-cathode in a vacuum tube. This electron is accelerated by a voltage of several keV onto a screen, where it causes some luminescence. A further amplification is possible, for example, with a channel plate in front of or behind this screen. The temporal resolution stems from a deflection voltage which changes linearly with time (often simply the linear part of a $\sin \omega t$ function is used) and which "streaks" the electrons along the screen, transforming a delay in time into a spatial separation. The spatially distributed luminescence pattern of the screen can be detected by an optical multichannel analyzer. An especially favorable configuration is attained if the entrance slit of the streak camera, which is imaged on the photo-cathode, is in the direction of the photon energy of a spectrometer (in Fig. 24.7 normal to the paper) and if a two-dimensional readout system is used behind the screen. This setup allows one to measure the signal intensity directly as a function of photon energy and of time $I(\hbar\omega, t)$. Streak cameras are designed either for the detection of single shots and events with a slow repetition rate, or for use in connection with mode-locked lasers, which often have repetition rates around 80 MHz, determined by the round-trip time of the laser resonator. A streak camera needs a trigger pulse usually several ns before the arrival of the optical signal. The time axis of a streak camera can be calibrated by using a variable optical delay line.

Direct measurements in the time domain become rather difficult below approximately 5 ps. For this time regime correlation techniques are generally used; these are either purely optical or electro-optical. The latter include the

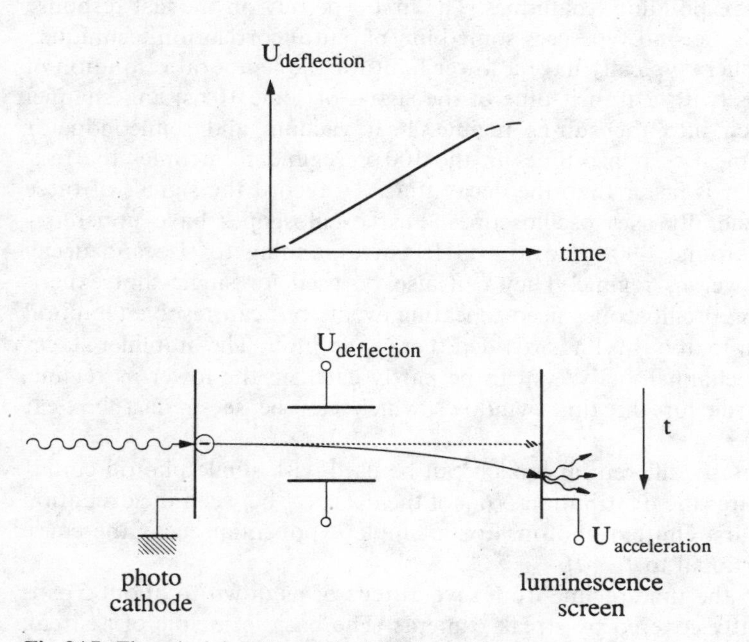

Fig. 24.7. The principle of the operation of a streak camera

so-called Auston switches [24.14]. We concentrate here, however, on the optical methods. A rather simple setup is the so-called Kerr shutter of Fig. 24.8a. A Kerr medium is placed between two crossed polarizers. Without electric field on the Kerr medium, the signal beam is not transmitted. To open the shutter, one uses the electric field of a short (typically ps) laser pulse which makes the Kerr medium birefringent and allows a fraction of the signal beam to be transmitted for a time given by the duration of the pulse and the relaxation time of the Kerr

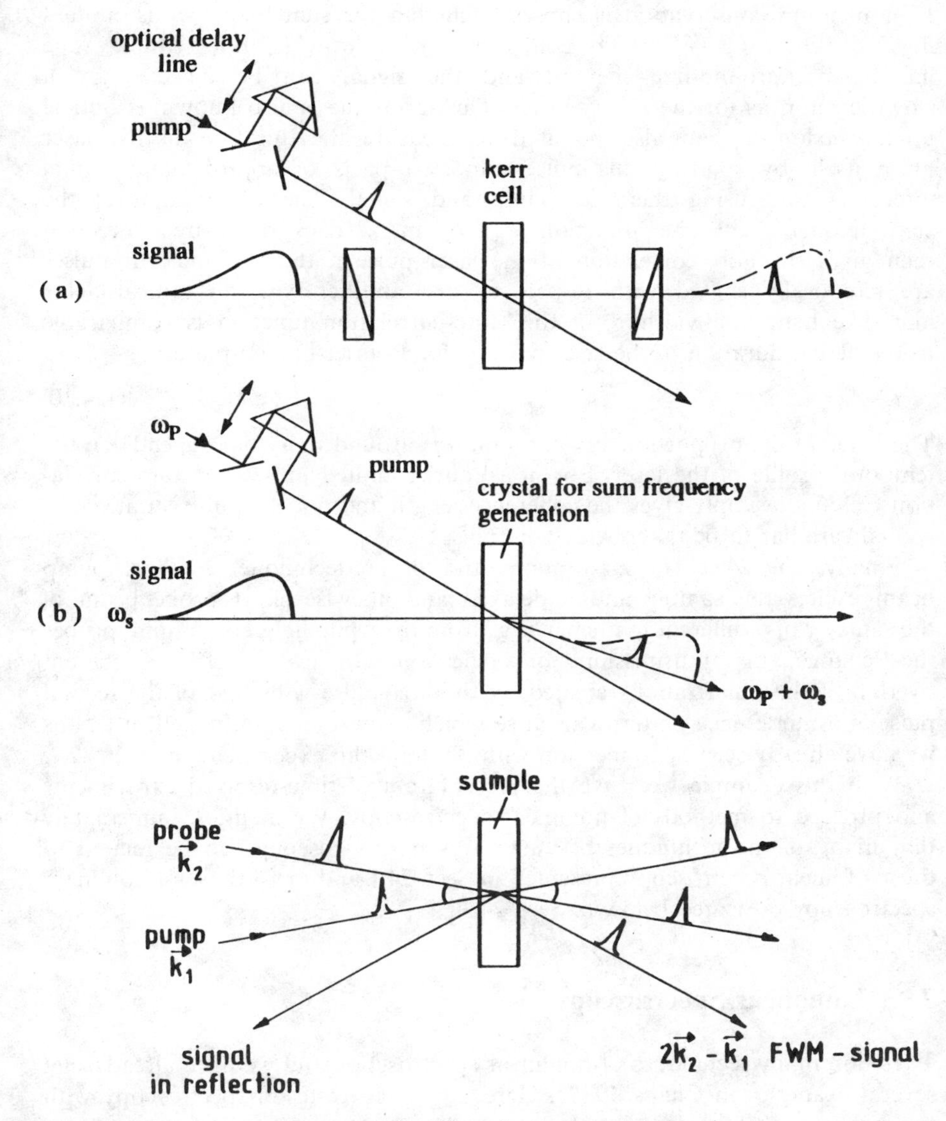

Fig. 24.8. Various methods for measuring light pulses in the (sub-) ps regime

effect. The latter quantity is usually in the sub-ps regime. The signal can now be detected with a slow (i.e., time-integrating) detector and the temporal evolution of the signal is tested by imposing a time shift between the pump-beam which opens the Kerr shutter and the signal beam using a variable optical delay line.

The next method (Fig. 24.8b) relies on frequency mixing (Sect. 18.1) in an optically nonlinear medium like KDP (Sodium dihydrogen phosphate): A signal beam with a certain temporal evolution and a frequency ω_s is mixed with a short laser pulse at ω_p and the sum frequency $\omega_s + \omega_p$ is detected. If momentum conservation is chosen such that the sum frequency is emitted in a direction $\boldsymbol{k}_p + \boldsymbol{k}_s$ which coincides neither with $\boldsymbol{k}_p$ nor with $\boldsymbol{k}_s$, one has as background-free method and the signal can be detected by a slow detector, as for the Kerr shutter. This technique is also known as optical up-conversion. It can also be used to measure the duration of the laser pulse itself, by splitting this pulse into two parts of approximately equal intensities and using them as pump and signal. The observation of the sum frequency in the direction $\boldsymbol{k}_p + \boldsymbol{k}_s$ in a background-free direction then gives the auto-correlation of the laser pulse if the two incident pulses are temporally shifted with respect to one another with an optical delay line. The temporal width τ_a of this auto-correlation function is comparable to the actual duration of the laser pulse τ_p for Fourier-limited pulses:

$$\tau_p = a\tau_a \,. \tag{24.10}$$

The value of the proportionality constant a is around unity but depends on the temporal profile of the laser. For non-Fourier-limited pulses the auto-correlation technique simply gives the coherence length and not the pulse duration, so special care has to be taken with such pulses.

Finally, Fig. 24.8c shows a pump and probe technique. A short pump beam excites the sample and a delayed and likewise short probe beam of the same or of a different frequency (e.g., from the white light continuum) probes the change in transmission or reflection. In cases of a coherent overlap of the polarizations created by the first pulse with that of the second pulse, an interference pattern can arise which causes some diffracted orders as we have already seen in connection with photon-echo experiments in Sect. 22.2.

With this example we leave the presentation of time-resolved experiments and proceed to methods of nonlinear spectroscopy. We mention immediately that many of the techniques of time-resolved spectroscopy can be related to those of linear spectroscopy presented in Sect. 24.1 and/or to those of nonlinear spectroscopy presented below in Sect. 24.3.

24.3 Nonlinear Spectroscopy

There are many techniques of nonlinear spectroscopy and we have already met several examples in Chaps. 19–22. Here, we concentrate on spectroscopy with the pump-and-probe beam technique and with laser-induced gratings. Both groups of methods can be related to the methods of temporal resolution mentioned above. The main idea of the pump-and-probe beam technique is to excite the sample with an intense and spectrally narrow laser beam and to probe changes of the transmission and/or reflection spectra with a weak, usually spectrally broad probe beam or alternatively with a weak spectrally narrow probe beam which is either scanned in frequency or tuned to match a narrow spectral structure of interest. The probe-beam spectrum is recorded twice; once without and once with the pump beam on. The difference between the two spectra (corrected for dark current and the luminescence caused by the pump beam if necessary) gives information about the changes of the transmision or absorption at a wavelength $\hbar\omega_{\mathrm{probe}}$ caused by excitation at $\hbar\omega_{\mathrm{exc}}$ with an intensity I_{exc}:

$$\Delta T(\hbar\omega_{\mathrm{probe}}, \hbar\omega_{\mathrm{exc}}, I_{\mathrm{exc}}), \quad \Delta\alpha(\hbar\omega_{\mathrm{probe}}, \hbar\omega_{\mathrm{exc}}, I_{\mathrm{exc}})\,. \tag{24.11a}$$

Similar data can be deduced for the reflectivity

$$\Delta R(\hbar\omega_{\mathrm{probe}}, \hbar\omega_{\mathrm{exc}}, I_{\mathrm{exc}})\,. \tag{24.11b}$$

If the sample under investigation has parallel surfaces and shows Fabry–Perot modes in transmission and/or reflection their shift with pump can also be used to deduce the change of the real part of the refractive index:

$$\Delta n(\hbar\omega_{\mathrm{probe}}, \hbar\omega_{\mathrm{exc}}, I_{\mathrm{exc}})\,, \tag{24.11c}$$

e.g., using the relations for the position of the Fabry-Perot maxima in Sect. 3.1.6. A crucial point for the pump-and-probe technique is the intensity of the probe beam. It must itself be so low that it introduces no changes of the optical properties, but still needs to be high enough to give a reasonable signal-to-noise ratio. A good choice is often to make the probe beam signal on the detector about half an order of magnitude stronger than that due to the luminescence arising from the pump.

The pump-and-probe beam technique can be used in a cw mode, pulsed under quasi-stationary conditions, i.e., with a pulse length τ_{p} comparable to or longer than the lifetime T_1 of the species under consideration, or for $\tau_{\mathrm{p}} \ll T_1$ as a method for time-resolved measurements as indicated above.

It is advisable to have the probe beam spatially and, for pulsed excitation, also temporally narrower than the pump (say by a factor two) and to probe only the central part of the pump (Fig. 24.9a). In this case the gradients which follow from temporal and/or spatial inhomogeneities and may cause lateral or longitudinal diffusion of the excited species can be minimized.

The duration of the pump pulse and its repetition rate should be low enough to avoid sample heating. This can be checked by measuring the probe beam spectrum without, during and appproximately 5 T_1 after the pump pulse or by varying the repetition rate with all other parameters unchanged.

Lensing effects should also be checked, i.e., the creation of a lateral refractive index profile by the pump which can lead to a (de-) focussing of the probe and consequently to a variation of its focussing onto the entrance slit of the spectrometer. Such lensing effects can be detected by observing the probe

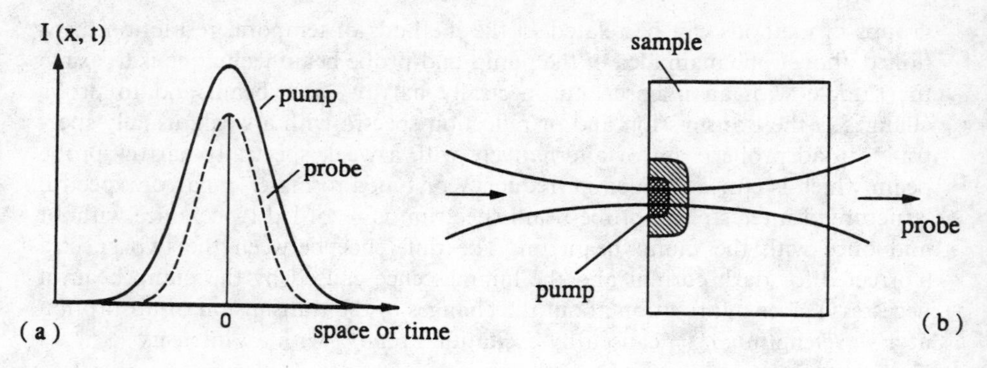

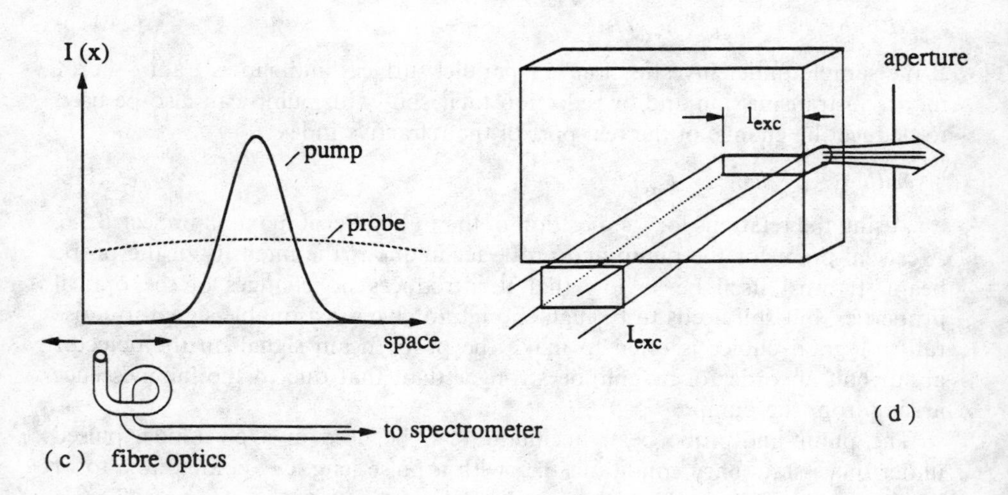

Fig. 24.9. Some aspects of pump-and-probe beam spectroscopy. For details see text

beam on a white screen about 0.5 m behind the sample. A lensing effect would result in a "blooming" of the probe beam. The stray light from the pump can be suppressed by choosing pump and probe polarized orthogonal to one another, provided the experiment allows such a configuration, or to use a non-collinear geometry of pump-and-probe beam and an aperture stop for the former.

Another crucial aspect is the homogeneity of the excitation with depth into the sample (Fig. 24.9b). If the pump light is in a spectral region of strong absorption, its penetration depth can be as low as 0.1 μm. With a typical diffusion length l_{diff} of excitons or electron–hole pairs in direct gap semiconductors of one or a few μm, only samples of comparable thickness can be excited homogeneously. If the sample is thicker, the inhomogeneity has to be taken into account, e.g., using a two layer model including an excited and an unexcited layer. To check whether the excited species reach the opposite side of the sample one can probe their influence on the excitonic reflection spectra at the front side which is pumped, and on the rear side. If this method is used for samples of different thicknesses around l_{diff} it even allows the diffusion length of the excited species to be measured. Two-photon excitation often results in a rather homogeneous excitation of samples with thicknesses up to one or several mm, but the maximum densities which can be reached by two-photon excitation are generally lower than those obtained under one-photon excitation. Furthermore, it is often important to excite small volumes of only a few μm dimension in all three directions of space to suppress or at least reduce unwanted stimulated emission.

A tuning of the pump energy $\hbar\omega_{exc}$ allows resonant excitation, e.g., in an exciton resonance, and even measurement of excitation spectra of some optical nonlinearities, e.g., the gain, which are sometimes very helpful for the identification and interpretation of optical nonlinearities [24.15].

If, in contrast to what was said above, the probe beam is chosen to be spatially considerably broader than the pump beam (Fig. 24.9c), one can perform laterally resolved experiments, e.g., by scanning with a glass fiber over an enlarged image of the sample. This type of experiment can give information about the lateral expansion or diffusion of the excited species [24.16].

We proceed now to consider some further modifications of pump-and- probe spectroscopy. If only one spectrally narrow beam impinges on the sample, and one measures its transmission as a function of its incident intensity, one can learn something about "self-renormalization" effects, i.e., changes of the absorption (or reflection) created by a beam at $\hbar\omega_{exc}$ at this very frequency, $\hbar\omega_{exc} = \hbar\omega_{probe}$ in (24.11).

If, on the other hand, an intense and spectrally broad beam is used simultaneously as pump and probe beam one does not have to worry about spatial and temporal coincidence, but violates the important rule $I_{exc} \gg I_{probe}$ given at the beginning of this section. The penalty that must be paid is that some nonlinearities like optical amplification cannot be detected.

A further variant of the pump-and-probe beam technique again uses only one intense, spectrally narrow pump beam and as the probe the luminescence excited by this beam. This method works well in direct gap materials and also automatically fulfills spatial and often temporal coincidence (depending on the ratio of τ_{pulse} and T_1). It is also known as luminescence-assisted two-photon spectroscopy (LATS) [24.17].

A further modification of the pump-and-probe technique involves the variation of the excitation stripe length [24.18]; see Fig. 24.9d. In this case the sample is excited in a stripe-like geometry with a width of about 20 μm and a variable length l_{exc}. The light emitted from the edge is measured as a function of l_{exc}. Most conveniently l_{exc} is modulated by a factor 2. The emission of the sample in the direction of observation is used as a probe beam. By integrating the spontaneous emission and its absorption or amplification along the excitation stripe one can deduce the absorption $\alpha(\hbar\omega)$ or gain spectra $g(\hbar\omega)$ under variation of

l_{exc} between l_{exc1} and l_{exc2} via

$$\frac{I_{lum}(l_{exc1}, \hbar\omega)}{I_{lum}(l_{exc2}, \hbar\omega)} = \frac{\exp[g(\hbar\omega)\, l_{exc1}] - 1}{\exp[g(\hbar\omega)\, l_{exc2}] - 1} . \tag{24.12}$$

Care has to be taken in this method that the excitation intensity is strictly independent of l_{exc} and, especially in gain measurements, that no gain saturation occurs, i.e., that the product $g(\hbar\omega)l_{exc} \lesssim 1$. A comparison of the gain spectra obtained with two different pairs of l_{exc} answers the question from the experimental side. Furthermore it should be noted that (24.12) relies on the assumption that only light propagating along the excitation stripe is detected, and none that leaves the stripe laterally. An aperture can help to fulfil this condition in the experiment (Fig. 24.9d).

A final application of the pump-and-probe technique is the so-called spectral hole burning. A laser is tuned over an inhomogeneously broadened absorption band. It excites and thus saturates some of the transitions. This results in a bleaching around the photon energy of the pump laser which shifts with $\hbar\omega_{exc}$.

Under favorable conditions the width $\Delta\hbar\omega$ of this "spectral hole" in the absorption band gives information about the T_2 time of the corresponding transition, via $\Delta\hbar\omega T_2 \approx \hbar$ [24.5, 19, 20]. Care has to be taken that only a small fraction of the transitions around $\hbar\omega_{exc}$ is bleached. If the pump laser is too strong it also bleaches spectrally neighboring transitions which have only a small absorption tail at $\hbar\omega_{exc}$. This so-called power broadening of the spectral hole falsifies the T_2 values deduced from the spectral width of the hole. More information on this technique may be found in [24.5, 19, 20].

We now leave the discussion of the pump-and-probe technique and proceed to the other widespread group of nonlinear optical techniques, which are based in some way or other on the formation of a grating caused by the interference of two coherent laser beams (see also Sect. 22.2).

The basic idea is the following: Two coherent laser beams of roughly equal intensities are allowed to interfere in the sample under investigation. This results in an interference pattern in one direction, i.e., in a periodic modulation of the light impinging on the sample (Fig. 24.10a). The period of this grating Λ depends on the wavelengths λ of the incident beams and on the angle Θ between them:

$$\Lambda = \frac{\lambda}{2\sin\frac{\Theta}{2}} = \frac{c\pi}{\omega}\frac{1}{\sin\frac{\Theta}{2}}, \tag{24.13a}$$

and the periodic lateral spatial modulation of the intensity in the interference pattern follows as

$$I(x) = I_1 + I_2 + 2(I_1 I_2)^{1/2} \cos\left(\frac{2\pi}{\Lambda}x\right) . \tag{24.13b}$$

If there is any optical nonlinearity in the sample, i.e., any change in the real and/or imaginary part of the complex index of refraction e.g. at the photon

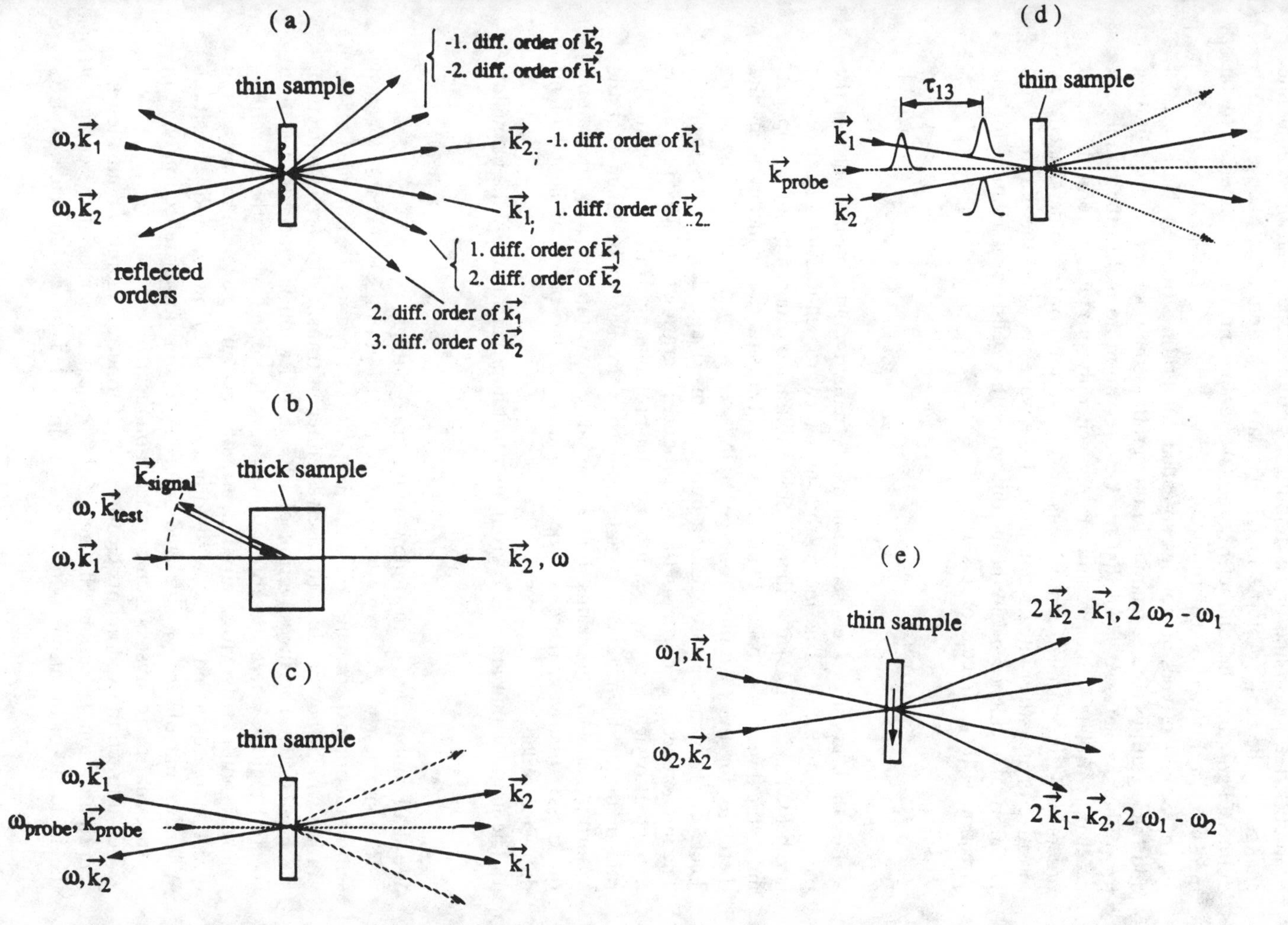

Fig. 24.10. Spectroscopy with laser-induced gratings. For details see text

energy of the incident laser, there will be a periodic modulation of the optical properties, i.e., phase and/or an amplitude grating, depending on whether the changes of the real or of the imaginary part of $\tilde{n}$ dominate. Such a grating diffracts light and the diffracted order(s) are the quantity of interest to be measured in this type of experiment.

There are two types of gratings: so-called thin gratings (Raman–Nath regime) and thick gratings (Bragg regime). In thin gratings the thickness d of the grating (or of the sample) is so small that the $\boldsymbol{k}$-conservation normal to the sample can be slightly violated. Such a thin grating generally produces several diffracted orders and the parallel components of the various orders differ exactly by (positive or negative) integer multiples of the wave vector of the grating, $2\pi/\Lambda$. For normal incidence of the beam this gives the well-known relation for the direction of the diffracted orders n

$$\sin \alpha_n = \frac{n\lambda}{\Lambda}\,, \quad n = 0, \pm 1, \pm 2 \ldots . \tag{24.14}$$

For thick gratings, $\boldsymbol{k}$-conservation has to remain valid, and not only for the parallel components. This means that in a forward self-diffraction experiment (see below) as in Fig. 24.10a, there is only a diffracted order of beam 1 propagating in the direction of beam 2 and vice versa. This situation is inconvenient and therefore, for thick gratings, one generally uses the geometry of Fig. 24.10b. There are two counter-propagating pump beams, 1 and 2, and a test beam, t. The interference, e.g., of beams 1 and t, produces a grating from which beam 2 is diffracted to yield a signal beam, s, counter-propagating to the test beam. The role of beams 1 and 2 can be interchanged. Care has to be taken that the absorption length for beams 1 and 2 is at least comparable to the sample thickness. Otherwise, there will be no spatial overlap between them and no diffracted order.

The limit for thin gratings is given, to a first approximation, by the inequality

$$d < \Lambda^2 \lambda^{-1} . \tag{24.15}$$

We shall concentrate in the following on thin gratings.

The simplest way to exploit the laser-induced gratings (LIG) are so-called self-diffraction experiments. Here the incident laser beams are diffracted off the grating which they produce themselves either in transmission or in relfection, see Fig. 24.10a. The investigation of LIG in reflection is especially useful in spectral regions where the absorption is very high and where the reflectivity changes significantly with excitation, as is the case in the exciton resonances of direct gap semiconductors with dipole-allowed band-to-band transition. Whenever one observes a diffracted order in transmission or reflection one can be sure that there must be some optical nonlinearity.

Self-diffraction experiments clearly give information mainly about the self-renormalization, i.e., about changes of the optical properties at the energy of the incident beam itself. Often various diffracted orders coincide in LIG experiments. In the case of self-diffraction, the diffracted order -1 of beam 1 coincides with the transmission (i.e., the zero order) of beam two and vice versa. In the direction of the $+1$ diffracted order of beam 1 we also have the $+2$ order of beam 2 and so on. Since the intensities of higher orders generally tend to be much weaker, they can be often neglected.

We now discuss some aspects of LIG that are not necessarily connected with self-diffraction, and then proceed to some more complex variants of LIG spectroscopy.

One usually measures the grating efficiency, η, i.e., the ratio of the intensity in a certain diffracted order to the total incident intensity as a function of photon energy. Some authors prefer a different definition of efficiency, e.g., by normalizing only with the intensity of one of the incident beams, which is diffracted into the desired direction. This procedure usually gives a value of η which is a factor two higher when both incident beams have roughly equal intensities.

Other authors consider the ratio of diffracted to transmitted intensity (i.e., of the zero order) as diffraction efficiency. This approach partly takes care of absorption effects, but the quantity obtained in this way is strictly speaking not an efficiency because it relates the quantities in two different output channels, while a true efficiency is always defined as the ratio of the output in a desired channel divided by the input.

If the input intensity is kept constant, e.g., while scanning the photon energy, it is often sufficient to give just the intensity of the diffracted order (in arbitrary units) e.g., to observe some resonances.

Indeed it is advisable to keep the intensity of the incident beams constant while scanning another parameter like $\hbar\omega_{\mathrm{exc}}$ or Λ, since the dependence of η on the incident intensity is rather strong.

If we consider the self-diffraction geometry of Fig. 24.10a and vary the intensities of the two incident beams simultaneously, e.g., by producing them from one beam with a beam splitter, we will find that the intensity of the first diffracted order I_{s} varies like

$$I_{\mathrm{s}1} \propto \eta I_1 \propto I^3 \quad \text{for} \quad I_1 \propto I_2 = I \, . \tag{24.16}$$

The cubic dependence follows immediately from the grating efficiency which, for weakly absorbing thin gratings and small changes of the complex index of refraction $\Delta\tilde{n}$, is given by [24.3]:

$$\eta \approx \left|\frac{\pi \Delta \tilde{n} d}{\lambda}\right|^2 = \left|\frac{\pi \Delta n d}{\lambda}\right|^2 + \left|\frac{\Delta \alpha d}{4}\right|^2 \tag{24.17}$$

and from $\Delta\tilde{n} \propto I$; see (18.3). This relation holds both for a coherent and an incoherent (or population) grating. The first diffracted order of a coherent grating is also known as a four-wave mixing process and is described by the $\chi^{(3)}$ contribution in (18.2, 13). The dependence of I_{s} on I according to (24.16) is also obvious in this case.

Values of the exponent below three indicate some saturation of the nonlinearity or an excitation-induced increase of absorption which can even lead to a switching down of the transmitted and diffracted intensities during the pulse; Sect. 23.4 and [24.21].

Though the description of incoherent or population gratings in terms of $\chi^{(3)}$ processes is in principle meaningless, some authors like to deduce a $\chi^{(3)}$ value from η, independent of the physical origin of the nonlinearity. In the following this construct will be called $\chi^{(3)}_{\text{eff}}$; it can be helpful when comparing the magnitudes of the nonlinearities of different physical origins and/or in different materials.

Formulas to convert η into $\chi^{(3)}_{\text{eff}}$ are given in [24.3, 22] and have forms such as

$$\chi^{(3)}_{\text{eff}} = \frac{19 n_0^2 \lambda \sqrt{\eta}}{\pi I_\text{p} d} \quad \text{or} \quad \chi^{(3)}_{\text{eff}} = \frac{8 c^2 n_0^2 \varepsilon_0 \alpha \sqrt{\eta}}{3 \omega I_\text{p} (1 - T_\text{p})}, \tag{24.18}$$

where n_0 is the refractive index at wavelength λ or frequency ω, I_p the (total) pump power for the grating, d the thickness of the grating, α the absorption coefficient, T_p the transmission at the pump wavelength, and ε_0 and c the usual constants.

Values of $\chi^{(3)}_{\text{eff}}$ are given by several authors in the otherwise rather out-dated electrostatic units (esu). The relation between esu units and those of the internationally recognized Système Internationale (SI) are

$$x^{(3)} \text{ in m}^2\,\text{V}^{-2} \mathrel{\hat{=}} x^{(3)} \text{ in esu} \,.\, 1.396 \cdot 10^{-8}. \tag{24.19}$$

Until now we have presented only the simplest method of LIG spectroscopy, namely self-diffraction. In its time-resolved version using two short ps pulses with a delay it is nevertheless already capable of measuring photon-echos and T_2 times as shown in Sect. 22.2.

The next step of sophistication is to write a grating with two coherent incident beams of equal frequency and to read it with a third beam as shown in Fig. 24.10c. This third beam can have the same frequency but be delayed in time, e.g., to measure the decay of the grating (see below) and/or it can have a different photon energy to probe the change of $\tilde{n}$ induced at $\hbar\omega_{\text{probe}}$ by illumination at $\hbar\omega_{\text{exc}}$. One can create a grating with $\hbar\omega_{\text{exc}} \gtrsim E_\text{g}$ in the strongly absorbing region, where the optical properties and thus the excitation conditions do not change very much under excitation, and read it in the transparent spectral region below the exciton energy ($\hbar\omega_{\text{probe}} < E_{\text{exc}}$) where essentially a phase-grating is probed.

If the third beam is not coplanar with the two beams producing the grating, its diffracted orders are also in another plane. This trick allows one to spatially separate self-diffracted orders of the two pump beams from the diffracted orders of the third (probe) beam.

Spectroscopy with LIG also allows one to determine the diffusion length l_{diff} of certain excited species in the following ways (see also Chap. 22). The efficiency of a thin coherent grating is roughly independent of Λ if all other parameters are kept constant. This is not usually true for incoherent or population gratings formed from some mobile excited particles like free carriers, excitons or phonons (e.g., via heat conductivity). In these cases the grating efficiency is influenced by the lifetime T_1 of e.g., the excitons and by their diffusion, which washes the grating out. T_1 is independent of Λ but the influence of diffusion depends on the ratio of Λ to l_{diff}. Thus it is possible to separate the two quantities via (24.20) by varying Λ with all other parameters unchanged.

In the case of quasi-stationary excitation ($\tau_p \gtrsim T_1$) one plots $\eta(\Lambda^{-1})$. This curve decays for increasing Λ^{-1}. As a rule of thumb one can say that η decays to one half for

$$\Lambda_{1/2} = 2\pi l_{\mathrm{diff}}. \tag{24.20a}$$

If one uses short pulses with $\tau_p \ll T_1$ one probes the decay of the grating efficiency with a third delayed probe beam, while the two pump beams arrive simultaneously (Fig. 24.10d). For $\tau_{13} > T_2$ one finds for the decay time constant τ_s of the signal intensity defined by

$$I_s \propto \mathrm{e}^{-t_{13}/\tau_s}, \tag{24.20b}$$

$$\frac{1}{\tau_s} = \frac{2}{T_1} + \frac{8\pi^2 D}{\Lambda^2}, \tag{24.20c}$$

under the assumption of classical diffusion with a density-independent diffusion constant. A plot of $\frac{1}{\tau_s} = f(\Lambda^{-2})$ gives with its slope the value of D and with the abscissa T_1.
It should be mentioned that these three beam techniques also allow the observation of photon-echos. In this case the time between the arrival of the two pulses 1 and 2 is varied in a range comparable to T_2, and the probe beam 3 arrives after a fixed delay time of several T_2. The photon-echo is is emitted with a delay $t_{1,2}$ after beam 3.

The configuration introduced in Fig. 24.10b for thick gratings is evidently also a variant of this three-beam LIG technique, which of course also works for thin samples.

The final technique that we want to mention uses two coherent laser beams of different frequencies ω_1 and ω_2 to write the grating. Coherence, i.e., a well-defined phase relation between the two beams which changes in this case linearly in time, can be achieved by the superposition of two very narrow laser beams of correspondingly long coherence times, pumped synchronously by one pump laser. The interference of two such laser beams results in a grating that moves laterally with a speed v_G given by

$$v_G = \frac{\omega_1 - \omega_2}{|\boldsymbol{k}_1 - \boldsymbol{k}_2|}. \tag{24.21}$$

The beams diffracted from such a moving grating are spectrally (Doppler-) shifted by integer multiples of $\omega_1 - \omega_2$. The same result is obtained if we use, instead of the wave picture, the particle picture or the $\chi^{(3)}$ notation with $\chi^{(3)}(\omega_s\colon \omega_1 + \omega_1 - \omega_2)$.

These moving gratings give information about the phase relaxation time and/or the lifetime T_1. The basic idea is the following: As long as the grating moves laterally with a speed that shifts it by one grating period Λ in a time much larger than T_2 or T_1, then the density of the coherent or incoherent species which form the grating can follow this motion adiabatically and the diffraction efficiency is high. If the lateral velocity of the grating increases, the density variation is smoothed out and the grating efficiency drops. If only one time constant T contributes one finds, to a first approximation

$$\eta \propto |\chi_{\text{eff}}^{(3)}|^2 \propto \frac{1}{1+(\omega_1-\omega_2)^2 T^2} . \tag{24.22}$$

Otherwise product terms with T_1 and T_2 have to be used [24.23]. See also (22.24). An example was given in Fig. 22.11.

The final aspect of laser-induced gratings to be discussed here is phase conjugation. This term means that some of the diffracted orders are phase conjugated or cum grano salis time reversed with respect to some of the incident beams. This effect is best explained in the configuration of Fig. 24.10b. In Fig. 24.11a we assume two coherent plane waves as pump beams 1 and 2. Further we assume that the wavefront of the probe beam is distorted in the way shown by the solid line. Due to the effect of phase conjugation, the diffracted signal beam

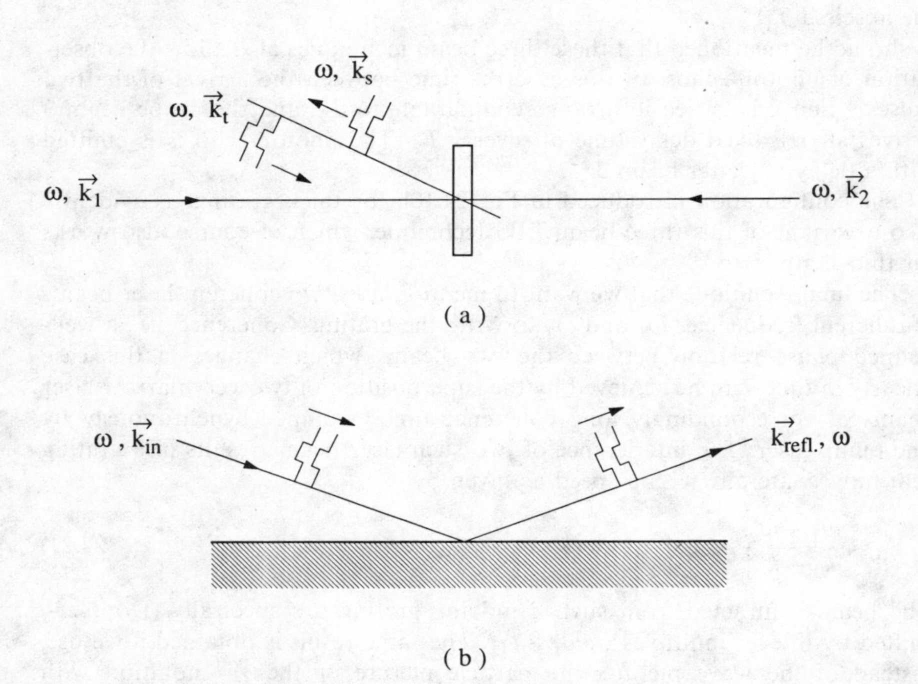

Fig. 24.11. Schematic drawing of the phenomenon of phase conjugation

has the opposite deformation, in contrast to the familiar case of reflection from an ordinary mirror (Fig. 24.11b). Another consequence of phase conjugation is that a divergent beam falling on a phase-conjugating mirror is reflected back as a convergent beam, whereas it would still be divergent in the case of a normal mirror. More information on the topic of phase conjugation and some of its possible (and impossible) applications can be found in [24.1, 3, 24] and references therein.

To conclude this section, we give some serious but not exhaustive warnings: never look directly into a laser beam, nor even into the stray light of it, if you are not absolutely sure that it is safe to do so. Wear safety goggles in the lab, whenever the laser is switched on. Avoid long-term exposure even of the skin to laser radiation, especially in the UV. If you want to see the exact shape or position of a laser spot, monitor it with a CCD camera, for example, behind a microscope and not by looking with your eye through it. Be careful with the high voltages present in many lasers (partly even after switch-off) and with the chemical risks of laser gases, dyes, and solvents. Read and obey the instruction manuals of your equipment before switching it on.

24.4 Problems

1. What is the blaze wavelength of a grating? How does it work? Why are the efficiencies for + and − first order different?

2. Which type of lens (plane-convex, bi-convex) would you choose to focus a parallel beam? And which to yield a 1:1 imaging?

3. Is it best to place a crystal polarizer (like a Glan-Thompson) in a parallel, a convergent, or a divergent beam?

4. Discuss the advantages and disadvantages of a spectrometer employing a grating or a prism as dispersive element.

5. What happens if you place an optical component with some coatings or some adhesive layers of at the focal point of a high-power laser?

6. What do you see when you look into a phase-conjugating mirror? Solve this problem by thinking, *do not* do it!

7. Have a good idea, which you like yourself, for an experiment on semiconductor optics. Convince some funding agency that your idea has great prospects, both in basic and applied research, and that it belongs to the best 10% in this universe and this century, and so acquire some 300 kilo units of ECU, US$ or another comparable currency to buy your equipment, pay overheads, hire coworkers etc. Set up your own experiments. If and when it works, please send me some comments or addenda for this or any other chapter based on your own experiences with Murphy's Laws [24.25].

25 Group Theory in Semiconductor Optics

In this last chapter, we shall give an outline of group theory, its connection to quantum mechanics, and its applications in semiconductor optics. We shall present only the most important aspects and rules and generally give no proofs of the various relations.

The information compiled here has been taken from different books on group theory, which are listed in [25.1–6]. A very good collection of tables is found in the book by Koster et al. [25.6], from which most of the Tables 25.6–15 are taken.

25.1 Introductory Remarks

Noether's theorem of Sect. 3.1.1, which we have already used several times in this book, states that a conservation law follows from every invariance of the Hamiltonian. The classic examples are the following.

- If H is invariant under an infinitesmally shift in time, then the total energy of the system described by this Hamiltonian is conserved.
- If H does not depend on a particular spatial coordinate x_i, i.e., if it is invariant under an infinitesimal translation along x_i, then the component of the momentum p_i in the direction x_i is conserved.
- If H has spherical symmetry (as for a single atom, for example) it is invariant with respect to an infinitesimal rotation $\mathrm{d}\phi$ around any axis, and a conservation law for the angular momentum follows for L^2 and for one component of $\boldsymbol{L}$, e.g., L_z.

In a crystal the first condition remains valid in the absence of explicitly time-dependent (perturbative) terms, i.e., energy is still conserved.

The second condition is no longer valid, but is replaced by invariance with respect to translations by integer multiples of the $\boldsymbol{a}_i$ (Sect. 9.2). We have already presented this subject and its consequences in some detail, e.g., in Sect. 3.1.3 and Sects. 9.1–5, 10.1, but we shall return to it briefly in this chapter too.

The third condition concerning infinitesimal rotations likewise no longer applies to crystals. Instead we have, depending on the crystal structure, only invariance with respect to rotations around 2-, 3-, 4-, and 6-fold symmetry axes. This means that strictly speaking $\boldsymbol{L}$ is not a good quantum number in crystals, but later, in connection with compatibility relations, we shall see that arguments based on L_z can still sometimes be used, if with considerable care.

In the following sections we give a short introduction to group theory (Sect. 25.2) and to representations and characters (Sect. 25.3). Then we present the crucial part, namely the connection between the Hamiltonian and group theory, in Sect. 25.4. Finally we give applications both of general nature and others which are more or less specific to semiconductors Sects. 25.5, 6. The aim of this chapter is not to give the reader a deep insight into group theory and its implications but rather a feeling and understanding of what the Γ_i mean that are met in bandstructures as in Sect. 10.6, and to enable him or her to calculate simple selection rules.

25.2 Abstract Group Theory

A group G is a set of elements $\{X_i\}$ and a connection between the elements which is often called "multiplication" with the following properties:

- Closure: if X_i and X_j are elements of $G(X_i, X_j \in G)$ then

$$X_i X_j = X_k \in G . \tag{25.1}$$

- Associative law:

$$X_i(X_j X_k) = (X_i X_j) X_k . \tag{25.2}$$

- There is a neutral or identity element E in G with

$$EX_i = X_i E = X_i \quad \text{for all} \quad X_i \in G . \tag{25.3}$$

- There is an inverse element X_i^{-1} for each $X_i \in G$ with

$$X_i^{-1} X_i = X_i X_i^{-1} = E . \tag{25.4}$$

The number g of different elements in a group gives the order of the group. There are finite and infinite groups.

If the commutative law holds, i.e., if

$$X_i X_j = X_j X_i \quad \text{for all} \quad X_i X_j \in G \tag{25.5}$$

the group is said to be Abelian.

Some examples of groups are:

- the positive and negative integers including zero $\{0, \pm 1, \pm 2 \cdots\}$ with normal addition as connection. They form an infinite Abelian group.
- the positive rational numbers p/q with $p = 0, 1, 2, 3 \ldots$ and $q = 1, 2, 3 \ldots$ with normal multiplication as connection. They again form an infinite Abelian group.
- all symmetry operations that transform an equilateral triangle into itself. The connection is to perform one operation after the other. The identity element is simply to leave the triangle unchanged. This group is generally

called D_3. It is not Abelian and is of order $g = 6$. We shall use it as a model example in the following.

The elements of D_3 are (Fig. 25.1) E, J: a rotation of $+120°$ around the axis through the center of the triangle and normal to it, K a rotation by $-120°$ around this axis, L a rotation by $\pm 180°$ around the axis a, M around b, and N around c. The elements L, M, N require that the upper and lower faces of the triangle are indistinguishable. For these elements the inverse is identical to the element itself, e.g., $L = L^{-1}$ since $LL = E$.

Alternatively there can be three mirror planes which contain the a-, b- and c-axes and are normal to the triangle. This latter case leads to the group C_{3v}. However, we choose the group D_3 as a model for abstract group theory due to its simplicity. For real semiconductors we will later consider the three most important point-groups T_d, O_h and C_{6v}. We should note here that different nomenclatures are used in group theory in order to confuse beginners. In particular, the notation used to describe molecules, (lattice) vibrations, and electronic states in solids are all different. We shall stick here to the notation used generally for electronic states in solids (Schönflies notation). "Translation tables" connecting the different notations are found in some of the books cited, and they can also be deduced from inspection of the character tables (see below).

The first important table which we encounter in group theory is the multiplication table. Table 25.1 gives the multiplication table for the group D_3. At the crossing point between the line with element X_i and the column with element X_j we find the element $X_i X_j$. In every line and in every column of the multiplication table every element of the group appears once and only once.

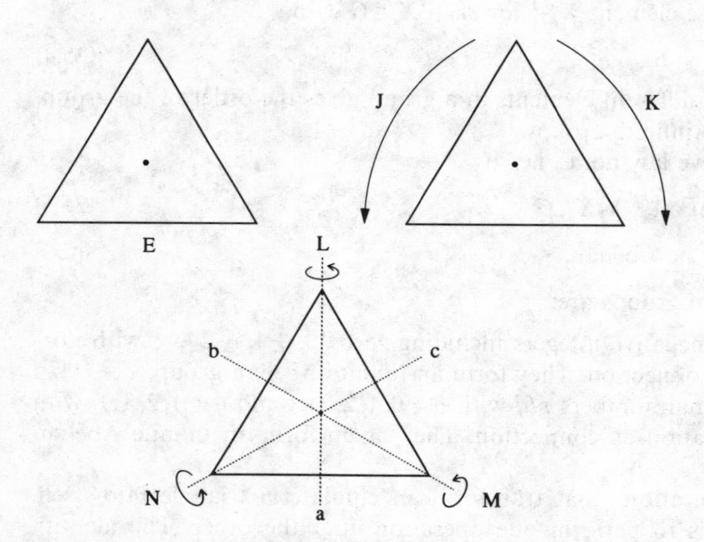

Fig. 25.1. The symmetry operations which map an equilateral triangle into itself

Table 25.1. Multiplication table of group D_3. We give the names of the elements and the operations corresponding to Fig. 25.1

	E	J $+120$	K -120	L $\pm 180_a$	M $\pm 180_b$	N $\pm 180_c$
E	E	J $+120$	K -120	L $\pm 180_a$	M $\pm 180_b$	N $\pm 180_c$
J $+120$	J $+120$	K -120	E	N $\pm 180^{\circ}_c$	L 180_a	M 180_b
K -120	K -120	E	J $+120$	M 180_b	N 180_c	L 180_a
L $\pm 180_a$	L $\pm 180_a$	M 180_b	N $+180_c$	E	J $+120°$	K $-120°$
M $\pm 180_b$	M $\pm 180_b$	N 180_c	L 180_a	K $-120°$	E	J $+120°$
N $\pm 180_c$	N $\pm 180_c$	L $\pm 180_a$	M 180_b	J $+120°$	K $-120°$	E

The fact that the multiplication table is not symmetric with respect to the main diagonal shows that the group is not Abelian.

An isomorphism is a one-to-one correspondence between the elements of two groups of the same order which is preserved during multiplication. Any two groups whose multiplication tables can be made identical are isomorphic, or, in an abstract sense, the same group. For example, the group of permutations of three different symbols (abc), (cab), (bca), (acb), (bac) and (cba) is isomorphic to D_3.

In contrast, a homomorphism is a many-to-one correspondence between the elements of two groups G and G' which is also preserved during multiplication. In this case $g \geqslant g'$ is a prerequisite.

A (proper) subgroup is a subset of the elements of G which itself forms a group. The whole group G and E alone are considered as improper subgroups. A subgroup of D_3 is $\{E, J, K\}$ as can be seen from an inspection of the multiplication table.

Two elements X_i and X_j of G are called conjugate to each other if there exists at least one element X_k in G with

$$X_k^{-1} X_i X_k = X_j \,. \tag{25.6}$$

In D_3, for example, the elements J and K are conjugate to each other since

$$N^{-1} J N = K \quad (\text{with } N^{-1} = N \text{ in } D_3) \,. \tag{25.7}$$

An element X_i from G for which

$$X_j^{-1} X_i X_j = X_i \tag{25.8}$$

for all X_j is called self-conjugate. The identity element E is always self-conjugate.

A class C_i is a subset of elements which are all conjugate to each other. A self-conjugate element always forms a class by itself.

In D_3 we have three classes $C_1 = \{E\}; C_2 = \{J.K\}$ and $C_3 = \{L, M, N\}$.

The outer or direct product of two groups $G = \{X_i\}$ and $G' = \{X'_i\}$ is again a group G'' which consists of all ordered pairs X_i, X'_j. The definition of the connection of the elements of G'' is

$$(X_i, X'_j)(X_k, X'_l) = (X_i X_k, X'_j X'_c), \tag{25.9}$$

and the order of G'' is $g \cdot g'$.

Any set of elements which fulfils the multiplication table of a group is called a representation of this group. The operations in Fig. 25.1 can be considered as a representation of D_3.

In a stricter sense, to which we shall stick from now on, we consider a representation to be a set of $(n \times n)$ matrices $(n = 1, 2, 3 \ldots)$ which fulfils the multiplication table with normal matrix multiplication as the connection.

25.3 Theory of Representations and of Characters

With the above definition we now consider matrix representations of groups. The name of a certain representation is given by a label α used as an index.

With $\mathbf{\Gamma}_\alpha$ we denote a set of g matrices that fulfil the multiplication table of group G. $\mathbf{\Gamma}_\alpha(X_i)$ is the matrix that represents the element X_i of G and $\mathbf{\Gamma}_\alpha(X_i)_{jk}$ is the element of this matrix in row j and column k.

The normal matrix multiplication then results in

$$\mathbf{\Gamma}(X_i X_j)_{kl} = \sum_h \mathbf{\Gamma}_\alpha(X_i)_{k,h} \mathbf{\Gamma}_\alpha(X_j)_{h,l} \,. \tag{25.10}$$

We introduce n_α to denote the dimension of the $(n_\alpha \times n_\alpha)$ matrices of a certain representation α; n_α is the same for all matrices of one representation $\mathbf{\Gamma}_\alpha$.

There is an infinite number of representations for one group. If $\mathbf{\Gamma}_\alpha$ is a representation and $\mathbf{X}$ is a non-singular matrix of the same dimension n_α then $\{\mathbf{X}^{-1} \mathbf{\Gamma}_\alpha \mathbf{X}\}$ is again a representation.

In the following we shall destill from this infinite number of representations some with special properties (the so-called irreducible representations) and present some of their characteristic properties.

Table 25.2 gives four examples of representations of D_3, two one-dimensional ones, a two-and a three-dimensional one. A representation $\mathbf{\Gamma}_\alpha$ is called reducible if a non-singular matrix $\mathbf{Y}$ can be found such that all matrices $\mathbf{Y}^{-1} \mathbf{\Gamma}_\alpha(\mathbf{X}_i) \mathbf{Y}$ acquire the same block form. This means there are square submatrices and otherwise zeros; see Fig. 25.2. From the properties of matrix multiplication it follows immediately that only submatrices at identical positions are connected with each other in multiplication. This means that the submatrices of all X_i which are located in the same position alone form a representation.

Table 25.2. Four matrix representations of the group D_3

	Elements of the group					
	E	J	K	L	M	N
Γ_1	1	1	1	1	1	1
Γ_2	1	1	1	-1	-1	-1
Γ_3	$\begin{pmatrix}1 & 0\\ 0 & 1\end{pmatrix}$	$\frac{1}{2}\begin{pmatrix}-1 & \sqrt{3}\\ -\sqrt{3} & -1\end{pmatrix}$	$\frac{1}{2}\begin{pmatrix}-1 & -\sqrt{3}\\ \sqrt{3} & -1\end{pmatrix}$	$\frac{1}{2}\begin{pmatrix}1 & -\sqrt{3}\\ -\sqrt{3} & -1\end{pmatrix}$	$\frac{1}{2}\begin{pmatrix}1 & \sqrt{3}\\ \sqrt{3} & -1\end{pmatrix}$	$\frac{1}{2}\begin{pmatrix}-1 & 0\\ 0 & 1\end{pmatrix}$
"Γ_4"	$\begin{pmatrix}1 & 0 & 0\\ 0 & 1 & 0\\ 0 & 0 & 1\end{pmatrix}$	$\begin{pmatrix}1 & 0 & 0\\ 0 & -\frac{1}{2} & -\frac{\sqrt{3}}{2}\\ 0 & -\frac{\sqrt{3}}{2} & -\frac{1}{2}\end{pmatrix}$	$\begin{pmatrix}1 & 0 & 0\\ 0 & -\frac{1}{2} & -\frac{\sqrt{3}}{2}\\ 0 & \frac{\sqrt{3}}{2} & -\frac{1}{2}\end{pmatrix}$	$\begin{pmatrix}-1 & 0 & 0\\ 0 & \frac{1}{2} & -\frac{\sqrt{3}}{2}\\ 0 & -\frac{\sqrt{3}}{2} & -\frac{1}{2}\end{pmatrix}$	$\begin{pmatrix}-1 & 0 & 0\\ 0 & \frac{1}{2} & \frac{\sqrt{3}}{2}\\ 0 & \frac{\sqrt{3}}{2} & -\frac{1}{2}\end{pmatrix}$	$\begin{pmatrix}-1 & 0 & 0\\ 0 & -1 & 0\\ 0 & 0 & 1\end{pmatrix}$

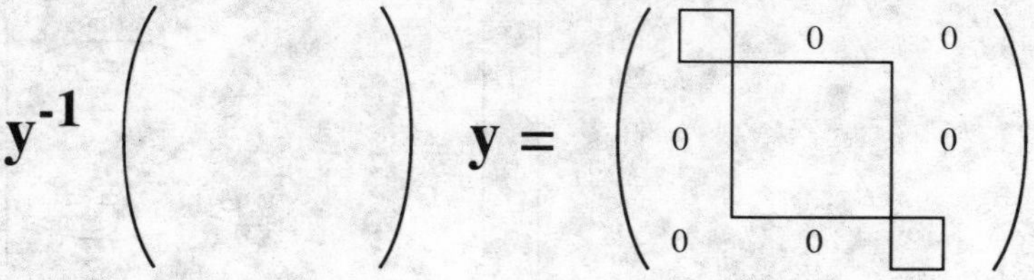

$$\mathbf{Y}^{-1}\begin{pmatrix} \\ \\ \\ \end{pmatrix}\mathbf{Y} = \begin{pmatrix} \square & 0 & 0 \\ 0 & \square & 0 \\ 0 & 0 & \square \end{pmatrix}$$

Fig. 25.2. The reduction of a reducible representation into irreducible ones

A representation $\boldsymbol{\Gamma}_\alpha$ which can be brought into such a block form of submatrices is called reducible. A representation $\boldsymbol{\Gamma}_\alpha$ for which no matrix $\mathbf{Y}$ can be found to bring all $\boldsymbol{\Gamma}_\alpha(X_i)$ into the same block form is called irreducible. The representations $\boldsymbol{\Gamma}_1$–$\boldsymbol{\Gamma}_3$ in Table 25.2 are irreducible; $\boldsymbol{\Gamma}_4$ is evidently reducible. It is already expressed in the desired block form.

We concentrate in the following on irreducible representations.

Two irreducible representations $\boldsymbol{\Gamma}_\alpha$ and $\boldsymbol{\Gamma}_\beta$ are called equivalent, if there exists a non singular matrix $\mathbf{S}$ with

$$\mathbf{S}^{-1}\boldsymbol{\Gamma}_\alpha(X_i)\mathbf{S} = \boldsymbol{\Gamma}_\beta(X_i) \quad \text{for all} \quad X_i \in G \tag{25.11}$$

With this definition of equivalence we arrive at a very important statement: There exist exactly as many non equivalent (=different) irreducible representations of a group as the group has classes C_i.

For D_3 we found three classes (see above). Consequently D_3 has three nonequivalent irreducible representations. These are the $\boldsymbol{\Gamma}_1$, $\boldsymbol{\Gamma}_2$ and $\boldsymbol{\Gamma}_3$ in Table 25.2. The irreducible representation in which all elements are represented by 1 is called the trivial representation and is always called $\boldsymbol{\Gamma}_1$. The representations of the identity element are always unit matrices with 1 on the main diagonal and zeros otherwise. For irreducible representations $\boldsymbol{\Gamma}_\alpha$ and $\boldsymbol{\Gamma}_\beta$ there exists an orthogonality relation

$$\sum_{X_i} \boldsymbol{\Gamma}_\alpha(X_i)_{kp}\boldsymbol{\Gamma}_\beta(X_i^{-1})_{ql} = \frac{g}{n_\alpha}\delta_{\alpha\beta}\delta_{kl}\delta_{pq} \tag{25.12}$$

with δ_{kl} and δ_{pq} being the Kroneckers symbol and

$$\delta_{\alpha\beta} = \begin{cases} 0 & \text{if } \boldsymbol{\Gamma}_\alpha \text{ and } \boldsymbol{\Gamma}_\beta \text{ are not equivalent} \\ 1 & \text{if } \boldsymbol{\Gamma}_\alpha \text{ and } \boldsymbol{\Gamma}_\beta \text{ are identical} \\ \neq 0 & \text{but undefined if } \boldsymbol{\Gamma}_\alpha \text{ and } \boldsymbol{\Gamma}_\beta \text{ are equivalent} \end{cases}$$

A characteristic quantity which is identical for all equivalent irreducible representations is the trace of the matrices

$$\chi_\alpha(X_i) = \mathrm{Tr}\,\boldsymbol{\Gamma}_\alpha(X_i) = \sum_j \boldsymbol{\Gamma}_\alpha(X_i)_{jj}\,. \tag{25.13}$$

These traces are called characters.

Two equivalent representations have the same characters since

$$\mathrm{Tr}\,\mathbf{\Gamma}_\alpha(X_i) = \mathrm{Tr}\,\mathbf{S}^{-1}\mathbf{\Gamma}_\alpha(X_i)\mathbf{S}\,. \tag{25.14}$$

We can thus write down the characters of all nonequivalent irreducible representations in a table. Since the characters of conjugate elements are the same, as seen from a comparison of (25.6) and (25.14), it is sufficient to give the characters for the different classes. This results in a square scheme of numbers. Table 25.3 gives as a first example the character table for the group D_3. It is identical, by the way, to that of the group C_{3v} mentioned above.

As already mentioned, $\mathbf{\Gamma}_1$ in the first line is always the trivial representation, while the first column gives the dimensionality n_α of the representation $\mathbf{\Gamma}_\alpha$.

The nomenclature $\chi_\alpha(C_i)$ means the character of the irreducible representation α for the class C_i.

Introducing the number h_i of elements of the class C_i, with a total number r of classes and of irreducible representations in the group, we find the following relations for the irreducible representations

$$\sum_{i=1}^{r} h_i\,\chi_\alpha(C_i)\,\chi^*_\beta(C_i) = g\delta_{\alpha\beta}\,, \tag{25.15a}$$

$$\sum_{\alpha=1}^{r} h_i\,\chi_\alpha(C_i)\,\chi^*_\alpha(C_j) = g\delta_{ij}\,. \tag{25.15b}$$

Two representations $\mathbf{\Gamma}_\alpha$ and $\mathbf{\Gamma}_\beta$ are equivalent if $\chi_\alpha(X_i) = \chi_\beta(X_i)$ for all X_i.

If a representation $\mathbf{\Gamma}$ is given, it contains the irreducible representation $\mathbf{\Gamma}_\alpha$ p_α times in the sense of a decomposition into block form:

$$\mathbf{\Gamma} = \sum_\alpha p_\alpha \mathbf{\Gamma}_\alpha \quad \text{with} \quad p_\alpha = \sum_{X_i} \chi(X_i)\,\chi^*_\alpha(X_i)\,. \tag{25.16}$$

This means that it is not necessary in practice to try out all possible non singular matrices since one can immediately deduce from (25.16) the decomposition of a given representation into irreducible ones. The reducible representation "$\mathbf{\Gamma}_4$" in Table 25.2 can consequently be decomposed into

$$\mathbf{\Gamma}_4 = \mathbf{\Gamma}_2 \oplus \mathbf{\Gamma}_3\,. \tag{25.17}$$

This process also explains the use of the symbol $\oplus$.

Table 25.3. The character table of the group D_3

	Classes of elements			
	D_3	E	$2C_3$	$3C_2$
Irreducible representations	Γ_1	1	1	1
	Γ_2	1	1	-1
	Γ_3	2	-1	0

The direct product $\mathbf{\Gamma}_\alpha \otimes \mathbf{\Gamma}_\beta$ of two representations $\mathbf{\Gamma}_\alpha$ and $\mathbf{\Gamma}_\beta$ with dimensions n_α and n_β, respectively, is a new representation of dimension $n_\alpha \cdot n_\beta$. The new set of matrices is obtained in the following way

$$\mathbf{\Gamma}_\alpha(X_i) \otimes \mathbf{\Gamma}_\beta(X_i) = \begin{pmatrix} \mathbf{\Gamma}_\alpha(X_i)_{11} \mathbf{\Gamma}_\beta(X_i) \ldots \mathbf{\Gamma}_\alpha(X_i)_{1n_\alpha} \mathbf{\Gamma}_\beta(X_i) \\ \vdots \\ \mathbf{\Gamma}_\alpha(X_i)_{n_{\alpha 1}} \mathbf{\Gamma}_\beta(X_i) \ldots \mathbf{\Gamma}_\alpha(X_i)_{n_\alpha n_\alpha} \mathbf{\Gamma}_\beta(X_i) \end{pmatrix} \tag{25.18}$$

The direct product of two representations is again a representation of the group. The direct product of two irreducible representations can be reducible or irreducible.

For the characters of the direct product one finds

$$\chi(\mathbf{\Gamma}_\alpha \otimes \mathbf{\Gamma}_\beta) = \chi_\alpha \cdot \chi_\beta \, . \tag{25.19}$$

The decomposition of the products of the irreducible representations of a group is usually given in tables:

$$\mathbf{\Gamma}_\alpha \otimes \mathbf{\Gamma}_\beta = \sum_\gamma g_{\alpha\beta\gamma} \mathbf{\Gamma}_\gamma \tag{25.20}$$

with

$$g_{\alpha\beta\gamma} = \frac{1}{g} \sum_{X_i} \chi_\alpha(X_i) \chi_\beta(X_i) \chi_\gamma^*(X_i) \, . \tag{25.21}$$

We give the direct products of the irreducible representations of D_3 in Table 25.4.

The criterion for a given representation to be irreducible is

$$\mathbf{\Gamma}_\alpha \quad \text{irred} \Leftrightarrow \sum_{X_i} |\chi_\alpha(X_i)|^2 = g \, . \tag{25.22}$$

A representation of a group is also a representation of each of its subgroups. An irreducible representation of a group can be a reducible or irreducible representation of the subgroup. Which possibility applies can either be checked with formula given above or it can be found in the tables. In Table 25.5 we give these so-called compatibility relations of the irreducible representations of D_3 with the irreducible representations of its two proper subgroups C_2 and C_3.

Table 25.4. Multiplication table for the irreducible representations of the group D_3

D_3	Γ_1	Γ_2	Γ_3
Γ_1	Γ_1	Γ_2	Γ_3
Γ_2	Γ_2	Γ_1	Γ_3
Γ_3	Γ_3	Γ_3	$\Gamma_1 + \Gamma_2 + \Gamma_3$

25.4 Hamilton Operator and Group Theory

In this section we derive the connections between the eigenfunctions and eigenstates of the stationary or time-independent Hamilton operator H and the irreducible representations of a group.

We first note that the symmetry operations which transform a system into itself form a group. For physical objects, like atoms, molecules and crystalline solids, there are essentially two types of symmetry operation. For one of them, at least one point is kept fixed. These "point operations" include reflection in a mirror plane, rotation around an axis, and inversion, i.e., transition from $\boldsymbol{r}=(x,y,z)$ to $-\boldsymbol{r}=(-x,-y,-z)$ and combinations of these. The groups that contain these operations are known as point groups. Except for the spherical or cylindrical symmetry, these are finite and usually non-Abelian groups. The other type of symmetry operations are translations. They exist only for systems which are ordered in a lattice. The translational group of a crystal, for example, is an infinite Abelian group. The combination of point group and translational group is known as space group.

We do not yet specify a particular type of group, but notice that an operator $P(X_i)$ which transforms the system into itself commutes with the Hamilton operator

$$P(X_i)H = HP(X_i)\,, \tag{25.23}$$

where X_i is a symmetry operation of the group. It is known that whenever an operator P commutes with H, there exists is a set of functions which are simultaneously eigenfunctions of H and of P. These eigenfunctions are called basis functions. They are given in the tables of Sect. 25.6. We also come back to them in Sect. 25.5.

The group which transforms the system into itself is also known as the group of H.

We consider now an eigenvalue E_α of H with eigenfunction(s) $\psi_{\alpha j}$; $j=1,\dots,m$. If the eigenvalue E_α is not degenerate one has $m=1$. For a degenerate eigenvalues $m>1$.

With

$$H\psi_{\alpha j} = E_\alpha \psi_{\alpha j} \tag{25.24}$$

Table 25.5. Compatibility table of the irreducible representations of the group D_3 with the irreducible representations of its proper subgroups C_2 and C_3

D_3	Γ_1	Γ_2	Γ_3
C_3	Γ_1	Γ_1	$\Gamma_2+\Gamma_3$
C_2	Γ_1	Γ_2	$\Gamma_1+\Gamma_2$

we also have

$$P(X_i)\,H\,\psi_{\alpha j} = H\,P(X_i)\,\psi_{\alpha j} = P(X_i)\,E_\alpha\,\psi_{\alpha j} = E_\alpha P(X_i)\psi_{\alpha j}\,, \tag{25.25}$$

since E_α is just a real number.

This means that together with $\psi_{\alpha j}$ also $P(X_i)\psi_{\alpha j}$ is an eigenfunction with eigenvalue E_α. Since the $\psi_{\alpha j}$ already include all possible eigenfunctions with eigenvalue E_α, the $P(X_i)\psi_{\alpha j}$ can only be linear combinations of the $\psi_{\alpha j}$.

This means

$$P(X_i)\,\psi_{\alpha k} = \sum_{j=1}^{m} \mathbf{\Gamma}_\alpha(X_i)_{jk}\,\psi_{\alpha j}\,. \tag{25.26}$$

If we perform this procedure for one symmetry operation X_i and for all $\psi_{\alpha k}$ we get a matrix of coefficients $\mathbf{\Gamma}_\alpha(X_i)_{jk}$. For a nondegenerate eigenvalue this is just a (complex) number of unit magnitude.

If we then perform the procedure for all symmetry operations X_i of the group of the Hamiltonian we get a set of matrices $\mathbf{\Gamma}_\alpha(X_i)$.
Now comes the crucial point: this set of matrices forms a representation of the group of H, and generally an irreducible representation.

If this is the case we can identify the eigenvalue E_α which belongs to the eigenstates $\psi_{\alpha j}$ with the label of the irreducible representation $\mathbf{\Gamma}_\alpha$, i.e., we say that the eigenvalue E_α has symmetry $\mathbf{\Gamma}_\alpha$ or the eigenstates $\psi_{\alpha j}$ of E_α transform according to $\mathbf{\Gamma}_\alpha$.

Since most physical systems have an infinite number of eigenstates but finite groups have only a finite number of non-equivalent irreducible representations, the name of one irreducible representation will occur many times as shown schematically in Fig. 25.3.

If the representation of the group produced by the eigenfunctions of one eigenenergy E_α is reducible, we speak of an accidental degeneracy. An example is shown in Fig. 25.3. This means there are actually two different eigenenergies with their sets of eigenfunctions which just happen to be equal for the specific parameters of H, but which could in principle be different.

The important point now is that group theory allows us to make many important predictions, for example, about selection rules or splitting of states under external perturbations, without even knowing the $\mathbf{\Gamma}_\alpha(X_i)_{ij}$ nor the eigenvalues or eigenfunctions. There is one drawback however group theory tells us only whether a certain matrix element is zero or not; it does not say anything about its magnitude if it is non zero. To get quantitative information we must either use physical arguments to deduce how small or large an effect can be, once group theory has told us that it exists, or we just have to calculate the matrix elements. But here group theory helps us again, by telling us which ones are zero.

Until now we have dealt only with real space. For electronic eigenstates we should also consider spin. Spin does not have a representation in real space. Since we do not want to indulge in a lot of mathematics, we simply note that

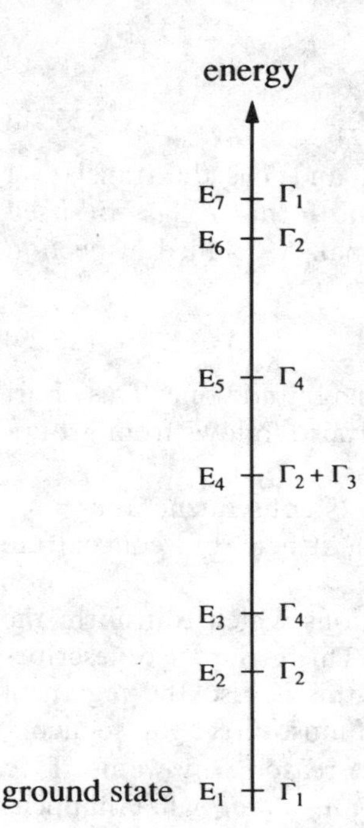

Fig. 25.3. The schematic relation of the eigenenergies E_α and the corresponding irreducible representation Γ_α

there exists a group $D_{1/2}$ which describes the behavior of particles with spin 1/2 under point operations (a translation does not affect spin).

To describe electronic wave functions with spin

$$\psi_{\alpha j} = \psi_{\alpha j} \cdot s \, , \tag{25.27}$$

we have to consider the "double group" which is formally just the direct product of $D_{1/2}$ and the corresponding point group $D_{1/2} \otimes$point group. This double group is again a group, possessing conjugate elements, classes, irreducible representations, and a character table. We shall finally end up with more classes and more Γ_α compared to the simple point group without spin. In the tables given at the end of this chapter for the three most important point groups of semiconductors one can see from the basis functions which ones are the additional Γ_α and classes. For vibrations (phonons) of molecules and of solids spin is not important. Therefore often only the simple and not the double groups are given in corresponding books.

25.5 Applications to Semiconductor Optics

The two groups that describe crystalline solids are, as already mentioned, the translation group and the point group.

The translation group is obviously Abelian, i.e.,

$$T(\boldsymbol{R}_1)\,T(\boldsymbol{R}_2) = T(\boldsymbol{R}_1 + \boldsymbol{R}_2) = T(\boldsymbol{R}_2 + \boldsymbol{R}_1) = T(\boldsymbol{R}_2)\,T(\boldsymbol{R}_1)\,, \tag{25.28}$$

where $\boldsymbol{R}_1$ and $\boldsymbol{R}_2$ are two translations of the lattice and T is the translation operator. Consequently every $\boldsymbol{R}_i$ is self-conjugate and forms a class by itself and all irreducible representations are one-dimensional. A detailed inspection of their properties leads to

$$T(\boldsymbol{R}_i)\,u(\boldsymbol{r}, k) = \mathrm{e}^{\mathrm{i}\boldsymbol{k}\boldsymbol{R}i} u(\boldsymbol{r}, \boldsymbol{k}) \tag{25.29}$$

since the prefactor must be linear in $\boldsymbol{R}_i$ and of the magnitude one. This short argument shows that the Bloch theorem of Sect. 10.1 also follows from group-theoretical considerations.

Since we have already treated this theorem and its consequences in some detail in Chaps. 9 and 10, we need not further pursue it here, but concentrate now on point groups.

The point group of a crystal contains all operations which transform the crystal into itself and leave at least one point fixed. This group also describes the eigenstates for $\boldsymbol{k} = 0$, i.e., at the Γ-point. Since this is also the region of $\boldsymbol{k}$-space that is important for the optical properties of most direct gap semiconductors, we stick in the following to $\boldsymbol{k} = 0$ (this is the reason why we use Γ to label the irreducible representations) and discuss only later what happens when we go to $\boldsymbol{k} \neq 0$. For the moment we note that a statement which is exact for $\boldsymbol{k} = 0$ will be almost correct in the close vicinity of the Γ-point.

We now leave the group D_3, which we used above for illustration purposes, and introduce the three point groups that are most important for the semiconductors, namely T_d, O_h, and C_{6v}.

In Sect. 25.6 we give the character tables and the basis functions of these three point groups. The remainder of the 32 point groups can be found in [25.5, 6]. Additionally we give the multiplication tables for the irreducible representations, their compatibility tables with the full rotation group, which is valid for spherical problems like atoms (see below), and with the proper subgroups. There we indicate examples of external perturbations which reduce the symmetry of the point group to that of its subgroups.

The classes with a bar come from the formation of the double group when we include spin and the same is true for the irreducible representations which contain noninteger $\boldsymbol{J}$ and J_z values in their basis functions $\phi(J,\ J_z)$. Basis functions are examples of functions that possess the symmetry properties of the irreducible representation. Basis functions S_x, S_y and S_z transform like x, y and z, however, without a change of sign under inversion and R means a spherically symmetric function like $x^2 + y^2 + z^2$. A superscript $^+$ or $^-$ appears where inversion C_i is an element of the group and where, consequently, parity is a good quantum number. Obviously $^+$ or $^-$ mean even and odd parity under inversion, respectively.

The information contained in these tables is sufficient to answer many of the questions addressed below.

C_{6v} is the point group of the hexagonal wurtzite structure and contains all symmetry elements of a pointed six-sided pencil, i.e., rotations by $\pm 60°$, $\pm 120°$ and $\pm 180°$ around the hexagonal axis and two sets of three equivalent mirror planes containing this axis.

T_d is the point group of the zinc-blende structure. It contains all symmetry operations which map a tetrahedron into itself. The classes of symmetry operations contain, apart from the identity element E (which is actually a rotation through 4π if spin is included), eight threefold axes c_3, three twofold axes c_2, six fourfold axes c_4 and six mirror planes σ_d. The inversion C_i is not an element of T_d.

Finally, O_h is the point group of the diamond lattices. It can be considered as the direct product of T_d and C_i. It contains all symmetry operations which transform a cube into itself. (The S_i in the classes are products of rotations with the inversion).

Having briefly introduced the most important point groups for semiconductor optics, we present the first important rule. If we have to calculate a transition matrix element M_{if} (see also Sect. 3.2)

$$M_{if} \propto \langle \overset{\Gamma_f}{f} | \overset{\Gamma_s}{H_s} | \overset{\Gamma_i}{i} \rangle = \int \psi_f^* H_s \psi_i \mathrm{d}\tau \,, \tag{25.30}$$

where we start from an initial state $|i\rangle$ of symmetry Γ_i and want to reach a final state $\langle f|$ of symmetry Γ_f with a perturbation operator H_s that transforms like Γ_s, then group theory tells us that this matrix element is zero or non-zero according to:

$$\langle f|H_s|i\rangle = \begin{cases} \neq 0 \text{ if } \Gamma_f \text{ is contained in the direct product } \Gamma_i \otimes \Gamma_s \text{ or} \\ \quad \text{if } \Gamma_1 \text{ is contained in } \Gamma_f \otimes \Gamma_s \otimes \Gamma_i; \\ \quad \text{These two statements are equivalent;} \\ 0 \text{ otherwise.} \end{cases} \tag{25.31}$$

This rule enables us, for example, to calculate the selection rules in crystals. For full spherical symmetry in atoms these rules read, for electric dipole transitions, $\Delta l = \pm 1$; $\Delta m = 0, \pm 1$. For crystals we have to know the symmetry of H_s. It can be often deduced from inspection of the basis functions. The dipole operator transforms in T_d like Γ_5, in O_h like Γ_5^-, and in C_{6v} like Γ_1 for $\boldsymbol{E} \| \boldsymbol{c}$ and like Γ_5 for $\boldsymbol{E} \perp \boldsymbol{c}$.

More information about the symmetries of other perturbations like magnetic fields, uniaxial stress etc. is found in [25.5–9] or in the tables of Sect. 26.6. We now use this knowledge to ask which states can be reached by one-and two-photon transitions in C_{6v} symmetry, starting from the crystal ground state, which has always symmetry Γ_1. For one-photon transitions, we have to consider matrix elements as in (25.30). Thus for C_{6v} we have

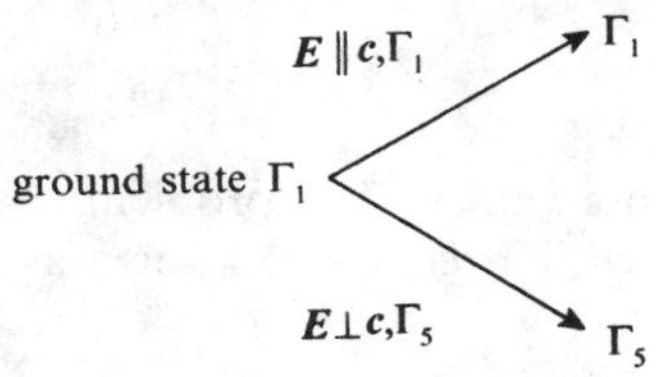

For two-photon transitions we have to consider expressions such as (see also Sect. 18.1):

$$M_{if} \propto \sum_{z_j} \frac{\langle f|H^{\mathrm{D},2}|z_j\rangle\langle z_j|H^{\mathrm{D},1}|i\rangle}{\hbar\omega - (E_{zj} - E_{\mathrm{i}})} + c.p. \tag{25.33}$$

and find the following selection rules with the help of the multiplication tables

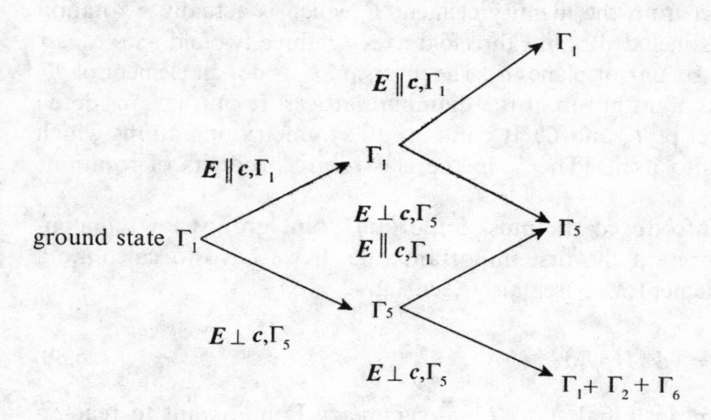

This means that in two-photon absorption we can reach states with symmetries $\Gamma_1, \Gamma_2, \Gamma_5$ and Γ_6 depending on the polarization.

Another field in which the vanishing or otherwise of a matrix element is of importance is perturbation theory. We assume that we have an unperturbed stationary Hamiltonian H with non-degenerate eigenstate E_n^0, ψ_n^0. We now apply a small stationary perturbation H_{s} to get a total Hamiltonian

$$H = H_0 + H_{\mathrm{s}}\,. \tag{25.35}$$

The perturbed eigenenergy E_n and wavefunction ψ_n are then given by

$$E_n = E_n^0 + \langle\psi_n^0|H_{\mathrm{s}}|\psi_n^0\rangle\,, \tag{25.36a}$$

and

$$\psi_n = \psi_n^0 + \sum_{k \neq n} \frac{\langle\psi_k^0|H_{\mathrm{s}}|\psi_n^0\rangle\psi_k^0}{E_n^0 - E_k^0}\,. \tag{25.36b}$$

Equation (25.31) tells us now which perturbations can change the eigenenergy and mix the states and which cannot. As stated earlier, group theory does not tell us how large this effect will be.

In the case of a degenerate level E_n^0 with $\psi_{n,i}^0$, perturbation theory leads to a secular equation given by the determinant

$$\det|\langle\psi_{n,i}^0|H_{\mathrm{s}}|\psi_{n,j}^0\rangle - E\delta_{ij}| = 0, \tag{25.37}$$

and again (25.31) can predict which matrix elements vanish. The problem of degenerate states has been developed even further with a technique known as "invariant expansion". A discussion of this technique is beyond the scope of this book but the reader may find an elaborate presentation in [25.7, 8].

Partly equivalent information is obtained if we consider that the perturbation reduces the symmetry of the problem, so that the perturbed problem corresponds to a subgroup of the unperturbed one. If the irreducible representation of E_n^0 is still irreducible in the subgroup, E_n may shift but does not split. If it decays into more than one irreducible representation in the subgroup, E_n^0 can split into a corresponding number of different levels under the influence of H_s. If some of the new eigenvalues still coincide we have accidental degeneracy.

To illustrate this fact, inspection of the compatibility relations C_{6v} shows that an electrostatic field $\boldsymbol{E} \perp \boldsymbol{c}(\Gamma_5)$ will result in a splitting of a Γ_5 level into two states of symmetries Γ_1 and Γ_2 of the group C_s.

Another case in which group theory is helpful concerns product wave functions. If, for example, we know that the spatial part of the hole wave function $\phi_h(\boldsymbol{r}_n)$ in T_d transforms like Γ_5 (this results from an atomic p-state as will be shown below) and we want to add spin σ we get

$$\psi = \sigma_n^{\Gamma_5}(\boldsymbol{r}_h)\cdot \boldsymbol{s} \,. \tag{25.38}$$

In order to discover the symmetry of the total wave function we just form the direct product of the symmetry of the spatial part (here Γ_5) and of the irreducible representation of spin 1/2. For T_d this reads

$$\Gamma_{\text{total}} = \Gamma_{5,\,\text{space}} \otimes \Gamma_{1/2} = \Gamma_5 \otimes \Gamma_6 = \Gamma_7 \oplus \Gamma_8 \,. \tag{25.39}$$

This means that the spin results in a splitting into two states of symmetries Γ_7 and Γ_8. From physical arguments we know that this is the spin–orbit splitting.

A very important example of product wave functions in semiconductors are exciton wave functions. As shown in Sect. 11.1 with (11.3), they are a product of the electron and hole wavefunctions and of the envelope function.

The possible symmetries of an exciton therefore result from the direct product of the symmetries of the electron, hole and envelope functions, i.e.,

$$\Gamma_{\text{exciton}} = \Gamma_{\text{el}} \otimes \Gamma_{\text{h}} \otimes \Gamma_{\text{env}} \,. \tag{25.40}$$

For the ground state (main quantum number $n_B = 1$) Γ_{env} is always Γ_1. So in T_d, for example, we find excitons formed with the hole in the Γ_7 or Γ_8 valance band and the electron in the Γ_6 conduction band

$$\Gamma_6 \otimes \Gamma_8 \otimes \Gamma_1 = \Gamma_3 \oplus \Gamma_4 \oplus \Gamma_5 \,, \tag{25.41a}$$

$$\Gamma_6 \otimes \Gamma_7 \otimes \Gamma_1 = \Gamma_2 \oplus \Gamma_5 \,. \tag{25.41b}$$

The Γ_5 is the singlet state, which can be reached from the ground state by an electric-dipole transition; the Γ_3, Γ_4 and Γ_2 states are dipole forbidden. Actually they are the triplet states with parallel electron and hole spins.

The number of possible exciton states increases rapidly for $n_B > 1$.

Going one step further, we come to the more complicated entities such as bound-exciton complexes or biexcitons. The procedure is basically as above, i.e.,

$$\Gamma_{\text{biex}} = (\Gamma_{\text{el}} \otimes \Gamma_{\text{el}})^{\pm} \otimes (\Gamma_{\text{h}} \otimes \Gamma_{\text{h}})^{\pm} \Gamma_{\text{env}}^{\mp\mp}. \tag{25.42}$$

However, one now has to consider that these systems contain partly indistinguishable fermions. Therefore the total wave-function must change sign under the exchange of two identical particles. This means that if the combination of the two electrons $(\Gamma_{\text{el}} \otimes \Gamma_{\text{el}})^{-}$ changes sign under exchange, then the envelope function must have even parity under this operation and vice versa. The same holds for the holes. The parity of the combinations can be seen from the coupling coefficients in tables like [25.6]. We do not go into details here but mention that the ground state of the biexciton always has an envelope of Γ_1 symmetry. The possible combinations for the excitons of (25.41a, b) are in the ground state Γ_1 and $\Gamma_1 \oplus \Gamma_3 \oplus \Gamma_5$ [25.8], respectively. In C_{6v} the biexciton containing two holes from the same valence band always has only symmetry Γ_1 in the ground state, but with a hole from the Γ_9 and another from a Γ_7 valence band it is possible to construct in addition Γ_5 and Γ_6 biexciton levels [25.8, 9].

The bands in semiconductors often still contain some information about the parent atomic orbitals, especially in the lattice periodic part $u_k(\boldsymbol{r})$ of the Bloch function. The atomic orbitals result from a spherically symmetric problem. Therefore it is reasonable to say a few words about this group and its compatibility relations with the point groups of semiconductors. In the full spherical rotation group, the system can be rotated around any aixs by any angle and is transformed into itself. It can be shown that all rotations by the same angle ϕ but around arbitrary axes are in the same class. The full rotation group can therefore be considered as a continuous group (Fig. 25.4) with rotation angle $0 \leqslant \phi < 2\pi$. Consequently there must be a correspondingly infinite number of irreducible representations Γ_l. Figure 25.4 gives the character of the representation. For the identity element E we get the dimensionality of Γ_l which is obviously $2l+1$. The basis functions are the spherical harmonics $Y_l^m(\phi, \theta)$.

The compatibility relations between the full rotation group for even and odd parity, called $D_{L_z}^{\pm}$ and $D_{J_z}^{\pm}$, and the three point groups T_d, O_h and C_{6v} are given in Sect. 25.6. These tables may be used as follows: In a tight-binding approximation we would assume that the uppermost valence band of CdS is formed from the filled 3p levels of S^{2-}, while the lowest conduction band comes from the empty 5s levels of Cd^{2+}. From the compatibility tables we learn that this results in the following band symmetries for C_{6v} at the Γ-point:

$$\text{CB:} \quad 5s \quad \text{levels} \Rightarrow D_{1/2} \Rightarrow \Gamma_7;$$

$$\text{VB:} \quad 3p \quad \text{levels} \Rightarrow \begin{cases} D_{3/2} \Rightarrow \Gamma_7 + \Gamma_9 \\ D_{1/2} \Rightarrow \Gamma_7. \end{cases} \tag{25.43}$$

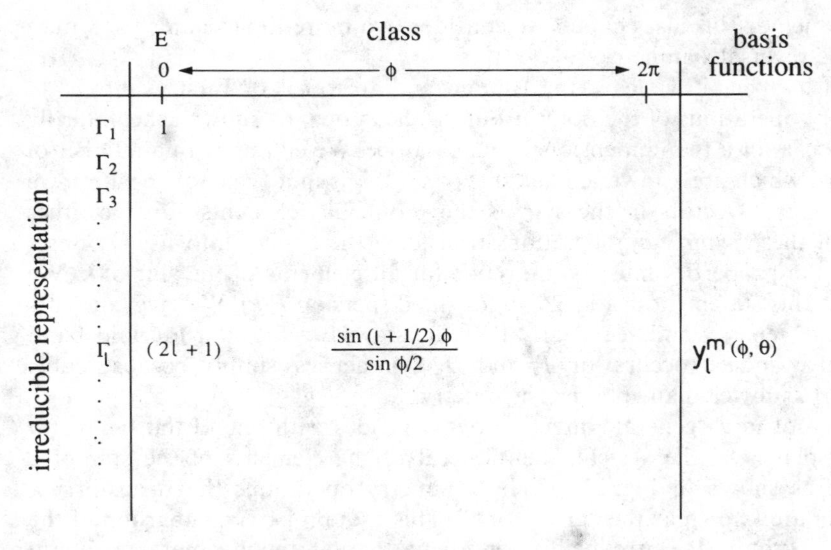

Fig. 25.4. The character table of the spherical rotation group

For T_d symmetry one finds similarly

$$\text{CB:} \quad \Gamma_6; \quad \text{VB:} \quad \Gamma_7 + \Gamma_8. \tag{25.44}$$

This implies a twofold splitting in the valence band of T_d symmetry, which comes from the spin–orbit interaction as already mentioned, and a threefold splitting in C_{6v} again arising from spin–orbit splitting and, in addition, from the hexagonal crystal field.

Group theory cannot tell us the magnitude of the splittings nor the ordering of the bands. It is found, however, that for T_d symmetry the Γ_8 state usually forms the upper valence band and the Γ_7 comes below (except in CuCl which has the reverse ordering) and in C_{6v} one usually has an uppermost valence band of symmetry Γ_9 and two Γ_7 levels below (except for ZnO, where the two upper bands are interchanged).

It is clear that angular momentum is no longer a good quantum number in solids. The good quantities are the Γ_i. With great care, however, one can sometimes still use arguments based on angular momentum. In the wurtzite structure the $\boldsymbol{c}$ axis is the quantization axis and J_z is to some extent usable for discussions. The p-orbitals forming the valence band have $l = 1$ and $m = 0, \pm 1$. The $m = 0$ states have symmetry Γ_1 (or Γ_2) the $m = \pm 1$ states have symmetry Γ_5. With the z-component of the spin of $\pm 1/2$ we can then produce $J_z = \pm 1/2\hbar$ and $J_z = \pm 3/2\hbar$ states corresponding to Γ_7 and Γ_9 respectively. For values of J_z larger than 3/2, however, the rather shaky approach above breaks down completely.

To conclude this last chapter we consider two more topics namely $\boldsymbol{k}$-values and time-reversal symmetry.

What happens if we leave the Γ point and go to $\boldsymbol{k} \neq 0$? First we apply all symmetry operations of the point group to the system as shown schematically in Fig. 25.5a for a two-dimensional square lattice. We end up with a number of $\boldsymbol{k}$-vectors, which are known as the "star of $\boldsymbol{k}$". If $\boldsymbol{k}$ is in a "general" position we get as many $\boldsymbol{k}$-vectors in the star as the group has elements. On the other hand, all these symmetry operations transform the system into itself. Consequently the properties must be the same for all elements of the star of $\boldsymbol{k}$. We illustrate this statement in Fig. 25.5b for the dispersion $E(\boldsymbol{k})$. We give contours of constant energy and see clearly that the star of $\boldsymbol{k}$ allows a fourfold band warping as indeed occurs for T_d and O_h symmetries, simply because cubic symmetry is lower than spherical symmetry.

If $\boldsymbol{k}$ is not in a "general" direction but coincides with one of the symmetry lines or planes of the system, then several of the elements of the star of $\boldsymbol{k}$ coincide as shown in Fig. 25.5c. The symmetry operations that transform $\boldsymbol{k}$ into itself are known as the "group of $\boldsymbol{k}$". This group of $\boldsymbol{k}$ is a subgroup of the full point group. The irreducible representations at the Γ-point are consequently also representations of the subgroup. If we start at the Γ-point with a certain level which transforms like Γ_α and move along some symmetry direction, the level Γ_α may or may not split, depending on whether Γ_α becomes a reducible representation or remains an irreducible one in the (sub) group of $\boldsymbol{k}$.

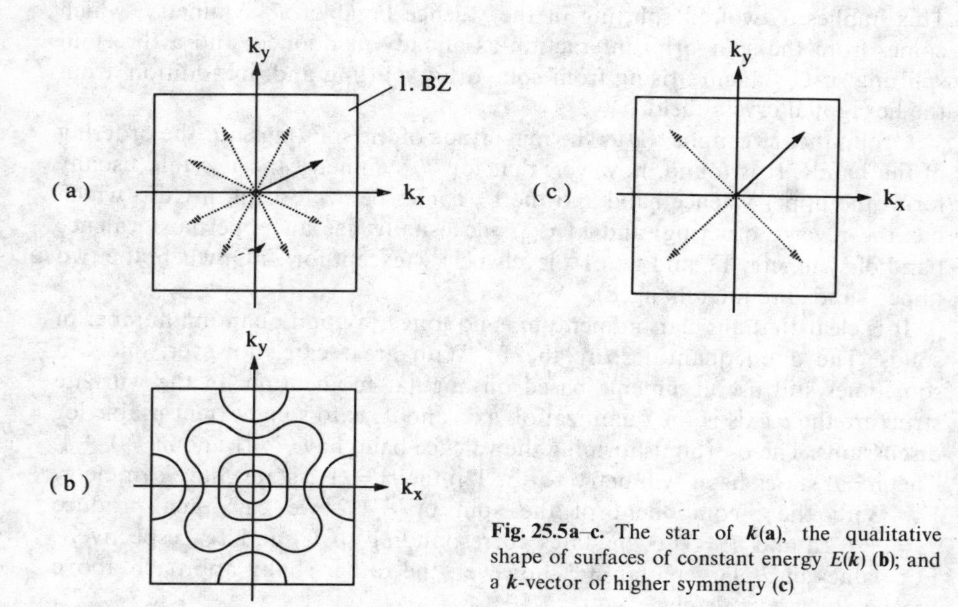

Fig. 25.5a–c. The star of $\boldsymbol{k}$(**a**), the qualitative shape of surfaces of constant energy $E(\boldsymbol{k})$ (**b**); and a $\boldsymbol{k}$-vector of higher symmetry (**c**)

A typical example is the splitting of the fourfold degenerate Γ_8 state in T_d symmetry for $\boldsymbol{k} \neq 0$ into two twofold degenerate states known as heavy and light-hole bands (Fig.25.6). The Γ_6 and Γ_7 states do not show such behavior.

For a general orientation of $\boldsymbol{k}$ all degeneracies can be lifted. In band structure theory one thus usually calculates $E(\boldsymbol{k})$ in directions of high symmetry (see, e.g., Figs. 10.9–12) to exploit the advantages of group theory and tries to extrapolate directions if necessary.

The final symmetry operation that we address is the invariance of a microscopic physical system under time reversal. If we neglect spin for the moment, time invariance has the following consequence for dispersion relations:

$$E(\boldsymbol{k}) = E(-\boldsymbol{k}), \tag{25.45}$$

even if the point group does not include the inversion. This phenomenon is known as Kramers degeneracy.

In a power expansion of the dispersion

$$E(\boldsymbol{k}) = \sum_{n=0}^{\infty} a_n \boldsymbol{k}^n \tag{25.46}$$

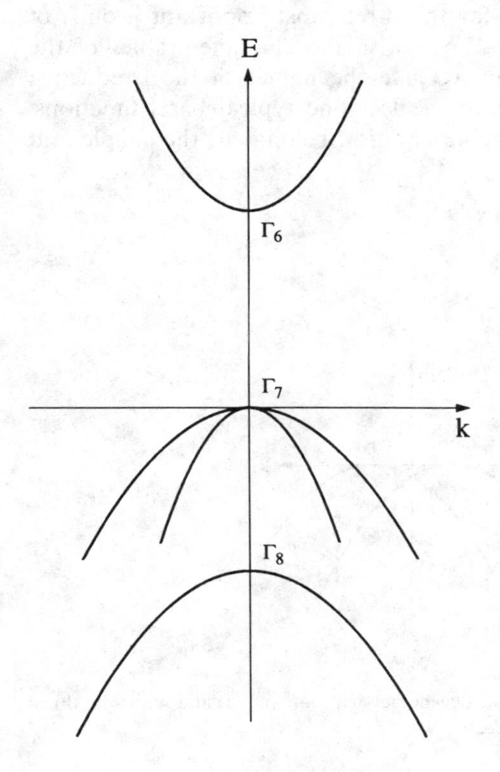

Fig. 25.6. The bandstructure around the Γ-point for T_d symmetry. Compare with Fig. 10.9

this permits only even values for n (Fig. 25.7a) If we now include the spin, (24.46) changes to

$$E(\boldsymbol{k},\uparrow)=E(-\boldsymbol{k},\downarrow)\,. \tag{25.47}$$

This means that time reversal flips the spin. This condition can be also fulfilled with odd powers in the expansion (25.46) and, in particular, allows terms linear in k (Fig. 25.7b).

Detailed group-theoretical investigations show that k-linear terms are possible for Γ_8 states T_d symmetry or for Γ_7 states $\boldsymbol{k}\perp\boldsymbol{c}$ in C_{6v} symmetry, but not for $\boldsymbol{k}\parallel\boldsymbol{c}$ and not for Γ_9 levels.

Indeed such k-linear terms are known; for example, for the B-exciton in CdS which contains a hole from the Γ_7 valence band (Fig. 14.4).

With this statement we conclude the text of the book. In the next two sections we give some tables and some problems. It would be a pleasure to us if the reader has learned to appreciate something of the beauty of semiconductor optics and of the excitement which this field offers to the scientists working in it.

25.6 Some Selected Group Tables

In this section, we give some tables for the three most important groups of semiconductors. We start in Tables 25.6–9 with the character tables of the point group C_{6v}. In the first column we give the names of the irreducible representations in various notations, in the last one typical basis functions. From them it becomes clear, which representation belongs to the simple and

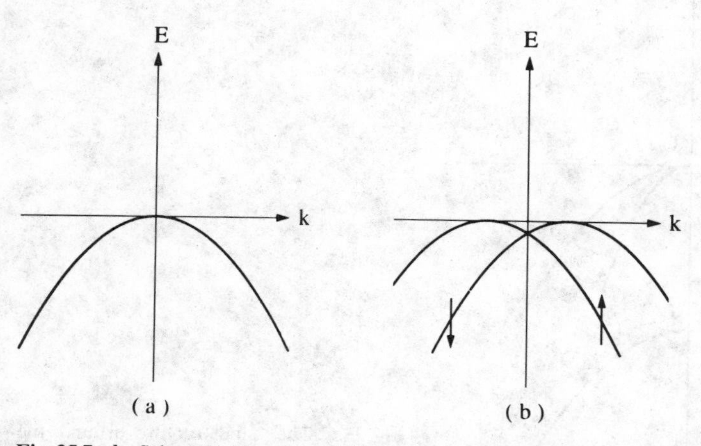

Fig. 25.7a, b. Schematic drawing of the Kramers degeneracy without spin (**a**) and with spin (**b**) as occurs e.g., for Γ_7 bands in C_{6v} symmetry for $\boldsymbol{k}\perp\boldsymbol{c}$

which one to the double group, including spin. The next tables give the multiplication tables of the irreducible representations and the combatibility relation with the subgroups and with the full rotation group. Tables 25.10–13 give the same information for T_d. For O_h we give only two tables for characters and the compatibility relation.

Table 25.6. Character table and basis functions of the point group C_{6v} (or 6mm). Below E_2 or Γ_6 are the additional irreducible representations of the double group which includes spin

C_{6v}		E	$\bar{E}$	$\frac{C_2}{\bar{C}_2}$	$2C_3$	$2\bar{C}_3$	$2C_6$	$2\bar{C}_6$	$\frac{3\sigma_d}{3\sigma_d}$	$\frac{3\sigma}{3\sigma_v}$		Basic functions
A_1	Γ_1	1	1	1	1	1	1	1	1	1	a	R or z
A_2	Γ_2	1	1	1	1	1	1	1	-1	-1	a	S_z
B_1	Γ_3	1	1	-1	1	1	-1	-1	1	-1	a	x^3-3xy^2
B_2	Γ_4	1	1	-1	1	1	-1	-1	-1	1	a	y^3-3yx^2
E_1	Γ_5	2	2	-2	-1	-1	1	1	0	0	a	$(S_x-\mathrm{i}S_y)$, $-(S_x+\mathrm{i}S_y)$
E_2	Γ_6	2	2	2	-1	-1	-1	-1	0	0	a	$\Gamma_3\times\Gamma_5$
	Γ_7	2	-2	0	1	-1	$\sqrt{3}$	$-\sqrt{3}$	0	0	c	$\phi(1/2,-1/2)$, $\phi(1/2,1/2)$
	Γ_8	2	-2	0	1	-1	$-\sqrt{3}$	$\sqrt{3}$	0	0	c	$\Gamma_7\times\Gamma_3$
	Γ_9	2	-2	0	-2	2	0	0	0	0	c	$\phi(3/2,-3/2)$, $\phi(3/2,3/2)$

Table 25.7. Multiplication table for the irreducible representations of the group C_{6v}

Γ_1	Γ_2	Γ_3	Γ_4	Γ_5	Γ_6	Γ_7	Γ_8	Γ_9	
Γ_1	Γ_2	Γ_3	Γ_4	Γ_5	Γ_6	Γ_7	Γ_8	Γ_9	Γ_1
	Γ_1	Γ_4	Γ_3	Γ_5	Γ_6	Γ_7	Γ_8	Γ_9	Γ_2
		Γ_1	Γ_2	Γ_6	Γ_5	Γ_8	Γ_7	Γ_9	Γ_3
			Γ_1	Γ_6	Γ_5	Γ_8	Γ_7	Γ_9	Γ_4
				$\Gamma_1+\Gamma_2+\Gamma_6$	$\Gamma_3+\Gamma_4+\Gamma_5$	$\Gamma_7+\Gamma_9$	$\Gamma_8+\Gamma_9$	$\Gamma_7+\Gamma_8$	Γ_5
					$\Gamma_1+\Gamma_2+\Gamma_6$	$\Gamma_8+\Gamma_9$	$\Gamma_7+\Gamma_9$	$\Gamma_7+\Gamma_8$	Γ_6
						$\Gamma_1+\Gamma_2+\Gamma_5$	$\Gamma_3+\Gamma_4+\Gamma_6$	$\Gamma_5+\Gamma_6$	Γ_7
							$\Gamma_1+\Gamma_2+\Gamma_5$	$\Gamma_5+\Gamma_6$	Γ_8
								$\Gamma_1+\Gamma_2$ $\Gamma_3+\Gamma_4$	Γ_9

Table 25.8. Compatibility table of the irreducible representations of the group C_{6v} with the irreducible representations of its proper subgroups

$C_{6v}: E(z)$	Γ_1	Γ_2	Γ_3	Γ_4	Γ_5	Γ_6	Γ_7	Γ_8	Γ_9
$C_6: H(z)$	Γ_1	Γ_1	Γ_4	Γ_4	$\Gamma_5+\Gamma_6$	$\Gamma_2+\Gamma_3$	$\Gamma_7+\Gamma_8$	$\Gamma_9+\Gamma_{10}$	$\Gamma_{11}+\Gamma_{12}$
C_{3v}	Γ_1	Γ_2	Γ_1	Γ_2	Γ_3	Γ_3	Γ_4	Γ_4	$\Gamma_5+\Gamma_6$
C_{2v}	Γ_1	Γ_3	Γ_2	Γ_4	$\Gamma_2+\Gamma_4$	$\Gamma_1+\Gamma_3$	Γ_5	Γ_5	Γ_5
$C_s: E(x): H(y)$	Γ_1	Γ_2	Γ_1	Γ_2	$\Gamma_1+\Gamma_2$	$\Gamma_1+\Gamma_2$	$\Gamma_3+\Gamma_4$	$\Gamma_3+\Gamma_4$	$\Gamma_3+\Gamma_4$
$C_s: H(x): E(y)$	Γ_1	Γ_2	Γ_2	Γ_1	$\Gamma_1+\Gamma_2$	$\Gamma_1+\Gamma_2$	$\Gamma_3+\Gamma_4$	$\Gamma_3+\Gamma_4$	$\Gamma_3+\Gamma_4$

Table 25.9. Compatibility table of the irreducible representations of the full rotation group of angular momentum $L_z \leqslant 6$ and $J_z \leqslant \frac{13}{2}$ with the irreducible representations of the group C_{6v}

D_0^+	Γ_1	D_0^-	Γ_2
D_1^+	$\Gamma_2+\Gamma_5$	D_1^-	$\Gamma_1+\Gamma_5$
D_2^+	$\Gamma_1+\Gamma_5+\Gamma_6$	D_2^-	$\Gamma_2+\Gamma_5+\Gamma_6$
D_3^+	$\Gamma_2+\Gamma_3+\Gamma_4+\Gamma_5+\Gamma_6$	D_3^-	$\Gamma_1+\Gamma_3+\Gamma_4+\Gamma_5+\Gamma_6$
D_4^+	$\Gamma_1+\Gamma_3+\Gamma_4+\Gamma_5+2\Gamma_6$	D_4^-	$\Gamma_2+\Gamma_3+\Gamma_4+\Gamma_5+2\Gamma_6$
D_5^+	$\Gamma_2+\Gamma_3+\Gamma_4+2\Gamma_5+2\Gamma_6$	D_5^-	$\Gamma_1+\Gamma_3+\Gamma_4+2\Gamma_5+2\Gamma_6$
D_6^+	$2\Gamma_1+\Gamma_2+\Gamma_3+\Gamma_4+2\Gamma_5+2\Gamma_6$	D_6^-	$\Gamma_1+2\Gamma_2+\Gamma_3+\Gamma_4+2\Gamma_5+2\Gamma_6$
$D_{1/2}^+$	Γ_7	$D_{1/2}^-$	Γ_7
$D_{3/2}^+$	$\Gamma_7+\Gamma_9$	$D_{3/2}^-$	$\Gamma_7+\Gamma_9$
$D_{5/2}^+$	$\Gamma_7+\Gamma_8+\Gamma_9$	$D_{5/2}^-$	$\Gamma_7+\Gamma_8+\Gamma_9$
$D_{7/2}^+$	$\Gamma_7+2\Gamma_8+2\Gamma_9$	$D_{7/2}^-$	$\Gamma_7+2\Gamma_8+\Gamma_9$
$D_{9/2}^+$	$\Gamma_7+2\Gamma_8+2\Gamma_9$	$D_{9/12}^-$	$\Gamma_7+2\Gamma_8+2\Gamma_9$
$D_{11/2}^+$	$2\Gamma_7+2\Gamma_8+2\Gamma_9$	$D_{11/12}^-$	$2\Gamma_7+2\Gamma_8+2\Gamma_9$
$D_{13/2}^+$	$3\Gamma_7+2\Gamma_8+2\Gamma_9$	$D_{13/2}^-$	$3\Gamma_7+2\Gamma_8+2\Gamma_9$

Table 25.10. Character table and basis functions for the irreducible representations of the point group T_d (or $\bar{4}3$m)

T_d		E	$\bar{E}$	$8C_3$	$8\bar{C}_3$	$3C_2$ $3\bar{C}_2$	$6S_4$	$6\bar{S}_4$	$6\sigma_d$ $6\bar{\sigma}_d$	Time reversal	Basis functions
A_1	Γ_1	1	1	1	1	1	1	1	1	a	R or xyz
A_2	Γ_2	1	1	1	1	1	-1	-1	-1	a	$S_xS_yS_z$
E	Γ_3	2	2	-1	-1	2	0	0	0	a	$(2z^2-x^2-y^2)$ $\sqrt{3}(x^2-y^2)$
T_1	Γ_4	3	3	0	0	-1	1	1	-1	a	S_x, S_y, S_z
T_2	Γ_5	3	3	0	0	-1	-1	-1	1	a	x, y, z
$\bar{E}_1$	Γ_6	2	-2	1	-1	0	$\sqrt{2}$	$-\sqrt{2}$	0	c	$\phi(1/2, -1/2)$ $\phi(1/2, 1/2)$
$\bar{E}_2$	Γ_7	2	-2	1	-1	0	$-\sqrt{2}$	$\sqrt{2}$	0	c	$\Gamma_6 \times \Gamma_2$
$\bar{E}_3$	Γ_8	4	-4	-1	1	0	0	0	0	c	$\phi(3/2, -3/2)$ $\phi(3/2, -1/2)$ $\phi(3/2, 1/2)$ $\phi(3/2, 3/2)$

Table 25.11. Multiplication table of the irreducible representations of the group T_d

Γ_1	Γ_2	Γ_3	Γ_4	Γ_5	Γ_6	Γ_7	Γ_8	
Γ_1	Γ_2	Γ_3	Γ_4	Γ_5	Γ_6	Γ_7	Γ_8	Γ_1
	Γ_1	Γ_3	Γ_5	Γ_4	Γ_7	Γ_6	Γ_8	Γ_2
		$\Gamma_1+\Gamma_2+\Gamma_3$	$\Gamma_4+\Gamma_5$	$\Gamma_4+\Gamma_5$	Γ_8	Γ_8	$\Gamma_6+\Gamma_7+\Gamma_8$	Γ_3
			$\Gamma_1+\Gamma_3+\Gamma_4+\Gamma_5$	$\Gamma_2+\Gamma_3+\Gamma_4+\Gamma_5$	$\Gamma_6+\Gamma_8$	$\Gamma_7+\Gamma_8$	$\Gamma_6+\Gamma_7+2\Gamma_8$	Γ_4
				$\Gamma_1+\Gamma_3+\Gamma_4+\Gamma_5$	$\Gamma_7+\Gamma_8$	$\Gamma_6+\Gamma_8$	$\Gamma_6+\Gamma_7+2\Gamma_8$	Γ_5
					$\Gamma_1+\Gamma_4$	$\Gamma_2+\Gamma_5$	$\Gamma_3+\Gamma_4+\Gamma_5$	Γ_6
						$\Gamma_1+\Gamma_4$	$\Gamma_3+\Gamma_4+\Gamma_5$	Γ_7
							$\Gamma_1+\Gamma_2+\Gamma_3+2\Gamma_4+2\Gamma_5$	Γ_8

Table 25.12. Compatibility table of the irreducible representations of T_d with the irreducible representations of its proper subgroups

T_d	Γ_1	Γ_2	Γ_3	Γ_4	Γ_5	Γ_6	Γ_7	Γ_8
T	Γ_1	Γ_1	$\Gamma_2+\Gamma_3$	Γ_4	Γ_4	Γ_5	Γ_5	$\Gamma_6+\Gamma_7$
O_{2d}	Γ_1	Γ_3	$\Gamma_1+\Gamma_3$	$\Gamma_2+\Gamma_5$	$\Gamma_4+\Gamma_5$	Γ_6	Γ_7	$\Gamma_6+\Gamma_7$
$C_{3v}: E(w)$	Γ_1	Γ_2	Γ_3	$\Gamma_2+\Gamma_3$	$\Gamma_1+\Gamma_3$	Γ_4	Γ_4	$\Gamma_4+\Gamma_5+\Gamma_6$
$S_1: H(z)$	Γ_1	Γ_2	$\Gamma_1+\Gamma_2$	$\Gamma_1+\Gamma_3+\Gamma_4$	$\Gamma_2+\Gamma_3+\Gamma_4$	$\Gamma_5+\Gamma_6$	$\Gamma_7+\Gamma_8$	$\Gamma_5+\Gamma_6+\Gamma_7+\Gamma_8$
$C_3: H(w)$	Γ_1	Γ_1	$\Gamma_2+\Gamma_3$	$\Gamma_1+\Gamma_2+\Gamma_3$	$\Gamma_1+\Gamma_2+\Gamma_3$	$\Gamma_4+\Gamma_5$	$\Gamma_4+\Gamma_5$	$\Gamma_4+\Gamma_5+2\Gamma_6$
$C_{2v}: E(z)$	Γ_1	Γ_3	$\Gamma_1+\Gamma_3$	$\Gamma_2+\Gamma_3+\Gamma_4$	$\Gamma_1+\Gamma_2+\Gamma_4$	Γ_5	Γ_5	$2\Gamma_5$
$C_s: E(v): H(v)$	Γ_1	Γ_2	$\Gamma_1+\Gamma_2$	$\Gamma_1+2\Gamma_2$	$2\Gamma_1+\Gamma_2$	$\Gamma_3+\Gamma_4$	$\Gamma_3+\Gamma_4$	$2\Gamma_3+2\Gamma_4$

Table 25.13. Compatability table of the irreducible representations of the full rotation group for $L_z \leqslant 6$ and $J_z \leqslant \frac{13}{2}$ with the irreducible representations of the point group T_d

D_0^+	Γ_1	D_0^-	Γ_2
D_1^+	Γ_4	D_1^-	Γ_5
D_2^+	$\Gamma_3+\Gamma_5$	D_2^-	$\Gamma_3+\Gamma_4$
D_3^+	$\Gamma_2+\Gamma_4+\Gamma_5$	D_3^-	$\Gamma_1+\Gamma_4+\Gamma_5$
D_4^+	$\Gamma_1+\Gamma_3+\Gamma_4+\Gamma_5$	D_4^-	$\Gamma_2+\Gamma_3+\Gamma_4+\Gamma_5$
D_5^+	$\Gamma_3+2\Gamma_4+\Gamma_5$	D_5^-	$\Gamma_3+\Gamma_4+2\Gamma_5$
D_6^+	$\Gamma_1+\Gamma_2+\Gamma_3+\Gamma_4+2\Gamma_5$	D_6^-	$\Gamma_1+\Gamma_2+\Gamma_3+2\Gamma_4+2\Gamma_5$
$D_{1/2}^+$	Γ_6	$D_{1/2}^-$	Γ_7
$D_{3/2}^+$	Γ_8	$D_{3/2}^-$	Γ_8
$D_{5/2}^+$	$\Gamma_7+\Gamma_8$	$D_{5/2}^-$	$\Gamma_6+\Gamma_8$
$D_{7/2}^+$	$\Gamma_6+\Gamma_7+\Gamma_8$	$D_{7/2}^-$	$\Gamma_6+\Gamma_7+\Gamma_8$
$D_{9/2}^+$	$\Gamma_6+2\Gamma_8$	$D_{9/2}^-$	$\Gamma_7+2\Gamma_8$
$D_{11/2}^+$	$\Gamma_6+\Gamma_7+2\Gamma_8$	$D_{11/2}^-$	$\Gamma_6+\Gamma_7+2\Gamma_8$
$D_{13/2}^+$	$\Gamma_6+2\Gamma_7+2\Gamma_8$	$D_{13/2}^-$	$2\Gamma_6+\Gamma_7+2\Gamma_8$

Table 25.14. Character table and basis functions of the point group O_h (or m3m) which contains the inversion and consequently has parity as a good quantum number

O_h	E	$\bar{E}$	$8C_3$	$8\bar{C}$	$3C_2$ $3\bar{C}_2$	$6C_4$	$6\bar{C}_4$	$6C_2'$ $6\bar{C}_2'$	I	$\bar{I}$	$8S_6$	$8\bar{S}_6$	$3\sigma_h$ $3\bar{\sigma}_h$	$6S_4$	$6\bar{S}_4$	$6\sigma_d$ $6\bar{\sigma}_d$	Time reversal	Basis functions
Γ_1^+	1	1	1	1	1	1	1	1	1	1	1	1	1	1	1	1	a	R
Γ_2^+	1	1	1	1	1	−1	−1	−1	1	1	1	1	1	−1	−1	−1	a	$(x^2-y^2)(y^2-z^2)(z^2-x^2)$
Γ_3^+	2	2	−1	−1	2	0	0	0	2	2	−1	−1	2	0	0	0	a	$(2z^2-x^2-y^2), \sqrt{3}(x^2-y^2)$
Γ_4^+	3	3	0	0	−1	1	1	−1	3	3	0	0	−1	1	1	−1	a	S_x, S_y, S_z
Γ_5^+	3	3	0	0	−1	−1	−1	1	3	3	0	0	−1	−1	−1	1	a	yz, xz, xy
Γ_1^-	1	1	1	1	1	1	1	1	−1	−1	−1	−1	−1	−1	−1	−1	a	$\Gamma_2^- \times \Gamma_2^+$
Γ_2^-	1	1	1	1	1	−1	−1	−1	−1	−1	−1	−1	−1	1	1	1	a	xyz
Γ_3^-	2	2	−1	−1	2	0	0	0	−2	−2	1	1	−2	0	0	0	a	$\Gamma_3^+ \times \Gamma_2^-$
Γ_4^-	3	3	0	0	−1	1	1	−1	−3	−3	0	0	1	−1	−1	1	a	x, y, z
Γ_5^-	3	3	0	0	−1	−1	−1	1	−3	−3	0	0	1	1	1	−1	a	$\Gamma_5^+ \times \Gamma_1^-$
Γ_6^+	2	−2	1	−1	0	$\sqrt{2}$	$-\sqrt{2}$	0	2	−2	1	−1	0	$\sqrt{2}$	$-\sqrt{2}$	0	c	$\phi(1/2, -1/2), \phi(1/2, 1/2)$
Γ_7^+	2	−2	1	−1	0	$-\sqrt{2}$	$\sqrt{2}$	0	2	−2	1	−1	0	$-\sqrt{2}$	$\sqrt{2}$	0	c	$\Gamma_6^+ \times \Gamma_2^+$
Γ_8^+	4	−4	−1	1	0	0	0	0	4	−4	−1	1	0	0	0	0	c	$\phi(3/2, -3/2), \phi(3/2), 3/2)$ $\phi(3/2, 1/2), \phi(3/2), 3/2)$
Γ_6^-	2	−2	1	−1	0	$\sqrt{2}$	$-\sqrt{2}$	0	−2	2	−1	1	0	$-\sqrt{2}$	$\sqrt{2}$	0	c	$\Gamma_6^+ \times \Gamma_1^-$
Γ_7^-	2	−2	1	−1	0	$-\sqrt{2}$	$\sqrt{2}$	0	−2	2	−1	1	0	$\sqrt{2}$	$-\sqrt{2}$	0	c	$\Gamma_6^+ \times \Gamma_2^-$
Γ_8^-	4	−4	−1	1	0	0	0	0	−4	4	1	−1	0	0	0	0	c	$\Gamma_8^+ \times \Gamma_1^-$

Table 25.15. The compatibility table of the irreducible representations of the group O_h and the irreducible representations of its proper subgroups

O^h	Γ_1^+	Γ_2^+	Γ_3^+	Γ_4^+	Γ_5^+	Γ_1^-	Γ_2^-	Γ_3^-	Γ_4^-	Γ_5^-	Γ_6^+	Γ_7^+	Γ_8^+	Γ_6^-	Γ_7^-	Γ_8^-
O	Γ_1	Γ_2	Γ_3	Γ_4	Γ_5	Γ_1	Γ_2	Γ_3	Γ_4	Γ_5	Γ_6	Γ_7	Γ_8	Γ_6	Γ_7	Γ_8
T_d	Γ_1	Γ_2	Γ_3	Γ_4	Γ_5	Γ_2	Γ_1	Γ_3	Γ_5	Γ_4	Γ_6	Γ_7	Γ_8	Γ_7	Γ_6	Γ_8
T_h	Γ_1^+	Γ_1^+	$\Gamma_2^+ + \Gamma_3^+$	Γ_4^+	Γ_4^+	Γ_1^-	Γ_1^-	$\Gamma_2^- + \Gamma_3^-$	Γ_4^-	Γ_4^-	Γ_5^+	Γ_5^+	$\Gamma_6^+ + \Gamma_7^+$	Γ_5^-	Γ_5^-	$\Gamma_6^- + \Gamma_7^-$
D_{4h}	Γ_1^+	Γ_3^+	$\Gamma_1^+ + \Gamma_3^+$	$\Gamma_2^+ + \Gamma_5^+$	$\Gamma_4^+ + \Gamma_5^+$	Γ_1^-	Γ_3^-	$\Gamma_1^- + \Gamma_3^-$	$\Gamma_2^- + \Gamma_5^-$	$\Gamma_4^- + \Gamma_5^-$	Γ_6^+	Γ_7^+	$\Gamma_6^+ + \Gamma_7^+$	Γ_6^-	Γ_7^-	$\Gamma_6^- + \Gamma_7^-$
D_{3d}	Γ_1^+	Γ_2^+	Γ_3^+	$\Gamma_2^+ + \Gamma_3^+$	$\Gamma_1^- + \Gamma_3^-$	Γ_1^-	Γ_2^-	Γ_3^-	$\Gamma_2^- + \Gamma_3^-$	$\Gamma_1^- + \Gamma_3^-$	Γ_4^+	Γ_4^-	$\Gamma_4^+ + \Gamma_5^- + \Gamma_6^-$	Γ_4^-	Γ_4^-	$\Gamma_4^- + \Gamma_5^- + \Gamma_6^-$
$C_{4h}: H(z)$	Γ_1^+	Γ_2^+	$\Gamma_1^+ + \Gamma_2^+$	$\Gamma_1^+ + \Gamma_3^+ + \Gamma_4^+$	$\Gamma_2^+ + \Gamma_3^+ + \Gamma_4^+$	Γ_1^-	Γ_2^-	$\Gamma_1^- + \Gamma_2^-$	$\Gamma_1^- + \Gamma_3^- + \Gamma_4^-$	$\Gamma_2^- + \Gamma_3^- + \Gamma_4^-$	$\Gamma_5^+ + \Gamma_6^+$	$\Gamma_7^+ + \Gamma_8^+$	$\Gamma_5^+ + \Gamma_6^+ + \Gamma_7^+ + \Gamma_8^+$	$\Gamma_5^- + \Gamma_6^-$	$\Gamma_7^- + \Gamma_8^-$	$\Gamma_5^- + \Gamma_6^- + \Gamma_7^- + \Gamma_8^-$
$C_{2h}: H(v)$	Γ_1^+	Γ_2^+	$\Gamma_1^+ + \Gamma_2^+$	$2\Gamma_2^+ + \Gamma_1^+$	$2\Gamma_1^+ + \Gamma_2^+$	Γ_1^-	Γ_2^-	$\Gamma_1^- + \Gamma_2^-$	$2\Gamma_2^- + \Gamma_1^-$	$2\Gamma_1^- + \Gamma_2^-$	$\Gamma_3^+ + \Gamma_4^+$	$\Gamma_3^+ + \Gamma_4^+$	$2\Gamma_3^+ + 2\Gamma_4^+$	$\Gamma_3^- + \Gamma_4^-$	$\Gamma_3^- + \Gamma_4^-$	$2\Gamma_3^- + 2\Gamma_4^-$
$C_{31}: H(w)$	Γ_1^+	Γ_1^+	$\Gamma_2^+ + \Gamma_3^+$	$\Gamma_1^+ + \Gamma_2^+ + \Gamma_3^+$	$\Gamma_1^+ + \Gamma_2^+ + \Gamma_3^+$	Γ_1^-	Γ_1^-	$\Gamma_2^- + \Gamma_3^-$	$\Gamma_1^- + \Gamma_2^- + \Gamma_3^-$	$\Gamma_1^- + \Gamma_2^- + \Gamma_3^-$	$\Gamma_4^- + \Gamma_5^-$	$\Gamma_4^- + \Gamma_5^-$	$\Gamma_4^+ + \Gamma_5^+ + 2\Gamma_6^+$	$\Gamma_4^- + \Gamma_5^-$	$\Gamma_4^- + \Gamma_5^-$	$\Gamma_4^+ + \Gamma_5^+ + 2\Gamma_6^+$
$C_{4v}: E(z)$	Γ_1	Γ_3	$\Gamma_1 + \Gamma_3$	$\Gamma_2 + \Gamma_5$	$\Gamma_4 + \Gamma_5$	Γ_2	Γ_4	$\Gamma_2 + \Gamma_4$	$\Gamma_1 + \Gamma_5$	$\Gamma_3 + \Gamma_5$	Γ_6	Γ_7	$\Gamma_6 + \Gamma_7$	Γ_6	Γ_7	$\Gamma_6 + \Gamma_7$
$C_{2v}: E(v)$	Γ_1	Γ_2	$\Gamma_1 + \Gamma_2$	$\Gamma_2 + \Gamma_3 + \Gamma_4$	$\Gamma_1 + \Gamma_3 + \Gamma_4$	Γ_3	Γ_4	$\Gamma_3 + \Gamma_4$	$\Gamma_1 + \Gamma_2 + \Gamma_4$	$\Gamma_1 + \Gamma_2 + \Gamma_3$	Γ_5	Γ_5	$2\Gamma_5$	Γ_5	Γ_5	$2\Gamma_5$
$C_{3v}: E(w)$	Γ_1	Γ_2	Γ_3	$\Gamma_2 + \Gamma_3$	$\Gamma_1 + \Gamma_3$	Γ_2	Γ_1	Γ_3	$\Gamma_1 + \Gamma_3$	$\Gamma_2 + \Gamma_3$	Γ_4	Γ_4	$\Gamma_4^+ + \Gamma_5^+ + \Gamma_6^+$	Γ_4	Γ_4^+	$\Gamma_4 + \Gamma_5 + \Gamma_6$

25.7 Problems

1. Calculate the symmetries of the states which can be reached by one- and two-photon transitions in point groups O_h and T_d.

2. Calculate the possible symmetries of the excitons with principal quantum number $n_B = 1, 2$ and 3 in the point groups O_h, T_d and C_{6v}.

3. Calculate the symmetries of the ground state (Γ_1 envelope) of D^0X, A^0X and D^+X centers in O_h, T_d, and C_{6v}, assuming that the radius of these complexes is so large that the carriers feel the full symmetry of the lattice and not only the nearest neighbors.

4. SnO_2 is a direct gap semiconductor with the band extrema at the Γ-point. It crystallizes in the point group D_{4h}. Which symmetries do you expect for the highest valance and the lowest conduction band. The lowest band-to-band transition is dipole forbidden. Why? Do you except dipole-allowed exciton transitions? Do you think that they will have a big oscillator strength?

5. The final problem: Find all errors in this book both in physics and in printing, collect all aspects which can be improved and all important phenomena which have been omitted and solve all problems. Send a corresponding listing to the author and help him by doing so to improve the next editions of this book—if there are any. Many thanks in advance for your help!

References

Chapter 1

1.1 H. Haug, S.W. Koch: *Quantum Theory of the Optical and Electronic Properties of Semiconductors* (World Scientific, Singapore 1990)
N. Peyghambarian, S.W. Koch, A. Mysyrowicz: *Introduction to Semiconductor Optics* (Prentice Hall, Englewood Cliffs, NJ 1993)

1.2 M. Ueta, H. Kazanki, K. Kobayashi, Y. Toyozawa, E. Hanamura: *Excitonic Processes in Solids*, Springer Ser. Solid State Sci. Vol. 60 (Springer, Berlin, Heidelberg 1986)

1.3 H. Haug, S. Schmitt-Rink: Prog. Quantum Electron. **9**, 3 (1984)
R. Zimmermann: *Many-Particle Theory of Highly Excited Semiconductors*, Teubner Texte Phys. Vol. 18 (Teubner, Leipzig 1988)

1.4 Landolt-Börnstein, New Series, Vol. III, 17a to i, and Vol. III, 22a and b, ed. by O. Madelung (Springer, Berlin, Heidelberg 1982–1989)

1.5 W. Demtröder: *Laserspektroskopie*, 2nd edn. (Springer, Berlin, Heidelberg 1991)
V.S. Letokhov and V.P. Chebotayev Nonlinear Laser Spectroscopy, Springer Series in Optical Sciences, Vol 4, Springer, Berlin (1977)

1.6 H. Kuzmany: *Festkörperspektroskopie* (Springer, Berlin, Heidelberg 1990)

1.7 The proceedings of the International School on "Atomic and Molecular Spectroscopy" are held every two years in Erice (Sicily), are edited by B. Di Bartolo (Boston) and are actually devoted to optical properties of solids. They are published by Plenum Press, New York. Some of the recent topics are:
a. *Collective Excitations in Solids* (1981), NATO ASI Ser. B **88** (1983)
b. *Energy Transfer Processes in Condensed Matter* (1983), NATO ASI Ser. B **114** (1984)
c. *Spectroscopy of Solid-State Laser Type Materials* (1985), Ettore Majorana Sci. Ser. **30** (1987)
d. *Disordered Solids: Structures and Processes* (1987), Ettore Majorana Int. Sci. Ser. **46** (1989)
e. *Advances in Nonradiative Processes in Solids* (1989), NATO ASI Ser. B **249** (1991)
f. *Optical Properties of Excited States in Solids* (1991), NATO ASI Ser. B **301** (1993)
g. *Nonlinear Spectroscopy of Solids: Advances and Applications* (1993) NATO ASI Ser. B, to be published

1.8 Some further series contain valuable information on semiconductor optics e.g. Modern Problems in Condensed Matter Sciences
V. M. Agranovich and A. A. Maradudin eds, North Holland, Amsterdam especially the volumes
Vol 1 Surface Polaritons, V. M. Agranovich and D.L. Mills eds.
Vol 2 Excitons, E. I. Rashba and M. D. Sturge eds.
Vol 6 Electron-Hole Droplets in Semiconductors, C. D. Jeffries and L.V. Keldysh eds.
Vol 9 Surface Excitations, V. M. Agranovich and R. Landon eds.
Vol 10 Electron-Electron Interactions in Disordered Systems, A. L. Efros and M. Pollak eds.
Vol 16 Non equilibrium Phonons in Nonmetallic Crystals, W. Eisenmenger and A. A. Kaplyanski eds.
Vol 22 Spin Waves and Magnetic Excitations, A. S. Borovik-Romanov and S.K. Sinha eds.
Vol 23 Optical Properties of Mixed Crystals, R.J. Elliott and I. P. Ipatova eds.

Vol 24 The Dielectric Function of Condensed Systems, L.V. Keldysh, D. A. Kirzhnitz and A. A. Maradudin eds.
The series Festkörperprobleme/Advances in Solid State Physics, Vol 1–34 (1962–1995) published by Vieweg (Braunschweig) contains invited contributions from the Spring Meeting of the German Physical Society.
The series Semiconductors and Semimetals (Vol 1–38) and Solid State Physics (Vol 1–48) are both published by Academic Press (Boston/San Diego) and contain many contributions on various aspects of semiconductor optics.

1.9 *Laser Spectroscopy of Solids*, 2nd edn. ed. by W.M. Yen, P.M. Seltzer Topics Appl. Phys. Vol. 49 (Springer, Berlin, Heidelberg 1986)
Laser Spectroscopy of Solids II, ed. by W.M. Yen Topics Appl. Phys., Vol. 65 (1989)

1.10 F. Wooten: *Optical Properties of Solids* (Academic, New York 1972)
F. Bassani, G.P. Parravicini: *Electronic States and Optical Transitions in Solids* (Pergamon, Oxford 1975)
P.T. Landsberg: *Recombination in Semiconductors* (Cambridge Univ. Press, Cambridge 1991)

Chapter 2

2.1 M. Abramowitz, J.A. Stegun (eds): *Pocket Book of Mathematical Functions* (Deutsch, Thun 1984)
H. Stöcker (ed.): *Taschenbuch Mathematischer Formeln und Moderner Verfahren* (Deutsch, Frankfurt 1992)
I.N. Bronstein, K.A. Semendjajew: *Taschenbuch der Mathematik*, 25th edn. (Teubner, Stuttgart 1991)

2.2 J.K. Furdyna, J. Kossut (eds): *Diluted Magnetic Semiconductors*, Semicond. Semimet. **25** (1988)
M.V. Ortenberg: *Festkörperprobleme*, Adv. Solid State Phys. **31**, 261 (1991)
O. Goede, W. Heimbrodt: Adv. Solid State Phys. **32**, 237 (1992)
D.R. Yakovlev: Adv. Solid State Phys. **32**, 251 (1992)

2.3 L.I. Schiff: *Quantum Mechanics*, 2nd edn. Adv. Solid State Phys. (McGraw-Hill, New York 1955)

2.4 A.L. Fetter, J.D. Walecka: *Quantum Theory of Many Particle Systems* (McGraw Hill, New York 1971)

2.5 H. Haken: *Quantenfeldtheorie des Festkörpers* (Teubner, Stuttgart 1973)
H. Haken: *Quantum Field Theory of Solids* (North Holland, Amsterdam 1976)
H. Haken, H.C. Wolf: *Atom- und Quantenphysik* (Springer, Berlin, Heidelberg 1980)

2.6 W. Greiner: *Theoretische Physik*, Vols. 1–10 (Deutsch, Frankfurt 1985)

2.7 P. Meystre, M. Sargent III: *Elements of Quantum Optics*, 2nd ed. (Springer, Berlin, Heidelberg 1992)
I. Abraham: In [1.7g]
B. Bowlby: In [1.7g]

2.8 P. Würfel, J. Phys C **15** 3967 (1982)
K. Schick, E. Daub, S. Finkbeiner and P. Würfel, *Appl. Phys.* A **54**, 109 (1992)

Chapter 3

3.1 M. Born, E. Wolf: *Principles of Optics* (Pergamon, Oxford 1977)
3.2 E. Hecht: *Optics*, 2nd edn. (Addison-Wesley, Reading 1987)
3.3 R.W. Pohl: *Optik und Atomphysik*, 13th edn. (Springer, Berlin, Heidelberg 1976)
3.4 D. Craig, A.K. Kar, J.G.H. Mathew, A. Miller: IEEE J. QE-**21**, 1363 (1985)
3.5 A. Thelen: *Design of Optical Interference Coatings* (McGraw Hill, New York 1989)
3.6 W. Kleber: *Einführung in die Kristallographie*, 8th edn. (VEB Verlag Technik, Berlin 1965)

3.7 V.M. Agranovich, V.L. Ginzburg: *Crystal Optics with Spatial Dispersion and Excitons.* Springer Ser. Solid-State Sci. **42** 2nd ed. (Springer, Berlin, Heidelberg 1984)
3.8 R. Claus, L. Merten, J. Brandmüller: *Light Scattering by Phonon Polaritons*, Springer Tracts Mod. Phys. Vol. 75 (Springer, Berlin, Heidelberg 1975)

Chapter 4

4.1 J.J. Hopfield, D.G. Thomas: Phys. Rev. **132**, 563 (1963)

Chapter 5

5.1 J. J. Hopfield Phys. Rev. **112**, 1555 (1958)
J.J. Hopfield, D.G. Thomas: Phys. Rev. **132**, 563 (1963)
5.2 R. Claus, L. Merten, J. Brandmüller: *Light Scattering by Phonon Polaritons*, Springer Tracts Mod. Phys. Vol. 75 (Springer, Berlin, Heidelberg 1975)
5.3 V.M. Agranovich, V.L. Ginzburg: *Crystal Optics with Spatial Dispersion and Excitons.* Springer Ser. Solid-State Sci. Vol. 42, 2nd edn., (Springer, Berlin, Heidelberg 1984)
5.4 H. Haug, S.W. Koch: *Quantum Theory of the Optical and Electronic Properties of Semiconductors.* (World Scientific, Singapore 1990)
N. Peyghambarian, S.W. Koch, A. Mysyrowicz: *Introduction to Semiconductor Optics* (Prentice Hall, Englewood Cliffs, NJ 1993)
5.5 M. Ueta, H. Kazanki, K. Kobayashi, Y. Toyozawa, E. Hanamura: *Excitonic Processes in Solids*, Springer Ser. Solid State Sci., Vol. 60 (Springer, Berlin, Heidelberg 1986)
5.6 P. Meystre, M. Sargent III: *Elements of Quantum Optics*, 2nd edn. (Springer, Berlin, Heidelberg 1992)
I. Abraham: In [1.7g]
B. Bowlby: In [1.7g]
5.7 G. Raithel, Phys. Bl. **50**, 1149 (1994) and references given therein
5.8 R. W. Pohl: *Optik und Atomphysik*, 13th edn. (Springer, Berlin, Heidelberg 1976)

Chapter 6

6.1 S.J. Pekar: Sov. Phys. Sol. State **4**, 953 (1962)
J.J. Hopfield and D. G. Thomas, Phys. Rev. **132**, 563 (1963)
G.S. Agarwal: Phys. Rev. B **10**, 1447 (1974)
see also Ref [5.1–4]
6.2 A. Stahl, Ch. Uhilein: *Festkörperprobleme XIX*, Adv. Solid State Phys. **159** (1979)
A. Stahl: Phys. Stat. Sol (b) **106**, 575 (1981)
I. Balslev: Phys. Rev. B **23** (1981)
A. Stahl, I. Balslev: Phys. Stat. Sol. (b) **111**, 531 (1982) ibid. **113**, 583 (1982)
6.3 J. Lagois, K. Hümmer: Phys. Stat. Sol. (b) **72**, 393 (1975)
K. Hümmer, P. Gebhardt: ibid. **85**, 271 (1978)
J. Lagois: Phys. Rev. B **23**, 5511 (1981)
T. Skettrup: Phys. Stat. Sol (b) **109**, 663 (1982)
6.4 P.Halevi, R. Fuchs: J. Phys. C **17**, 3869 and 3889 (1984)
6.5 W. Stössel, H.J. Wagner: Phys. Stat. Sol. (b) **89**, 403 (1978)

I. Broser, M. Rosenzweig, R. Broser, M. Richard, E. Birkicht: ibid. **90**, 77 (1978)
I. Broser, M. Rosenzweig: ibid. **95**, 141 (1979); Phys. Rev. B **22**, 2000 (1980)
6.6 Y. Onodera: J. Phys. Soc. Jpn. **51**, 2194 (1982)
T. Shigenari, X.Z. Lu, H.Z. Cummins: Phys. Rev. B **30**, 1962 (1984)
B. Hönerlage, R. Lévy, J.B. Grun, C. Klingshirn, K. Bohnert: Phys. Rep. **124**, 161 (1985)
V. Ya Reznichenko, M.I. Strashnikova and V.V. Cherny Phys. Stat. Sol b **152**, 675 (1989) and **167**, 311 (1991)
6.7 M. Matsushito, J. Wicksted, H.Z. Cummins: Phys. Rev. B **29**, 3362 (1983)
6.8 M. Rosenzweig: Excitonische Polaritonen – optische Eigenschaften räumlich dispersiver Medien Dissertation, Berlin (1982)

Chapter 7

7.1 V.M. Agranovich, D.L. Mills (eds.): *Surface Polaritons* Mod. Prob. Condens. Mat. Sci., Vol. 1, (North Holland, Amsterdam 1982)
7.2 A. Otto: *Festkörperprobleme*/Adv. Solid State Phys. **14**, 1 (1974)
7.3 J. Lagois: Oberflächenpolaritonen, Habilitation Thesis, Erlangen (1981) and references therein

Chapter 8

8.1 F. Wooten: *Optical Properties of Solids* (Academic, New York 1972)
8.2 abc Physik. (Brockhaus, Leipzig 1972)
8.3 H. Kuzmany: *Festkörperspektroskopie* (Springer, Berlin, Heidelberg 1990)
8.4 R.W. Pohl: *Optik und Atomphysik*, 13th edn. (Springer, Berlin, Heidelberg 1976)
8.5 K. Hümmer: Excitonische Polaritonen in einachsigen Kristallen Habilitation Thesis, Erlangen (1978)

Chapter 9

9.1 O. Madelung: *Introduction to Solid State Theory*, Springer Ser. Solid-State Sci. Vol. 2, 2nd edn. (Springer, Berlin, Heidelberg 1981)
9.2 Ch. Kittel: *Introduction to Solid State Physics*, 6th edn. (Wiley, New York 1986)
Ch. Kittel: *Einführung in die Festkörperphysik*, 7th edn. (Oldenbourg, München 1988)
9.3 H. Ibach, H. Lüth: *Festkörperphysik*, 3rd edn. (Springer, Berlin, Heidelberg 1991)
9.4 N.W. Ashcroft, N.D. Mermin: *Solid State Physics* (Holt, Rinehart and Winston, New York 1976)
9.5 B. Di Bartolo (ed.): *Collective Excitations in Solids*, NATO ASI Series B **88** (Plenum, New York 1983)
9.6 H. Haug, S.W. Koch: *Quantum Theory of the Optical and Electronic Properties of Semiconductors* (World Scientific, Singapore 1990)
9.7 H. Haken: *Quantenfeldtheorie des Festkörpers* (Teubner, Stuttgart 1973)
9.8 Landolt-Börnstein, New Series, Group III Vol.17 a and b, O. Madelung, ed. by M. Schulz, H. Weiss (Springer, Berlin, Heidelberg 1982)
9.9 M.A. Herman, H. Sitter: *Molecular Beam Epitaxy* (Springer, Berlin, Heidelberg 1989)
B.R. Pamplin (ed.): *Molecular Beam Epitaxy* (Pergamon, Oxford 1980)
M. Razeghi: *The MOCVD Challenge* (Hilger, Bristol 1989) and references therein
9.10 T. Ruf, J. Spitzer, V. F. Sapega, V. I. Belitsky, M. Cardona and K. Ploog Festkörperprobleme/Advances in Solid State Physics **34**, (1994) in press 9.11 M. Cardona, ibid, in press

Chapter 10

10.1 J.M. Ziman: *Principles of the Theory of Solids*, 2nd edn. (Cambridge Univ. Press, Cambridge 1992)
J.M. Ziman: *Prinzipien der Festkörpertheorie* (Deutsch, Frankfurt 1975)

10.2 O. Madelung: *Introduction to Solid State Theory*, Springer Ser. Solid-State Sci. Vol. 2 (Springer, Berlin, Heidelberg 1978)

10.3 N.W. Ashcroft, N.D. Mermin: *Solid State Physics* (Holt, Rinehart and Winston, New York 1976)

10.4 M.L. Cohen, J.R. Chelikowsky: *Electronic Structure and Optical Properties of Semiconductors*, Springer Ser. Solid-State Sci. Vol. 75 (Springer, Berlin, Heidelberg 1988)

10.5 R.W. Cahn, P. Haasen, E.J. Kramer (eds.): *Electronic Structure and Properties of Semiconductors*: Materials Science and Technology, Vol. 4 (VCH, Weinheim 1991)

10.6 C. Klingshirn: In *Laser Spectroscopy of Solids II*, ed. by W.M. Yen, Topics Appl. Phys. p. 201, Vol.65 (Springer, Berlin, Heidelberg 1989)

10.7 Landolt-Börnstein, New Series Group III, Vol.17, O. Madelung, ed. by M. Schulz, H. Weiss (Springer, Berlin, Heidelberg 1982)

10.8 H. Ibach, H. Lüth: *Festkörperphysik*, 2nd edn. (Springer, Berlin, Heidelberg 1987)

10.9 J. Pollmann, H. Büttner: Phys. Rev. B **16**, 4480 (1977)
H.R. Trebin: Phys. Stat. Sol. (b) **92**, 601 (1979)
B. Pertzsch, U. Rössler: ibid. **101**, 197 (1980)
M. Matsuura, H. Büttner: Solid State Commun. **36**, 81 (80)

10.10 D.R. Yakovlev: *Festkörperprobleme*/Adv. Solid State Phys. **31**, 261 (1991)

10.11 O. Madelung: *Grundlagen der Halbleiterphysik*, Heidelberger Taschenbücher, Vol. 71 (Springer, Berlin, Heidelberg 1970)

10.12 U. Rössler: Phys. Rev. **184**, 733 (1969)

10.13 G. Bastard, J.A. Brum: IEEE J. QE-**22**, 1625 (1986)

10.14 C. Weisbuch: In *Semicond. Semimet*, Vol. 24 (Academic, London 1987)

10.15 U. Rössler, F. Malcher, A. Ziegler: In NATO ASI series B **183**, 219 (Plenum, New York 1988)

10.16 E.E. Mendez, K.v. Klitzing (eds.): *Physics and Applications of Quantum Wells and Superlattices*, NATO ASI Ser., Vol. 170 (Plenum, New York 1987)

10.17 T.C. McGill, C.M. Sotomayor Torres, W. Gebhardt (eds.): *Growth and Optical Properties of Wide-Gap II-VI Low Dimensional Semiconductors*, NATO ASI Ser. B, Vol. 200 (Plenum, New York 1988)

10.18 M.A. Herman, H. Sitter: *Molecular Beam Epitaxy*, Springer Ser. Mater. Sci. Vol. 7 (Springer, Berlin, Heidelberg 1989)

10.19 E.O. Göbel, K. Ploog: Progr. Quantum Electron. **14**, 289 (1990)

10.20 G. Abstreiter, J. Brunner, F. Meier, U. Menczigar, J. Nützel, R. Schorer and D. Többen Proc. 21th ICPS, Beijing (1992), Ping Jiang and Hou–Zhi Zeng eds., p 827, (World Scientific, Singapore 1993) G. Abstreiter Physica Scripta T **49**, A + B, 42 (1993) P. Vogl, M. M. Rieger, J. A. Majewski and G. Abstreiter, ibid, p 476.

10.21 M.A. Haase, J. Qui, J.M. De Puydt, H. Cheng: Appl. Phys. Lett. **59**, 1272 (1991)
H. Jeon, J. Ding, W. Patterson, A.V. Nurmikko, W. Xie, D.C. Grillo, M. Kobayashi, R.L. Gunshor: ibid. p. 3619

10.22 F. Daiminger, L. V. Butov, Ch. Greus, J. Straka and A. Forchel Proc. 21th ICPS, Beijing (1992), Ping Jiang and Hou-Zhi Zeng eds p1293 (World Scientific, Singapore 1993)
K. Ploog, R. Nötzel and O. Brandt ibid. p 1297
M. Illing, G. Bacher, A. Forchel, A. Waag, Th. Litz and G. Landwehr J. Crystal Growth **138**, 638 (1994).

10.23 R. Nötzel, N.N. Ledenstow, L. Däweritz, M. Hohenstein, K. Ploog: Phys. Rev. B **45**, 3507 (1992)

10.24 Ch. Flytzains, D. Ricard, Ph. Roussignol: NATO ASI Ser. B **194**, 181 (1988)
A.I. Ekimov et al.: J. Opt. Soc. Am. B **10**, 100 (1993)

10.25 A. Henglein, Chem. Rev. **89**, 1861 (1989)
M. A. Reed, Scientific American, January issue p 98 (1993)
M.G. Bawendi, M. L. Steigerwald and L. E. Brus, Annual Rev. Phys. Chem. **41**, 477 (1990)
C.B. Murray, D.J. Norris and M.G. Bawendi, J. Am. Chem. Soc. **115**, 8706 (1993)
10.26 N. Peyghambarian, B. Fluegel, D. Hulin, A. Migus, M. Joffre, A. Antonetti, S.W. Koch, M. Lindberg: IEEE J. QE-**25**, 2516 (1989)
S.H. Park, R.A. Morgan, Y.Z. Hu, M. Lindberg, S.W. Koch, N. Peyghambarian: J. Opt. Soc. Am. B **7**, 2097 (1990)
10.27 L.E. Brus: Appl. Phys. A **53**, 465 (1991)
A.I. Ekimov: Phys. Scr. T **39**, 217 (1991)
L. Banyai and S.W. Koch Semiconductor Quantum Dots, World Scientific Series on Atomic, Molecular and Optical Physics, Vol 2 (World Scientific, Singapore 1993)
U. Woggon, Optical Properties of Semiconductor Quantum Dots, Habilitation Thesis, Kaiserslautern (1995)
U. Woggon and S.V. Gaponenko, Phys. Stat. Sol. b (1995) in press
10.28 F. Henneberger, J. Puls, A. Schülzgen, V. Jungnickel, Ch. Spiegelberg: Adv. Solid State Phys. *Festkörperprobleme* **32**, 279 (1992)
10.29 U. Woggon, S. Gaponenko, W. Langbein, A. Uhrig, C. Klingshirn: Phys. Rev. B **47**, 3684 (1993)
10.30 S.M. Sze: *Physics of Semiconductor Devices*, 2nd edn. (Wiley, New York 1981)
10.31 K.J. Ebeling: *Integrierte Optoelektronik*, 2nd edn. (Springer, Berlin, Heidelberg 1992)
10.32 W. Schröter (ed.): *Electronic Structure and Properties of Semiconductors*, Materials Science and Technology, Vol. 4 (VCH, Weinheim 1991)
10.33 C.Y. Fong, I.P. Batra, S. Ciraci (eds.): *Properties of Impurity States in Superlattice Semiconductors*, NATO ASI Ser. B Vol. 183 (Plenum, New York 1988)
10.34 B. Di Bartolo (ed.): *Energy Transfer Processes in Condensed Matter*, NATO ASI Ser. B **114** (Plenum, New York 1984)
10.35 B. Di Bartolo (ed.): *Spectroscopy of Solid State Laser-Type Materials*, Ettore Majorana Int'l Sci, Ser. Phys. Sci. Vol. 30 (Plenum, New York 1987)
10.36 N.F. Mott, E.A. Davies: *Electronic Processes in Non-Crystalline Materials*, 2nd edn. (Clarendon, Oxford 1979)
10.37 J.M. Ziman: *Models of Disorder* (Cambridge Univ. Press, Cambridge 1979)
10.38 R. Zallen: *Physics of Amorphous Solids* (Wiley, New York 1983)
10.39 B.I. Shklovskii, A.L. Efros: *Electronic Properties of Doped Semiconductors*, Springer Ser. Solid-State Sci. Vol. 45 (Springer, Berlin, Heidelberg 1984)
B. Kramer, G. Bergmann, Y. Bruynserade (eds.): *Localization, Interaction and Transport Phenomena*, Springer Ser. Solid-State Sci., Vol. 61 (Springer, Berlin, Heidelberg 1985)
10.40 D.M. Finlayson (ed.): *Localization and Interaction.* (SUSSP, Edinburgh 1986)
10.41 B. Di Bartolo (ed.): *Disordered Solids, Structures and Processes*, Ettore Majorana Int'l Sci. Ser. Phys. Sci. **46** (Plenum, New York 1989)
10.42 S.D. Baranovskii, A.L. Efros: Sov. Phys. Semicond. **12**, 1328 (1978)
10.43 S. Permogorov, A. Reznitskii: J. Lumin **52**, 201 (1992)
10.44 A.A. Klochikhin, S.G. Ogloblin: Sov. Phys. JETP **73**, 1122 (1991) and Phys. Rev. B **48**, 3100 (1993)

Chapter 11

11.1 S. Nikitine: Progr. Semicond. **6**, 233, 269 (1962)
R.S. Knox: *Theory of Excitons*, Solid State Phys., Suppl 5 (Academic, New York 1963)
11.2 C.G. Kuper, G.D. Whitfield (eds.): *Polarons and Excitons* (Plenum, New York 1963)
11.3 R.J. Elliot: In [11.2] p. 269; Phys. Rev. **108**, 1384 (1957)

11.4 M.M. Denisov, V.P. Makarov: Phys. Stat. Sol. (b) **56**, 9 (1973)
11.5 K. Cho (ed.): *Excitons*, Topics Curr. Phys. Vol. 14 (Springer, Berlin, Heidelberg 1979)
11.6 E.I. Rashba, M.D. Sturge (eds.): *Excitons*, Mod. Probl. Cond. Mat. Sci. Vol. 2 (North Holland, Amsterdam 1982)
11.7 D. Bimberg: *Festkörperprobleme*/Adv. Solid State Phys., **17**, 195 (1977)
R.G. Ulbrich, C. Weisbuch: ibid. **18**, 217 (1978)
U. Rössler: ibid. **19**, 77 (1979)
A. Stahl, Ch. Uihlein: ibid. **19**, 159 (1979)
D. Fröhlich: ibid. **21**, 363 (1981)
11.8 C. Klingshirn, H. Haug: Phys. Rep. **70**, 315 (1981)
11.9 B. Hönerlage, R. Lévy, J.B. Grün, C. Klingshirn, K. Bohnert: Phys. Rep. **124**, 161 (1985)
11.10 G.H. Wannier: Phys. Rev. **52**, 191 (1937)
11.11 J. Frenkel: Phys. Rev. **37**, 1276 (1931); Phys. Z. Sowjetunion **9**, 158 (1936)
11.12 H. Haug, S.W. Koch: *Quantum Theory of the Optical and Electronic Properties of Semiconductors* (World Scientific, Singapore 1990)
N. Peyghambarian, S.W. Koch, A. Mysyrowicz: *Introduction to Semiconductor Optics* (Prentice Hall, Englewood Cliffs, NJ 1993)
11.13 E. Hanamura, H. Haug: Phys. Rep. **33c**, 209 (1977)
11.14 R. Lévy, C. Klingshirn, E. Ostertag, Vu Duy Phach, J.B. Grun: Phys. Stat. Sol. (b) **77**, 381 (1976)
N. Peyghambarian, L.L. Chase, A. Mysyrowicz: Phys. Rev. B **27**, 2325 (1983)
J.P. Wolfe, A. Mysyrowicz: Sci. Am. **250** (3) p 70 (1984)
11.15 A. Mysyrowicz, D.P. Trauernicht, J.P. Wolfe, H.-R. Trebin: Phys. Rev. B **27**, 2562 (1983)
H. Haug, H.H. Kranz: Z. Phys. B **53**, 151 (1983)
D. Fröhlich, K. Reimann, R. Wille: Europhys. Lett. **3**, 853 (1987)
11.16 H. Haken: *Halbleiterprobleme*, IV, 1 (1955); Nuovo Cimento, **3**, 1230 (1956)
11.17 M. Krause, H.-E. Gumlich, U. Becker: Phys. Rev. B **37**, 6336 (1988)
R.D. Carson, S.E. Schnatterly: Phys. Rev. Lett. **59**, 319 (1987)
11.18 R.C. Miller, D.A. Kleinmann, W.T. Tsang, A.C. Gossard: Phys. Rev. B **24**, 1134 (1981)
R.L. Greene, K.K. Bajaj: ibid. **29**, 1807 (1984)
E.S. Koteles, J.Y. Chi: ibid. **37**, 6332 (1988)
11.19 M. Shinada, S. Sugano: J. Phys. Soc. Jpn. **21**, 1936 (1966)
T. Ogawa, T. Takagahara: Phys. Rev. B **44**, 8138 (1991)
11.20 See [10.24–29]
11.21 J.R. Haynes: Phys. Rev. Lett. **4**, 361 (1960)
B. Hönerlage, U. Schröder: Phys. Rev. B **16**, 3608 (1977)
11.22 G. Blattner, C. Klingshirn, R. Helbig, R. Meinl: Phys. Stat. Sol. (b) **107**, 105 (1981)
J. Gutowski: NATO ASI Ser. B, **200**, p. 139 (1989) and references therein
11.23 M.L.W. Thewalt: Solid State Commun. **25**, 513 (1978) and references therein
P.J. Dean, D.C. Herbert, D. Bimberg, W.J. Choyke: Phys. Rev. Lett. **37**, 1635 (1976)
11.24 P.J. Dean, D.C. Herbert: In *Excitons*, ed. by K. Cho Topics Curr. Phys., Vol.14 (Springer, Berlin, Heidelberg 1979) p. 55
11.25 C.Y. Fong, I.P. Batra, S. Ciraci (eds.): *Properties of Impurity States in Superlattice Semiconductors*, NATO ASI Ser. B, Vol.183 (Plenum, New York 1988)
11.26 See, e.g. [10.39–44]
11.27 H. Schwab et al.: Phys. Stat. Sol. (b) **172**, 479 (1992)
U. Siegner et al.: Phys. Rev. B **46**, 4564 (1992)
A. Reznitsky, S.D. Baranovskii, A. Tsekoun and C. Klingshirn, Phys. Stat sol. b **184**, 159 (1994)
11.28 J. Hegarty, M.D. Sturge: J. Opt. Soc. Am. B **2**, 1143 (1985)
H. Kalt et al. Physica B **191**, 90 (1993)
11.29 R. Cingolani: Phys. Scr. T **49** B, 470 (1993)
R. Cingolani, H. Lage, L. Tapfer, H. Kalt, D. Heitmann and K. Ploog Phys. Rev. Lett. **67**, 891 (1991), see also ref. [10.22]

Chapter 12

12.1 H. Haug, S.W. Koch: *Quantum Theory of the Optical and Electronic Properties of Semiconductors* (World Scientific, Singapore 1990)
12.2 I. Egri: Phys. Rep. **119**, 363 (1985)
12.3 O. Madelung: *Introduction to Solid State Theory*, Springer Ser. Solid-State Sci. Vol. 2 (Springer, Berlin, Heidelberg 1978)
12.4 R. Höpfel, G. Lindemann, E. Gomik, G. Stangl, AC. Gossard and W. Wiegmann Surf. Science **113**, 118 (1982)
D. Heitmann ibid. **170**, 332 (1986)
A. Pinczuk, and J.M. Worlock, Physica **117/118** B, 637 (1983)
M. Helm, P. England, E. Colas, F. De Rosa and S.J. Allen Phys. Rev. Lett. **63**, 74 (1989).
Th. Egeler, Festkörperprobleme/Advances in Solid State Physics **31**, 315 (1991).
U. Meskt ibid. **30**, 70 (1990)

Chapter 13

13.1 D.C. Reynolds, C.W. Litton, T.C. Collins, E.N. Frank: In *Proc. 10th Int'l Conf. Physics of Semiconductors*, Cambridge, M (1970) p. 519
13.2 H. Kuzmany: *Festkörperspektroskopie* (Springer, Berlin, Heidelberg 1989)
13.3 R.M. Martin, T.C. Damen: Phys. Rev. Lett. **26**, 86 (1971)
13.4 I. Broser, M. Rosenzweig, R. Broser, E. Beckmann, E. Birckicht: J. Phys. Soc. **49**, Suppl. A, 401 (1980)
13.5 R. Claus, L. Merten, J. Brandmüller: *Light Scattering by Phonon-Polaritons*, Springer Tracts Modern Phys. Vol.75 (Springer, Berlin, Heidelberg 1975), R. Claus, Phys. Stat. Sol. b **100**, 9 (1980)
13.6 Ch. Henry, J.J. Hopfield: Phys. Rev. Lett. **15**, 964 (1965)
13.7 R.L. Schmidt, K. Kunc, M. Cardona, H. Bilz: Phys. Rev. B **20**, 3345 (1979)
13.8 C. Colvard, R. Merlin, M.V. Klein, A. Gossard: Phys. Rev. Lett. **45**, 298 (1980)
13.9 N. Marshall, B. Fischer: Phys. Rev. Lett. **28**, 811 (1972)
13.10 W.G. Spitzer: *Festkörperprobleme*/Adv. Solid State Phys. **11**, 1 (1971)
13.11 G.W. Spitzer, H.Y. Fan: Phys. Rev. **106**, 882 (1957)
13.12 N. Marshall, B. Fischer, H.-J. Queisser: Phys. Rev. Lett. **27**, 95 (1971)
13.13 H. Nather, L.G. Quagliano: Solid State Commun. **50**, 75 (1984)
13.14 U. Nowak, W. Richter, G. Sachs: Phys. Stat. Sol. (b) **108**, 131 (1981)
13.15 W. Kütt, Festkörperprobleme/Advances in Solid State Physics **32**, 133 (1992)
13.16 M. Cardona, P. Etch egoin, H.D. Fuchs and P. Molinàs-Mata, J. Phys. Condens. Matter **5A**, 61 (1993)
M. Cardona, Festkörperprobleme/Advances in Solid State Physics **34**, (1994) in press
13.17 T. Ruf, J. Spitzer, V.F. Sapega, V.I. Belitsky, M. Cardona and K. Ploog and Festkörperprobleme/Advances in Solid State Physics **34**, (1994) in press
B. Jusserand and M. Cardona, Light Scattering in Solids V , Topics in Applied Physics **66**, 49, Springer Heidelberg (1989)
13.18 G. Borstl, H.J. Falge and A. Otto, Surface and Bulk Phonon Modes Observed by Attenuated Total Reflection, Springer Tracts in Modern Physics **74**, (1974)
13.19 O. Madelung, Introduction to Solid State Theory, 2nd ed, Springer Series in Solid State Sciences **2** (Springer, Berlin 1981)
13.20 See ref [12.4]

Chapter 14

14.1 H. Haug, S.W. Koch: *Quantum Theory of the Optical and Electronic Properties of Semiconductors* (World Scientific, Singapore 1986)

14.2 See e.g. ref [5.1] or [6.1, 2]
14.3 W. Ekardt, K. Lösch, D. Bimberg: Phys. Rev. B **20**, 3303 (1979)
14.4 T. Skettrup: Phys. Stat. Sol. b **109**, 663 (1982)
J. Lagois: Phys. Rev. B **23**, 5511 (1981)
14.5 G. Blattner, G. Kurtze, G. Schmieder, C. Klingshirn: Phys. Rev. B **25**, 7413 (1982)
14.6 J. Lagois, K. Hümmer: Phys. Stat. Sol. (b) **72**, 393 (1975)
14.7 I. Broser, M. Rosenzweig: Phys. Rev. B **22**, 2000 (1980)
W. Stößel, H.J. Wagner: Phys. Stat. Sol. (b) **89**, 403 (1978)
14.8 R. Ruppin: Phys. Rev. B **29**, 2232 (1984)
14.9 B. Hönerlage, R. Lévy, J.B. Grun, C. Klingshirn, K. Bohnert: Phys. Rep. **124**, 161 (1985)
C. Klingshirn: NATO ASI Ser. B **301**, 119 (1993)
14.10 W. Maier, G. Schmieder, C. Klingshirn: Z. Phys. B **50**, 193 (1983)
14.11 M. Rosenzweig: Dissertation Thesis TU Berlin (1982)
14.12 K. Bohnert, G. Schmieder, C. Klingshirn: Phys. Stat. Sol. (b) **98**, 175 (1980)
14.13 K. Hümmer: Dissertation Habilitation Thesis, Erlangen (1978)
14.14 R.L. Weiher, W.C. Tait: Phys. Rev. B **5**, 623 (1972)
14.15 M. Fiebig, D. Fröhlich, Ch. Pahlke-Lerch: Phys. Stat. Sol. (b) **177**, 187 (1993)
14.16 J. Voigt, F. Spiegelberg, M. Senoner: Phys. Stat. Sol. (b) **91**, 189 (1979)
14.17 R. Ulbrich: In *Materials Science and Technology* ed. by W. Schröter, Vol.4 (VCH, Weinheim 1991) p. 65
14.18 F. Urbach: Phys. Rev. **92**, 1324 (1953)
14.19 W. Martienssen: J. Phys. Chem. Solids **2**, 257 (1957), ibid. **8**, 294 (1959)
14.20 D. Dutton: Phys. Rev. **112**, 785 (1958)
14.21 M.V. Kurik: Phys. Stat. Sol (a) **8**, 9 (1971)
14.22 F. Spiegelberg, E. Gutsche, J. Voigt: Phys. Stat. Sol. (b) **77**, 233 (1976)
14.23 H. Sumi, Y. Toyozawa: J. Phys. Soc. Jpn. **31**, 342 (1971)
14.24 J.D. Dow, D. Redfield: Phys. Rev. **85**, 94 (1972)
14.25 J.G. Liebler, S. Schmitt-Rink, H. Haug: J. Lumin. **34**, 1 (1985)
14.26a R. Kuhnert, R. Helbig, K. Hümmer: Phys. Stat. Sol. (b) **107**, 83 (1981)
14.26b J. Voigt and F. Spiegelberg Phys. Stat. Sol. b **30**, 659 (1968)
J. Puls and J. Voigt ibid. b **94**, 199 (1979)
14.26c H. Venghaus, S. Suga and K. Cho, Phys. Rev. **B16**, 4419 (1977)
W. Dreybrodt, K. Cho, S. Suga, F. Willmann and Y. Niji, Phys. Rev. **B21**, 4692 (1980) and references therein
14.27 S. Permogorov: In *Excitons*, ed. by E.I. Rashka, M.D. Sturge, Mod. Probl. Cond. Mat. Sci., Vol. 2, (North Holland, Amsterdam 1982) p. 177
14.28 C. Klingshirn: Phys. Stat. Sol. (b) **71**, 547 (1975)
14.29 V.A. Kiselev, B.S. Razbirin, I.N. Uraltsev: Phys. Stat. Sol. (b) **72**, 161 (1975) and references therein
14.30 I.V. Makarenko, I.N. Uraltsev, V.A. Kiselev: Phys. Stat. Sol. (b) **98**, 773 (1980)
14.31 T. Mita, N. Nagasawa: Solid State Commun. **44**, 1003 (1984)
14.32 R.G. Ulbrich, C. Weisbuch: *Festkörperprobleme*/Adv. Solid State Phys. **28**, 217 (1978)
14.33 J. Wicksted, M. Matsushita, H.Z. Cummins, T. Shigenari, X.Z. Lu: Phys. Rev. B **29**, 3350 (1984)
M. Matsushita, J. Wicksted, H.Z. Cummins: ibid. **29**, 3362 (1984)
14.34 B. Sermage, G. Fishman: Phys. Rev. Lett. **43**, 1043 (1979)
G. Winterling, E.S. Koteles, M. Cardona: Phys. Rev. Lett. **39**, 1286 (1977)
E. Koteles, G. Winterling: ibid. **44**, 948 (1980)
I. Broser, M. Rosenzweig: Solid State Commun. **36**, 1027 (1980)
14.35 I. Broser, R. Broser, E. Beckmann, E. Birkicht: Solid State Commun. **39**, 1209 (1981)
14.36 M.V. Lebedev, M.I. Strashnikova, Y.B. Timofeev, V.V. Chernyi: JETP Lett. **39**, 366 (1984)
14.37 Y. Masumoto, Y. Unuma, Y. Tanaka, S. Shionoya: J. Phys. Soc. Jpn. **47**, 1844 (1979)
14.38 R.G. Ulbrich, G.W. Fehrenbach: Phys. Rev. Lett. **43**, 963 (1979)
Y. Segawa, Y. Aoyagi, S. Namba: J. Phys. Soc. Jpn. **52**, 3664 (1983)

T. Itho, P. Lavallard J. Reydellet and C. Benoit à la Guillaume Solid State Commun. **37**, 925 (1981)
14.39 D. Fröhlich: In [1.7g] and references therein
14.40 C. Klingshirn, H. Haug: Phys. Rep. **70**, 315 (1981)
14.41 D. Fröhlich, E. Mohler, P. Wiesner: Phys. Rev. Lett. **26**, 554 (1971)
14.42 B. Hönerlage, R. Lévy, J.B. Grun, C. Klingshirn, K. Bohnert: Phys. Rep. **124**, 161 (1985)
14.43 F. Beerwerth, D. Fröhlich: Phys. Rev. Lett. **55**, 2603 (1985); ibid. **57**, 1344 (1986)
14.44 J. Lagois, B. Fischer: *Festkörperprobleme*/Adv. Solid State Phys. **18**, 197 (1978)
14.45 D.L. Mills, V.M. Agranovich (eds.): *Surface Polaritons*, Mod. Probl. Cond. Mat. Sci., Vol.1 (North Holland, Amsterdam 1981)
V.M. Agranovich, R. Loudon (eds.): *Surface Excitations*, ibid., Vol. 9 (1984)
14.46 Landolt-Börnstein, New Series, Vol. 17b, O. Madelung, M. Schulz, H. Weiss (eds.): (Springer, Berlin, Heidelberg 1981)
14.47 Ch. Uihlein, D. Fröhlich, R. Kenklies: Phys. Rev. B **23**, 2731 (1981)
14.48 D. Fröhlich, K. Kulik, B. Uebbing, A. Mysyrowicz, V. Langer, H. Stolz, W. von der Osten: Phys. Rev. Lett. **67**, 2343 (1991)
14.49 O. Madelung: *Grundlagen der Halbleiterphysik*, Heidelberger Taschenbücher, Vol.71 (Springer, Berlin, Heidelberg 1970)
14.50 H. Schweizer, A. Forchel, A. Hangleiter, S. Schmitt-Rink, J.P. Löwenau, H. Haug: Phys. Rev. Lett. **51**, 698 (1983)
14.51 G.A. Thomas, M. Capizzi: In *Proc. 13th Int'l Conf. Phys. Semicond.*, ed. by F.G. Fumi, Rome (1976) p. 915
14.52 S. Schmitt-Rink, D.S. Chemla, D.A.B. Miller: Adv. Phys. **38**, 89 (1989)
14.53 E.O. Göbel, K. Ploog: Prog. Quantum Electron. **14**, 289 (1990)
14.54 R. Cingolani, K. Ploog: Adv. Phys. **40**, 535 (1990)
R. Cingolani: Phys. Scripta **T 494**, 470 (1993)
14.55 D.S. Chemla: Private communication
14.56 J.J. Le Pore: J. Appl. Phys. **51**, 6441 (1980)
14.57 D. Oberhauser: Dissertation, Kaiserslautern (1992)
14.58 L.C. Andreani, F. Bassani: Phys. Rev. B **41**, 7536 (1990) and **45**, 6023 (1992)
F. Tassone, F. Bassani, L.C. Andreani: Il Nuovo Cimento **12D** 1673 (1990) and **140** 1241 (1992) in press
R. Atanasov, F. Bassani, V.M. Agranovich: Phys. Rev. B
The contributions to Intern Conf. on Nonlinear Optics and Excitation Kinetics (NOEKS) IV Gosen (1994) by L. C. Andteani and by D. C. Citrin Phys. Stat. Sol. b (1995) in press
14.59 K.P. O'Donnel, P.J. Parbrook, F. Yang, X. Chem, D.J. Irvine, C. Traeger-Cowan, B. Henderson: J. Cryst. Growth **117**, 497 (1992)
14.60 W. Sack, D. Oberhauser, K.P. O'Donnel, P.J. Parbrook, P.J. Wright, B. Cockayne, C. Klingshirn: J. Lumin. **53**, 409 (1992)
14.61 H. Tuffigo, N. Magnea, H. Mariette, A. Wasiela, Y.M. d'Aubiné: Phys. Rev. B **43**, 14629 (1991)
14.62 J. Crystal Growth **59** (1982); **72** (1985); **86** (1986); **101** (1990); **117** (1992); **138**, (1994): Semicond. Sci. Technol. **6** issue 9A (1991); Adv. Mater. Opt. Electron. **3** Issues 1–6 (1993) Mat. Science Forum (1995) in press
14.63 H. Kalt, J.H. Collet, Le Si Dang, J. Cibert, S.D. Baranovskii, R. Saleh, M. Umlauff, K.P. Geyzers, M. Heuken, C. Klinghsirn: Physica B **191**, 90 (1993)
14.64 R. Hellmann, A. Pohlmann, D.R. Yakovlev, A. Waag, R.N. Bicknell-Tassius, G. Landwehr: In *Proc. 21st Int'l Conf. Phys. Semicond.*, ed. by P. Jiang, H.Z. Zheng (World Scientific, Singapore 1993) p. 1008
D.R. Yakovlev, W. Ossau, A. Waag, R.N. Bicknell-Tassius, G. Landwehr, K.V. Kavokin, A.V. Kavokin, I.N. Uraltsev, A. Pohlmann: ibid. p. 1136
14.65 J. Cibert: et al. Phys. Scr. T **49** B, 487 (1993)
C. Klingshirn, W. Langbein, H. Kald, M. Hetterich and M. Grün J. Crystal Growth **138**, 191 (1994)

14.66 See [10.20] and references therein

14.67 F. Daiminger, L.V. Butov, Ch. Greus, J. Straka, A. Forchel: In *Proc. 21st Int'l Conf. Phys. Semicond.*, Beijing (1992), ed. by P. Jiang, H.-Z. Zheng (World Scientific, Singapore 1993) p.1293
K.H. Ploog, R. Nötzel, O. Brandt: ibid. p. 1297

14.68 Y. Hirugama, S. Tarucha, Y. Suyiki, H. Okamoto: Phys. Rev. B **37**, 2774 (1988)

14.69 *Physics of Nanostructures Proc. 38th Scottish Universities Summer School in Physics*, St. Andrews (1991) (SUSSP, Edinburgh 1992)
U. Merkt: *Festkörperprobleme*/Adv. Solid State Phys. **30**, 77 (1990)

14.70 A.I. Ekimov, Al.L. Efros: Acta Phys. Polnica A **79**, 5 (1991)
Ch. Flytzanis: In [1.7g]
R. Reisfeld: In [1.7g]
U. Woggon: In [1.7g]
References [10.24–29] and references therein

14.71 U. Woggon, S. Gaponenko, W. Langbein, A. Uhrig, C. Klingshirn: Phys. Rev. B **47**, 3684 (1993)
S.V. Gaponenko, U. Woggon, M. Saleh, W. Langbein, A. Uhrig, M. Müller, C. Klingshirn: J. Opt. Soc. Am. B **10**, 1947 (1993)

14.72 S.W. Koch: Phys. Bl. **46**, 167 (1990)

14.73 F. Henneberger, U. Woggon, J. Puls, Ch. Spiegelberg: Appl. Phys. B **46**, 19 (1988)

14.74a A.I. Ekimov, et al. J. Opt. Soc. Am. B **10**, 100 (1993)

14.74b S.W. Koch et al., J. Crystal Growth **117**, 592 (92)

14.75 O. Wind et al.: Adv. Mat. Opt. El **3**, 89 (1993)

14.76 G.H. Doehler, IEEE J. QE-**22** 1682 (1986)
S. Malzer, N. Linder, K.H. Gulden, A. Hoefler, P. Kiesel, M. Kneissl, X. Wu, J.S. Smith, G.H. Doehler: Phys. Stat. Sol. (b) **173**, 459 (1992)
N. Linder, F. Gabler, K.H. Gulden, P. Kiesel, M. Kneissl, P. Diehl, G.H. Doehler: Appl. Phys. Lett. **62**, 1916 (1993)

14.77 H. Kuzmany: *Festkörperspektroskopie* (Springer, Berlin, Heidelberg 1990) and references therein

14.78 M. Cardona, G. Harbeke: Phys. Rev. **137** A, 1467 (1965)
M. Cardona, M. Weinstein, G.A. Wolff: Phys. Rev. A. **140**, 633 (1965)
C. Janowitz, O. Günther, G. Jungk, R.L. Johnson, P.V. Santos, M. Cardona, W. Faschinger and H. Sitter, Phys. Rev. B **50**, 2181 (1994)

Chapter 15

15.1 J.R. Haynes: Phys. Rev. Lett. **4**, 361 (1960)
B. Hönerlage, U. Schröder: Phys. Rev. B **16**, 3608 (1977)

15.2 R. Helbig: Habilitation Thesis, Erlangen (1976)

15.3 H. Schrey: Dissertation, Karlsruhe (1979)

15.4 J. Shah, R.F. Leheny, W.F. Brinkmann: Phys. Rev. B **10**, 659 (1974)
A.F. Dite, V.I. Revenko, V.B. Timofeev: Sov. Phys. Solid State **16**, 1273 (1974)

15.5 C. Klingshirn: Phys. Stat. Sol. (b) **71**, 547 (1975)

15.6 G. Blattner, C. Klingshirn, R. Helbig, R. Meinl: Phys. Stat. Sol. (b) **107**, 105 (1981)

15.7 J. Gutowski: Solid State Commun. **58**, 523 (1986) and references therein

15.8 M.L.W. Thewalt: Solid State Commun. **25**, 513 (1978)

15.9 K. Bohnert, G. Schmieder, S. El-Dessouki, C. Klingshirn: Solid State Commun. **27**, 295 (1978)

15.10 G. Blattner: Diploma Thesis, Karlsruhe (1979)

15.11 K. Cho (ed.): *Excitons*, Topics Curr. Phys. Vol. 14 (Springer, Berlin, Heidelberg 1979) and Refs. [11.22–25]

15.12 R. Kuhnert, R. Helbig: J. Lumin. **26**, 203 (1981)

15.13 See Ref. [1.7]

15.14 Landolt-Börnstein, New Series Group III, O. Madelung, M. Schulz, H. Weiss (eds.): Vol.17 (Springer, Berlin, Heidelberg 1982)
15.15 B. Di Bartolo (ed.): *Optical Properties of Ions in Solids.* (Plenum, New York 1975)
15.16 K.H. Hellwege: *Einführung in die Festkörperphysik*, 2nd edn. (Springer, Berlin, Heidelberg 1981)
15.17 D.L. Wood, J. Tauc: Phys. Rev. B **5**, 3144 (1972)
15.18 P.C. Taylor: In *Laser Spectroscopy of Solids II*, ed. by W.M. Yen, Topics Appl. Phys. Vol. 65 (Springer, Berlin, Heidelberg 1989)
15.19 S. Permogorov, A. Reznitsky: J. Lumin. **52**, 201 (1992) and references therein
15.20 C. Klingsh.rn: NATO ASI Series B **301**, 119 (1993)
15.21 U. Siegner: et al. Phys. Rev. B **46**, 4564 (1992)
15.22 H. Schwab: et al. Phys. Stat. Sol. b **172**, 479 (1992)
15.23 A.A. Klochikin, S.G. Ogloblin: Sov. Phys. JETP **73**, 1122 (1991)
15.24 B.I. Shklovskii, A.L. Efros: *Electronic Properties of Doped Semiconductors*, Springer Ser. Solid-State Sci. Vol. 45 (Springer, Berlin, Heidelberg 1984)
15.25 D.M. Finlayson (ed.): *Localization and Interaction*, 31st Scottish Universities Summer School in Physics (SUSSP, Edinburgh 1986)
15.26 See Ref. [1.7d], [10.42–44] or [11.27]

Chapter 16

16.1 W. Voigt, Magneto- und Elektrooptik (Teubner, Leipzig, 1908)
16.2a K. Cho (ed.): *Excitons*, Topics Curr. Phys. Vol. 14 (Springer, Berlin, Heidelberg 1979)
16.2b B. Hönerlage, R. Lévy, J.B. Grun, C. Klingshirn, K. Bohnert: Phys. Rep. 124, 161 (1985)
16.3 M. Cardona: *Modulation Spectroscopy* (Academic, New York 1969)
B.O. Seraphin (ed.): *Modulation Spectroscopy*, (North-Holland, Amsterdam 1973)
16.4 G. Landwehr (ed.): *Application of High Magnetic Fields in Semiconductor Physics*, Lecture Notes Phys. Vol. 177 (Springer, Berlin, Heidelberg 1982)
16.5 G. Blattner, G. Kurtze, G. Schmieder, C. Klingshirn: Phys. Rev. B **25**, 7413 (1982)
16.6 H.W. Hölscher, A. Nöthe, Ch. Uihlein: Physica B **117/118** 2379 (1982)
16.7 G. Blattner, C. Klingshirn, R. Helbig, R. Meinl: Phys. Stat. Sol. (b) **107**, 105 (1981)
16.8 C. Klingshirn: In Ref. [16.4] p.214
16.9 J. Gutowski: NATO ASI Ser. B **200**, 139 (Plenum, New York 1989)
J. Gutowski: Solid State Commun. **58**, 523 (1986) and references therein
16.10 H. Hümmer: Phys. Stat. Sol. (b) **56**, 249 (1973)
16.11 D. Oberhauser, K.H. Pantke, W. Langbein, V.G. Lyssenko, H. Kalt, J.M. Hvam, G. Weimann, C. Klingshirn: Phys. Stat. Sol (b) **173**, 53 (1992)
16.12 S. Schmitt-Rink: *Festkörperprobleme*/Adv. Solid State Phys. **31**, 243 (1992)
16.13 T. Rappen, G. Mohs, M. Wegener: Phys. Stat. Sol. (b) **173**, 77 (1992)
16.14 E. Gutsche, H. Lange: Phys. Stat. Sol. (b) **22**, 229 (1967)
N. Hase, M. Onuka: J. Phys. Soc. Jpn. **28**, 965 (1970)
O.W. Madelung: Z. Phys. **249**, 12 (1971)
O.W. Madelung: Dissertation, Erlangen (1976)
M.V. Lebedev and V.G. Lysenko, Sov. Phys. Sol. State **24**, 1721 (1982) and Phys. Stat. Sol. b **161**, 395 (1990)
16.15 L. Schultheis, J. Lagois: Phys. Rev. B **29**, 6784 (1984)
L. Schultheis, C.W. Tu: Phys. Rev. B **32**, 6978 (1985)
16.16 Y. Hamakawa, T. Nishino: In *Optical Properties of Solids, New Developments*, ed. by B.O. Seraphim (North-Holland, Amsterdam 1976) p. 255
16.17 D.A.B. Miller, D.S. Chemla, T.C. Damen, A.C. Gossard, W. Wiegmann, T.H. Wood, C.A. Burrus: Phys. Rev. B **32**, 1043 (1985)
D.A.B. Miller, D.S. Chemla, S. Schmitt-Rink: Phys. Rev. B **33**, 6976 (1986)
16.18 I. Bar-Joseph, C. Klingshirn, D.A.B. Miller, D.S. Chemla, U. Koren, B.I. Miller: Appl. Phys. Lett. **50**, 1010 (1987)

16.19 F. Haché, D. Ricard, C. Flytzanis: Appl. Phys. Lett. **55**, 1504 (1989)
16.20 H. Rossmann, A. Schülzgen, F. Henneberger, M. Müller: Phys. Stat. Sol. b **159**, 287 (1990)
16.21 S. Nomura, T. Kobayashi: Phys. Rev. B **45**, 1305 (1992)
16.22 U. Woggon, S.V. Bogdanov, O. Wind, K.-H. Schlaad, H. Pier, C. Klingshirn, P. Chatziagoraston, H.P. Fritz, Phys. Rev. B **48**, 1979 (1993)
16.23 D.W. Langer, R.N. Euwema, K. Era, T. Koda: Phys. Rev. B **2** 4005 (1970)
16.24 H.-R. Trebin, U. Rössler, R. Ranvaud: Phys. Rev. B **20**, 686 (1979)
R. Ranvaud, H.-R. Trebin, U. Rössler, F.H. Pollak: Phys. Rev. B **20**, 701 (1979)
16.25 Chr. Solbrig: Z. Phys. **211**, 429 (1968)
16.26 Mater. Sci. Technol. **4** (1991)
16.27 R. People, J.C. Bean: Appl. Phys. Lett. **47**, 322 (1985)
B.W. Dodson, J.Y. Tsao: Appl. Phys. Lett. **51**, 1325 (1987)
16.28 H.P. Wagner, H. Leiderer: *Festkörperprobleme*/Adv. Solid State Phys. **32**, 221 (1992)
H.P. Wagner, S. Lanken, K. Wolf, D. Lichtenberger, W. Kuhn, P. Link, W. Gebhardt: J. Lumin. **52**, 41 (1992)
16.29 W. Gebhardt: Mater. Sci. Eng. B **11**, 1 (1992)
16.30 U. Becker, H. Gießen, F. Zhou, Th. Gilsdorf, J. Loidolt, M. Müller, M. Grün, C. Klingshirn: J. Crystal Growth **125**, 384 (1992)
H. Gießen: Diploma Thesis, Kaiseslautern (1991)

Chapter 17

17.1 S. Nudelman, S.S. Mitra (eds.): *Optical Properties of Solids* (Plenum, New York 1969)
17.2 F. Abeles (ed.): *Optical Properties of Solids* (North-Holland, Amsterdam 1972)
17.3 F. Wooten: *Optical Properties of Solids* (Academic, New York 1972)
17.4 F. Bassani, G.P. Parravicini: *Electronic States and Optical Transitions in Solids* (Pergamon, Oxford 1975)
17.5 H. Haug, S.W. Koch: *Quantum Theory of the Optical and Electronic Properties of Semiconductors* (World Scientific, Singapore 1990)
17.6 H. Kuzmany: *Festkörperspektroskopie* (Springer, Berlin, Heidelberg 1991)
17.7 P.T. Landsberg: *Recombination in Semiconductors* (Cambridge Univ. Press, Cambridge 1991)
17.8 N. Peyghambarian, S.W. Koch, A. Mysyrowicz: *Introduction to Semiconductor Optics* (Prentice Hall, Englewood Cliffs, NJ 1993)

Chapter 18

18.1 Maria Göppert-Mayer, Annalen der Physik, **9**, 273 (1931)
18.2 P.A. Franken, J.F. Ward: Rev. Mod. Phys. **35**, 23 (1963)
M. Bass, P.A. Franken, J.F. Ward: Phys. Rev. **138**. A, 534 (1965)
18.3 N. Bloembergen: *Nonlinear Optics* (Benjamin, New York 1965)
Y.R. Shen: *The Principles of Nonlinear Optics* (Wiley, New York 1984)
D.L. Mills: *Nonlinear Optics* (Springer, Berlin, Heidelberg 1991)
H.J. Eichler, P. Günter, D.W. Pohl: *Laser Induced Dynamic Gratings*, Springer Ser. Opt. Sci. Vol. 50 (Springer, Berlin, Heidelberg 1986)
P.N. Butcher, and D. Cotter, The Elements of Nonlinear Optics, Cambridge Studies in Modern Optics Vol 9 (Cambridge University Press, Cambridge 1990) and Ref [17.5–8]
18.4 L.S. Brown and T.W.B. Kibble Phys. Rev. **133**, A705 (1964) and T.W.B. Kibble, ibid. **150**. A, 1060 (1966)
18.5 C. Klingshirn, H. Haug: Phys. Rep. **70**, 315 (1981)
18.6 S. Schmitt-Rink, D.A.B. Miller, D.S. Chemla: Adv. Phys. **38**, 9 (1989)

18.7 E.O. Göbel, K. Ploog: *Progr. Quantum Electron.* **14**, 289 (1990)
R. Cingolani, K. Ploog: Adv. Phys. **40**, 535 (1991)
18.8 K.-H. Pantke and J.M. Hvam, Intern. Journ. of Modern Physics **B8**, 73 (1994)

Chapter 19

19.1 D. Fröhlich: *Festkörperprobleme*/Adv. Solid State Phys. 77 (1981); Ref. [17.g]
19.2 C. Klingshirn, H. Haug: Phys. Rep. **70**, 315 (1981)
C. Klingshirn, Adv. Materials for Optics and Electronics **3**, 103 (1994) and ref. given therein
19.3 J.M. Hvam: Solid State Commun. **12**, 95 (1973); Phys. Stat. Sol. (b) **63**, 511 (1974)
19.4 C. Klingshirn: Phys. Stat. Sol. b **71**, 547 (1975)
19.5 E.O. Göbel, K.L. Shaklee, R. Epworth: Solid State Commun. **17**, 1185 (1975)
19.6 B. Hönerlage, C. Klingshirn, J.B. Grun: Phys. Stat. Sol. b **78**, 599 (1976)
19.7 M. Ueta, H. Kazanaki, K. Kobayashi, Y. Toyozawa, E. Hanamura: *Excitonic Processes in Solids*, Springer Ser. Solid-State Sci. **60** (Springer, Berlin, Heidelberg 1986)
19.8 R. Cingolani, K. Ploog: Adv. Phys. **40**, 535 (1991)
19.9 O. Akimoto, E. Hanamura: J. Phys. Soc. Jpn. **33**, 1537 (1972); Solid State Commun. **10**, 253 (1972)
W.F. Brinkmann, T.M. Rice, B. Bell: Phys. Rev. B **8**, 1570 (1972)
W.T. Huang: Phys. Stat. Sol b **60**, 309 (1973)
19.10 B. Hönerlage, R. Lévy, J.B. Grun, C. Klingshirn, K. Bohnert: Phys. Rep. 124, 161 (1985)
19.11 V.B. Timofeev: In *Excitons*, ed. by E.I. Rashba, M.D. Sturge (North-Holland, Amsterdam 1982) Chap.9, p. 349
19.12 D.A. Kleinmann: Phys. Rev. B **28**, 871 (1983)
19.13 R.T. Phillips, D.J. Lovering, G.J. Denton, G.W. Smith: Phys. Rev. B **45**, 4308 (1992); Phys. Rev. Lett. **68**, 1880 (1992)
19.14 S. Bar-Ad, I. Bar Joseph: Phys. Rev. Lett. **66**, 2591 (1991); Phys. Rev. Lett. **68**, 349 (1992)
19.15 D. Oberhauser, K.H. Pantke, W. Langbein, V.G. Lyssenko, H. Kalt, J.M Hvam, G. Weimann, C. Klingshirn: Phys. Stat. Sol b **173**, 53 (1992)
19.16 K.-H. Pantke, D. Oberhauser, V.G. Lyssenko, J.M. Hvam: Phys. Rev. B **47**, 2413 (1993)
K.H. Pantke and J.M. Hvam Intern. Journ. of Modern Physics B **8**, 73 (1994)
19.17 U. Woggon, O. Wind, W. Langbein, O. Gogolin, C. Klingshirn: J. Lumin. **59**, 135 (1994)
19.18 Y.Z. Hu, M. Lindberg, S.W. Koch: Phys. Rev. B **42**, 1713 (1990)
19.19 A. Uhrig, L. Banyai, Y.Z. Hu, S.W. Koch, C. Klingshirn, N. Neuroth: Z. Phys. B **81**, 385 (1990)
19.20 L. Banyai and S.W. Koch, Semiconductor Quantum Dots World Scientific Series on Atomic, Molecular and Optical Physics, Vol 2 (World Scientific, Singapore, 1993)
U. Woggon, Optical Properties of Semiconductor Quantum Dots, Habilitation Thesis, Kaiserslautern (1995), to be published by Springer, Berlin
19.21 U. Woggon: In Ref [1.7 g]
S.V. Gponenko, U. Woggon, M. Saleh, W. Langbein, A. Uhrig, M. Müller, C. Klingshirn: J. Opt. Soc. Am. B **10**, 1947 (1993)
19.22 R. Zimmermann: *Festkörperprobleme*/Adv. Solid State Phys. **30**, 295 (1990)
19.23 H. Haug, L. Banyai (eds.): Optical Switching in Low Dimensional Solids, NATO ASI Ser. B **194** (1988)
19.24 S. Schmitt-Rink, D.S. Chemla, D.A.B. Miller: Adv. Phys. **38**, 89 (1989)
19.25 L.V. Keldysh: Phys. Stat. Sol. b **173**, 119 (1992)
19.26 C. Klingshirn: Semicond. Sci. Technol. **5**, 457, 1006 (1990)
19.27 E. Hanamura, H. Haug: Phys. Rep. **33c**, 209 (1977)
19.28 A. Mysyrowicz, D.P. Trauernicht, J.P. Wolfe, H.-R. Trebin: Phys. Rev. B **27**, 2562 (1983)
H. Haug, H.H. Kranz: Z. Phy. B **53**, 151 (1983)
D. Fröhlich, K. Reimann, R. Wille: Europhys. Lett. **3**, 853 (1987)

19.29 R. Lévy, C. Klingshirn, E. Ostertag, Vu Duy Pach, J. B. Grun: Phys. Stat. Sol B **77**, 381 (1976)
19.30 A. Mysyrowicz: Phys. Rev. B **27**, 2325 (1983)
J.P. Wolfe, A. Mysyrowicz: Sci. Am. **250** (3), 70 (1984)
19.31 W.D. Johnyston Jr., K.L. Shaklee: Solid State Commun. **15**, 73 (1974)
19.32 L.V. Butov, A. Zrenner, G. Abstreiter, G. Böhm and G. Weimann Phys. Rev. Lett. **73**, 304 (1994)
19.33 C. Klingshirn, J. Grohs, M. Wegener: In *Nonlinear Dynamics in Solids*, ed. by H. Thomas (Springer, Berlin, Heidelberg 1992) p. 88 and references therein

Chapter 20

20.1 T.M. Rice: Solid State Phys. **32**, 1 (1977)
J.C. Hensel, T.G. Phillips, G.A. Thomas: Solid State Phys. **32**, 88 (1977)
20.2 C. Klingshirn, H. Haug: Phys. Rep. **70**, 315 (1981)
20.3 S.W. Koch: *Dynamics of First-Order Phase Transitions in Equilibrium and Nonequilibrium Systems*. Lecture Notes Phys. Vol. 207 (Springer, Berlin, Heidelberg 1984)
H. Haug, S. Schmitt-Rink: Prog. Quantum Electron. **9**, 3 (1984)
20.4 D. Bimberg in Landolt-Börnstein New Series, Group III Vol 17 i, 297 (1985)
20.5 C.D. Jeffries, L.V. Keldysch (eds.): Electron–Hole Droplets in Semiconductors, Mod. Probl. Cond. Mat. Sci., Vol. 6 (North-Holland, Amsterdam 1985)
20.6 R. Zimmermann: *Many Particle Theory of Highly Excited Semiconductors*, Teubner Texte Phys. (Teubner, Leipzig 1988)
20.7 H. Haug, S.W. Koch: *Quantum Theory of Optical and Electronic Properties of Semiconductors* (World Scientific, Singapore 1990)
20.8 R. Cingolani, K. Ploog: Adv. Phys. **40**, 535 (1990)
R. Cingolani: Physica Scripta T **49** B, 470 (1994)
20.9 C. Klingshirn: In Ref. [1.8g]
20.10 H. Kalt: Optical Properties of III–V Semiconductors: the Influence of Multi-Valley Band Structures Habilitation Thesis, Kaiserslautern (1993) to be published (Springer, Berlin 1995)
M. Kalt: *Festkörperprobleme*/Adv. Solid State Phys. **32**, 145 (1992)
H. Kalt: J. Lumin **60/61**, 262 (1994)
20.11 P. Debye, E. Hückel: Phys. Z. **24**, 185, 305 (1923)
20.12 L.H. Thomas: Proc. Camb. Phil. Soc. **23**, 542 (1927)
E. Fermi: Z. Phys. **48**, 73 (1928)
20.13 N.F. Mott: *Metal–Insulator Transitions* (Taylor, Francis, London 1974)
20.14 P. Vashista, R.K. Kalia: Phys. Rev. B **25**, 6492 (1982)
R. Zimmermann: Phys. Stat. Sol. b **146**, 371 (1988)
20.15 D.S. Chemla, I. Bar-Joseph, J.M. Kuo, T.Y. Chang, C. Klingshirn, G. Livescu, D.A.B. Miller: IEEE J.QE-**24**, 1664 (1988)
20.16 C. Klingshirn: *Festkörperprobleme*/Adv. Solid State Phys. **30**, 335 (1990)
20.17 E. Hanamura, H. Haug: Phys. Rep. **33C**, 209 (1977)
20.18 M. Combescot: Solid State Commun. **30**, 81 (1979)
M. Combescot, C. Benoit à la Guillaume: ibid. **46**, 579 (1983)
20.19 H. Stolz, R. Zimmermann, G. Röpke: Phys. Stat. Sol. b **105**, 585 (1981);
G. Manzke, V. May, K. Henneberger: ibid **125**, 693 (1984)
20.20 G.W. Fehrenbach, W. Schäfer, J. Treusch, R.G. Ulbrich: Phys. Rev. Lett. **49**, 1281 (1982)
20.21 R.W. Martin: Solid State Commun. **19**, 373 (1976); ibid. **22**, 523 (1977)
20.22 H. Kalt, A.L. Smirl, T.F. Bogges: Phys. Stat. Sol. b **150**, 895 (1988)
H. Fieseler, R. Schwabe, J. Staehli: ibid. **159**, 411 (1990)
S. Gürtler, R. Schwabe, J.L. Stäehli, K. Unger: ibid. **173**, 441 (1992)
P.P. Paskov: Europhys. Lett. **20**, 143 (1992)
H. Kalt, M. Rinker: Phys. Rev. B **45**, 1139 (1992)

20.23 A. Forchel, B. Laurich, G. Moersch, W. Schmid, T.L. Reinecke: Phys. Rev. Lett. **46**, 678 (1981)
20.24 G.A. Thomas, T.M. Rice, J.C. Hensel: Phys. Rev. Lett. **33**, 219 (1974)
T.L. Reinecke, S.C. Ying: ibid. **35**, 311 (1975)
20.25 R.M. Westervelt, J.C. Culbertson, B.S. Black: Phys. Rev. Lett. **42**, 269 (1979)
20.26 M. Greenstein, J.P. Wolfe: Solid State Commun. **33**, 309 (1980)
20.27 R.S. Markiewicz, J.P. Wolfe, C.D. Jeffries: Phys. Rev. B **15**, 1988 (1977)
20.28 V.G. Lyssenko, V.I. Revenko, T.G. Tratas, V.B. Timofeev: Sov. Phys. JETP **41**, 163 (1975)
20.29 K. Bohnert, M. Anselment, G. Kobbe, C. Klingshirn, H. Haug, S.W. Koch, S. Schmitt-Rink, F.F. Abraham: Z. Phys. B **42**, 1 (1981)
F.A. Majumder, H.-E. Swoboda, K. Kempf, C. Klingshirn: Phys. Rev. B **32**, 2407 (1985)
20.30 M.S. Skolnick, K.J. Nash, P.E. Simmonds, D.J. Mowbray, T.A. Fisher, M.K. Saber, D.M. Whittaker, S.J. Bass, R.S. Smith: In *21st Int'l. Conf. on Physics of Semiconductors* , Beijing 1992, ed. by Ping Jiang, Hou-Zhi Zheng (World Scientific, Singapore 1993) p.41
20.31 H. Kalt, K. Leo, R. Cingolani, K. Ploog: Phys. Rev. B. **40**, 12017 (1989)
M. Combescot, Ch. Tanguy: In *21st Int'l. Conf. on Physics of Semiconductors* , Beijing 1992, ed. by Ping Jiang, Hou-Zhi Zheng (World Scientific, Singapore 1993) p. 141
20.32 H.E. Swoboda, F.A. Majumder, V.G. Lyssenko, C. Klingshrin, L. Banyai: Z. Phys. B. **70**, 341 (1988)
20.33 D.J. Hagan, H.A. MacKenzie, H.A. Al Attar, W.J. Firth: Opt. Lett. **10,** 187 (1985)
20.34 F.A. Majumder, S. Shevel, V.G. Lyssenko, H.-E. Swoboda, C. Klingshirn: Z. Phys. B **66**, 409 (1987)
20.35 C. Klingshirn, M. Kunz, F.A. Majumder, D. Oberhauser, R. Renner, M. Rinker, H.-E. Swoboda, A. Uhrig, C. Weber: NATO ASI Ser. B **200**, 167 (1989)
20.36 M. Capizzi, A. Frova, S. Modesti, A. Selloni, J.L. Staehli, M. Guzzi: Helv. Phy. Acta **58**, 272(1985)
20.37 C. Klingshirn, Ch. Weber, D.S. Chemla, D.A.B. Miller, J.E. Cunningham, C. Ell, H. Haug: NATO ASI Seri. B **194**, 353 (1989)
K.-H. Schlaad, Ch. Weber, D.S. Chemla, J. Cunningham, C.V. Hoof, G. Borghs, G. Weimann, W. Schlapp, H. Nickel, C. Klingshirn: Phys. Stat. Sol. b **159**, 173 (1990); Phys. Rev. B **43**, 4268 (1991)
20.38 V.D. Kulakovskii, E. Lach, A. Forchel, D. Grützmacher: Phys. Rev. B **40**, 8087 (1989)

Chapter 21

21.1 C. Klingshirn, H. Haug: Phys. Rep. **70**, 315 (1981)
C. Klingshirn: In Ref. [1.7c] p.485; Adv. Mater. Opt. Electron., **2**(1993)
21.2 W.W. Chow, S.W. Koch and M. Sargent III, Semiconductor Laser Physics (Springer, Berlin, 1994)
21.3 N.G. Basov, O.V. Bogdankevich: In *Proc 7th Int'l Conf. Phys. Semicond.*, Paris, 1964 (Dunod, Paris 1964) p. 225
21.4 J. Bille: *Festkörperprobleme*/Adv. Solid State Phys. **13**, 111 (1973)
21.5 R.B. Bhargava: J. Cryst. Growth **117**, 894 (1992)
A.S. Nasibov, V.I. Kozlovsky, P.V. Reznikov, Ya.K. Skasysisky, Yu.M. Popov: ibid. **117**, 1040 (1992)
21.6 M.S. Brodin, K.A. Dmitrenko, L.V. Taranenko, S.G. Shevel: Sov. Phys. Techn. Phys. **9**, 1852 (1983)
21.7 S.M. Sze: *Physics of Semiconductor Devices*, 2nd ed. (Wiley, New York 1981)
K.J. Ebeling: *Integrierte Optoelectronic*, 2nd edn. (Springer, Berlin, Heidelberg 1992)
21.8 A.V. Nurmikko, R.L. Gunshor: Phys. Stat. Sol. b **173**, 291 (1992) and references therein and in [14.62]
H. Kalt, M. Umlaff, W. Petri, F.A. Majumder and C. Klingshirn Materials Science Forum (1995) in press

21.9 W. Wünstel, C. Klingshirn: Opt. Commun. **32**, 269 (1980)
21.10. V.S. Dneprovskii, V.I. Klimov, D.K. Okorokov, Yu.V. Vandyshev: Phys. Stat. Sol. b **173**, 405 (1992)
21.11 Y. Masumoto, T. Kawamura, K. Era: Appl. Phys. Lett. **62**, 225 (1993)
P. Faller, B. Kippelen, B. Hönerlage, R. Lévy: Opt. Mater. **2**, 39 (1993)
U. Woggon, O. Wind, W. Langbein and C. Klingshirn Japanes J. of Appl. Physics **34**, Supp1, 232 (1995)
21.12 B. Hönerlage, R. Lévy, J.B. Grun, C. Klingshirn, K. Bohnert: Phys. Rep. **124**, 161 (1985)
21.13 F.A. Majumder, S. Shevel, V.G. Lyssenko, H.-E Swoboda, C. Klingshirn: Z. Phys. B **66**, 409 (1987)
21.14 See Refs. [20.10,22]
21.15 A. Wörner, H. Kalt, R. Westphäling: Proc. 22th ICPS, Vancouver (1994) in press
21.16 D.G. Thomas, J.J. Hopfield: J. Appl Phys. **33**, 3243 (1962)
21.17 C. Klingshirn: Phys. Stat. Sol. b **71**, 547 (1975)

Chapter 22

22.1 See the two conference series mentioned in the preface
22.2 Ultrafast Phenomena VIII, eds.: J.-L. Martin, A. Migus, G.A. Mourou and A.H. Zewail, Springer Ser. Chem. Phys. **55**, (Springer, Berlin, Heidelberg 1993) and previous Vols. 4, 14, 23, 38, 46, 48, 53 of this series, and Ultrashort Processes in Condensed Matter W.E. Bron ed., NATO ASI Series B **314**, (1993)
22.3 Proceedings of Int'l Workshops on "Nonlinear Optics and Excitation Kinetics in Semiconductors"
NOEKS I: Phys. Stat. Sol. b **146**, 311–391; **147**, 699–756 (1988)
NOEKS II: Phys. Stat. Sol. b **159**, 1–484 (1990)
NOEKS III: Phys. Stat. Sol. b **173**, 1–478 (1992)
NOEKS IV: Phys. Stat. Sol. b (1995) in press
22.4 H. Kalt, J.H. Collet, Le Si Dang, J. Cibert, S.D. Baranovskii, R. Saleh, M. Umlauff, K.P. Geyzers, M. Heuken, C. Klingshirn: Physica B **191**, 90 (1993)
22.5 A. Haug: *Festkörperprobleme* XII, 411 (1975) J. Lumin. **20**, 173 (1979) and references therein
22.6 V.S. Letokhov, V.P. Chebotayev: *Nonlinear Laser Spectroscopy*, Springer Ser. Opt. Sci., Vol. 4 (Springer, Berlin, Heidelberg 1977)
22.7 W. Demtröder: *Laserspektroskopie* 2nd edn. (Springer, Berlin, Heidelberg 1991)
22.8 F. Bloch: Phys. Rev. **70**, 460 (1946)
R.P. Feynman, F.L. Vornon, R.W. Hellwarth: J. Appl. Phys. **28**, 49 (1957)
22.9 See Refs. [1.7f, g]
22.10 R.L. Shoemaker: In *Coherence and Laser Spectroscopy* (Plenum, New York 1978) p. 197
Y.R. Shen: *The Principles of Nonlinear Optics* (Wiley, New York 1984)
22.11 M. Wegener, D.S. Chemla, S. Schmitt-Rink, W. Schäfer: Phys. Rev. A **44**, 2124 (1991)
22.12 C. Dörnfeld, J.M. Hvam: IEEE J. QE-**25**, 904 (1989)
22.13 J. Kuhl, A. Honold, L. Schultheis, Ch.W. Tu: *Festkörperprobleme*/Adv. Solid State Phys. **29**, 157 (1989)
22.14 D. Oberhauser, K.-H. Pantke, W. Langbein, V.G. Lyssenko, H. Kalt, J.M. Hvam, G. Weimann, C. Klingshirn: Phys. Stat. Sol. b **173**, 53 (1992)
K.-H. Pantke, D. Oberhauser, V.G. Lyssenko, J.M. Hvam, G. Weimann: Phys. Rev. B **47**, 2413 (1993)
22.15 Ch. Weber, U. Becker, R. Renner, C. Klingshirn: Z. Phy. B **72**, 379 (1988)
22.16 K.-H. Pantke, V.G. Lyssenko, B.S. Razbirin, H. Schwab, J. Erland, J.M. Hvam: Phys. Stat. Sol. b **173**, 69 (1992)
K.-H. Pantke, J.M. Hvam: J. Mod. Phy. B **8**, 73 (1994)
22.17 K.-H. Pantke, V.G. Lyssenko, B.S. Razbirin, J. Erland, J.M. Hvam: Proc 21th ICPS (1992) p 129

22.18 H. Stolz: *Festkörperprobleme*/Adv. Solid State Phy. **32**, 219 (1991)
22.19 H. Schwab, C. Dörnfeld, E.O. Göbel, J.M. Hvam, C. Klingshirn, J. Kuhl, V.G. Lyssenko, F.A. Majumder, G. Noll, J. Nunenkamp, K.-P. Pantke, R. Renner, A. Retznitsky, U. Siegner, H.-E. Swoboda, Ch. Weber: Phys. Stat. Sol. b **172**, 479 (1992)
22.20 U. Siegner, D. Weber, E.O. Göbel, D. Bennhardt, V. Henckeroth, R. Saleh, S.D. Baranovskii, P. Thomas, H. Schwab, C. Klingshirn, J.M. Hvam, V.G. Lyssenko: Phys. Rev. B **46**, 4564 (1992)
22.21 H. Schwab, C. Klingshirn: Phys. Rev. B **45**, 6938 (1992)
22.22 D. Oberhauser, K.-H. Pantke, J.M. Hvam, G. Weimann, C. Klingshirn: Phys. Rev. B **47**, 6827 (1993)
22.23 J.M. Hvam: In Ref. [1.7g]
22.24 D. Fröhlich, A. Kulik, B. Uebbing, A. Mysyrowicz, V. Langer, H. Stolz, W. von der Osten: Phys. Rev. Lett. **67**, 2343 (1991)
22.25 K.-H. Pantke, J. Schillak, B.S. Razbirin, V.G. Lyssenko, J.M. Hvam: Phys. Rev. Lett. **70**, 327 (1993)
J. Erland, B.S. Razbirin, V.G. Lyssenko, K.-H. Pantke and J.M. Hvam, J. Crystal Growth **138**, 800 (1994)
22.26 S. Bar.Ad, I. Bar-Joseph: Phys. Rev. Lett. **66**, 2591 (1991), ibid. **68**, 349 (1992)
22.27 E.O. Göbel, M. Koch, J. Feldmann, G.v. Plessen, T. Meier, A. Schulze, P. Thomas, S. Schmitt-Rink, K. Köhler, K. Ploog: Phys. Stat. Sol. b **173**, 21 (1992); Phys. Rev. Lett. **69**, 3633 (1992)
M. Koch, D. Weber, J. Feldmann, E.O. Göbel, T. Meier, A. Schulze, P. Thomas, S. Schmitt-Rink and K. Ploog Phys. Rev. B **47**, 1532 (1993)
22.28 T. Yajima, H. Souma: Phys. Rev. A **17**, 309, 324 (1978)
22.29 U. Woggon et al. in Ref. 1.7g and J. Crystal Growth **138**, 976 and 988 (1994)
22.30 U. Woggon, S. Gaponenko, W. Langbein, A. Uhrig, C. Klingshirn: Phys. Rev. B **47**, 3684 (1993)
22.31 R.M. Macfarlane: In Refs. [1.7f p 399 and 1.7g]
22.32 P. Wiesner, U. Heim: Phys. Rev. B **11**, 3071 (1975)
22.33 D.S. Chemla, S. Schmitt-Rink, D.A.B. Miller: In *Optical Nonlinearities and Instabilities in Semiconductors*, ed. by H. Haug (Academic, New York 1988) p.83
22.34a J. Feldmann, G. Peter, E.O. Göbel, P. Dawson, K. Moore, C. Foxon, R.J. Elliot, B. Devaud, F. Clérot, N. Roy, K. Satzke, B. Sermage, D.S. Katzer: Phys. Rev. Lett. **67**, 2355 (1991)
22.34b L.C. Andreani, Solid State Commun. **77**, 641 (1991)
22.35 D. Oberhauser, W. Sack, K.P. O'Donnel, P.J. Parbrook, P.J. Wright, B. Cockayne, C. Klingshirn: Superlat. Microstruct. **9**, 107 (1991); J. Lumin. **53**, 409 (1992)
22.36 F. Bloch: Z. Phys. **52**, 555 (1928)
22.37 C. Zener: Proc. Royal. Soc. London A **145**, 523 (1934)
22.38 E. Spenke: *Elektronische Halbleiter* 2nd edn. (Springer, Berlin, Heidelberg 1965)
22.39 J. Feldmann: Festkörperprobleme/Adv. Solid State Phys. **32**, 81 (1992)
22.40 K. Leo: ibid. p. 97
22.41 H.G. Roskos: ibid. **34** (1994) in press
22.42 K. Victor, H.G. Roskos, Ch. Waschke: J. Opt. Soc. Am. B (1994) **11**, 2470 (1994)
22.43 P. Leisching, P.H. Bolivar, W. Beck, Y. Dhaibi, F. Brüggemann, R. Schwedler, H. Kurz, K. Leo, K. Köhler: Phys. Rev. B **50**, 14 389 (1994)
22.44 S.W. Koch, J. Feldmann: Private communications
22.45 S. Bar-Ad, I. Bar-Joseph: Phys. Rev. Lett. **66**, 2491 (1991); ibid. **68**, 349 (1992)
S. Bar-Ad, I. Bar-Joseph, Y. Levinson, H. Strikman: Phys. Rev. Lett. **72**, 776 (1994)
L.V. Butor, V.D. Kulakovskii, G.E.W. Bauer, A. Forchel, D. Grützmacher: Phys. Rev. B **46**, 12765 (1992)
22.46 R. Höpfel, G. Lindemann, E. Gornik, G. Stangl, A.C. Gossard, W. Wiegmann: Surf. Sci. **113**, 118 (1982) and see Ref [12.4]

M. Helm, P. England, E. Colas, F. DeRosa, S.J. Allen: Phys. Rev. Lett. **63**, 74 (1989)
U. Merkt: *Festkörperprobleme*/Adv. Solid State Phy. **27**, 109 (1987); ibid. **30**, 77 (1990)
D. Heitmann: Surf. Sci. **170**, 332 (1986) and see Ref. [12.4]

Chapter 23

23.1 Ch. M. Bowden, M. Ciftan, H.R. Robl eds.: *Optical Bistability* (Plenum, New York 1980)
Ch. M. Bowden, H.M. Gibbs, S.L. McCall (eds.): *Optical Bistability II* (Plenum, New York 1984)
H.M. Gibbs, P. Mandel, N. Peyghambarian, S.D. Smith (eds.): *Optical Bistability III* Springer Proc. Phys. Vol.8 (Springer, Berlin, Heidelberg 1986)
W. Firth, N. Peyghambarian, A. Tallet (eds.): *Optical Bistability IV* J. Phys. (Paris) **49**, C2 suppl. au no. 6 (1988)
23.2 H.M. Gibbs: *Optical Bistability, Controlling Light with Light* (Academic, New York 1985)
23.3a P. Mandel, S.D. Smith, B. Wherett (eds.): *From Optical Bistability Towards Optical Computing*, (North-Holland, Amsterdam 1987)
23.3b Optical Information Technology, S.D. Smith and R.F. Neale eds., (Springer, Berlin 1993)
23.4a H. Haug, L. Banyai (eds.): *Optical Switching in Low-Dimensional Solids*, NATO ASI Ser. B **194**, (1988)
23.4b H. Haug (ed.): *Optical Nonlinearities and Instabilities in Semiconductors* (Academic, New York 1988)
23.5 F. Henneberger: Phys. Stat. Sol. **137**, 371 (1986)
23.6 B.S. Wherett, F.A.P. Tooley (eds.): *Optical Computing* (IOP, Bristol 1989)
23.7 H. Thomas (ed.): *Nonlinear Dynamics in Solids* (Springer, Berlin, Heidelberg 1992)
23.8 J. Grohs, F. Zhou, H. Ißler, C. Klingshirn: Int'l J. Bifurcation Chaos **2**, 861 (1992)
23.9 M. Wegener, C. Klingshirn, S.W. Koch, L. Banyai: Semicond. Sci. Technol. **1**, 366 (1986)
23.10 C. Klingshirn, J. Grohs, M. Wegener: In Ref. [23.6] p. 88
C. Klingshirn: In Ref. [1.7e] p.529
23.11 D.A.B. Miller: J. Opt. Soc. Am. **B1**, 857 (1984)
23.12 C. Klingshirn: In Ref. [23.3] p.252
23.13 M. Wegener, C. Klingshirn: Phys. Rev. A **35**, 1740, 4247 (1987)
23.14 S.W. Koch in Ref. 23.4b p 273
23.15 J. Grohs, S. Apanasevich, H. Issler, M. Kuball, J. Steffen, C. Klingshirn: SPIE Proc. **1807**, 192 (1992)
J. Grohs, M. Müller, A. Schmidt, A. Uhrig, C. Klingshirn, H. Bartelt: Opt. Commun. **78**, 77 (1990)
U. Zimmermann, K.-H. Schlaad, G. Weimann, C. Klingshirn: In Ref. [23.3b] p 232
23.16 J. Grohs, H. Issler, C. Klingshirn: Opt. Commun. **86**, 183 (1991)
23.17 D.A.B. Miller: In Ref. [23.6] pp.55, 71
D.S. Chemla, D.A.B. Miller, S. Schmitt-Rink: In Ref. [23.4b] p 83 and ref. therein
23.18 F. Forsmann, D. Jäger, W. Niessen: Opt. Commun. **62**, 193 (1987)
23.19 A. Witt, M. Wegener, V.G. Lyssenko, C. Klingshirn, G. Wingen, Y. Iyechika, D. Jäger, G. Müller-Vogt, H. Sitter, H. Heinrich, H.A. Mackenzie: IEEE J. QE-**24**, 2500 (1988)
W. Kazukauskas, J. Grohs, C. Klingshirn, G. Wingen, D. Jäger: Z. Phys. B **79**, 149 (1990)
23.20 H.J. Eichler, V. Glaw, A. Kummrow, V. Penschke, S. Boede: SPIE Proc. **1017**, 90 (1988)
T. Brand, H.J. Eichler, B. Smandek: SPIE Proc. **1017**, 200 (1988)
23.21 S.V. Bogdanov, V.G. Lyssenko: Sov. Tech. Phys. Lett. **14**, 270 (1988)
23.22 G.H. Doehler: IEEE J. QE-**22**, 1682 (1986)
N. Linder, T. Gabler, K.-H. Gulden, P. Kiesel, M. Kneissl, P. Diehl, G.H. Doehler: Appl. Phys. Lett. **62**, 1916 (1993)
23.23 J. Cibert et al. Phys. Scr. T **49**, B 487 (1993)

23.24 W. Langbein, H. Kalt, M. Hetterich, M. Grün C. Klingshirn, J. Crystal Growth **138**, 559 (1994)
23.25 S.M. Sze: *Physics of Semiconductor Devices*, 2nd edn. (Wiley, New York 1981)
K.J. Ebeling: *Integrierte Optoelektronik*, 2nd edn. (Springer, Berlin, Heidelberg 1992)
23.26 H. Haken: *Synergetics, an Introduction*, Springer Ser. Syn., Vol. 1 and subsequent volumes (Springer, Berlin, Heidelberg 1983)
23.27 S.W. Koch: *Dynamics of First-Order Phase Transitions in Equilibrium and Nonequilibrium Systems*, Lecture Notes Phys. Vol. 207 (Springer, Berlin, Heidelberg 1984)
23.28 H.G. Schuster: Deterministic Chaos, 2nd edn. (Physik Verlag, Weinheim 1988)
23.29 K. Ikeda, H. Daido, O. Akimoto: Phys. Rev. Lett. **45**, 709 (1980); **48**, 617 (1982)
J.V. Moloney: Phys. Rev. A **33**, 4061 (1986)
23.30 M. Lindberg, S.W. Koch, H. Haug: J. Opt. Soc. Am. B **3**, 751 (1986)
I. Galbraith, H. Haug: ibid. **4**, 1116 (1987)
23.31 M. Wegener, C. Klingshirn, G. Müller-Vogt: Z. Phys. B **68**, 519 (1987)

Chapter 24

24.1 Y.R. Shen: *The Principles of Nonlinear Optics* (Wiley, New York 1984)
24.2 *Laser Spectroscopy of Solids*, 2nd edn; ed. by W.M. Yen, P.M. Selzer, Topics Appl. Phys. Vol. 49 (Springer, Berlin, Heidelberg 1986)
W.M. Yen (ed.): *Laser Spectroscopy of Solids II*, Topics Appl. Phys. Vol. 65 (Springer, Berlin, Heidelberg 1989)
24.3 J.P. Eichler, P. Güter, D.W. Pohl: *Laser-Induced Gratings*, Springer Ser. Opt. Sci. Vol. 50 (Springer, Berlin, Heidelberg 1986)
24.4 H. Kuzmany: *Festkörperspektroskopie* (Springer, Berlin, Heidelberg 1989)
24.5 W. Demtröder: *Laserpektroskopie*, 2nd edn. (Springer, Berlin, Heidelberg 1991)
24.6 W. Stößel: *Fourier Optik*, (Springer, Berlin, Heidelberg 1993)
24.7 Y.H. Pao: *Optoacoustic Spectroscopy and Detection* (Academic, New York 1977)
W.B. Jackson, N.M. Amer, A.C. Boccara, D. Fournier: Appl. Opt. **20**, 1333 (1981)
D. Bimberg, T. Wolf, J. Böhrer: NATO ASI Ser. B **249**, 577 (1991)
24.8 I.N. Uraltsev, E.L. Ichenko, P.S. Kop'ev, V.P. Kochereshko, D.R. Yakovlev: Phys. Stat. Sol. b **150**, 673 (1988)
24.9 J. Voigt, F. Spiegelberg: Phys. Stat. Sol. b **30**, 659 (1968)
J. Voigt, E. Ost: ibid, **33**, 381 (1969)
24.10 J.J. Le Pore: J. Appl. Phys. **51**, 6441 (1980)
24.11 D.G. Seiler, D. Heiman, R. Feigenblatt, R.L. Aggarwal, B. Lax: Phys. Rev. B **25**, 7666 (1982)
D.G. Seiler, D. Heiman, B.S. Wherett: Phys. Rev. B **27**, 2355
24.12 I. Bar-Joseph, C. Klingshirn, D.A. Miller, D.S. Chemla, U. Koren, B.I. Miller: Appl. Phys. Lett. **50**, 1010 (1987) and references therein
24.13 W.H. Knox, R.S. Knox, J.F. Hoose, R.N. Zare: Opt. Photon. News, April (1990) p. 44
24.14 D.H. Auston: Appl. Phys. Lett. **26**, 101 (1975)
J. Rosenzweig, C. Moglestue, A. Axmann, J. Schneider, A. Hülsmann, M. Lambsdorff, J. Kuhl, M. Klingenstein, H. Leier, A. Forchel: SPIE Proc. **1362**, 168 (1990)
24.15 K. Bohnert, M. Anselment, G. Kobbe, C. Klingshirn, H, Haug, S.W. Koch, S. Schmitt-Rink, F.F. Abraham: Z. Phys. **42**, 1 (1981)
24.16 F.A. Majumder, H.-E. Swoboda, K. Kempf, C. Klingshirn: Phys. Rev. B **32**, 2407 (1985)
24.17 H. Schrey, V. Lyssenko, C. Klingshirn: Solid State Commun. **32**, 897 (1979)
J.M. Hvam, G. Blattner, M. Reuscher, C. Klingshirn: Phys. Stat. Sol. b **118**, 179 (1983)
24.18 C. Benoit à la Guillaume, J.M. Debever, F. Slavan: Phys. Rev. **177**, 567 (1969)
K.L. Shaklee, R.E. Nahory, R.F. Leheny: J. Lumin. **7**, 284 (1973)
J.M. Hvam: J. Appl. Phys. **49**, 3124 (1978)

24.19 V.S. Lotokhov, V.P. Chebotayev: *Nonlinear Laser Spectroscopy*, Springer Ser. Opt. Sci. Vol. 4 (Springer, Berlin, Heidelberg 1977)
24.20 R. Macfarlane: In Ref. [1.17f, g]
24.21 Ch. Weber, U. Becker, R. Renner, C.Klingshirn: Appl. Phys. B **45**, 113 (1988)
24.22 P. Horan, W. Blau, H. Byrne, P. Berglund: Appl. Opt. **29**, 31 (1990)
24.23 T. Yajima, H. Souma: Phys. Rev. A **17**, 309, 324 (1978)
24.24 H. Rajbenbach, J.P. Huignard: In *Optical Computing*, ed. by B.S. Wherett, F.A.P. Tooley (SUSSP, Edinburgh 1989) p. 133
V.V. Shkunov, B. Ya Zel'dovich: Sci. Am. **253**, p. 40 (1985)
D.M. Pepper: Sci. Am. **254**, 56 (1986)
24.25 A. Bloch (ed.): *Murphy's Laws*, Books I–III, (Price/Stern/Sloan, Los Angeles 1986)

Chapter 25

25.1 V. Heine: *Group Theory in Quantum Mechanics* (Pergamon, Oxford 1960)
25.2 R.S. Knox, A. Gold: *Symmetry in the Solid State* (Benjamin, New York 1964)
25.3 W.A. Tinkham: *Group Theory and Quantum Mechanics* (McGraw-Hill, New York 1964)
25.4 H.W. Streitwolf: *Gruppentheorie in der Festkörperphysik* (Akad. Verl Leipzig 1967)
25.5 W. Ludwig, C. Falter: *Symmetries in Physics*, Springer Ser. Solid-State Sci. Vol. 64 (Springer, Berlin, Heidelberg 1988)
25.6 G.F. Koster, J.O. Dimmock, R.G. Wheeler, H. Statz (eds.): *Properties of the Thirty-Two Point Groups*, (MIT, Cambridge 1963)
25.7 K. Cho: Phys. Rev. B **14**, 4463 (1976)
25.8 B. Hönerlage, B. Lévy, J.B. Grun, C. Klingshirn, K. Bohnert: Phys. Rep. **124**, 161 (1985)
25.9 O. Goede: Phys. Stat. Sol. b **81**, 235 (1977)

Subject Index

Springer-Verlag and the Environment

Printing: Mercedesdruck, Berlin
Binding: Buchbinderei Lüderitz & Bauer, Berlin